Lecture Notes in Computer Science 1888

Edited by G. Goos, J. Hartmanis, and J. van Leeuwen

Springer
Berlin
Heidelberg
New York
Barcelona
Hong Kong
London
Milan
Paris
Singapore
Tokyo

Gerald Sommer Yehoshua Y. Zeevi (Eds.)

Algebraic Frames for the Perception-Action Cycle

Second International Workshop, AFPAC 2000
Kiel, Germany, September 10-11, 2000
Proceedings

Springer

Series Editors

Gerhard Goos, Karlsruhe University, Germany
Juris Hartmanis, Cornell University, NY, USA
Jan van Leeuwen, Utrecht University, The Netherlands

Volume Editors

Gerald Sommer
Universität Kiel
Institut für Informatik und Praktische Mathematik, Kognitive Systeme
Preussestr. 1-9, 24105 Kiel, Germany
E-mail: gs@ks.informatik.uni-kiel.de

Yehoshua Y. Zeevi
Technion – Israel Institute of Technology
Faculty of Electrical Engineering
Technion City, Haifa 32000, Israel
E-mail: zeevi@ee.technion.ac.il

Cataloging-in-Publication Data applied for

Die Deutsche Bibliothek - CIP-Einheitsaufnahme

Algebraic frames for the perception-action cycle : second international workshop ; proceedings / AFPAC 2000, Kiel, Germany, September 10 - 11, 2000. Gerald Sommer ; Yehoshua Y. Zeevi (ed.). - Berlin ; Heidelberg ; New York ; Barcelona ; Hong Kong ; London ; Milan ; Paris ; Singapore ; Tokyo : Springer, 2000
(Lecture notes in computer science ; Vol. 1888)
ISBN 3-540-41013-9

CR Subject Classification (1998): I.3.5, I.2.9-10, I.5, I.6, I.4

ISSN 0302-9743
ISBN 3-540-41013-9 Springer-Verlag Berlin Heidelberg New York

Springer-Verlag Berlin Heidelberg New York
a member of BertelsmannSpringer Science+Business Media GmbH

Printed in Germany

Typesetting: Camera-ready by author, data conversion by DA-TeX Gerd Blumenstein
Printed on acid-free paper SPIN: 10722492 06/3142 5 4 3 2 1 0

Preface

This volume presents the proceedings of the 2nd International Workshop on Algebraic Frames for the Perception and Action Cycle. AFPAC 2000. held in Kiel, Germany, 10–11 September 2000. The presented topics cover new results in the conceptualization, design, and implementation of visual sensor-based robotics and autonomous systems. Special emphasis is placed on the role of algebraic modelling in the relevant disciplines, such as robotics, computer vision, theory of multidimensional signals, and neural computation. The aims of the workshop are twofold: first, discussion of the impact of algebraic embedding of the task at hand on the emergence of new qualities of modelling and second, facing the strong relations between dominant geometric problems and algebraic modelling.

The first workshop in this series, AFPAC'97. inspired several groups to initiate new research programs, or to intensify ongoing research work in this field, and the range of relevant topics was consequently broadened, The approach adopted by this workshop does not necessarily fit the mainstream of worldwide research-granting policy. However, its search for fundamental problems in our field may very well lead to new results in the relevant disciplines and contribute to their integration in studies of the perception–action cycle.

The background of the workshop is the design of autonomous artificial systems following the paradigm of behavior-based system architectures. The perception–action cycle constitutes the framework in which the designer has to make sure that robust, stable, and adaptive system behavior will result. The mathematical language used to shape this frame is crucial for getting system features such as the ones mentioned above or, in addition, semantic completeness and in some cases linearity. By semantic completeness we mean a representation property which is purpose-oriented in its nature rather then the traditional mathematical meaning of the term of completeness. While linearity is, without any restriction, a useful system property, most of the problems we have to handle turn out to be nonlinear. We learn from the approach of this workshop that this is not a matter of fate which traditionally results in non-complete, approximating linearizations. Instead, various problems can be algebraically transformed into linear and, thus, complete ones. The reader can identify this approach in several contributions related to multidimensional signal processing, neural computing, robotics, and computer vision.

This volume includes 7 invited papers and 20 regular papers. The invited papers are contributed by members of the program committee. Regretably, not all of them were able to present a talk or to contribute a paper to the proceedings. We wish, however, to thank all of them for their careful reviewing of the contributed papers. All authors of papers presented in this volume contributed to important aspects relevant to the main theme of the workshop. Our thanks go to all the authors of the invited and contributed papers for the high quality of their contributions and *for* their cooperation.

We thank the Christian-Albrechts-Universitt Kiel for hosting the workshop and the industrial sponsors for their financial support. Special thanks to the Deutsche Forschungsgemeinschaft (DFG) which, by awarding grant no. 4851/223/00, made it possible to invite selected speakers. Last but not least the workshop could not have taken place without the extraordinary commitment of the local organizing committee.

Kiel and Haifa, June 2000 Gerald Sommer and Yehoshua Y. Zeevi

Workshop Co-Chairs

G. Sommer, Germany
Y. Y. Zeevi, Israel

Program Committee

Y. Aloimonos, USA
L. Dorst, The Netherlands
J. O. Eklundh, Sweden
G. M. Granlund, Sweden
D. Hestenes, USA
J. J. Koenderink, The Netherlands
V. Labunets, Russia
J. Lasenby, UK
H. Li, China
H. Ritter, Germany
J. M. Selig, UK
G. Sommer, Germany
Y. Y. Zeevi, Israel

Invited Speakers

Y. Aloimonos, USA
L. Dorst, The Netherlands
G. Granlund, Sweden
D. Hestenes, USA
J. J. Koenderink, The Netherlands
V. Labunets, Russia
J. Lasenby, UK
H. Li, China
H. Ritter, Germany
J. M. Selig, UK

Organizers

D. Grest, Technical Support
N. Krger, Program
C. Perwass, Finance
B. Rosenhahn, Local
G. Sommer, Chair
F. Maillard, Secretary

Table of Contents

Analyzing Action Representations

Y. Aloimonos and C. Fermüller

Computer Vision Laboratory, Center for Automation Research, Institute for Advanced Computer Studies, and the Department of Computer Science, University of Maryland, College Park, MD 20742-3275, USA
yiannis@cfar.umd.edu,
WWW home page: http://www.cfar.umd.edu/~yiannis/research.html

Abstract. We argue that actions represent the basic seed of intelligence underlying perception of the environment, and the representations encoding actions should be the starting point upon which further studies of cognition are built. In this paper we make a first effort in characterizing these action representations. In particular, from the study of simple actions related to 3D rigid motion interpretation, we deduce a number of principles for the possible computations responsible for the interpretation of space-time geometry. Using these principles, we then discuss possible avenues on how to proceed in analyzing the representations of more complex human actions.

1 Introduction and Motivation

During the late eighties, with the emergence of active vision, it was realized that vision should not be studied in a vacuum but in conjunction with action. Adopting a purposive viewpoint makes visual computations easier by placing them in the context of larger processes that accomplish tasks. This new framework, known by a variety of names such as active, purposive, animate or behavioral vision, has contributed a wealth of new technical results and has fueled a variety of new application areas. Several groups constructed active vision systems and studied the basics of many visual competences, but for the most part, this study of the perception/action coupling concentrated on problems related to navigation. A flurry of activity produced a veritable cornucopia of algorithmic approaches to 3D motion estimation, 3D shape recovery, tracking, and motion and scene segmentation. As a matter of fact, one can say that this is the most mature area in the field of computer vision. This was, perhaps, not surprising since all these problems are related to the most basic action of all, one that transcends all systems with vision, that of self-motion; although the perception/action cycle surrounding the basic action of self-motion is not yet fully understood (we will discuss later some of the intricacies involved), it is clear that the problem is one of geometry and statistics awaiting the right modeling for its complete resolution. But an intelligent system with perception is not just a system that understands its movement, and the shape and movement of other objects in its field of view. It has a large number of capabilities related to recognition,

G. Sommer and Y. Y. Zeevi (Eds.): AFPAC 2000, LNCS 1888, pp. 1-21, 2000.

reasoning, planning, imagination and learning. Despite the fact that the active vision framework was embraced as the right tool in dealing with the recovery of 3D information from image sequences, that was not the case when it came to higher-level problems such as recognition, planning and reasoning. Lacking a foundation, not much progress has been made in those higher-level problems. For example, with the exception of simple objects in controlled environments, not much progress has been made in the field of object recognition. All new proposals during the past decade have followed some advance in the structure from motion problem.[1] There are, of course, many reasons for this, and some of them will become clear in the course of our exposition.

The goal of this paper is twofold, one methodological and the other technical. Our intent is to put forward the thesis that action represents the basic seed of intelligence, underlying perception of our environment, which is our major interest, recognition and other high-level processes such as communication. If this is the case, action representations are very basic components of an intelligent system. What are they? What could their nature and kind be? How can we build them? What does it mean to understand them? Most important, what techniques should we employ in order to obtain them? Answering these questions is our second goal and we approach it using only computational arguments. Before we begin, we need to discuss other efforts attempting to answer such kinds of questions.

2 Intelligence, Learning and the Stratification of Reality

Slowly but steadily, the paradigm that has dominated the study of cognition and the fields surrounding it for the past century is starting to be challenged. The dominant approach has been to view mental states as having meaning in terms of symbol systems like natural language. The emerging new approach is referred to as embodied intelligence or the sensorimotor theory of cognition which proposes that higher cognition makes use of the same structures as those involved in sensorimotor activity. This view is highly exemplified in recent efforts in AI to achieve intelligence through building robots and having them "develop" through interaction with their environment. Although there exist interesting points in the contemporary sensorimotor theories of cognition, and the projects of AI engineers have merits of their own, there exist two fundamental problems with them that we explain here.

The belief that smaller systems develop through interaction with their environment and become integrated to form new and more advanced systems owes its appeal to its intended mimicking of evolution. It makes, however, an implicit assumption, namely that evolution is evolution in degree and not evolution in kind. By evolution in kind we mean the introduction of new components having

[1] When it became clear that 3D information could be extracted from point and line correspondences, researchers concentrated on pose recovery in the polyhedral world. When view interpolation was achieved for some cases, recognition approaches considering objects as collections of 2D views appeared.

qualitatively different characteristics from the existing components. This leads to a strange kind of one-way relationship, both between the whole and its constituent parts within the system, and between higher systems and their primitive ancestors. One can define this relationship by saying that the whole is its parts, and continues to be so even if, as the result of the introduction of new subsystems (kind), it acquires a number of additional characteristics in the course of its evolution. The subsidiary systems themselves do not gain any higher characteristics and may even lose some in the process of simplification. The one-way relationship consists in the system as a whole possessing all the characteristics of its component parts, but none of these parts possesses the characteristics specific to the whole. Similarly, every system possesses most of the characteristics of its primitive ancestors, but even complete knowledge of a system's characteristics will not make it possible to predict those of its more highly developed descendants [17]. Thus, we can explain a system only if we accept the present structures in its body as our data.[2]

The second problem, related to the first, is best explained by utilizing the ideas of the philosopher Nicolai Hartmann [15]. Like all processes in life, those of acquiring and storing relevant information take place on many different levels and are interlinked at many points. The world in which we live, according to Hartmann, is built of different strata, each with its own existential categories that distinguish it from other strata. "There are in the hierarchy of existence certain phenomenal realities whose fundamental differences our minds fail to bridge .... Any true and accurate theory of categories must have as much regard for these gaps as for the existential relationships that bridge them." These relationships, however, only transcend in a unilateral manner the divisions between the four great strata of existence—the inorganic, organic, cognitive and conscious. The principles of existence and laws of nature that govern inorganic matter apply with equal validity to higher strata. But Hartmann insists that the differences between the higher and lower strata are far from restricted to the distinctions between inorganic, organic, cognitive and conscious. "The higher elements that make up the world are stratified in a similar way to the world itself," he wrote. This means that each step leading from a system of a lower order to one of a higher order is the same in nature and complexity as the coming into existence of life itself. This viewpoint about the stratification of existence can guard us against common mistakes in studying cognition. If we neglect the categories and laws that are exclusive properties of the higher strata or even deny their existence, we commit the fallacy of an upward transgression of the limits imposed upon us by the stratification of the real world. It is not possible to explain the laws and processes proper to higher levels in terms of categories derived from lower levels. Another common mistake results from transgressing the boundaries between levels of existence in the opposite direction. This is described by Hartmann

[2] Irrational Residue: The number of historical causes one would need to know to fully explain why an organism is as it is may not be infinite but it is sufficiently great to make it impossible to trace all causal chains to their end. As Polanyi pointed out [19], a higher animal cannot be reduced to its simpler ancestors.

as follows: The base of the entire world picture is then chosen on the level of conscious experience—the level on which man experiences his own subjective life—and from there the principle is extended downwards to the lower level of reality.

What does this all mean for us who wish to gain an understanding of perceptual mechanisms and, in particular, how action representations make up an intelligent system? There are several lessons we can obtain. It is a mistake, while seeking unitary explanatory principles behind the world, to try to explain lower and more primitive systems on the basis of principles applicable only to higher systems and vice versa. Similarly, the attempts of some psychologists and behavioral scientists to achieve intelligence through learning are at fault. This is true for both primitive systems that are incapable of it and systems of more advanced organisms which are not only incapable of being modified by learning but whose phylogenetic program makes them resistant to all modification.

The only chance at understanding intelligent systems seems to be to study them by accepting their present structures as our data, and work towards understanding the different components. But here is the tricky part. The components we study should be such that they do not require other components in order to be understood, or if they do, all interrelated components should be studied together. Otherwise it appears that there is no hope for understanding. And finally, our only tools should be the ones of the physical sciences, geometry, statistics and computation.

3 Understanding Actions, Objects

Our task is to increase our understanding of how to piece together the components of an intelligent system with vision. Our thesis is that action is fundamental in this regard. It is of course irresistible to attempt to arrive at a unitary world image and try to explain the diversity of the world (and intelligent systems) on the basis of ontological and phenomenal principles of one single kind. According to our previous analysis this would be a mistake after all. But our goals are much more modest. We want to uncover principles underlying the workings of the perceptual system and in particular shed light on the nature of action representations.

Our basic thesis is that the content of the mind is organized in the form of a model of the world. That model contains objects and their relationships, events and the interaction of the system with its environment. In building, maintaining, manipulating and working with that model we should distinguish three items: the world itself, the system (its body) and the system's interaction with the world. What could the best candidate be for building a foundation on the basis of which this complex model can be understood?

Recall that it is trivially true that understanding a thing is relating it to something already understood; if we continue along this chain, this means that there must be something understood in terms of itself. This entity must be something with which a system is intimately familiar in a non-symbolic way. We conclude that the best candidate for this is the agent's own activity.

There is a number of basic actions that are the basis of acquiring models of the shape and motion of the environment. These actions are related to the system's understanding of its own motion and they are universal, as all systems must have them in some form. The next sections are devoted to the computational mechanisms underlying the coupling of such actions with perception. By taking a look at these actions, on the one hand we can give an example and elaborate on what it means to understand these actions. On the other hand, since the computations involved in these basic actions form the foundation for many other computational processes we gain insight into the nature of space-time representations. That is, it allows us to deduce a number of principles on what the shape and motion models of the environment might be and how they could be computed. The computational principle of feedback loops is emerging as a foundational theme.

Next we are interested in the very large repertoire of more complex actions through which an agent interacts with his environment. Since understanding these actions should serve as a foundation for all other understanding, we should not presuppose understanding of other things. We merely assume that the agent understands its own body. Equipped with this knowledge and the foundation of space-time perception delivered from basic actions, we can provide a characterization of action as: *One understands an action if one is able to imagine performing the action with images that are sufficient for serving as a guide in actual performance.* This characterization provides two things. First, it shows that it is the ability to imagine an action which constitutes understanding it. Imagining it amounts to representing the constitutive components of the action as such, in a form that is capable of guiding the performance of the action. Because the components of the action form the content of the agent's mental state, the agent can be said to understand it. Second, this sort of understanding produces the feeling of confidence that one understands an action, because the understanding consists in a mental state of which the agent can be aware. Moving along this line of thought, understanding anything amounts to knowing the possible actions one might perform in relation to that thing and being aware of that understanding is consciously imagining those actions. Thus, we can say that one has an understanding of a perceived object if one is able to imagine incorporating the object into the performance of an action that is already understood, with an image sequence sufficient for serving as a guide in actual performance.

Thus, the ability to intentionally perform an action, manifested in the ability to imagine performing it, is a way of understanding the action that presupposes no other understanding. Similarly, an object that is intentionally incorporated into the performance of an (understood) action is itself understood in relation to that action. In moving from the understanding of actions to the understanding of objects we have crossed an important boundary. Understanding actions does not require a concept of the objective external world; that is how it can be a foundation for such a concept. But understanding objects does require such a concept.

It is not our intention to delve further into a philosophical inquiry about understanding objectivity and the mind/world distinction. Our purpose here is to achieve an understanding of action representations. Having a crisp characterization of what it means to understand an action we will be ready in Sect. 4 to investigate constraints on such representations. We will not delve into object recognition issues but the framework is in place to carry this investigation further into higher-level problems.

But remaining truthful to the principles of embodied intelligence, we cannot study in a technical sense actions unless we commit ourselves to a particular body. Of course the same is true for the basic actions and the perception of shape and motion, but in that case the dependence lies mostly in image acquisition (eye design) and computational capacity; in our days of extreme computational power these issues are not that important. But when it comes to higher-level actions and interaction with the world, it is important to choose a body. A failure to do so will turn our study completely philosophical. As such a body, we choose the human body.

4 Basic Actions

The most basic actions are the ones which can be based on the most simple model of 3D motion, and this is the model of instantaneous rigid motion. As living organisms move through the environment their eyes undergo a rigid motion. The visual system has to recognize and interpret this rigid motion from the visual input, that is, the sequence of images obtained by its eyes. This immediately gives rise to two basic actions: first, the capability of estimating one's self-motion or egomotion; second, the detection of objects in the static scene which move themselves, that is, the detection of independently moving objects. Similarly, the motion of various objects is rigid or can be closely approximated by one. Thus a third action arising from this model is the estimation of rigid object motion. This action, however, is much more difficult than the estimation of self-motion. One reason is that objects usually cover a small field of view and thus less data is available for the estimation. Another reason is that objects lie outside the system's body and thus there is less information from other sensors. In contrast, for self-motion estimation biological systems—in addition to vision—also use inertial sensors (such as those in our ears). As a result, most systems only possess the capability to partially estimate rigid object motion. Even rather simple organisms understand the approaching of the enemy, or can estimate how long it will take to intercept an object (the estimation of time to contact). A somewhat more sophisticated but still partial estimation would be a rough estimate of the object's translation or rotation.

The visual input is a sequence of images, a video. Aside from how the images are formed (that is, the properties of the eye or camera), the images depend on the spatiotemporal geometry, that is, the layout of the scene and the relative 3D motion between observer and scene, and the optics, that is, the light sources and the reflectance properties of the surfaces. By considering only changes of image patterns, an image representation of the movement of the 3D scene points is obtained. This representation, a vector field, ideally is only due to the geometry that is the relative 3D motion between the eye and the scene and the shape of the scene. If the relative 3D motion between the observer and parts of the scene is rigid, then segmentation amounts to the localization of those parts of the flow field corresponding to the same rigid motion and motion estimation amounts to decoding the 3D motion parameters. The shape of the scene determines the remaining components of the vector field and it can be decoded after motion estimation.

The interpretation of motion information, probably because of its well-defined geometric nature, has received a lot of attention in the computer vision literature. A methodology emerged under the hood of which most studies were conducted, which defined rigid motion interpretation as the "structure from motion" problem and it was considered to be carried out with three consecutive computational steps. First the exact movement of image points is estimated, either in the form of the optical flow field if images densely sampled in time are considered, or in the form of correspondences of single image points if images further apart in time are used. The optical flow field or correspondences are approximations of the projections of the 3D motion field which represents the movement of scene points. The second computational step consists in estimating the 3D rigid motion from the optical flow field, and the third step amounts to estimating the 3D structure of the scene usually from multiple flow fields using the estimates of the 3D motion.

With this framework it has been considered that exact reconstruction of space-time geometry is possible—it is just a matter of using the right sophisticated tools—and it can be addressed in a purely bottom-up manner computationally. We'll argue that this cannot be. Throughout this and the next section we'll provide computational arguments supporting our view and elaborate on what the computations should be.

First, optical flow or correspondence cannot be estimated accurately using only the image data, that is, before performing estimations about the 3D geometry. The local image data defines only one component of the image velocity. This is the so-called aperture problem illustrated in Fig. 1: Local image data usually only provides one-dimensional spatial structures. From the movement of these linear features between image frames the movement of single image points cannot be determined, but only the flow component perpendicular to the linear feature; this is the so-called normal flow. In order to estimate the two-dimensional image flow, a model of the flow field is needed; then normal measurements in different directions within local neighborhoods can be combined. Here lies the problem: In order to model the flow field, additional information is necessary. It is easy to model the parts of the image which correspond to smooth scene patches by using smoothness assumptions of different kinds, but there are discontinuities in the flow field, and their locations must be known prior to modeling the flow

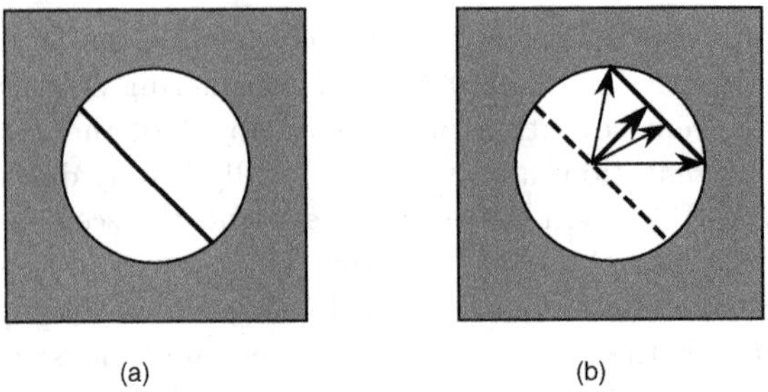

Fig. 1. Aperture problem: (a) Line feature observed through a small aperture at time t. (b) At time $t + \delta t$ the feature has moved to a new position. It is not possible to determine exactly where each point has moved to. From local measurements only the flow component perpendicular to the line feature can be computed

field. The discontinuities in the flow field are due either to independently moving objects in the scene or to scene surfaces at different depths. On the other hand, the discontinuities cannot be detected unless an estimate of the flow field is available. Thus the problems of estimating the flow field and locating the discontinuities are inherently coupled and the situation seems to present itself as a chicken and egg problem. Discontinuities provides an even more severe problem than usually considered. Most of the literature considers the idealized problem of estimating the motion for one rigidly moving observer in a static environment. Thus all the discontinuities they deal with are due to depth. Even if the locations of the discontinuities in the flow field are known, a system would need additional information about the 3D motion and scene to distinguish the motion discontinuities from the pure scene discontinuities. There also has been a lot of work in the literature on motion segmentation, but these studies consider the problem separate from 3D motion estimation: either they attempt a localization of the discontinuities on the basis of flow field information only, or they assume the 3D motion to be known. It is clear that for a real system the two problems of 3D motion estimation and independent motion detection have to be considered together.

Besides the problem due to discontinuities there are also statistical difficulties with the estimation of optical flow. As will be elaborated on later, the estimation of optical flow from noisy normal flow measurements even within neighborhoods of smooth flow is very difficult. To avoid bias and obtain very accurate flow, theoretically one would need detailed models of the flow field, and these in turn can only be obtained from 3D information; specifically, this means the discontinuities, the shape of the scene and the 3D motion.

Since accurate optical flow estimation bottom up seems to be infeasible, the question is, is it possible to perform 3D motion estimation using as input normal flow measurements? In the past we have studied this question, and we have developed two constraints that allow to extract 3D geometry directly flow normal flow. One constraint that relates image motion to 3D motion only, and a second one which also involves the scene.

5 Relating Image Measurements to 3D Geometry

Consider a system moving with rigid motion in a static environment. The motion is described by an instantaneous translational velocity $\mathbf{t}$ and a rotational velocity $\boldsymbol{\omega}$. Each scene point $\mathbf{R} = (X, Y, Z)$ measured with respect to a coordinate system $OXYZ$ fixed to the nodal point of the system's eye where Z is the optical axis has velocity $\dot{\mathbf{R}} = -\mathbf{t} - \boldsymbol{\omega} \times \mathbf{R}$.

We consider the image to be formed on a plane orthogonal to the Z axis at distance f (focal length) from the nodal point. Thus through perspective projection image points $\mathbf{r} = [x, y, f]$ are related to scene points $\mathbf{R}$ by $\mathbf{r} = \frac{\mathbf{R}f}{\mathbf{R}\cdot\hat{\mathbf{z}}}$, where $\hat{\mathbf{z}}$ is a unit vector in the direction of the Z axis. The projection of the 3D motion field on the image gives the image motion field

$$\dot{\mathbf{r}} = \frac{1}{(\mathbf{R}\cdot\hat{\mathbf{z}})}(\hat{\mathbf{z}} \times (\mathbf{t} \times \mathbf{r})) + \frac{1}{f}\hat{\mathbf{z}} \times (\mathbf{r} \times (\boldsymbol{\omega} \times \mathbf{r})) = \frac{1}{Z}\mathbf{u}_{\mathrm{tr}}(t) + \mathbf{u}_{\mathrm{rot}}(\omega) \quad (1)$$

where Z is used to denote the scene depth $(\mathbf{R}\cdot\hat{\mathbf{z}})$ and $\mathbf{u}_{\mathrm{tr}}, \mathbf{u}_{\mathrm{rot}}$ are the direction of the component of the flow due to translation and the component of the flow due to rotation, respectively. Due to the coupling of Z and $\mathbf{u}_{\mathrm{tr}}$ in this relationship only the direction of translation (the Focus of Expansion, FOE, or focus of contraction, FOC, depending on whether the observer is approaching or moving away from the scene), scaled depth and the three rotational parameters can be obtained. The direction of the axis of rotation will be denoted as AOR. The normal flow $\mathbf{v_n}$ amounts to the projection of the flow $\dot{\mathbf{r}}$ on the local gradient direction. If $\mathbf{n}$ is a unit vector denoting the orientation of the gradient, the value of the normal flow, v_n amounts to

$$v_n = \dot{\mathbf{r}} \cdot \mathbf{n}$$

The first constraint, the "pattern constraint," uses as its only physical constraint the fact that the scene lies in front of the camera and thus all the depth values have to be positive. This allows the relation of the sign of normal flow measurements to the directions of the translation and the rotation, that is, the FOE and AOR. The basis lies in selecting groups of measurements of normal flow along pre-defined directions (orientation fields) which form patterns in the image plane, and these patterns encode the 3D motion [4], [5], [6], [7].

There are two classes of orientation fields. The first class are called copoint fields. Each copoint field parameterized by $\mathbf{s}$, is defined as the unit vectors perpendicular to a translation flow field with translation $\mathbf{s}$. These unit vectors are in the direction $\mathbf{v}_{\mathrm{cp}}(\mathbf{s}, \mathbf{r}) = \hat{\mathbf{z}} \times \mathbf{u}_{\mathrm{tr}}(\mathbf{s}) = \hat{\mathbf{z}} \times (\hat{\mathbf{z}} \times (\mathbf{s} \times \mathbf{r}))$ and they are perpendicular to the bundle of lines passing through $\mathbf{s_0}$ (the point where $\mathbf{s}$ intersects the image plane) (Fig. 2a).

Now consider a normal flow field due to rigid motion $(\mathbf{t}, \boldsymbol{\omega})$ and consider all the normal flow vectors of this field which are in direction $\mathbf{v}_{\mathrm{cp}}(\mathbf{s}, \mathbf{r})$, that is,

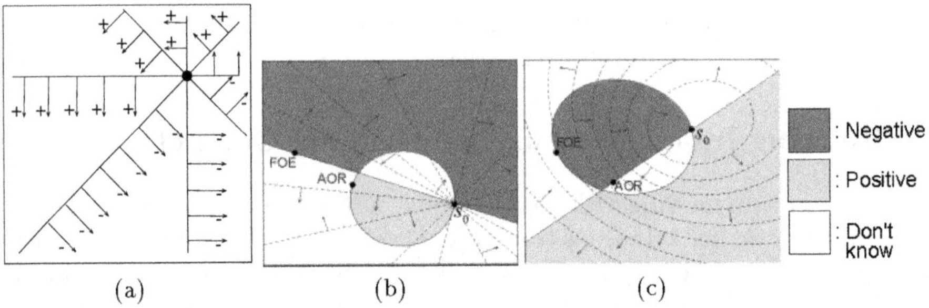

Fig. 2. (a) Copoint field filled with qualitative motion measurements. (b) copoint vectors and copoint pattern. (c) coaxis vectors and patterns

the vectors of value $\frac{\mathbf{v}_{cp}}{\|\mathbf{v}_{cp}\|} \cdot \dot{\mathbf{r}}$. The translational components of these vectors are separated by the conic $(\boldsymbol{\omega} \times \mathbf{r}) \cdot (\mathbf{s} \times \mathbf{r}) = \mathbf{0}$ into an area with negative values and an area with positive values. Similarly, the rotational components of the same vectors are separated by the line $(\mathbf{t} \times \mathbf{s}) \cdot \mathbf{r} = \mathbf{0}$ into areas of positive and negative values. The rigid motion field is the sum of the translational and the rotational flow field; thus the areas above need to be superimposed. Where both the translational and rotational components are positive the combined flow is positive and where both the translational and rotational components are negative the combined flow is negative. In the remaining areas the sign is undetermined. The areas of different sign form a pattern in the image plane as shown in Fig. 2b. For every direction $\mathbf{v}_{cp}$ parameterized by $\mathbf{s}$ there is a different pattern and the intersection of the different patterns provides the 3D motion.

The second class of patterns is defined by the orientation field of coaxis vectors, which are the unit vectors perpendicular to a certain rotational flow field, that is, the coaxis vectors parameterized by $\mathbf{s}$ are in direction $\mathbf{v}_{cA} = \hat{\mathbf{z}} \times (\hat{\mathbf{z}} \times (\mathbf{r} \times (\boldsymbol{\omega} \times \mathbf{r})))$. In this case the translational components are separated by a line and the rotational components are separated by a cone into positive and negative values and give patterns such as that shown in Fig. 2c. Using these pattern constraints the problem of self-motion estimation amounts to recognizing particular patterns in the image plane. The system would compute the normal flow and then for a number of orientation fields find the patterns which are consistent with the sign of the estimated normal flow vectors. The intersection of the possible patterns provides the solution, usually in the form of possible areas for the FOE and AOR, as most often there is not enough data to provide an exact localization of the 3D motion.

The question now is, is it possible at all to have a technique that accurately computes the 3D rigid motion parameters? The answer is no, if the data available is from a limited field of view. This statement is based on an error analysis conducted in [8], [9]. Estimating 3D motion amounts to minimizing some function. If the pattern constraints are used, this function is based on negative depth values and if the classical epipolar constraint requiring optical flow is used, this function is based on the distances of the flow vectors from the epipolar lines. The correct 3D motion is defined as the minim um in the functions. A topographic analysis of the expected values showed that for all these functions the minimum lies within a valley, and locating the exact position of the minim um within this valley under noisy conditions generally is not possible.

If the 3D motion is known, the depth of the scene can be computed from the flow field. If $\hat{\mathbf{t}}$ and $\hat{\boldsymbol{\omega}}$ are the estimates of the direction of translation and the rotation the estimated depth $\hat{Z}$ is derived from (1) as

$$\hat{Z} = \frac{\mathbf{u}_{\mathrm{tr}}(\hat{\mathbf{t}}) \cdot \mathbf{n}}{\dot{\mathbf{r}} \cdot \mathbf{n} - \mathbf{u}_{\mathrm{rot}}(\hat{\boldsymbol{\omega}}) \cdot \mathbf{n}} \tag{2}$$

If the estimates $\hat{\mathbf{t}}$ and $\hat{\boldsymbol{\omega}}$ correspond to the correct values of the actual rigid motion $\mathbf{t}$ and $\boldsymbol{\omega}$, then $\hat{Z}$ will give a correct scene depth estimate, but what if there are errors in the estimates? Then the estimated depth will be a distorted version of the actual depth [2], [3], [8], [11].

Substituting for $\dot{\mathbf{r}}$ in (2) the actual motion parameters the estimated depth can be written as

$$\hat{Z} = Z \cdot \frac{\mathbf{u}_{\mathrm{tr}}(\hat{\mathbf{t}}) \cdot \mathbf{n}}{\mathbf{u}_{\mathrm{tr}}(\mathbf{t}) \cdot \mathbf{n} - \mathbf{Z}\mathbf{u}_{\mathrm{rot}}(\delta\boldsymbol{\omega}) \cdot \mathbf{n}},$$

with $\delta\boldsymbol{\omega}$ denoting the estimation error in rotation $\delta\boldsymbol{\omega} = \hat{\boldsymbol{\omega}} - \boldsymbol{\omega}$, or to make clear the distortion as $\hat{Z} = Z \cdot D$, where $D = \frac{\mathbf{u}_{\mathrm{tr}}(\hat{\mathbf{t}}) \cdot \mathbf{n}}{\mathbf{u}_{\mathrm{tr}}(\mathbf{t}) \cdot \mathbf{n} - \mathbf{Z}\mathbf{u}_{\mathrm{rot}}(\delta\boldsymbol{\omega}) \cdot \mathbf{n}}$ is the multiplicative distortion factor to express how wrong depth estimates result from inaccurate 3D motion values. Note that D also depends on $\mathbf{n}$. For different directions, $\mathbf{n}$, the distortion factor takes on different values, ranging from $-\infty$ to $+\infty$.

This concept of space distortion forms the basis of the second constraint relating normal flow to 3D geometry, the "depth variability constraint." The idea is, instead of form ulating smoothness constraints on the 2D optical flow field as is usually done, to relate the smoothness of scene patches and the 3D motion directly to normal flow. To be more specific, if we estimate the depth from the flow field corresponding to a smooth scene patch using the correct 3D rigid motion, we will compute a smooth scene patch. If we use an erroneous 3D motion estimate, then the obtained depth function, assuming that normal flow vectors in different directions are available, will be rugged. Not only do incorrect estimates of motion parameters lead to incorrect depth estimates, but the distortion is such that the worse the motion estimate, the more lik ely depth estimates are obtained that locally vary more than the correct ones. Thus one has to define functions which express the variability of depth values and search for the 3D motion which minimizes these functions. Locally for the flow values $\dot{\mathbf{r}}_i$ in direction $\mathbf{n}_i$ within a region R we define depth variability measures as

$$\theta_0\left(\hat{\mathbf{t}}, \hat{\omega}, R\right) = \sum W_i \left(\dot{\mathbf{r}}_i \cdot \mathbf{n}_i - \mathbf{u}_{\mathrm{rot}}(\mathbf{u}) \cdot \mathbf{n}_i - \frac{1}{\hat{Z}}\left(\mathbf{u}_{\mathrm{tr}}\left(\hat{\mathbf{t}}\right) \cdot \mathbf{n}\right)\right)^2$$

where $1/\hat{Z}$ is the depth estimate minim izingθ_0, which is obtained by modeling the depth within the region using a parametric model, and W_i are weights that can be chosen differently. Solving for the minimum $1/\hat{Z}$ and substituting back we obtain θ_1. The sum of all θ_1 in all regions, θ_2, is the global function which we are interested in minimizing.

To algorithmically utilize these constraints one has to proceed in iterations as follows. Split the image into patches and perform a search in the space of translational directions. For every candidate translation estimate the best rotation using the depth variability measure, then perform a depth segmentation using the image data and the candidate motion parameter, and finally evaluate the depth variability measure taking into account the segmentation. The solution is found as the 3D motion minimizing the global depth variability measure.

In summary of the above discussion a number of conclusions can be drawn regarding the possible computations of space-time geometry.

- It is not possible to reliably estimate the 3D rigid motion very accurately if the field of view is limited. This is true even for the simplified case of one camera moving in a static environment. If there is less data available—as it is when one considers motion of objects covering a small field of view—the estimation is much complicated and all we can expect are bounds on the motion parameters.
- It is also not possible for a real-time system to derive accurate depth estimates. Considering the cue of motion, this follows from the fact that 3D motion cannot be estimated accurately. As a result of this one obtains a distorted depth map, but it also has been shown experimentally for other cues that human space perception is not exactly Euclidean [16].
- Since reconstruction is not possible the information describing the models must be encoded in the form of patterns in the flow field.
- The computations carrying out the space-time geometry interpretation must be implemented as feedback processes. This is true even for the simple, basic actions as will be elaborated on in the next section, and for more complicated actions the interplay between bottom-up and top-down processes becomes extremely important.

6 Feedback

The interpretation of motion fields is a difficult computational problem. From the raw images only, it is not possible to compute a very accurate 2D motion field, and from erroneous motion field measurements it is not possible to perform good 3D motion estimation, and, in the sequel, scene estimation. To estimate good flow we need information about the space-time geometry, and to estimate the geometry well we need at least the flow field to be segmented accurately.

This seemingly chicken and egg situation calls for a simultaneous estimation of image motion and 3D motion and scene geometry, and this could only be implemented through an iterative or feedback process. The whole interpretation process should take the following form: First, the system estimates approximate image velocity by combining normal flow measurements. The representation of these estimates could be in the form of qualitative descriptions of local flow field patches or bounds on the flow values, but it should allow for finding the more easily detected discontinuities in the flow field. Using the flow computed in this way or maybe even normal flow measurements within segmented areas an estimate of 3D motion is derived and a partial 3D shape model of the scene computed. Subsequently, the computed information about the space-time geometry is fed back to use image measurements from larger regions to perform better flow estimation, more accurate discontinuity localization, and in the sequel the estimates of 3D motion and structure are improved. Naturally the whole interpretation process has to be dynamic. While the system computes structure and motion from one flow field, the geometry changes and new images are taken. Thus the system also has to be able to make predictions and relate the 3D information computed earlier in time to images acquired later.

There are a number of ways these feedback processes could be implemented in a system. From a computational perspective we can ask for an optimal way, and this will depend on the computational power of the system and the accuracy of the estimates needed. There are many efforts in the biological sciences, in fields such as psychophysics, neurophysiology and anatomy, to figure out the structure and processes which exist in human and other primate brains—this structure is generally referred to as the motion pathway; but many of these studies do not consider that the sole purpose of motion processing is to interpret the 4D space-time geometry. We'll next elaborate on two visual illusions that in our opinion demonstrate that the human visual system must perform feedback processes to interpret image motion.

6.1 The Ouchi Illusion

The striking Ouchi illusion shown in Fig. 3 consists of two black and white rectangular checkerboard patterns oriented in orthogonal directions—a background orientation surrounding an inner ring. Small retinal motions, or slight movements of the paper, evince a segmentation of the inset pattern, and motion of the inset relative to the surround. Our explanation lies in the estimation of differently biased flow vector fields in the two patterns which in turn give rise to two different 3D motion estimates. When the system feeds back these motion estimates it performs a segmentation of the image using the flow fields and the static image texture, and because of the difference in 3D motion one pattern appears to move relative to the other.

The bias is easily understood by considering the most simple model of image motion estimation. Local image measurements define at a point one component of the image flow, for example, if the image measurements are the spatial derivatives E_x, E_y and the temporal derivative E_t of the image intensity function, E, at point $\mathbf{r}_i$ the flow $\dot{\mathbf{r}} = (u, v, 0)$ is constrained by the equation

$$E_{x_i} u + E_{y_i} v E_{t_i} = E_{t_i} \tag{3}$$

We consider additive zero mean independent noise in the derivatives of E. Let n be the number of indices i to which (3) applies, which for convenience we write in matrix form as

$$\mathbf{E}_s \mathbf{u} = \mathbf{E}_t \tag{4}$$

where $\mathbf{E}_s$ is the n by 2 matrix incorporating the spatial derivatives E_{x_i}, E_{y_i}, $\mathbf{E}_t$ is an n by 1 matrix incorporating the E_{t_i}, and $\mathbf{u}$ is the vector (u, v).

If (4) is solved by standard least squares estimation the solution is

$$\mathbf{u} = \left(\mathbf{E}_s{}^{\top}\mathbf{E}_s\right)^{-1} \mathbf{E}_t{}^{\top}\mathbf{u}$$

and this solution, if there are errors in the spatial derivatives, is biased. The expected value, $E(\mathbf{u})$, of $\mathbf{u}$ amounts to

$$E(\mathbf{u}) = \mathbf{u}' - n\sigma_s^2 \left(\mathbf{E}_s{}'^{\top}\mathbf{E}_s{}'\right)^{-1} \mathbf{u}',$$

where primes are used here to denote actual values and σ_s^2 is the variance of the noise in the spatial derivatives. This expected value of $\mathbf{u}$ is an underestimate in length and in a direction closer to the majority of gradient directions in the patch.

It has been shown in [13], [14] that it is not only gradient based techniques and least squares estimation that produce bias, but other techniques (frequency-domain, correlation techniques, and different estimation procedures) as well. Correcting for the bias would require knowledge of the noise parameters and these are difficult to obtain. In the case of the Ouchi illusion the 3D motion, which is caused by small eye movements or rapid motion of the paper, changes very quickly so the system does not have enough data available to acquire the statistics of the noise. In more natural situations, however, the system can reduce the bias problem if it performs feedback processes. This way, after some estimates of motion and structure have been obtained, image measurements from larger image regions can be considered to acquire more accurate statistical noise sampling.

6.2 *Enigma* and Variants

The static figure, *Enigma*, painted by Leviant and shown in Fig. 4, consists of radial lines emanating from the center of the image and interrupted by a set of concentric, uniformly colored rings. Upon extended viewing of *Enigma*, most humans perceive illusory movement inside the rings which keeps changing direction. Based on a study of positron emission tomography, researchers in neurophysiology [20], [21] have argued that higher level processes of motion interpretation in the human brain are responsible for the perception of this illusion. To be more specific, these researchers compared the cerebral blood flow during the viewing of *Enigma* and a similar reference image which does not give rise to an illusion, and they found the values comparable in the early motion processing area V1, but found significantly higher values for *Enigma* in V5 (or MT)—a later area in visual processing.

The spatial texture in *Enigma* is closely related to the patterns described in Sect. 5 that form constraints for 3D motion estimation. This lets us hypothesize that the illusion essentially is due to higher level activity. It is due to the

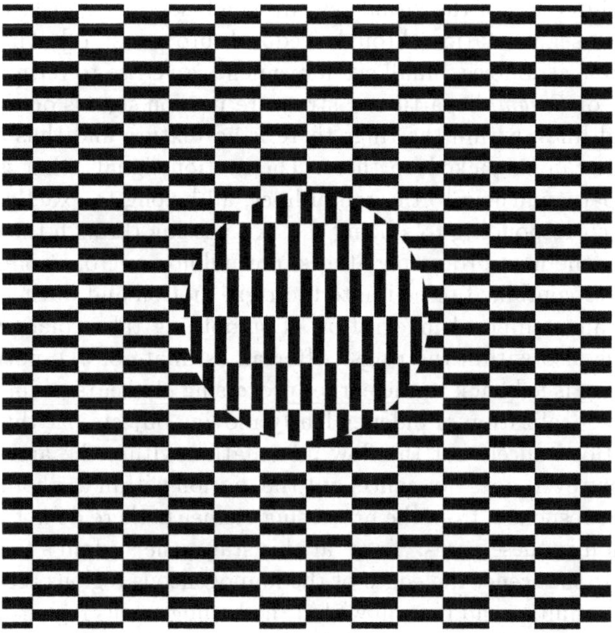

Fig. 3. A pattern similar to one by Ouchi [18]

Fig. 4. *Enigma*, giving rise to the Leviant illusion: Fixation at the center results in perception of a rotary motion inside the rings

particular architecture of the visual system which solves the tasks of 3D motion estimation and segmentation through feedback processes [12]. The explanation in more detail is as follows. Small eye movements give rise to retinal motion signals. Because of *Enigma*'s particular structure motion signals occur only in the areas of radial lines; within the rings there is no texture and thus also no early motion signals can be picked up there. The motion signals, that is the normal flow vectors perpendicular to the radial lines, all belong to exactly one copoint field, the one corresponding to s being the axis passing through the center of the image. During the next processing stages, normal flow vectors from increasingly larger areas are combined to estimate the rigid motion and during these stages some spatial and temporal integration of smoothing takes place which causes motion signals also within the rings. After an estimate of 3D motion is found, this information is fed back to the earlier processing elements which then have to perform exact localization of independently moving objects. At this stage, temporal integration cancels out the flow information within the areas covered by the rays. It, however, does not cancel out the motion within the homogeneous regions since the responses from the processing elements there do not contradict an existing motion in these areas. Since, at this stage, the system's task is to accurately segment the scene, the edges of the homogeneous regions are perceived as motion discontinuities and motion within the homogeneous regions is reinforced. The defining feature of *Enigma* is that all the spatial gradients in its patterns excite only one class of vector fields involved in 3D motion estimation—a copoint field. To test our hypothesis we created figures based on the same principle as *Enigma*—that is with black and white rays giving rise to normal flow vectors corresponding to exactly one coaxis or copoint vector field and homogeneous areas perpendicular to these rays. For an example, see Fig. 5. We found that it is exactly these patterns giving rise to illusory motion. Other figures based on the same principle (rays interrupted by perpendicular homogeneous areas) which do not correspond to the vector fields involved in 3D motion estimation have been found not to give rise to an illusion, thus supporting the hypothesis.

7 Complex Actions

One may be able to derive a simple procedure for recognizing some action based on the statistics of local image movement. There exists a variety of such techniques in the current literature mostly applied to human movement or facial expressions, but according to our account of what it means to understand an action, representations that enable this sort of recognition are simply inadequate. What we need are representations that can visualize the action under consideration. Let us summarize a number of key points that are important in figuring out the nature of action representations:

1. Action representations should be view independent. We are able to recognize and visualize actions regardless of viewpoint.

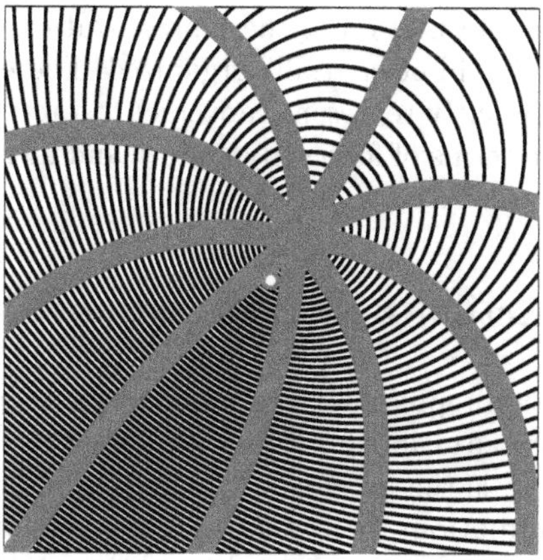

Fig. 5. Example of an illusory movement figure based on a coaxis field. Here the focal length is chosen to be about equal to the size of the image. Thus, the illusion is best experienced at a short distance

2. Action representations capture dynamic information which is manifested in a long image sequence. Put simply, it is not possible to understand an action on the basis of a small sequence of frames (viewpoints).
3. It is the combination of shape and movement that makes up an action representation.

A number of issues become clear from the impositions of the constraints listed above. *First*, to understand human action, a model of the human body should be available. *Second*, the movement of the different parts of the body relative to each other and to the environment should be recovered in a way that allows matching. *Third*, since action amounts to the movement of 3D space, action representations should contain information about both components. Let us further examine those issues in order. Acquiring a sophisticated model of the human body, although of extreme practical importance, is not essential to our discussion. We would like to caution, however, against learning approaches. It is probably an impossible task to learn such a model from visual data in a reasonable time. It is quite plausible that humans are born with such a model available to them. After all, it is a model of themselves. Later, we will describe technological solutions to this question. The second issue deserves more attention, because it introduces constraints. The previous section discussed the difficulties in acquiring 3D motion and shape, when normal motion fields were available in the whole visual field. From this we conclude that it is impossible to acquire exact 3D motion information from a very small part of the visual field, that is, it is not feasible to accurately estimate the 3D motion of the different rigid parts of the body from image sequences of it. Whatever the action representation is, it appears very hard to achieve recognition in a bottom-up fashion. The concept of feedback explained in the previous section acquires here a major importance. Recognizing action should be mostly a top-down process. Finally, the third issue reflects the need to incorporate both shape and motion in the representation. This means that any motion representation should be spatially organized. There are, however, various options here. Do we explicitly represent movement in terms of motion fields, or do we do it implicitly in terms of forces, dynamics, and inertial constraints? Both options have their pros and cons.

8 How Could Action Representations Be?

To gain insights on action representations we consider them in a hierarchy with two more kinds of representations and the properties of the possible mappings among them. *First* there is the image data, that is, videos of humans in action. Considering the cue of motion, then our image data amounts to a sequence of normal flow fields computed from the videos. The second kind of representations are intermediate descriptions encoding information about 3D space and 3D motion, estimated from the input (video). These representations consist of a whole range of descriptions of different sophistication encoding partially the space-time geometry, and they are scene dependent. Finally, we have the action representations themselves, which are scene independent.

Before we start considering the relationships in this hierarchy, let us introduce a novel and very useful tool in the study of action. It is a tool because it can help us in this study, but at the same time it amounts to a theoretical intermediate representation that combines the advantages of view independence with the capture of the action as a whole. It is a theoretical tool in the same way that the image motion field is a theoretical construct. An intermediate representation for the specific action in view is then a sequence of evolving 3D motion fields and it is the most sophisticated intermediate description that could be obtained. Acquiring this representation is no simple matter, but it can be achieved by employing a very large number of viewpoints. We have established in our laboratory a multi-camera network (sixty-four cameras, Kodak ES-310, providing images at a rate of eighty-five frames per second; the video is collected directly on disk—the cameras are connected by a high-speed network possessing sixteen dual processor Pentium 450s). By observing human action from all viewpoints we are able to create a sequence of 3D motion fields [10]. See Fig. 6 that displays a schematic description of our facility. The solution requires a calibration of the camera network [1] and proceeds by carving the rays in space using image intensity and image motion values.

Let us now consider the mapping from the second level of the hierarchy (intermediate descriptions and, in particular, 3D motion fields) to the scene independent action representations (top level). This mapping should be such that it extracts from a specific action quantities of a generic character common to all actions of the same type. These quantities most probably take the form of spatiotemporal patterns in four dimensions.

One way of obtaining such patterns is to perform statistics on a large enough sample. Considering, for example, a particular action (e.g., walking), we can obtain data in the multi-camera laboratory described before for a large number

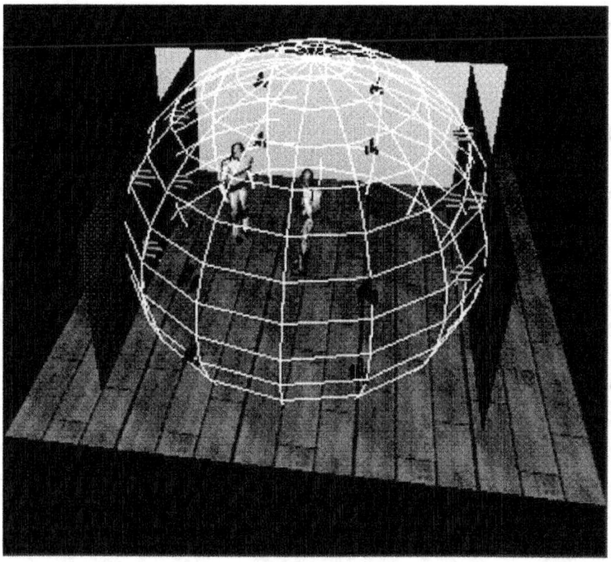

Fig. 6. A negative spherical eye

of individuals. In each case we can obtain a 3D motion field and thus have at our disposal a large number of 3D motion fields. A number of statistical techniques, such as principal component analysis, can reduce the dimensionality of this space and describe it with a small number of parameters, which parameters will in turn constitute a representation for the generic action under consideration.

Another way of obtaining these patterns would be to employ a geometric approach as well. Since a human action consists of a large number of different rigid motions, one possibility is to obtain these rigid motions from the 3D motion field, encode their relationships, and study invariances related to symmetry, and geometric quantities in space-time (angles, velocities, accelerations, periodicity, etc.).

Considering now the mappings from the top level to the image data and to the intermediate representations we can gain further insight. For the sake of recognition, action descriptions will have to be projected on the image and matched against the intermediate descriptions and the image data. But, as explained in previous sections, these intermediate descriptions are errorful and constitute distorted versions of the scene. In addition, there is variability among actions of the same type. In trying to perform this matching in space we will have to revise both the action representations and the nature of the intermediate descriptions so that the matching can be facilitated. In actual fact we may have to develop specific spatiotemporal patterns in the image data for this goal and these patterns will constitute some form of action representations for quick recognition. To conclude, the interplay among the mappings between the introduced hierarchy levels seems to be a feasible and structured way for building powerful action descriptions. This constitutes our current research efforts.

9 Conclusions

Being truthful to Hartmann's and Lorenz's approach regarding evolution and the structure of intelligent systems, we wish to study intelligent systems as they actually are, using the present structures in their bodies as our data. Considering representations for a number of important complex actions, we have two options about their origin: either the system learned them or it was born with them. We find the argument about the innate character of (at least some) action representations very appealing, because humans have retained the same basic structure and have been performing, for the most part, the same repertoire of actions for a very long time. It makes good sense to build these representations using all available knowledge and technology. This paper offered a number of ways for accomplishing this goal.

References

1. P. Baker and Y. Aloimonos. Complete calibration of a multi-camera network. In *Proc. IEEE Workshop on Omnidirectional Vision*, pages 134–141, Los Alamitos, CA, 2000. IEEE Computer Society.
2. T. Brodský, C. Fermüller, and Y. Aloimonos. Structure from motion: Beyond the epipolar constraint. *International Journal of Computer Vision*, 2000. In press. Also available as CfAR Technical Report CAR-TR-911.
3. L. Cheong, C. Fermüller, and Y. Aloimonos. Effects of errors in the viewing geometry on shape estimation. *Computer Vision and Image Understanding*, 71:356–372, 1998.
4. C. Fermüller and Y. Aloimonos. Direct perception of three-dimensional motion from patterns of visual motion. *Science*, 270:1973–1976, 1995.
5. C. Fermüller and Y. Aloimonos. Qualitative egomotion. *International Journal of Computer Vision*, 15:7–29, 1995.
6. C. Fermüller and Y. Aloimonos. Vision and action. *Image and Vision Computing*, 13:725–744, 1995.
7. C. Fermüller and Y. Aloimonos. On the geometry of visual correspondence. *International Journal of Computer Vision*, 21:223–247, 1997.
8. C. Fermüller and Y. Aloimonos. Ambiguity in structure from motion: Sphere versus plane. *International Journal of Computer Vision*, 28:137–154, 1998.
9. C. Fermüller and Y. Aloimonos. Observability of 3D motion. *International Journal of Computer Vision*, 2000. In press.
10. C. Fermüller, Y. Aloimonos, P. Baker, R. Pless, J. Neumann, and B. Stuart. Multi-camera networks: Eyes from eyes. In *Proc. IEEE Workshop on Omnidirectional Vision*, pages 11–18, Los Alamitos, CA, 2000. IEEE Computer Society.
11. C. Fermüller, L. Cheong, and Y. Aloimonos. Visual space distortion. *Biological Cybernetics*, 77:323–337, 1997.
12. C. Fermüller, R. Pless, and Y. Aloimonos. Families of stationary patterns producing illusory movement: Insights into the visual system. *Proc. Royal Society, London B*, 264:795–806, 1997.

13. C. Fermüller, R. Pless, and Y. Aloimonos. The Ouchi illusion as an artifact of biased flow estimation. *Vision Research*, 40:77–96, 2000.
14. C. Fermüller, D. Shulman, and R. Pless. The statistics of optical flow. Technical Report CAR-TR-928, Center for Automation Research, University of Maryland, 1999.
15. N. Hartmann. *Der Aufbau der Realen Welt.* de Gruyter, Berlin, 1964.
16. J. J. Koenderink and A. J. van Doorn. Relief: Pictorial and otherwise. *Image and Vision Computing*, 13:321–334, 1995.
17. K. Lorenz. *Behind the Mirror: A Search for a Natural History of Human Knowledge.* Harcourt Brace Jovanovich, New York, 1977.
18. H. Ouchi. *Japanese and Geometrical Art.* Dover, 1977.
19. M. Polanyi. *Personal Knowledge: Towards a Post-Critical Philosophy.* The University of Chicago Press, Chicago, 1958.
20. S. M. Zeki. The cortical *enigma*: A reply to Professor Gregory. *Proc. Royal Society, London B*, 257:243–245, 1994.
21. S. M. Zeki. Phenomenal motion seen through artificial intra-ocular lenses. *Proc. Royal Society, London B*, 260:165–166, 1995.

The Systems Theory of Contact

Leo Dorst and Rein van den Boomgaard

Research Institute for Computer Science, University of Amsterdam
Kruislaan 403, 1098 SJ Amsterdam, The Netherlands
{leo,rein}@wins.uva.nl

Abstract. The sense of touch and the capability to analyze potential contacts is important to many interactions of robots, such as planning exploration, handling objects, or avoiding collisions based on sensing of the environment. It is a pleasant surprise that the mathematics of touching and contact can be developed along the same algebraic lines as that of linear systems theory. In this paper we exhibit the relevant spectral transform, delta functions and sampling theorems. We do this mainly for a piecewise representation of the geometrical object boundary by Monge patches, i.e. in a representation by (umbral) functions. For this representation, the analogy with the linear systems theory is obvious, and a source of inspiration for the treatment of geometric contact using a spectrum of directions.

1 Kissing Contact

1.1 Intuition and Simplification

When two objects touch (Fig.), they are locally tangent. More precisely, at the point of contact they have tangent planes with a common attitude and location, but opposite orientation. Such a contact is called 'kissing'.

Kissing is a local property. It is often made impossible when the objects would penetrate each other at other locations along their boundary. This makes the operation hard to analyze (for instance, it becomes non-differentiable), and one often does not get further than a lattice-theoretic (and hence rather qualitative) analysis [ ]. To go beyond that, we must regularize the operation, and boldly decide to ignore those overlaps elsewhere – we analyze the mathematics of local kissing first, and consider the exclusion of situations in which penetration occurs as a worry for later (not treated in this paper).

Two objects are involved in the kissing, $\mathcal{A}$ and $\overline{\mathcal{B}}$, and obviously only their boundaries are of concern to the operation. To characterize the result of the kissing of the boundaries, it is customary to consider that result as a boundary itself. We denote it by $\mathcal{A} \otimes \overline{\mathcal{B}}$. This boundary exists in the configuration space of the objects $\mathcal{A}$ and $\overline{\mathcal{B}}$, but to simplify the analysis we will bring it down to the 'task space' of locations in which $\mathcal{A}$ and $\overline{\mathcal{B}}$ are defined as objects. This is done as follows.

Both $\mathcal{A}$ and $\overline{\mathcal{B}}$ are rigid bodies, characterized as a set of points in their own frame of reference, i.e. as a set of (translation) vectors in the vector space V^m in

G. Sommer and Y. Y. Zeevi (Eds.): AFPAC 2000, LNCS 1888, pp. 22– , 2000.

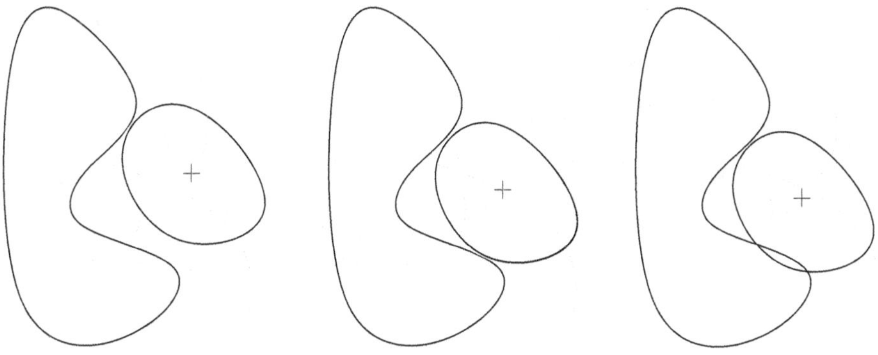

Fig. 1. Two objects in contact in 2-dimensional space. The rightmost example shows how penetration is 'beyond the double kiss', changing one of the kisses into penetration; but it makes the curve of kissing positions differentiable (dashed)

which the objects reside. These sets need not be connected: $\mathcal{A}$ could represent all the obstacles in a scene, $\overline{\mathcal{B}}$ could be the moving robot. Consider the kissing from the point of view of $\mathcal{A}$, i.e. keep $\mathcal{A}$ stationary and move $\overline{\mathcal{B}}$ around her. The various configurations of the rigid body $\overline{\mathcal{B}}$ can be denoted by a translation T and a rotation R, as $TR\overline{\mathcal{B}}$. The space spanned by the characterizing parameters of T and R is called configuration space; it is 3-dimensional if $\mathcal{A}$ and $\overline{\mathcal{B}}$ are 2-dimensional objects in V^2, and 6-dimensional for 3-dimensional $\mathcal{A}$ and $\overline{\mathcal{B}}$ in V^3. The configurations at which kissing occurs determine some curved hypersurface in this space. This surface is actually a boundary since it has an obvious inside and outside: we can (locally) freely move into a kissing configuration, but are not allowed locally to penetrate the object beyond the kissing contact point. Obviously, the rotations make for complicated boundaries in the configuration space. It is therefore customary to focus on translational motions, analyzing those for a fixed rotation. We do the same.

Since the configuration space of translations of objects is obviously of the same dimensionality as the sets of translation vectors defining the object points, the boundaries of $\mathcal{A}$, $\overline{\mathcal{B}}$ and $\mathcal{A} \otimes \overline{\mathcal{B}}$ may all be considered to reside in the same 'task space' of the robot. We could refer to $\mathcal{A} \otimes \overline{\mathcal{B}}$ as a 'kissing boundary', and we will treat it as the boundary of an object of the same reality as $\mathcal{A}$ and $\overline{\mathcal{B}}$.

1.2 Geometrical Objects and Boundary Representation

For robotics, the boundaries of interest are those of actual geometrical objects as they occur in the 'real world'. These can be characterized in many different ways: approximately as polyhedra, exactly as parametrized oriented hypersurfaces. In all cases, there should be a way of retrieving the set of oriented tangent planes at each point of the boundary. Most representations do this implicitly, by specify-

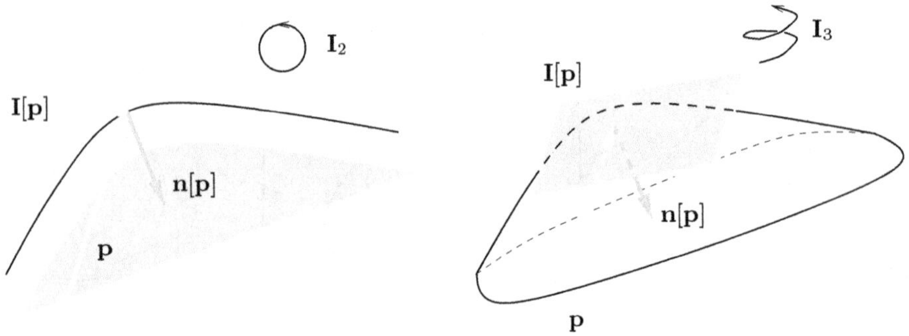

Fig. 2. The relative orientation of **I** and **n** in 2-dimensional Euclidean space (left) and 3-dimensional Euclidean space (right) with right-handed orientation convention

ing the boundary facets or points and providing the possibility of differencing or differentiating them to obtain the tangent information. It will be convenient for us to view a point of a boundary as a set of (position, tangent space)-pairs. This is a local representation of the surface, independent of whether other points are known. It is realistic in robotics, where the complete boundary information may not be available due to partial sensing. We denote position by $\mathbf{p}$, and characterize the tangent hyperplane at $\mathbf{p}$ by $\mathbf{I}[\mathbf{p}]$. We will represent it computationally using geometric algebra [,] as an $(m-1)$-blade, but you can read $\mathbf{I}[\mathbf{p}]$ as a symbolic notation if you are unfamiliar with that framework. We will also characterize it, dually, by the more classical *inward pointing normal vector* $\mathbf{n}[\mathbf{p}]$ (although this involves introducing a metric to specify perpendicularity, whereas kissing is actually an affine, and even conformal, property.) We then associate orientation of the tangent plane and the direction of the inward pointing normal in a consistent manner. In 2-dimensional space V^2, we use the convention that the inward pointing normal is achieved by an counterclockwise turn of the tangent vector $\mathbf{I}$, if the spatial area element has counterclockwise orientation. In 3-dimensional space, we will relate the orientation of $\mathbf{n}$ and the oriented tangent plane $\mathbf{I}$ by a left-hand screw relationship if the space has a right-handed volume element $\mathbf{I}_3$ (see Fig.). (In m-dimensional space with pseudoscalar (hypervolume element) $\mathbf{I}_m$, we may generalize this rule to $\mathbf{n}[\mathbf{p}] = -\mathbf{I}[\mathbf{p}]/\mathbf{I}_m$.)

To keep things simple, we will take smooth surfaces with a unique tangent plane at every point; tangent cones would require too much administration, obscuring the essence of this paper.

If we have a tangent plane at a location $\mathbf{p}$ (let's refer to that as an 'off-set tangent plane'), it is convenient to encode this using homogeneous coordinates, or more generally using the homogeneous model of Euclidean geometry provided by geometric algebra [ ]. This is an embedding of Euclidean m-space in a vector space of 1 dimension higher. We denote the extra dimensional direction by $\mathbf{e}_0$, with reciprocal $\mathbf{e}^0$, so that $\mathbf{e}_0 \cdot \mathbf{e}^0 = 1$. Both $\mathbf{e}_0$ and its reciprocal are orthogonal to any vector of the space with pseudoscalar $\mathbf{I}_m$, so $\mathbf{e}_0 \cdot \mathbf{I}_m = 0$ and $\mathbf{e}^0 \cdot \mathbf{I}_m = 0$. The homogeneous model embeds the *point* at location $\mathbf{p}$ as the *vector*

$$\mathbf{e}_0 + \mathbf{p}$$

in $(m+1)$-dimensional space. (This is thus very similar to homogeneous coordinates, where $\mathbf{p} = (p_1, p_2)^T$ is embedded as $(1, p_1, p_2)^T$.) The offset tangent plane at location $\mathbf{p}$ with tangent $\mathbf{I}[\mathbf{p}]$ is represented in the homogeneous model by the m-blade:

$$(\mathbf{e}_0 + \mathbf{p}) \wedge \mathbf{I}[\mathbf{p}] \tag{1}$$

(If you are not familiar with geometric algebra, you may view this computational expression as a mathematical shorthand; you should still be able to follow the reasoning that follows, although you will miss out on its computational precision.) It is more convenient to consider the position as a function of the tangent rather than vice versa, so that we will treat eq.( ) as:

$$\mathcal{R}[\mathbf{I}] = (\mathbf{e}_0 + \mathbf{p}[\mathbf{I}]) \wedge \mathbf{I}. \tag{2}$$

Note that this re-indexing from $\mathbf{I}[\mathbf{p}]$ to $\mathbf{p}[\mathbf{I}]$ requires an 'inversion of the derivative' which is conceptually straightforward. Yet to achieve it, for instance when the object boundary has been given analytically, requires a functional inversion which may not have a closed-form solution. We will not worry about these algorithmic issues for now. The inversion also leads to multi-valuedness, since the same tangent hyperplane attitude $\mathbf{I}$ may occur at several positions (if the object is not convex). We take all computations to be implicitly overloaded on these multi-valued outcomes, rather than messing up our equations with some index or set notation.

In our dual representation of the tangent plane by a normal vector, we need to take the dual of eq.( ), and characterize both $\mathbf{p}$ and the representation as a function of $\mathbf{n}$ rather than $\mathbf{I}$. We denote them by $\mathbf{p}[\mathbf{n}]$ and $\mathcal{R}[\mathbf{n}]$. We obtain by dualization:

$$\mathcal{R}[\mathbf{n}] = (\mathbf{e}_0 + \mathbf{p}[\mathbf{n}]) \cdot (\mathbf{e}^0 \mathbf{n}) = \mathbf{n} - \mathbf{e}^0 (\mathbf{p}[\mathbf{n}] \cdot \mathbf{n}) = \mathbf{n} + \mathbf{e}^0 \sigma[\mathbf{n}] \tag{3}$$

where we defined the *support function*

$$\sigma[\mathbf{n}] \equiv -\mathbf{p}[\mathbf{n}] \cdot \mathbf{n},$$

which gives the signed distance of the tangent plane to the origin, positive when the origin is (locally viewed) on the 'inside'. The support function for non-convex objects is multi-valued, since there may be more than one location where the inward pointing normal equals $\mathbf{n}$.

The representation in eq.( ) is now, geometrically, the support function plotted in the $\mathbf{e}^0$-direction, as a function of $\mathbf{n}$. Since $\mathbf{n}$ denotes the directions of normal vectors, it parametrizes the *Gaussian sphere* of directions, the range of the Gauss map. The distance function is a function on the Gauss sphere. But the geometrization of this function using geometric algebra offers advantages in computation, since duality becomes simply taking the orthogonal complement (through division by $\mathbf{e}_0 \mathbf{I}_m$).

Example: In the case of a solid *sphere* of radius ρ centered around the origin, the representation has support function ρ at each $\mathbf{n}$; therefore the representation is $\mathcal{R}[\mathbf{n}] = \mathbf{n} + \mathbf{e}^0\rho$. Note that this also applies when ρ is negative; this gives a *spherical hole.* A *point* at the origin (which is a sphere of radius 0) has as representation $\mathcal{R}[\mathbf{n}] = \mathbf{n}$. A point at location $\mathbf{q}$ has as representation $\mathcal{R}[\mathbf{n}] = \mathbf{n} - \mathbf{e}^0(\mathbf{q}\cdot\mathbf{n})$.

The object representations based on the multi-valued support function is invertible. Intuitively, you can see this: from a specification of the tangent planes of all points around $\mathbf{p}$, one should be able to reconstruct $\mathbf{p}$ as the intersection of these differentially different planes. You should think of this reconstruction of a surface from its tangent planes as the computation of a *caustic.* In geometric algebra, this is the *meet* of the neighboring tangent planes, and it can be dually computed as the *join.* To determine it, we need the derivative of $\mathcal{R}$ (or if you prefer, the derivative of the support function on the Gauss map); the dual of this derivative tangent plane to the representation is proportional to $\mathbf{p}$. This is described in [ ] using differential geometry in the 2-dimensional case, and in [ ] using geometric algebra for the m-dimensional case. Computable formulas result, but repetition of those here would be a bit involved and not lead to much insight. (We'll see a simpler inversion formula for the two dimensional case in section).

1.3 Kissing Objects

When two objects $\mathcal{A}$ and $\overline{\mathcal{B}}$ kiss, they have (locally) opposite tangents, $\mathbf{I}$ and $-\mathbf{I}$ at the contact point. This contact generates a point on the resulting boundary which is the addition of the two position vectors (see Fig. a). In the system of $\mathcal{A}$, this is $\mathbf{p}_\mathcal{A}[\mathbf{I}] - \mathbf{p}_{\overline{\mathcal{B}}}[-\mathbf{I}]$. It makes the mathematics more convenient (fewer signs to keep track of) if we see this as a basic operation between $\mathcal{A}$, and an object $\mathcal{B}$ which is the point-mirrored version of $\overline{\mathcal{B}}$, defined by:

$$\mathbf{p}_\mathcal{B}[\mathbf{I}] \equiv -\mathbf{p}_{\overline{\mathcal{B}}}[-\mathbf{I}].$$

Then the result of the kissing of $\mathcal{A}$ onto $\overline{\mathcal{B}}$ is identical to the operation of Huygens *wave propagation* using $\mathcal{A}$ as primary wave front, and $\mathcal{B}$ as propagator (see Fig. b). We denote this operation of wave propagation by $\breve{\oplus}$. In [ ] it was called *tangential dilation,* since it is the 'dilation' operation of mathematical morphology specialized to the boundaries of objects. The dilation is the Minkowski sum of the sets of points representing the objects, and is therefore operation on their volume; but this may be reconstructed from the much more tractable tangential dilation of their boundaries.

So the kissing of $\mathcal{A}$ by an object $\overline{\mathcal{B}}$ is the tangential dilation of $\mathcal{A}$ with the point-mirrored object $\mathcal{B}$ (and vice versa):

$$\mathcal{A}\breve{\oplus}\mathcal{B} \equiv \mathcal{A} \otimes \overline{\mathcal{B}}.$$

¿From now on, we treat the tangential dilation only, but this identity makes all results transferable to kissing.

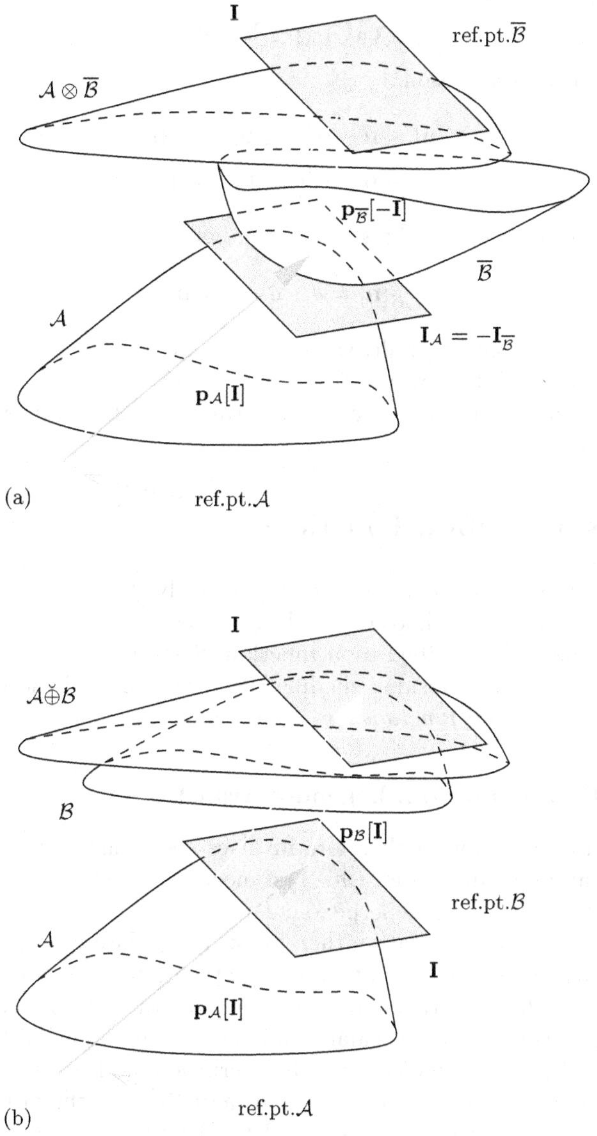

Fig. 3. (a) Kissing of boundaries, $\mathbf{p}_{\mathcal{A}\otimes\overline{\mathcal{B}}}(\mathbf{I}) = \mathbf{p}_{\mathcal{A}}(\mathbf{I}) - \mathbf{p}_{\overline{\mathcal{B}}}(-\mathbf{I})$. (b) Propagation or tangential dilation of boundaries, corresponding to (a) and generating the same boundary point; $\mathbf{p}_{\mathcal{A}\breve{\oplus}\mathcal{B}}(\mathbf{I}) = \mathbf{p}_{\mathcal{A}}(\mathbf{I}) + \mathbf{p}_{\mathcal{B}}(\mathbf{I})$

The resulting point(s) of $\mathcal{A}\breve{\oplus}\mathcal{B}$ as a consequence of the tangential dilation of point $\mathbf{p}_{\mathcal{A}}$ with tangent $\mathbf{I}$ on $\mathcal{A}$ and a point $\mathbf{p}_{\mathcal{B}}$ with tangent $\mathbf{I}$ on $\mathcal{B}$ is then the point $\mathbf{p}_{\mathcal{A}} + \mathbf{p}_{\mathcal{B}}$. The tangent of the result is also $\mathbf{I}$ (as may be seen by local, first-order variation of $\mathbf{p}_{\mathcal{A}}$ and $\mathbf{p}_{\mathcal{B}}$.) For the representation this addition has as its consequence:

$$\mathcal{R}_{\mathcal{A}\breve{\oplus}\mathcal{B}}[\mathbf{I}] = \mathbf{e}_0\mathbf{I} + (\mathbf{p}_\mathcal{A}[\mathbf{I}] + \mathbf{p}_\mathcal{B}[\mathbf{I}]) \wedge \mathbf{I} \tag{4}$$

and for the dual representation

$$\begin{aligned}\mathcal{R}_{\mathcal{A}\breve{\oplus}\mathcal{B}}[\mathbf{n}] &= \mathbf{n} - \mathbf{e}^0(\mathbf{p}_\mathcal{A}[\mathbf{n}] + \mathbf{p}_\mathcal{B}[\mathbf{n}])\cdot\mathbf{n}\\ &= \mathbf{n} - \mathbf{e}^0(\sigma_\mathcal{A}[\mathbf{n}] + \sigma_\mathcal{B}[\mathbf{n}]).\end{aligned}$$

So, *under tangential dilation, the support functions add up*:

$$\sigma_{\mathcal{A}\breve{\oplus}\mathcal{B}}[\mathbf{n}] = \sigma_\mathcal{A}[\mathbf{n}] + \sigma_\mathcal{B}[\mathbf{n}]. \tag{5}$$

(Since support functions are multi-valued, this may involve multiple additions, and a more proper notation would be to indicate this overload of the addition as a Minkowski sum of the set of values, but this is again administrative rather than insightful.)

2 Objects as Umbral Functions

It is interesting to see what happens locally. To study this, we denote the object by a Monge patch, i.e. we choose some local hyperplane relative to which the object surface may be described by a function. Such a function should still be endowed with a notion of 'inside'; shading the points 'under' the function then leads to the name *umbral function* (i.e. a function with a shadow)[ ].

2.1 Patch Representation; Legendre Transform

The Monge patch description therefore involves choosing a function direction, denoted by $\mathbf{e}$, and introducing an $(m-1)$-dimensional vectorial coordinate $\mathbf{x}$ to describe the position in a plane perpendicular to this direction (so $\mathbf{x}\cdot\mathbf{e} = 0$). The consequences for the representation then follow immediately from this.

In our treatment of objects by Monge patches, we are mostly interested in the case $m = 2$, since this paper will demonstrate the analogy with the linear filtering of 1-dimensional scalar functions. So we may introduce $\mathbf{e}_2$ as the special 'function' direction $\mathbf{e}$ of the patch description, and $\mathbf{e}_1$ as the orthogonal coordinate direction for a coordinate x. Thus a point $\mathbf{p}$ on the object boundary can be written as a function $\mathbf{p}\colon \mathbb{R} \to V^2$ defined by $\mathbf{p}(x) = x\mathbf{e}_1 + f(x)\mathbf{e}_2$ with $f\colon \mathbb{R} \to \mathbb{R}$ locally encoding the boundary.

The tangent plane now has a direction which is obtained by differentiation of $\mathbf{p}(x)$ to x, and proper orienting. It is denoted by the tangent vector $\mathbf{I}[x] = -\mathbf{e}_1 - f'(x)\mathbf{e}_2$, or dually by the normal vector $\mathbf{n}[x] = -\mathbf{I}[x]/\mathbf{I}_2 = f'(x)\mathbf{e}_1 - \mathbf{e}_2$. Then we compute the dual representation of this patch as:

$$\begin{aligned}\mathcal{R}[\mathbf{n}[x]] &= \mathbf{n}(x) - \mathbf{e}^0\,\mathbf{p}[x]\cdot\mathbf{n}[x]\\ &= f'(x)\mathbf{e}_1 - \mathbf{e}_2 - \mathbf{e}^0(x\mathbf{e}_1 + f(x)\mathbf{e}_2)\cdot(f'(x)\mathbf{e}_1 - \mathbf{e}_2)\\ &= f'(x)\mathbf{e}_1 - \mathbf{e}_2 + \mathbf{e}^0(f(x) - xf'(x)).\end{aligned} \tag{6}$$

This is a spatial curve which may be parametrized by its $\mathbf{e}_1$ coordinate. Let us denote that coordinate by ω:

$$\omega \equiv f'(x).$$

Then we need the value of x where the slope equals ω to rewrite the curve in terms of ω. If f' is invertible, this is $x = f'^{-1}(\omega)$ and the curve becomes

$$\mathcal{R}[\mathbf{n}[x]] = \omega\mathbf{e}_1 - \mathbf{e}_2 + \mathbf{e}^0 \left(f(f'^{-1}(\omega)) - f'^{-1}(\omega)\,\omega \right).$$

But this is a clumsy notation for computations. We can improve the formulation by introducing a notation for the 'stationary value of a function'

$$\text{stat}_u[g(u)] = \{g(u_*) \mid g'(u_*) = 0\}.$$

This is multi-valued, since there may be several extrema, of different magnitudes. (We will also encounter functions with multi-valued derivatives and in that case we should read $g'(u_*) = 0$ in the definition above as 'the derivative of g at u_* contains 0', so $g'(u_*) \ni 0$.) With this 'stat' operation, we can encode the definition of ω nicely in the stat expression specifying the $\mathbf{e}^0$-component, for we observe that

$$\text{stat}_u[f(u) - u\,\omega] = \{f(u_*) - u_*\omega \mid f'(u_*) - \omega = 0\}$$

is precisely the value we need. Therefore we rewrite

$$\begin{aligned} \mathcal{R}_f[\omega] &= \omega\mathbf{e}_1 - \mathbf{e}_2 + \mathbf{e}^0\,\text{stat}_u[f(u) - u\,\omega] \\ &= \omega\mathbf{e}_1 - \mathbf{e}_2 + \mathcal{L}[f](\omega)\,\mathbf{e}^0, \end{aligned}$$

where we defined the *(extended) Legendre transform* of f by:

$$\mathcal{L}[f](\omega) \equiv \text{stat}_u[f(u) - u\,\omega] = \{f(u_*) - u_*\omega \mid f'(u_*) = \omega\}. \tag{7}$$

The dual representation of eq.( ) shows that in the plane with $\mathbf{e}_2$-coordinate equal to -1, parallel to the $(\mathbf{e}_1 \wedge \mathbf{e}^0)$-plane, it is the (extended) Legendre transform as a function of ω.

For Monge patches and umbral functions, the Legendre transform $\mathcal{L}[f]$ thus plays the role of the support function σ in the purely geometrical framework. It denotes the support (a measure of the distance to the origin) of the tangent hyperplane with slope ω, through its intercept with the $\mathbf{e}_2$-axis, see Fig. a,b.

Since the support function representation of objects is invertible, so is the intercept characterization of these supports by the Legendre transform. And indeed *the (extended) Legendre transform is invertible*, through

$$\mathcal{L}^{-1}[F](x) = \text{stat}_\nu[F(\nu) + x\,\nu], \tag{8}$$

[1] We call it 'extended' since the Legendre transform does not usually admit multi-valuedness, being applied to purely convex or concave functions only; but the basic transform principle is so similar that it hardly justifies a new name. We originally introduced it under the name 'slope transform' [ ][ ], but now prefer honoring Legendre.

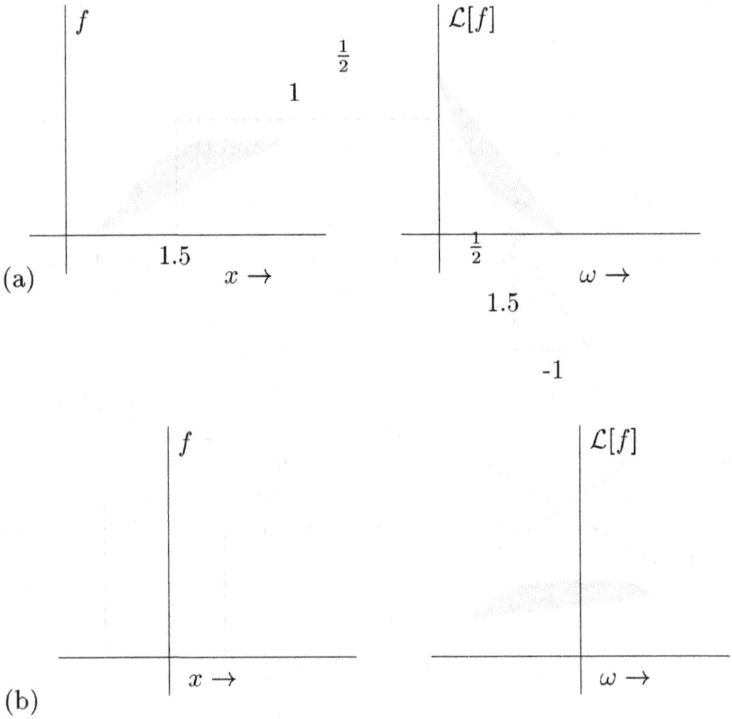

Fig. 4. (a) The Legendre transform for a convex function. (b) The multi-valued extended Legendre transform for a non-convex function

in the sense that

$$\mathcal{L}^{-1}[\mathcal{L}[f]] = f.$$

This is easily shown (if you are willing to take some properties of the multivalued 'stat' operator for granted):

$$\begin{aligned}
&\mathcal{L}^{-1}[\mathcal{L}[f]](x) = \\
&= \text{stat}_\nu[\text{stat}_u[f(u) - \nu\, u] + x\, \nu] \\
&= \text{stat}_\nu[\{f(u_*) + \nu(x - u_*) \mid f'(u_*) - \nu = 0\}] \\
&= \{f(u_*) + \nu_*(x - u_*) \mid f'(u_*) - \nu_* = 0 \text{ and } x - u_* = 0\} \\
&= \{f(x) + \nu_*(x - u_*) \mid \nu_* = f'(u_*) \text{ and } x = u_*\} \\
&= f(x)
\end{aligned}$$

(a singleton set results, so we drop the set notation).

[2] To do all this properly using multi-valued functions is a bit of a pain, and requires making sure that one keeps track of the various branches of the Legendre transform caused by different convex/concave portions of the original functions. This takes

2.2 Tangential Dilation of Umbral Functions

In the object representations by support functions, we saw that the tangential dilation is additive. This additivity of the support functions under tangential dilation refers to a translational shift of the tangents. It transfers directly to additivity of the intercepts of umbral functions, i.e. to Legendre transforms. Since the Legendre transform is invertible, we can use this to define what we mean by tangential dilation of umbral functions:

$$\mathcal{L}[f \breve{\oplus} g](\omega) \equiv \mathcal{L}[f](\omega) + \mathcal{L}[f](\omega). \tag{9}$$

Inversion of this gives the formula for $f \breve{\oplus} g$ as an operation on umbral functions:

$$\begin{aligned}(f \breve{\oplus} g)(x) &= \mathcal{L}^{-1}[\mathcal{L}[f] + \mathcal{L}[g]](\omega) \\ &= \mathrm{stat}_\nu[\mathcal{L}[f](\nu) + \mathcal{L}[g](\nu) + \nu\, x] \\ &= \mathrm{stat}_\nu\, \mathrm{stat}_u\, \mathrm{stat}_v[f(u) + g(v) + (x-u-v)\,\nu]] \\ &= \mathrm{stat}_u\, \mathrm{stat}_v\, \mathrm{stat}_\nu[f(u) + g(v) + (x-u-v)\,\nu]] \\ &= \mathrm{stat}_u\, \mathrm{stat}_v[\{f(u) + g(v) + (x-u-v)\,\nu_* \mid x-u-v = 0\}] \\ &= \mathrm{stat}_u[f(u) + g(x-u)].\end{aligned}$$

Therefore we obtain as the actual definition of tangential dilation of umbral functions:

$$(f \breve{\oplus} g)(x) = \mathrm{stat}_u[f(u) + g(x-u)]. \tag{10}$$

You may verify that this has indeed the desired properties at the corresponding points with coordinates x_f, x_g and $x_{f \breve{\oplus} g}$:

$$\begin{aligned} x_{f \breve{\oplus} g} &= x_f + x_g \\ (f \breve{\oplus} g)(x_{f \breve{\oplus} g}) &= f(x_f) + g(x_g) \\ (f \breve{\oplus} g)'(x_{f \breve{\oplus} g}) &= f'(x_f) = g'(x_g) \end{aligned} \tag{11}$$

In this function description, the tangential dilation has been studied extensively, though almost exclusively with an interest in the globally valid dilation corresponding to an actual collision process, i.e. not permitting intersection of the boundary functions. That globalization is achieved by a replacement of the 'stat' in eq.( ) by supremum and/or infimum operations. The literature on convex analysis [ ] and mathematical morphology [ ] is then relevant to its study.

2.3 Convolution and the Fourier Transform

It is interesting to compare the tangential dilation of (umbral) functions with the definition of their *convolution*, the basic operation of linear systems theory.

administration rather than essential mathematics. It can be avoided by describing the boundaries as parametrized curves rather than as functions – but this would lose the obvious similarity to the Fourier transform. It is all just a matter of choosing the most convenient representation of an algebraic intuition which is the same in all cases, and we will not worry about such details.

Convolution of two functions $f : \mathbb{R} \to \mathbb{R}$ and $g : \mathbb{R} \to \mathbb{R}$ is defined as:

$$(f * g)(x) = \int du\, [f(u)g(x-u)].$$

Under the Fourier transform defined as

$$\mathcal{F}[g](\omega) = \int_u du\, g(u)e^{-i\omega u},$$

this becomes multiplicative:

$$\mathcal{F}[f * g](\omega) = \mathcal{F}[f](\omega) \times \mathcal{F}[g](\omega).$$

It is good to remember why this is the case, to understand the role of the correspondence between the transformation formula and the convolution operation. The basic observation is that the convolution has *eigenfunctions*, i.e. functions which do not change their form more than by a multiplicative factor when used in convolution. These are the complex exponentials, each defined by an *amplitude* a and a *frequency* ω:

$$e_\omega(x) = ae^{i\omega x}.$$

Then we obtain for the convolution of f by an eigenfunction:

$$\begin{aligned}(f * e_\omega)(x) &= \int du\, [f(u) \times ae^{i\omega(x-u)}] \\ &= \left(\int du\, f(u)e^{-i\omega u}\right) \times (ae^{i\omega x}) \\ &= \mathcal{F}[f](\omega) \times e_\omega(x).\end{aligned}$$

The eigenfunction is unchanged in frequency, but its amplitude changes by a multiplicative amount $\mathcal{F}[f](\omega)$, which only depends on f and ω. Thus *the Fourier transform is the (multiplicative) eigenvalue of the eigenfunction of the convolution.*

To take advantage of this, it is essential to be able to decompose an arbitrary function using eigenfunctions, for if this is done the involved convolution operation becomes a simple frequency-dependent multiplication of transforms. This is where the Fourier transform transform plays a second role, for it is precisely that decomposition.

This is demonstrated in the theory of Fourier transforms, which shows that (under certain weak conditions) f can be reconstructed from knowing its spectrum $\mathcal{F}[f]$, by

$$f(x) = \frac{1}{2\pi} \int_\nu d\nu\, \mathcal{F}[f](\nu) \times e^{ix\nu}.$$

So the Fourier transform plays two roles in linear systems theory:

- The Fourier transform is a way of characterizing a signal f by a spectrum $\mathcal{F}[f]$ of *(complex) amplitudes* as a function of *frequency* in an invertible manner;

- such a spectral characterization of two signals f and g is just right for convolution: *multiplicative combination* ($\mathcal{F}[f] \times \mathcal{F}[g]$) produces the spectrum of the *convolution* ($f * g$) of the two signals.

Those properties form the basis of linear systems theory, and its powerful spectral representation.

2.4 The Legendre Transform as Spectrum

We observe that the Legendre transform conforms to a similar pattern. Indeed, the concept of 'eigenfunction' is also valid to its analysis, although this should now be interpreted in an additive sense. The eigenfunctions of dilation are then the straight boundaries characterized by *slope* and *intercept*:

$$e_\omega(x) = a + \omega\, x,$$

since dilating a straight boundary those by any function f, a straight boundary with the same slope will again be the result (although if f is not convex, this may be a set of straight boundaries; apparently, amplitudes should be permitted to be multi-valued, but we knew that already). The additive eigenvalue of the line is computed as:

$$\begin{aligned}(f \check{\oplus} e_\omega)(x) &= \mathrm{stat}_u[f(u) + a + \omega(x-u)] \\ &= \mathrm{stat}_u[f(u) - \omega\, u] + a + \omega\, x \\ &= \mathcal{L}[f](\omega) + e_\omega(x).\end{aligned}$$

The additive eigenvalue of e_ω is thus precisely the Legendre transform of f. The analogy with the linear theory is strong indeed:

- The Legendre transform can characterize an umbral function invertibly by a 'spectrum' of *(multi-valued) intercepts* as a function of the *slope*, see eq.( );
- the *additive combination* of such spectra $\mathcal{L}[f] + \mathcal{L}[g]$ is the spectrum of the *tangential dilation* of the umbral functions f and g, see eq.( ).

Therefore the essential properties of a Fourier transform *vis-à-vis* convolution of signals hold for the Legendre transform *vis-à-vis* tangential dilation of umbral functions. We may therefore expect that a lot of concepts which have proved useful in linear systems theory will be useful techniques in dealing with tangential dilation, and have applications to touching, collision computation and analysis of wave propagation.

3 Systems Theories

The comparison with the eigenfunctions of the convolution and tangential dilation shows how we should want to interpret the results. There is a conserved parameter which can be used to index the eigenfunctions – in the case of convolution it is a *frequency*, in the case of tangential dilation a *slope* or *tangent* direction. Then there is a second parameter, the *amplitude* or *intercept* or *support vector* which changes according to simple arithmetic (multiplication or addition) completely determined by the function g used to dilate with.

Now that we have a very similar algebraic structure to the systems theory of linear convolution, we should be able to develop it along similar lines. We sketch this development; again the precise notation in healthy mathematics has not caught up (as was the case with Fourier transformations and delta functions, which only later obtained their embeddings in the mathematics of distribution theory), but we may be confident that it will be fixed in time (by proper mathematicians rather than by us).

3.1 Delta Functions

In linear systems theory, the delta function (or impulse) is a convenient concept to describe sampling. It is simply the identity function of convolution:

$$(f * \delta)(x) = f(x), \quad \text{for all } x$$

This implies that its spectrum should be the multiplicative identity. Under the Fourier transform, the equation transforms to $\mathcal{F}[f](\omega) \times \mathcal{F}[\delta](\omega) = \mathcal{F}[f](\omega)$, so that $\mathcal{F}[\delta](\omega) = 1$, independent of ω. Reverse transformation (not elementary) yields the familiar delta function:

$$\delta(x) = \frac{1}{2\pi} \int d\nu \; e^{i\nu\, x} = \begin{cases} 0 & \text{if } x \neq 0 \\ \text{`1'} & \text{if } x = 0 \end{cases}$$

(Here the '1' is used as a shorthand to denote that it is not the value of $\delta(x)$ which is 1 at $x = 0$, but its integral. Yet most systems engineers are used to thinking of the value as 1, especially since they often work with the Kronecker delta.) This function is sketched in Fig. a.

We can make the same construction in the tangential dilation case, now demanding:

$$(f \breve{\oplus} \delta)(x) = f(x)$$

and after Legendre transformation this yields $\mathcal{L}[f](\omega) + \mathcal{L}[\delta](\omega) = \mathcal{L}[f](\omega)$, so that $\mathcal{L}[\delta](\omega) = 0$ independent of ω. We may invert this to:

$$\begin{aligned} \delta(x) &= \text{stat}_\nu [0 + \nu\, x] \\ &= \{\nu_* x \mid x = 0\} \\ &= \begin{cases} \{0\} & \text{if } x = 0 \\ \emptyset & \text{if } x \neq 0 \end{cases} \\ &= \begin{cases} 0 & \text{if } x = 0 \\ -\infty & \text{if } x \neq 0 \end{cases} \end{aligned}$$

The final step converts the set notation to the umbral function notation where an 'empty' function value is represented by $-\infty$ (this has no shadow, so it represents no object point).

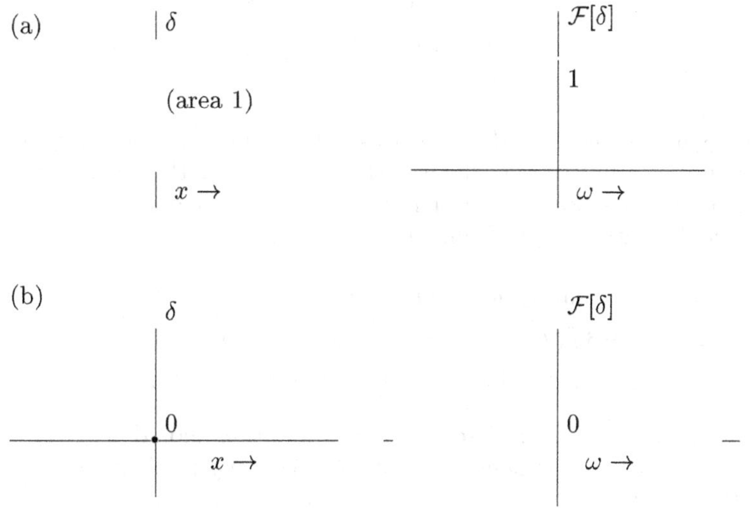

Fig. 5. Delta functions in linear theory (a) and contact theory (b), and their corresponding transforms

The delta function for dilation is sketched in Fig. b. As an object, the delta-function is clearly the *point at the origin.* Being tangentially dilated by a point therefore reveals the shape of f (as does colliding with a point at the origin).

3.2 Representation Using Delta-Functions

In linear filtering, the representation of a signal as a weighted sum of eigenfunctions is basic; but when moving towards sampling, another mode of representation is important. This is the decomposition of a function by means of delta-functions of appropriate magnitude.

$$f(x) = (f * \delta)(x) = \int du\ f(u) \times \delta(x-u)$$

Thus $f(u)$ is like the multiplicative amplitude of the delta-function shifted to u.

We can rewrite umbral functions in a similar manner, though now the appropriate delta-functions have amplitudes which are additive rather than multiplicative:

$$f(x) = (f \breve{\oplus} \delta)(x) = \mathrm{stat}_u[f(u) + \delta(x-u)].$$

Let us verify this by rewriting the Legendre transform reconstruction formula into a formula containing a δ-function:

$$\begin{aligned} f(x) &= \mathrm{stat}_\nu[\mathcal{L}[f](\nu) + x\,\nu] \\ &= \mathrm{stat}_\nu\, \mathrm{stat}_u[f(u) + (x-u)\,\nu] \end{aligned}$$

$$= \text{stat}_u \, \text{stat}_\nu [f(u) + (x-u)\,\nu]$$
$$= \text{stat}_u [f(u) + \text{stat}_\nu [(x-u)\,\nu]]$$
$$= \text{stat}_u [f(u) + \delta(x-u)].$$

It is interesting to see how the stat operator provides the 'scanning' of all u, picking out $x = u$, in precisely the way the integral does this in the linear case.

Working out the 'stat' operator in the final expression gives a clue on how to treat the derivative of the δ functions:

$$\begin{aligned} f(x) &= \text{stat}_u [f(u) + \delta(x-u)] \\ &= \{f(u_*) + \delta(x-u_*) \mid f'(u_*) - \delta'(x-u_*) = 0\} \\ &= \{f(u_*) + \delta(x-u_*) \mid f'(u_*) \ni \delta'(x-u_*)\} \end{aligned}$$

Since this should be identical to $f(x)$, the condition should be equivalent to stating that $u_* = x$, independent of what the function f is. Therefore $\delta'(v)$ must apparently be defined to be non-zero only when $v = 0$, and then to contain all slopes; this feels perfectly reasonable.

3.3 Translations

The basic operation in the representation by delta-functions is a translation of the signal. Fortunately, such an (abscissa) *translation is both a convolution and a tangential dilation.* So this important operation is part of both systems theories.

In linear systems, translation over t is represented by convolution with the shifted delta-function $\delta_t(x) \equiv \delta(x-t)$, of which the Fourier transform is $e^{-i\omega t}$:

$$\mathcal{F}[\delta_t](\omega) = \int du \; \delta(u-t) e^{-i\omega u} = \int dv \; \delta(v) e^{-i\omega v} \times e^{i\omega t} = e^{-i\omega t}.$$

This gives

$$f_t(x) \equiv f(x-t) \underset{\mathcal{F}^{-1}[\cdot]}{\overset{\mathcal{F}[\cdot]}{\rightleftharpoons}} \mathcal{F}[f_t](\omega) = \mathcal{F}[f](\omega) \times e^{-i\omega t}$$

In contact systems, translation over t is represented as tangential dilation by the shifted delta-function $\delta_t(x) \equiv \delta(x-t)$, of which the Legendre transform is $-\omega t$:

$$\mathcal{L}[\delta_t](\omega) = \text{stat}_u [\delta(u-t) - \omega\, u] = \text{stat}_v [\delta(v) - \omega\, v] - \omega t = -\omega t$$

This gives

$$f_t(x) \equiv f(x-t) \underset{\mathcal{L}^{-1}[\cdot]}{\overset{\mathcal{L}[\cdot]}{\rightleftharpoons}} \mathcal{L}[f_t](\omega) = \mathcal{L}[f](\omega) - \omega t$$

Note that the computations in both frameworks run completely analogously.

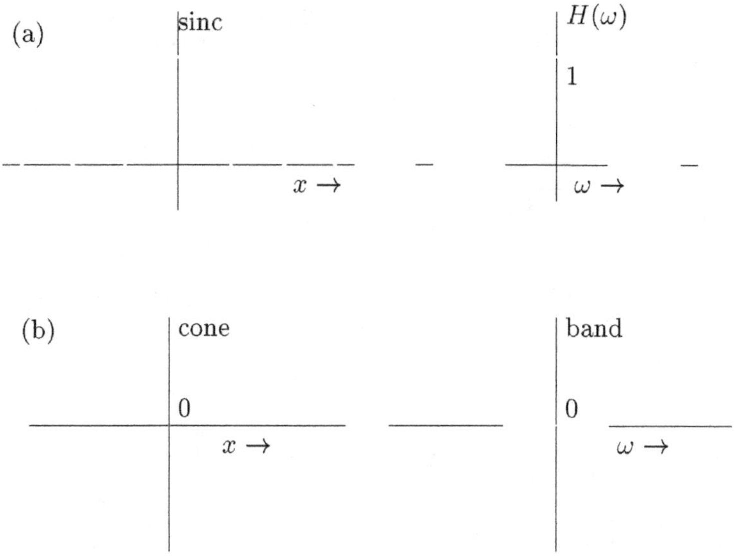

Fig. 6. Band limitations functions in linear theory (a) and contact theory (b), and their corresponding inverse transforms

3.4 Bandwidth Limitation

In linear systems theory, band limitation is achieved through multiplying the spectrum by the ideal bandpass filter

$$H(\omega) = \begin{cases} 1 & \text{if} \ \ |\omega| \leq \omega_0 \\ 0 & \text{if} \ \ |\omega| > \omega_0 \end{cases}$$

This inverse Fourier transform yields the famous 'sinc'-function:

$$\mathcal{F}^{-1}[H](x) = \frac{\omega_0}{\pi}\text{sinc}(\frac{\omega_0 x}{\pi}) \equiv \begin{cases} \frac{\sin \omega_0 x}{\pi x} & \text{if} \ \ x \neq 0 \\ 1 & \text{if} \ \ x = 0 \end{cases}$$

Convolution with this function indeed leads to a limitation of the spectrum of a signal to the frequencies between $-\omega_0$ and ω_0, limiting the bandwidth, see Fig. a.

In tangential dilation, we can achieve bandwidth limitation in a similar manner. We now need to make an additive bandpass filter, which should be defined as

$$H(\omega) = \begin{cases} 0 & \text{if} \ \ |\omega| \leq \omega_0 \\ -\infty & \text{if} \ \ |\omega| > \omega_0 \end{cases}$$

Now the inverse Legendre transform (most easily done pictorially) gives a 'cone'-function:

$$\mathcal{L}^{-1}[H](x) = -|\omega_0 x|$$

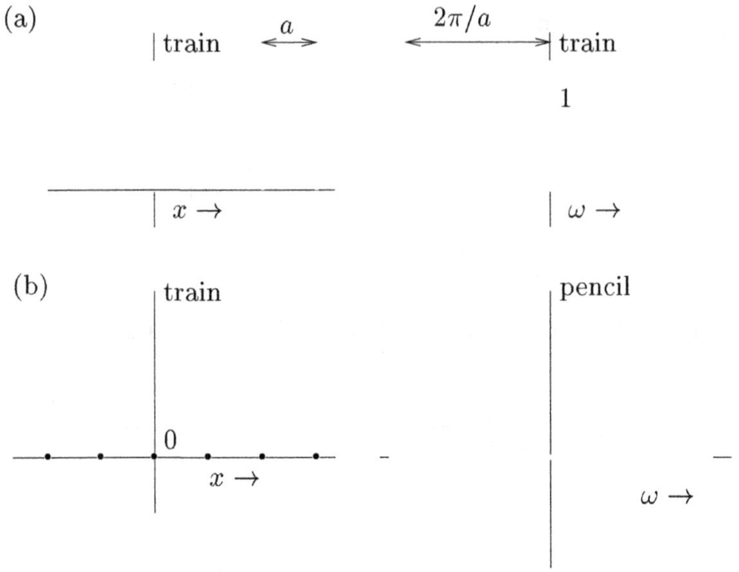

Fig. 7. Impulse trains in linear theory (top) and contact theory (bottom), and their corresponding transforms

Tangential dilation with this function produces a slope-limited function, see Fig. b. Those functions are known as Lipschitz-functions, and their theoretical importance as been recognized in theoretical developments in the mathematical morphology on umbral functions [ ]. Since dilation of a function with a cone results in a kind of clamping, filling in the locally concave parts while passing the local maxima unchanged, it is relevant to the analysis of envelopes in signal processing [ ].

3.5 Sampling

In linear systems theory, there are various theorems on sampling. The most surprising of those is that certain signals can be reconstructed completely after sampling.

Commonly sampling is described primarily in the spatial domain, as the multiplication by a train of impulses. The Fourier transform of this train is a train of impulses in the frequency domain, with a separation reciprocal to the separation in the spatial domain, see Fig. a.

$$\sum_{k\in\mathbb{Z}} \delta(x - ka) \underset{\mathcal{F}^{-1}[\cdot]}{\overset{\mathcal{F}[\cdot]}{\rightleftharpoons}} \sum_{k\in\mathbb{Z}} \delta(\omega - \frac{2\pi}{a}k) \tag{12}$$

For tangential dilation, the impulse train is a sum of shifted delta functions, of which the Legendre transform is a 'star' of lines in the slope domain, see Fig. 7b:

$$\sum_{k=-\infty}^{\infty} \delta(x - ka) \underset{\mathcal{L}^{-1}[\cdot]}{\overset{\mathcal{L}[\cdot]}{\rightleftarrows}} \quad \mathrm{stat}_u[\sum_{k\in\mathbb{Z}} \delta(u - ka) - \omega\, u]$$
$$= \{\sum_{k\in\mathbb{Z}} \delta(u_* - ka) - \omega\, u_* \mid \sum_{k\in\mathbb{Z}} \delta'(u_* - ka) = \omega\}$$
$$= \{-\omega\, ak \mid k \in \mathbb{Z}\}. \tag{13}$$

This is a 'star', a union of linear functions through the origin, see Fig. b.

3.6 Reconstruction after Sampling

In linear filtering, the spatial sampling of a function f is done through *multiplication* by the impulse train (which is a *sum* of delta functions). In the frequency domain, this leads to a *convolution* of the Fourier transform of this train and the spectrum $\mathcal{F}[f]$, leading to a sequence of *additively* overlapping copies of the original spectrum $\mathcal{F}[f]$ (see Fig. b). If those do not overlap, which happens if the original spectrum was band-limited and the sampling was sufficiently fine-grained (at least twice the highest frequency, this is the Nyquist criterion) one can reconstruct the original signal by multiplication of its spectrum by the band-pass filter, see Fig. c. In the spatial domain this is done through convolution of the sampled signal by the sinc-function corresponding to the band limitation.

In tangential dilation, if we use the train of delta functions (made as a *sum* of δ-functions) for sampling, this is done by an *addition* operation. This becomes a *tangential dilation* in the slope domain, by the star of lines which is the Legendre transform of the sum of impulses by eq.( ). This gives a smeared out spectrum over the directions present in the star, see Fig. b. The original spectrum is now *not* retrievable by a single global bandpass filter. Rather, we have to select an appropriate portion around each slope to approximate the original spectrum, i.e. we have to *add a local bandpass filter* of an appropriate width. This effectively approximates the spectral curve with a piecewise linear, locally tangent curve, with slopes takes from the slopes $-ak$ prescribed by the transform of the impulse train. Transforming back to the umbral function domain, this implies tangential dilation by the inverse legendre transform of the local bandpass filters, which are cones. This tangential dilation yields precisely a *first order interpolation* of the sample points, see Fig. c! (The carefully chosen intersection slopes in the slope domain to select the right portions correspond to the slopes required to connect the discrete points in the spatial domain.) Therefore linear interpolation is analyzable within the new systems theory of contact.

Exact reconstruction of a signal from its sampled version is now only possible if the original sample points were the vertices of a polygon (we have assumed above that the points were equidistant, for simplicity; this is not required, as the reader may check). Then the linear interpolation of the sample points obviously retrieves the original polygonal umbral function.

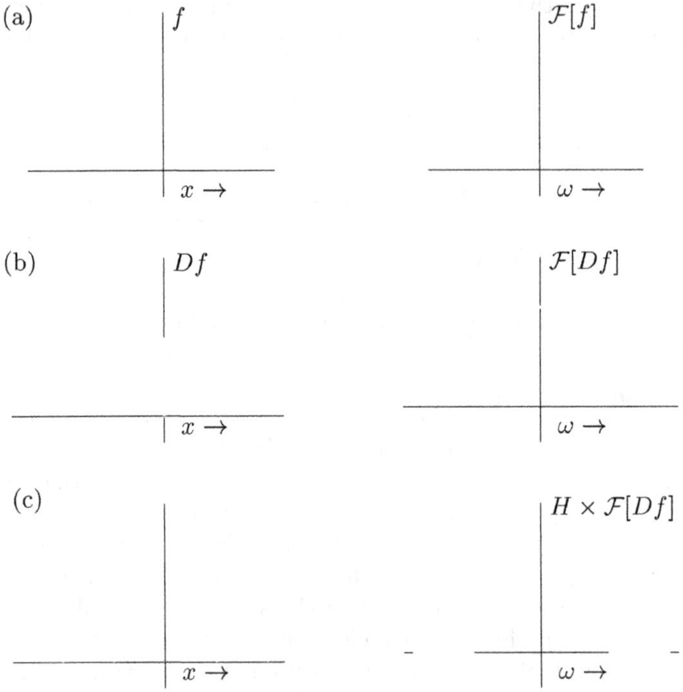

Fig. 8. Linear systems theory: (a) Original function and spectrum (b) The effect of sampling (c) Nyquist reconstruction using convolution

This is doable if one has only one polygonal umbral function, but when combining two such functions, it would be unusual for both to have their vertices at the same x-coordinates; one would then need to take the union of the sample locations. This is the counterpart of the Nyquist criterion for the sampling of tangential dilation operations – it is rather different in form, and seems less practical in its consequences.

3.7 Discretization of Convolution and Tangential Dilation

An important way of considering discretization is not as sampling of the original signals, but as a discretization of their relevant algebraic combination. One then desires a discretization method which 'commutes' with the basic operation, in a natural manner. Denoting discretization by D, we demand in the linear theory of convolution

$$D(f * g) = (Df) * (Dg),$$

and in the contact theory of tangential dilation

$$D(f \breve{\oplus} g) = (Df)\breve{\oplus}(Dg).$$

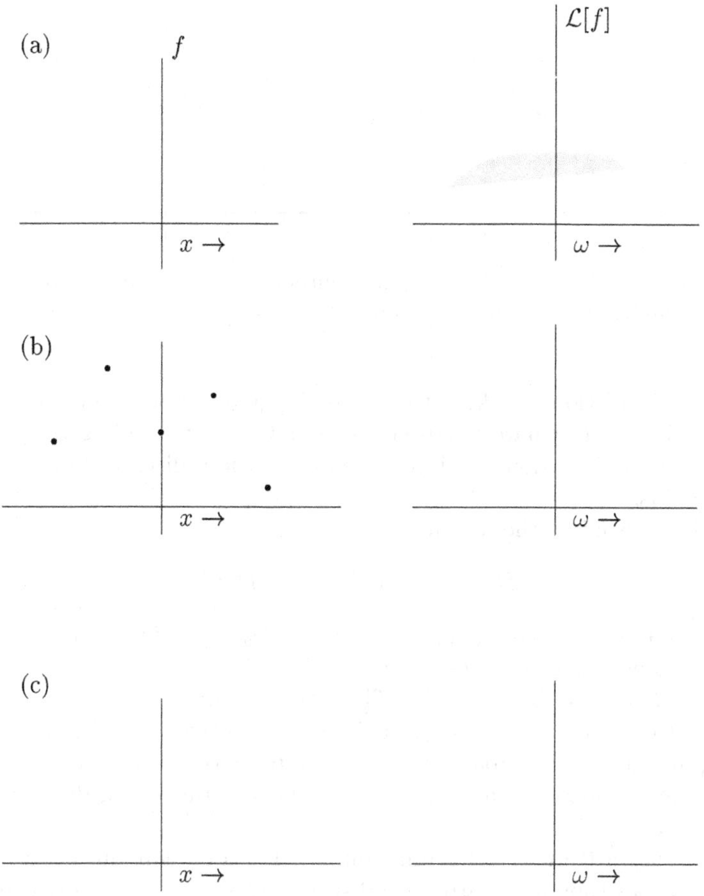

Fig. 9. Contact systems theory: (a) Original function and spectrum (b) The effect of sampling (c) Piecewise linear reconstruction in tangential dilation

Such discretizations are more easily designed in the spectral domain, where the combination operation is a simply arithmetical:

$$\mathcal{F}[D(f * g)] = \mathcal{F}[Df] \times \mathcal{F}[Dg].$$

We will only consider the case when D is itself a convolution, since then we can use the associativity to re-arrange terms. So let $Df = d * f$ for some function d. Then the Fourier transforms are simple: of the left hand side, it is $\mathcal{F}[d * (f * g)] = \mathcal{F}[d] \times \mathcal{F}[f] \times \mathcal{F}[g]$, and of the right hand side: $\mathcal{F}[d * f] \times \mathcal{F}[d * g] = \mathcal{F}[d]^2 \times \mathcal{F}[f] \times \mathcal{F}[g]$. For general signals, equality implies $\mathcal{F}[d]^2 = \mathcal{F}[d]$, so $\mathcal{F}[d]$ is a projection of functions. Any selection on the spectrum using a sum of non-overlapping bandpass filters has this property. Taking for example a train of delta functions in the frequency domain, this implies a convolution by a widely spaced train

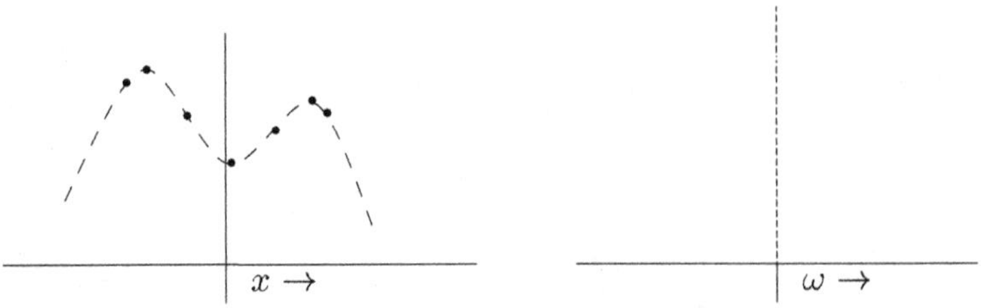

Fig. 10. Commutative discretization in contact systems theory: tangent discretization using a limited slope spectrum

of spatial delta functions, making the original signal periodic but retaining its continuity. It is therefore hardly a discretization. Only by sampling using a pulse train does one obtain a doubly discrete representation, discrete (and periodic) in both domains.

By similar reasoning, the demand

$$\mathcal{L}[D(f\breve{\oplus}g)] = \mathcal{L}[Df] + \mathcal{L}[Dg].$$

may be satisfied by a tangential dilation D. So let $Df = d\breve{\oplus}f$ for some function d. Then the left hand side gives $\mathcal{L}[d\breve{\oplus}(f\breve{\oplus}g)] = \mathcal{L}[d]+\mathcal{L}[f]+\mathcal{L}[g]$, and the right hand side $\mathcal{L}[d\breve{\oplus}f]+\mathcal{L}[d\breve{\oplus}g] = 2\mathcal{L}[d]+\mathcal{L}[f]+\mathcal{L}[g]$. We thus find a sampling $Df = d\breve{\oplus}f$ which needs to satisfy $2\mathcal{L}[d] = \mathcal{L}[d]$. This is an additive selector, and again a sum of non-overlapping bandpass filters has the property (or union, if you prefer the set notation). Since it is a sum, it corresponds to tangential dilation in the spatial domain.

If we now use a train of delta functions in the slope domain, i.e. a *discrete spectrum of slopes* (not necessarily equidistant), this leads to a polyhedral approximation of the spatial domain using those slopes in a star of linear functions (the derivation is analogous to that in section). This gives lots of infinite lines tangent to the function in the spatial domain. Selecting the correct portions of those in the spatial domain implies that we do a linear interpolation in the slope domain, leading to a *linear tangent approximation in the spatial domain*, using the selected spectrum of tangents, see Fig. .

So this structural discretization produces a polyhedral approximation which correctly represents the collection of offset tangent hyperplanes. To reconstruct the object, one also needs to store information on which hyperplanes are neighbors; one thus obtains a network of tangent hyperplanes. This is a discretization method which –by design– works well for contact: collision with the discretization is discretization of the collision.

[3] This is the full story for convex functions. For functions with inflection points, separate branches results for each convex and concave section, but there are ways to connect those – we will not go into this here. It is interesting to note that if such a function is coarsely sampled (so that concave portions are skipped), one obtains a linear tangent discrete approximation to the *convex hull* of the umbral function.

It can be shown that this leads to a simple algorithm which can be performed fully in the spatial domain, in which one is permitted to do a sorting on directions and add the contributions of the directions; not only as intercepts, but also as line lengths, since those are additive as well. Such representations were first introduced in [ ], though without the backing of a systems theory which suggests the equal-slope-spectrum sampling method to retain closure of the representation.

3.8 Filtering; Touch Sensing

Just as the Fourier transform gives us a way to analyze filtering for convolutions (i.e. optical blurring), the Legendre transform gives this possibility for touch sensors (or collision testing).

We would have liked to include a section here about the counterpart of Wiener filtering in stochastic signal processing, and other techniques to estimate unknown transfer functions from a statistically sufficiently rich collection of input signals. This would for instance permit the estimation of the unknown atomic probe used in scanning tunneling microscopy [ ], to 'de-dilate' it from the measurements to obtain a good estimate of the actual atomic surface that was being scanned.

However, for such applications one would really need the full theory which incorporates the impossibility of intersecting contacts, also in its statistical aspects. We have not looked into this. In its present form, the analogous filtering techniques might be useful to the inversion of waves (as in seismic migration studies), but even that is conjectural.

4 Towards a Systems Theory of Kissing

The analogy of the Monge patch descriptions of contact to linear systems theory have led us to consider tangential dilation as a systems theory. We will now briefly investigate whether this also provides insights to the fuller geometrical representation of objects and their interactions through contact, and how we should discretize those. The results are preliminary, but promising.

4.1 Gauss Sphere as Direction Spectrum

The principle of the slope spectrum is that the function may be considered as a collection of tangent planes, characterized by a support function, in an invertible manner.

We have indicated in section that we can do this for arbitrary objects, in a coordinate-free manner, by the support function σ as a function of the tangent direction characterization by $\mathbf{n}$ or $\mathbf{I}$. This is the *extended Gauss sphere* representation of the 'Gauss sphere' of directions. We therefore view the *Gauss sphere as the spectrum of directions*, and the support function on it as the *amplitude* indicating the strength of presence of that direction in the object. This support

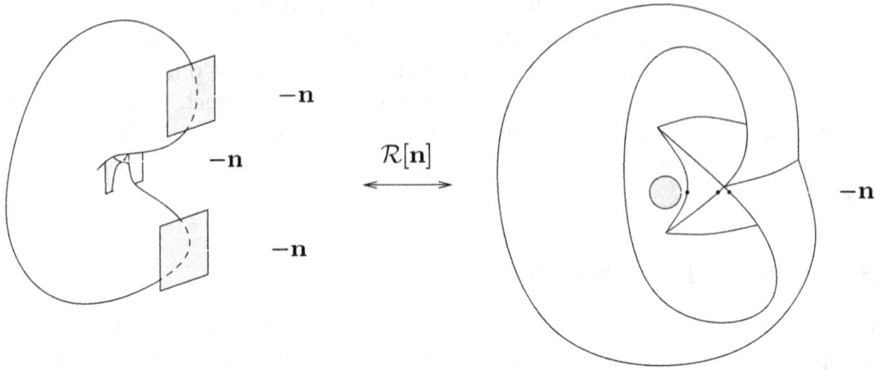

Fig. 11. The support function of an object sketched as a function on the Gaussian sphere of directions. For convenience in depiction, the outward pointing normal $-\mathbf{n}$ is drawn, rather than $\mathbf{n}$

function is, of course, multi-valued. We sketch this in Fig. ; the geometric algebra computations are easier to perform than to sketch...

Combining two objects in a dilation operation may be done as addition of their direction spectra. Whether or not this is practical depends on finding a 'fast direction transform'; but viewing the operation in its natural spectral description should provide analytic insight in the dilation and collision operations.

4.2 Point as Delta-Object

The 'delta object' of the dilation operation is an object which is the additive identity for all directions. It thus has as representation: $\mathcal{R}[\mathbf{n}] = \mathbf{n} + \mathbf{e}^0 \times 0 = \mathbf{n}$. This is a point at the origin. And indeed, if a boundary $\mathcal{A}$ is being dilated with (or collided by) a point object, the point describes the boundary $\mathcal{A}$.

4.3 Discretization of Collision Computations

We have seen in the umbral function description that a discretization which commutes with the dilation is obtained by choosing a (discrete) spectrum of directions, and encode the object as a network of offset tangent hyperplanes.

This is also true for geometrical objects. If all objects involved in the dilation are represented using this same limited spectrum of directions, then the dilation is simply computable through the addition of the support functions, without any additional interpolation. Again, *discretization of the dilation equals dilation of the discretization equals addition of the spectrum of support functions*, see Fig. . This is a potentially cheap way of evaluating collisions to any desired accuracy.

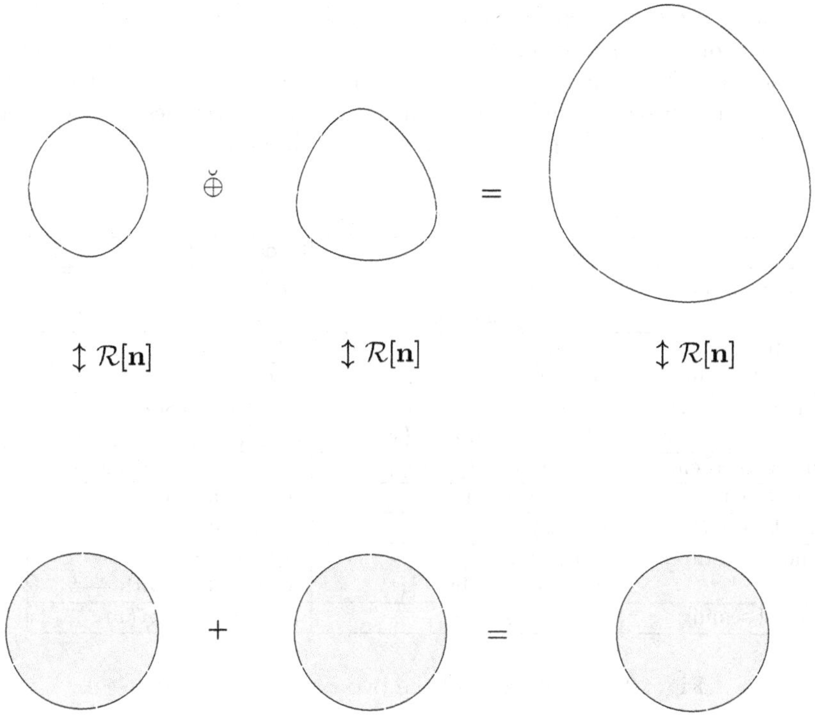

Fig. 12. Contact of objects: the tangent hyperplane supports are additive. (a) Spatial representation, the objects have been discretized using a icosahedral spectrum of directions; (b) Support functions on the Gauss sphere add up

5 Summary

We have seen that various operations which have the nature of collision or wave propagation can be analyzed using a 'systems theory'. This is based on the realization that the boundaries of the objects involved can be decomposed according to a spectral transform, and that in terms of this spectral transform the combination operation (wave propagation or collision) becomes a simple arithmetic operation, namely addition. The geometric intuition behind both is the same: an object boundary can be indexed as the support of its tangent planes, as a function of their direction. The spectral transform is the 'support function on the Gauss sphere' for objects, and the 'Legendre transform' for functions.

Especially the Legendre transform formulation shows that the resulting algebraic structure is very similar to the linear systems theory of convolution, with the Fourier transform as the spectrum (see Fig. , after [ ]). We have shown how that analogy gives useful techniques with sensible results for the analysis of collision/propagation of functions. The techniques should be transferable to the interaction of the boundaries of objects in space. We believe that this will lead to analytical insights which will be the foundation in the design of efficient algorithms for wave propagation, robotic collision, and object growing.

	contact systems	**linear systems**
signal combination	$(f \breve{\oplus} g)(x) =$ $\text{stat}_u\ [f(x) + g(x-u)]$	$(f * g)(x) =$ $\int du\, f(u) \times g(x-u)$
eigenfunctions	ωx	$e^{-\imath\omega x}$
spectral parameter	orientation	frequency
canonical transform	Legendre transform: $\mathcal{L}[f](\omega) = \text{stat}_x\ [f(x) - \omega x]$	Fourier transform: $\mathcal{F}[f](\omega) = \int dx\, f(x) e^{-\imath\omega x}$
central theorem	$\mathcal{L}[f \breve{\oplus} g] = \mathcal{L}[f] + \mathcal{L}[g]$	$\mathcal{F}[f * g] = \mathcal{F}[f] \times \mathcal{F}[g]$
delta function	point at origin	unit area at origin
translation $f(x-a)$	$\mathcal{L}[f](\omega) - \omega a$	$\mathcal{F}[f](\omega) e^{-\imath\omega a}$
band-limitation	Lipschitz: propagation by $-\|x\|$	Nyquist/Shannon: convolution with $\frac{\sin x}{x}$
natural scaling	umbral $af(x/a)$	proportional $af(x)$

Fig. 13. Comparison of the two systems theories

References

1. van den Boomgaard, R., Smeulders, A. W. M.: The morphological structure of images: the differential equations of morphological scale space. IEEE Trans. on Pattern Analysis and Machine Intelligence **16** (1994) 1101–1113
2. Dorst, L.: Objects in contact: boundary collisions as geometric wave propagation. In: Geometric Algebra: A Geometric Approach to Computer Vision, Quantum and Neural Computing, Robotics and Engineering, E. Bayro-Corrochano, G. Sobczyk, eds. Birkhäuser (2000) Chapter 17, 355–375
3. Dorst, L., van den Boomgaard, R.: Morphological signal processing and the slope transform. Signal Processing **38** (1994) 79–98 , ,
4. Dorst, L., van den Boomgaard, R.: The Support Cone: a representational tool for the analysis of boundaries and their interactions. IEEE PAMI **22(2)** (2000) 174–178
5. Ghosh, P. K.: A Unified Computational Framework for Minkowski Operations. Comput. & Graphics, **17(4)** (1993) 357–378
6. Hawkes, P. K.: The evolution of electron image processing and its potential debt to image algebra. Journal of Microscopy **190(1-2)** (1998) 37–44
7. Heijmans, H. J. A. M.: Morphological Image Operators. Academic Press, Boston (1994)

8. Heijmans, H. J. A. M., Maragos, P.: Lattice calculus of the morphological slope transform. Signal Processing **59** (1997) 17–42
9. Hestenes, D.: The design of linear algebra and geometry. Acta Applicandae Mathematicae **23** (1991) 65–93
10. Lasenby, J., Fitzgerald, W. J., Doran, C. J. L., Lasenby, A. N.: New Geometric Methods for Computer Vision. Int. J. Comp. Vision **36(3)** (1998) 191–213
11. Maragos, P.: Slope transforms: theory and application to nonlinear signal processing. IEEE Transactions on Signal Processing **43(4)** (1995) 864–877
12. R. T. Rockafellar, Convex analysis, Princeton University Press (1972)

An Associative Perception-Action Structure Using a Localized Space Variant Information Representation

Gösta H. Granlund

Computer Vision Laboratory, Department of Electrical Engineering,
Linköping University, SE-581 83 Linköping, Sweden
gosta@isy.liu.se

Abstract. Most of the processing in vision today uses spatially invariant operations. This gives efficient and compact computing structures, with the conventional convenient separation between data and operations. This also goes well with conventional Cartesian representation of data.

Currently, there is a trend towards context dependent processing in various forms. This implies that operations will no longer be spatially invariant, but vary over the image dependent upon the image content.

There are many ways in which such a contextual control can be implemented. Mechanisms can be added for the modification of operator behavior within the conventional computing structure. This has been done e.g. for the implementation of adaptive filtering.

In order to obtain sufficient flexibilility and power in the computing structure, it is necessary to go further than that. To achieve sufficiently good adaptivity, it is necessary to ensure that sufficiently complex control strategies can be represented. It is becoming increasingly apparent that this can not be achieved through prescription or program specification of rules. The reason being that these rules will be dauntingly complex and can not be be dealt with in sufficient detail.

At the same time that we require the implementation of a spatially variant processing, this implies the requirement for a spatially variant information representation. Otherwise a sufficiently effective and flexible contextual control can not be implemented.

This paper outlines a new structure for effective space variant processing. It utilises a new type of localized information representation, which can be viewed as outputs from band pass filters such as wavelets. A unique and important feature is that convex regions can be built up from a single layer of associating nodes. The specification of operations is made through learning or action controlled association.

1 Introduction

Most of the processing in vision today uses spatially invariant operations. This gives efficient and compact computing structures, with the conventional convenient separation between data and operations. This also goes well with conventional Cartesian representation of data.

G. Sommer and Y. Y. Zeevi (Eds.): AFPAC 2000, LNCS 1888, pp. 48– , 2000.

Currently, there is a trend towards context dependent processing in various forms. This implies that operations will no longer be spatially invariant, but vary over the image dependent upon the image content.

There are many ways in which such a contextual control can be implemented. Mechanisms can be added for the modification of operator behavior within the conventional computing structure. This has been done e.g. for the implementation of adaptive filtering [ ].

In order to obtain sufficient flexibilility and power in the computing structure, it is necessary to go further than that. To achieve sufficiently good adaptivity, it is necessary to ensure that sufficiently complex control strategies can be represented. It is becoming increasingly apparent that this can not be achieved through prescription or program specification of rules. The reason being that these rules will be dauntingly complex and can not be be dealt with in sufficient detail.

At the same time that we require the implementation of a spatially variant processing, this implies the requirement for a spatially variant information representation. Otherwise a sufficiently effective and flexible contextual control can not be implemented[ ].

Most information representation in vision today is in the form of iconic arrays, representing the pattern of intensity and color or some function of this, such as edges, lines, convexity, etc. This is advantageous and easily manageable for stereotypical situations of images having the same resolution, size, and other typical properties. Increasingly, various demands upon flexibility and performance are appearing, which makes the use of array representation less attractive.

The increasing use of actively controlled and multiple sensors requires a more flexible processing and representation structure. The data which arrives from the sensor(s) is often in the form of image patches of different sizes, rather than frame data in a regular stream. These patches may cover different parts of the scene at various resolutions. Some such patches may in fact be image sequence volumes, at a suitable time sampling of a particular region of the scene, to allow estimation of the motion of objects [ ]. The information from all such various types of patches has to be combined in some suitable form in a data structure.

The conventional iconic array form of image information is impractical as it has to be searched and processed every time some action is to be performed. It is desirable to have the information in some partly interpreted form to fulfill its purpose to rapidly evoke actions. Information in interpreted form, implies that it should be represented in terms of content or *semantic* information, rather than in terms of array values. Content and semantics implies *relations* between units of information or symbols. For that reason it is useful to represent the information as relations between objects or as *linked objects.* The discussion of methods for representation of objects as linked structures will be the subject of most of this paper, but we can already observe how some important properties of a desirable representation relate to shortcomings of conventional array representations:

- An array implies a given size frame, which can not easily be extended to incorporate a partially overlapping frame

- Features of interest may be very sparse over parts of an array, leaving a large number of unused positions in the array
- A description of additional detail can not easily be added to a particular part of an array

The following sections of this paper outline a new structure for effective space variant processing. It utilises a new type of localized information representation. The specification of operations is made through learning or action controlled association.

2 Channel Information Representation

A continuous representation of similarity requires that we have a *metric* or distance measure between items. For this purpose, information is in the associative structure expressed in terms of a *channel representation*[,]. See Figure .

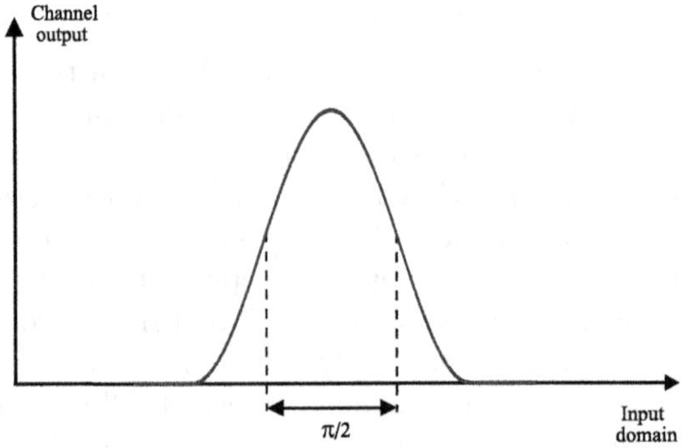

Fig. 1. Channel representation of some property as a function of match between filter and input pattern

Each channel represents a particular property measured at a particular position of the input space. We can view such a channel as the output from some band pass filter sensor for some property /citeg78a. An appropriate object evokes an output from the activated channel, corresponding to the match between the object presented and the properties of the filter, characterizing the passband of the channel. This resembles the function of biological neural feature channels. There are in biological vision several examples available for such properties; edge and line detectors, orientation detectors, etc [,].

If we view the channel output as derived from a band pass filter, we can establish a measure of *distance* or *similarity* in terms of the parameters of this filter. See Figure . For a conventional, linear simple band pass filter, the phase distance between the flanks is a constant $\pi/2$. Different filters will have different band widths, but we can view this as a standard unit of similarity or distance, with respect to a particular channel filter.

2.1 Sequentially Ordered Channels

There are several envelope functions with the general appearance of Figure , such as Gaussian and trigonometric functions. Functions which are continuous and have continuous derivatives within the resolution range are of interest. For the introductory discussion of channel representation, we assume the representation of a single scalar variable x, as an ordered one-dimensional sequence of band pass function envelopes x_k, which represent limited intervals, say $k - \frac{3}{2} \leq x \leq k + \frac{3}{2}$, of a scalar variable x. A class of functions which has some attractive properties for analysis is

$$x_k(x) = p_k(x) = \begin{cases} \cos^2(\frac{\pi}{3}(x-k)) & \text{if } k - \frac{3}{2} \leq x \leq k + \frac{3}{2} \\ 0 & \text{otherwise} \end{cases} \tag{1}$$

The scalar variable x can be seen as cut up into a number of local but partially overlapping intervals, $k - \frac{3}{2} \leq x \leq k + \frac{3}{2}$, where the center of each interval corresponds to $x = k$. It should be observed that we use the notation of x without subscript for the scalar variable and x_k with subscript for the channel representation of scalar variable x. The channel output signals which belong to a particular set are bundled together, to form a vector which is represented in boldface:

$$\mathbf{x} = [x_1 \ x_2 \ \ldots \ x_k \ \ldots \ x_K]^T \tag{2}$$

We assume for conceptual simplicity that the numbers k are consecutive integers, directly corresponding to the numbers of consecutive channels. This allows a more consistent treatment and a better understanding of mechanisms. We are obviously free to scale and translate the actual input variable in any desired way, as we map it onto the set of channels. An actual scalar variable ξ can be scaled and translated in the desired way

$$x = scale \cdot (\xi - translation) \tag{3}$$

to fit the interval spanned by the entire set of channels $\{x_k\}$. We will later see how other nonlinear scaling transformations can be made.

With each channel center representing consecutive integers, the distance between two adjacent channels in terms of the variable x is one unit. From Equation it is apparent that the distance in terms of angle is $\frac{\pi}{3}$ or 60°. We will in subsequent discussions refer to this as the typical channel distance of $\frac{\pi}{3}$ or 60°.

In Figure we have a one-dimensional set of 13 sequentially ordered channels. The position of each channel is indicated by the dashed lines. It is designed to provide a channel representation of scalars within a range $0 \leq x \leq 10$. To provide a continuous representation at the boundaries of this interval, the set of channels is padded with an extra channel at each end. Components from these channels are required to perform a reliable reconstruction back to a scalar value from the vector representation. In order to start adapting ourselves to the major purpose of processing of spatial data, we can view Figure as a one-dimensional image

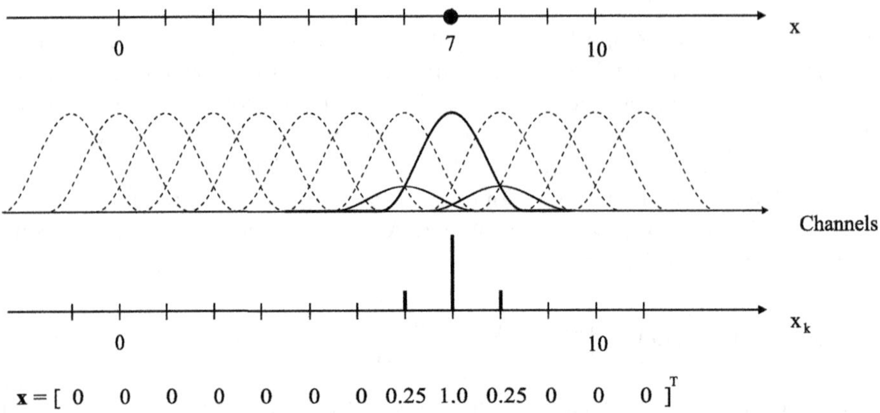

Fig. 2. Channel representation of a scalar $x = 7$

with a single simple object in the form of a dot. The channels are scaled to unit resolution between the filter centers, and the center values correspond to values

$$\mathbf{k} = [\ -1\ \ 0\ \ 1\ \ 2\ \ 3\ \ 4\ \ 5\ \ 6\ \ 7\ \ 8\ \ 9\ \ 10\ \ 11\ ]^T \tag{4}$$

If the set of channels is activated by a scalar $x = 7$, represented by a point at position $x = 7$, we will obtain a situation as indicated in Figure . We assume that the output of a channel is given by the position of the point $x = 7$ within its band pass function, according to Equation . The channels activated are indicated by the solid line curves. The scalar $x = 7$ will produce the vector $\mathbf{x}$ as indicated in Figure .

Below are a few additional examples which hopefully will shed some light on the representation, in particular at the boundaries. We still assume a set of 13 channels which are used to represent scalar values in the interval between 0 and 10.

$$\begin{array}{lcl}
x = 0.0 & \Rightarrow & \mathbf{x} = [\,0.25\ \ 1.0\ \ 0.25\ \ 0\ \ 0\ \ 0\ \ 0\ \ 0\ \ 0\ \ 0\ \ 0\ \ 0\ \ 0\,]^T \\
x = 3.73 & \Rightarrow & \mathbf{x} = [\,0\ \ 0\ \ 0\ \ 0\ \ 0.52\ \ 0.92\ \ 0.06\ \ 0\ \ 0\ \ 0\ \ 0\ \ 0\ \ 0\,]^T \\
x = 9.0 & \Rightarrow & \mathbf{x} = [\,0\ \ 0\ \ 0\ \ 0\ \ 0\ \ 0\ \ 0\ \ 0\ \ 0\ \ 0.25\ \ 1.0\ \ 0.25\ \ 0\,]^T \\
x = 10.0 & \Rightarrow & \mathbf{x} = [\,0\ \ 0\ \ 0\ \ 0\ \ 0\ \ 0\ \ 0\ \ 0\ \ 0\ \ 0\ \ 0.25\ \ 1.0\ \ 0.25\,]^T
\end{array} \tag{5}$$

We can clearly see the necessity for padding with extra channels at the boundaries. Under the conditions stated earlier, we have the following values of x_k within an interval $k - \frac{3}{2} \leq x \leq k + \frac{3}{2}$:

$$
\begin{aligned}
x_k(k-\tfrac{3}{2}) &= \cos^2(-\tfrac{\pi}{2}) = 0 \\
x_k(k-1) &= \cos^2(-\tfrac{\pi}{3}) = 0.25 \\
x_k(k) &= \cos^2(0) \quad = 1 \\
x_k(k+1) &= \cos^2(\tfrac{\pi}{3}) \quad = 0.25 \\
x_k(k+\tfrac{3}{2}) &= \cos^2(\tfrac{\pi}{2}) \quad = 0
\end{aligned} \tag{6}
$$

In relation to this, it can be shown that

$$\sum_k x_k(x) = 1.5 \qquad \text{if} \quad -\frac{1}{2} \leq x \leq K - \frac{1}{2} \tag{7}$$

where K is the last channel used for padding. This consequently gives a margin of $1/2$ outside the second last channel. This means that the sum of all channel contributions over the entire channel set from the activation by a single scalar x is 1.5, as long as x is within the definition range of the entire set. Related properties are:

$$x_k(k-1) + x_k(k) + x_k(k+1) = x_{k-1}(k) + x_k(k) + x_{k+1}(k) = 1.5 \tag{8}$$

Most components of $\mathbf{x}$ are zero, with only two or three non-zero components representing the scalar value x as discussed earlier.

An array may be activated by more than one value or stimulus. In Figure we have two scalars, at $x = 1$ and $x = 7$. It is apparent that as the difference between the two scalars decreases, there is going to be overlap and interference between the contributions. This indicates a need to worry about proper resolution, like for any sampling process. Still, the representation gives us the possibility to keep track of multiple events within a single variable, without their superimposing, something which a Cartesian representation does not allow.

2.2 Two-Dimensional Channels

Most of the information we want to deal with as input is two-dimensional, or possibly of even higher dimensionality. For that reason we will extend the definition to two dimensions, x and y:

$$
p_{kl}(x,y) = \begin{cases} \begin{cases} \cos^2(\frac{\pi}{3}\sqrt{(x-k)^2+(y-l)^2}\,) \\ \text{if} \quad k-\frac{3}{2} \leq x \leq k+\frac{3}{2},\ l-\frac{3}{2} \leq y \leq l+\frac{3}{2} \end{cases} \\ 0 \qquad \text{otherwise} \end{cases} \tag{9}
$$

The arrangement of sequential integer ordering with respect to k and l is similar to the one-dimensional case. The output from a channel is now dependent upon the distance, $d = \frac{\pi}{3}\sqrt{(x-k)^2+(y-l)^2}$ from the center of a particular channel at position (k,l) in the array.

As we will see later, good functionality requires that there are several non-zero outputs generated from a sensor array. As we in this case are dealing with point objects, this requires that there is an overlap between the transfer functions

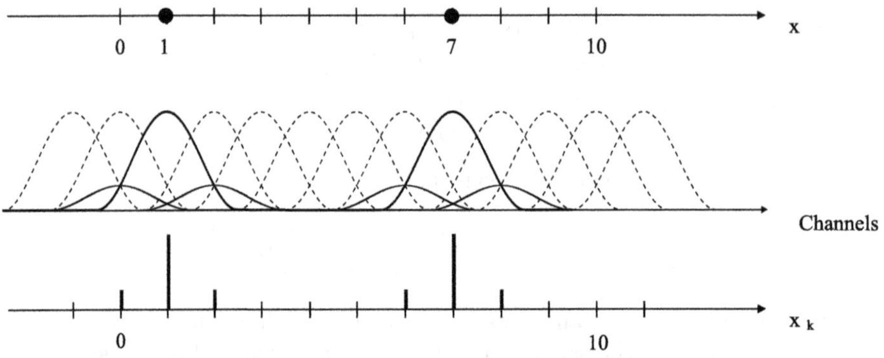

$\mathbf{x} = [\ 0\ \ 0.25\ \ 1.0\ \ 0.25\ \ 0\ \ \ 0\ \ \ 0\ \ 0.25\ \ 1.0\ \ 0.25\ \ 0\ \ \ 0\ \ \ 0\ ]^T$

Fig. 3. Channel representation of two scalars at $x = 1$ and $x = 7$

of the different detectors. When the object is a line, or other spatially extended object, no overlap is required. Rather we will see that receptive fields of sensors normally only have to cover parts of the array.

So far, we have only dealt with the position dependent component of the channel band pass function. Generally, there is as well a component dependent upon some property of the sensor, such as dominant orientation, color, curvature, etc. Equation will then take on the general form:

$$p_{klm}(x,y,\phi) = p_{kl}(x,y)p_m(\phi) = \begin{cases} \begin{cases} \cos^2(\frac{\pi}{3}\sqrt{(x-k)^2+(y-l)^2}\,)\,p_m(\phi) \\ \text{if } k-\frac{3}{2} \le x \le k+\frac{3}{2},\ l-\frac{3}{2} \le y \le l+\frac{3}{2} \end{cases} \\ 0 \qquad \text{otherwise} \end{cases} \tag{10}$$

As this property is often modular and e.g. representing an angle, it has been given an argument ϕ. Because the use of modular channel sets is not restricted to this application, we will give it a somewhat more extensive treatment.

3 Modular Channel Sets

There are several situations where it is desirable to represent a modular or circular variable, such as angle, in a channel representation. There are two different cases of interest:

- Modular channel distance $\frac{\pi}{3}$
- Modular channel distance $\frac{\pi}{4}$

Of these, we will only deal with the first one:

3.1 Modular Channel Distance $\frac{\pi}{3}$

The structure easiest to deal with has a channel distance of $\frac{\pi}{3}$, similarly to the earlier treatment. In this case, the least complex structure contains three channels in a modular arrangement:

$$\phi_m(\phi) = p_m(\phi) = \begin{cases} \cos^2(\phi - m\frac{\pi}{3}) & \text{if} \quad | \phi - m\frac{\pi}{3} | \leq \frac{\pi}{2} \qquad m = 0, 1, 2; \\ 0 & \text{otherwise} \end{cases} \tag{11}$$

The scalar variable ϕ will be represented by a vector

$$\boldsymbol{\phi} = [\phi_0 \ \phi_1 \ \phi_2]^T = \{\phi_m\} \qquad m = 0, 1, 2 \tag{12}$$

As earlier, we use the notation of ϕ without subscript for the scalar variable and ϕ_m with subscript for the channel representation of scalar variable ϕ. The channels which belong to a particular set are bundled together, to form a vector which is represented in boldface, to the extent that this type font is available.

The modular arrangement implies that as the scalar argument increases from $\frac{2\pi}{3}$ it will not activate a fourth channel but map back into channel 1, which is equivalent to the dashed channel curve in Figure . This is the minimum number of channels which will provide a continuous representation of a modular variable. It is for example useful for the representation of orientation of lines and edges. It can be shown that this is the minimum number of filter components which give an unambiguous representation of orientation in two dimensions [ ]. If we view the distance between adjacent channels to $\frac{\pi}{3}$ or 60° like in the earlier discussion, this implies that the total modulus for 3 channels is π or 180°. This is well suited for representation of "double angle" [ ] features such as the orientation of a line. If it is desired to represent a variable with modulus 2π or 360°, the variable ϕ can be substituted by $\phi/2$ in Equation above. Any different desired modulus can be scaled accordingly. There are several different ways to express the scaling. In this presentation we have tried to maintain the argument in terms of the $\cos^2()$ function as a reference.

Assuming a resolution of a 10 to 20 levels per channel, this will give a total resolution of 3° to 6° given modulus 180° and a total resolution of 6° to 12° given modulus 360°. This is sufficient for many applications. The modular arrangement is illustrated in Figure .

If a higher resolution is desired, more channels can be added in the modular set as required. There are several ways to express the scaling, such as in constant modulus or in constant channel argument. The way selected here is in terms of constant argument of the $\cos^2()$ function. This gives a variable modulus for the entire system, but makes it easy to keep track of the type of system. The generalized version becomes:

$$\phi_m(\phi) = p_m(\phi) = \begin{cases} \cos^2(\phi - m\frac{\pi}{3}) & \text{if} \quad | \phi - m\frac{\pi}{3} | \leq \frac{\pi}{2} \quad m = 0, 1, \ldots, M-1 \\ 0 & \text{otherwise} \end{cases} \tag{13}$$

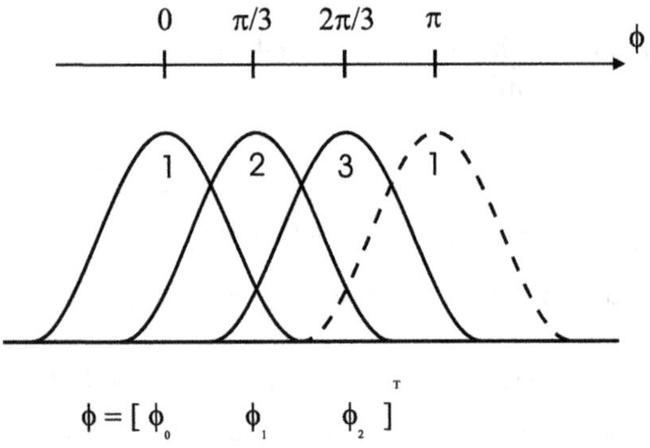

Fig. 4. Three component modular channel vector set

The scalar variable ϕ will be represented by a vector

$$\phi = [\phi_0 \; \phi_1 \; \ldots \phi_{M-1}]^T = \{\phi_m\} \quad m = 0, 1, \ldots, M-1 \quad \text{modulus} \quad M\frac{\pi}{3} \tag{14}$$

4 Variable Resolution Channel Representation

In the preceding discussion we have assumed a constant or linear mapping and resolution for the variable in question to the channel vector representation. There are however several occasions where a nonlinear mapping is desired.

4.1 Logarithmic Channel Representation

In many cases it will be useful to have a representation whose resolution and accuracy varies with respect to the value of the variable. As an example, we can take the estimated distance z to an object, where we typically may require a constant relative accuracy within the range.

We can obtain this using a logaritmic mapping to the channel representation.

$$z_k(z) = \begin{cases} \cos^2(\frac{\pi}{3}(\,^b\log(z - z_0) - k)) & \text{if} \quad k - \frac{3}{2} \leq \,^b\log(z - z_0) \leq k + \frac{3}{2} \\ 0 & \text{otherwise} \end{cases} \tag{15}$$

It is convenient to view the process of scaling as a mapping to the integer vector set. There are two cases of scaling which are particularly convenient to use:

- One octave per channel. This can for example be achieved by using a mapping $x = {}^2\log(z - z_0)$, where z_0 is a translation variable to obtain the proper scaling for z.
- One decade per two channels. This can for example be achieved by using a mapping $x = {}^{10}\log(z - z_0)/2$, where z_0 is a translation variable to obtain the proper scaling for z.

4.2 Arbitrary Function Mapping

A mapping with an arbitrary function $x = f(z)$ can be used, as long as it is strictly monotonous. It is possible to employ such a function to obtain a variable resolution in different parts of a scene, dependent upon the density of features or the required density of actions.

4.3 Foveal Arrangement of Sensor Channels

A non-uniform arrangement of sensors with a large potential is the foveal structure. See Figure . A foveal window, has a high density of sensors with a small scale near the center of the window. More peripherally, the density decreases at the same time as scale or size of sensors increases. This is similar to the sensor arrangement in the human retina.

The low level orientation outputs from the sensor channels will be produced from the usual procedures. They should have a bandwidth, which corresponds to the size of the sensor channel field, as illustrated in Figure . This implies a representation of high spatial frequencies in the center and low frequencies in the periphery.

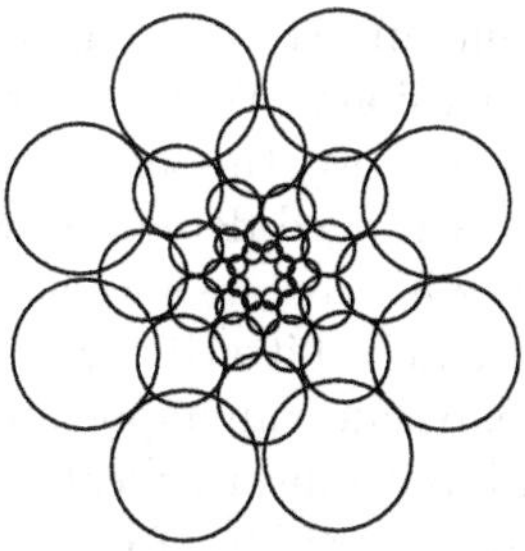

Fig. 5. Foveal arrangement of channels in sensor window

As the computing structure can easily deal with a non-uniform arangement of sensors, there is a great deal which speaks in favor of a foveal arrangement of sensors. It provides a high resolution at the center of the visual field. While the lower resolution towards the periphery does not provide detailed information, it is sufficient to relate the high resolution description to its background, as well as to guide the attentive search mechanism to regions of interest.

5 Arrangement of Channels

The abstract function of this associative structure is to produce a mapping between a set of arbitrarily arranged channels for feature variables, and a set of sequentially ordered channels for response variables. This constitutes a process of recognition.

We assume two distinctly different categories of channel representations:

1. Sequentially ordered channels for response variables
2. Arbitrarily arranged channels for sensor and feature variables

What we have been dealing with so far, can be said to imply the first category of response variables. This reflects the fundamental property that response states are defined along, at least locally, one-dimensional spaces. We assume an availability of consecutive, sufficiently overlapping channels which cover these spaces.

In general, there is no requirement for a regular arrangement of channels, be it on the input side or on the output side. The requirement of an orderly arrangement comes as we need to interface the structure to the environment, e.g. to determine its performance. We typically want to map the reponse output channel variables back into scalar variables in order to compare them with the reference. The mapping back into scalars is greatly facilitated by a regular arrangement.

5.1 Arbitrarily Arranged Sensor Channels

For sensor channels, we assume an arrangement which is typically two-dimensional, or in general multi-dimensional. While the response space is assumed to be one-dimensional as described above, the sensor or feature space is assumed to be populated with arbitrarily arranged detectors, where we have no guarantee for overlap or completeness. See Figure . As we will see, there is no problem for the associative structure to use an arbitrarily arranged array of input channels, as long as it is stable over time, because an important part of the learning process is to establish the identity of input sensor or feature channels.

The preferential orientation sensitivity of a sensor is indicated as a line segment, and the extent of the spatial sensitivity function is indicated by the size of the circle. As indicated in this figure, detectors for orientation may typically have no overlap, but rather be at some distance from each other. The reason is that an expected object, such as a line, has an extent, which makes it likely that it will still activate a number of sensor channels

Like any other analysis procedure, this one will not be able to analyze an entire image of say $512 \cdot 512$ elements in one single bite. It is necessary to limit the

Fig. 6. Example of random arrangement of orientation detectors over space

size of a processing window onto the image. We assume that a sensor map window contains $40 \cdot 40 = 1.6 \cdot 10^3$ orientation detectors, distributed as a two-dimensional array. Each orientation detector contains a combination of an edge and a line detector to produce a quadrature bandpass output. Detectors are assumed to be distributed such that we only have one detector for some preferred orientation within some neighborhood. This will give a lower effective resolution with respect to orientation over the array, corresponding to around $20 \cdot 20 = 400$ orientation detectors with a full orientation range. Detectors will have to be distributed in an arrangement such that we do not have the situation that there are only detectors of a particular orientation along a certain line, something which may happen with certain simple, regular arrangements.

Given no overlap between sensors, the reader may suspect that there will be situations, where an applied line will not give an output from any sensor. This is true, e.g. when a line is horizontal or vertical in a regular array. It is however no problem to deal with such situations, but we will leave out this case from the present discussion.

The channel representation has an elegant way to represent the non-existence of information, which is something totally different from the value 0. This is very different from the representation in a common Cartesian array, where all positions are assumed to have values, which are as well reliable. The channel representation does not require such a continuity, neither spatially, nor in terms of magnitude. This allows for the creation of a more redundant representation.

6 Feature Vector Set for Associative Machinery

A sensor channel will be activated depending upon how the type of stimulus matches, and how its position matches. The application of a line upon an array as indicated in Figure , will evoke responses from a number of sensors along the lenghth of the line.

All sensor channels which we earlier may have considered as different vector sets, with different indices, will now be combined into a *single* vector. We can obviously concatenate rows or columns after each other for an array such as in Figure ; we can freely concatenate vectors from different sensor modalities one after the other. We will in the ensuing treatment for simplicity assume that *all* sensor channels to be considered, are bundled together into a *single sensor channel vector set*:

$$\mathbf{x} = [x_1 \; x_2 \; \dots \; x_K]^T \qquad k = 1, \dots, K \tag{16}$$

We can see each sensor channel as an essentially independent wire, carrying a signal from a band pass filter, describing some property at some position of the image. It is assumed that we have a set of such sensor channels within some size *frame of interpretation*, which is a subset or window onto the image to be be interpreted, which is substantially smaller than the entire image. The *frame of interpretation* may in practise contain somewhere between 10^2 to 10^4 sensor channels, which is equivalent to the dimensionality K of $\mathbf{x}$, dependent upon the problem and available computational resources. This vector is very sparse however, because most sensors do not experience a matching stimulus, which gives the vector a density, typically from 10^{-1} to 10^{-3}.

A particular length of line at a particular position and orientation, will produce a stimulation pattern reflected in the vector $\mathbf{x}$ which is unique. As we will see next, this vector can be brought into a form such that it can be associated with the state vectors (length, orientation, position) related to it.

The set of features used for association, derives from the above mentioned sensor channels, as illustrated in Figure .

We will use the notation:

- Sensor channel vector set: $\mathbf{x} = [x_1 \; x_2 \; \dots \; x_K]^T = \{x_k\} \qquad k = 1, \dots, K$
- Feature channel vector set: $\mathbf{a} = [a_1 \; a_2 \; \dots \; a_H]^T = \{a_h\} \qquad h = 1, \dots, H$

The sensor channel vector $\mathbf{x}$ is an arbitrary but fixed one-dimensional arrangement of the outputs from the two-dimensional sensor channel array. The sensor channel vector $\mathbf{x}$ forms the basis for the *feature channel vector*, $\mathbf{a}$, which is to be associated with the response state. The feature vector can contain three different functions of the sensor vector:

1. **Linear components** This is the sensor channel vector itself, or components thereof. This component will later be denoted simply as $\mathbf{x}$.

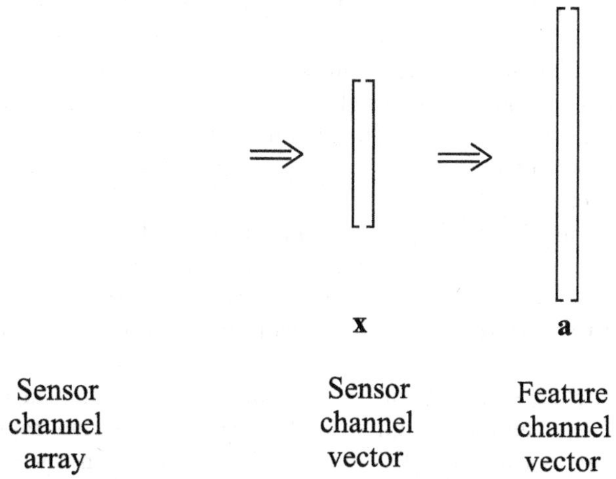

Fig. 7. Illustration of steps in going from sensor array to feature vector

2. **Autocovariant components** These are product components of type $(x_1x_1,\ x_2x_2,\ \dots\ , x_kx_k)$, which are the diagonal elements of the covariance matrix. The corresponding vector containing these components will be denoted as $\mathbf{xx}^T_{auto}$.
3. **Covariant components** These are product components of type $(x_1x_2,\ x_1x_3,\ \dots\ , x_{k-1}x_k)$, which are the off-diagonal elements of the covariance matrix. The corresponding vector containing these components will be denoted as $\mathbf{xx}^T_{cov}$.

The feature vector used for association will be:

$$\mathbf{a} = \begin{bmatrix} \mathbf{x} \\ \mathbf{xx}^T_{auto} \\ \mathbf{xx}^T_{cov} \end{bmatrix} \tag{17}$$

of which in general only the last covariant components will be present. Experiments indicate that the covariant feature components are the most descriptive as they describe coincidences between events, but the existence of the others should be kept in mind for various special purposes such as improved redundancy, low feature density, etc.

From this stage on, we will not worry about the sensor channel vector set $\mathbf{x}$, and only use feature channel vector set, $\mathbf{a}$. For that reason you will see some of the indices recycled for new tasks, which will hopefully not lead to any confusion.

Before we go into the rest of the associative structure, and how this feature vector is used, we will recognize the fact that we can recover the conventional scalar meaning of data expressed as a channel vector.

7 Reconstruction of Scalar Value From Channel Vectors

It should first be made clear that this computing structure is intended for consistent use of information represented as channel signals as explained earlier. Input to a computing unit will have the channel representation, as will normally the output. The output from one unit or computing stage will be used as input to another one, etc.

As a system of this type has to interface to the external world, input or output, requirements become different. For biological systems there are sensors available which do give a representation in this form, as well as that output actuators in the form of muscle fibers can directly use the channel signal representation.

For technical systems, it will be necessary to provide interfaces which convert between the conventional high resolution cartesian signal representation and the channel representation. We have in the introduction discussed how this is accomplished for input signals. We will now look at how this can be done for output signals as well. Output signals which will be used to drive a motor or similar device, or for visualization of system states.

The output from a single channel u_k of a response vector $\mathbf{u}$, will not provide an unambiguous representation of the corresponding scalar signal u, as there will be an ambiguity in terms of the position of u with respect to the center of the activated channel u_k. This ambiguity can be resolved in the combination with adjacent channel responses within the response vector $\mathbf{u} = \{u_k\}$. By using a sufficiently dense representation in terms of channels, we can employ the knowledge of a particular similarity or distance between different channel contributions.

It can be shown that if the distance between adjacent channels is 60° or less, we can easily obtain an approximative reconstruction of the value of u as a linear phase. Reconstruction of the scalar value u_e which corresponds to a particular response vector $\mathbf{u}$, formally denoted as *sconv*:

$$u_e = sconv(\mathbf{u}) = sconv(\{u_k\}) \qquad k = 1, \ldots, K \tag{18}$$

can, given the earlier discussion, be implemented in several ways. We will however leave out the details of the computation in this context.

8 System Structure for Training

The general aspects of training are obviously related to the current large field of Neural Networks [ ]. Training of a system implies that it is exposed to a succession of pairs of samples of a feature vector $\mathbf{a}$ and a corresponding response vector $\mathbf{u}$.

There are several ways in which this training can be done, but typically one can identify a training structure as indicated in Figure .

The *Pseudorandom Training Sequencer* supplies transformation parameters of a training pattern to the system. The most characteristic property of this is that the output variables are guaranteed to vary continuously. The training

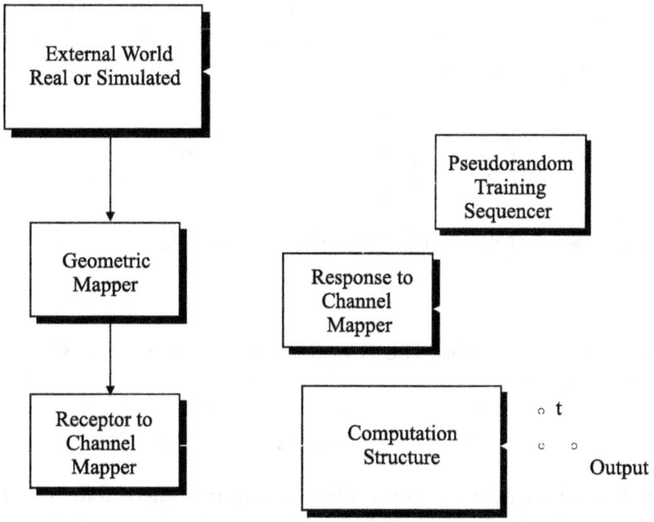

Fig. 8. System structure as set up for training, in interaction with the environment

variables have to cover the space over which the system is expected to operate. The Pseudorandom Training Sequencer is expected to produce its output in conventional digital formats.

The training data is input to the *External World Simulator* or interface. There it is generating the particular transformations or modes of variation for patterns, that the system is supposed to learn. This can be the generation of movements of the system itself, which will modify the precepts available

The *Geometric Mapper* will in the general case produce a two-dimensional projection from a three-dimensional world, such as to implement a camera.

The *Receptor to Channel Mapper* will in the general case convert an image projected onto it, into a parallel set of channels, each channel describing some property according to the discussion earlier.

The training data, representing transformations to the input pattern, is also supplied to the *Response to Channel Mapper*, where it is converted from conventional Cartesian format to the channel representation of the response state, as discussed earlier. This information is supplied directly to the output side of the associative computation structure.

8.1 Basic Training Procedure

The basic training procedure is to run the Pseudorandom Training Sequencer to have it vary its output. This will have an effect onto the external world model in that something changes. The response variables out from the Pseudorandom Training Sequencer will also be fed to the output of the Computing Structure.

We will in this discussion assume *batch mode training*, which implies that pairs of corresponding feature vectors $\mathbf{a}$ and response vectors $\mathbf{u}$ are obtained for each one of N samples, which form matrices $\mathbf{A}$ and $\mathbf{U}$:

$$\begin{cases} \mathbf{U} = [\mathbf{u}_1 \; \mathbf{u}_2 \; \ldots \; \mathbf{u}_n \; \ldots\ldots \; \mathbf{u}_N] \\ \mathbf{A} = [\mathbf{a}_1 \; \mathbf{a}_2 \; \ldots \; \mathbf{a}_n \; \ldots \; \mathbf{a}_N] \end{cases} \tag{19}$$

These matrices are related by the linkage matrix $\mathbf{C}$:

$$\mathbf{U} = \mathbf{C}\mathbf{A} \tag{20}$$

From this matrix equation, the coupling or linkage matrix $\mathbf{C}$ can be solved, superficially expressed as

$$\mathbf{C} = \mathbf{U}/\mathbf{A} \tag{21}$$

The feature matrix $\mathbf{A}$ may contain tens of thousands of features, represented by tens of thousands of samples. This implies that the method of solution has to be chosen carefully, in order not to spend the remaining part of the millenium solving the equation.

There are now very fast numerical methods available for an efficient solution of such systems of linear equations. These methods utilize the sparsity of the $\mathbf{A}$ and $\mathbf{U}$ matrices; i.e. the fact that most of the elements in the matrices are zero. Although this is an important issue in the use of the channel representation, it is a particular and well defined problem, which we will not deal with in this presentation. One of the methods available is documented in a Ph.D. Thesis by Mikael Adlers: Topics in Sparse Least Squares Problems [ ].

8.2 Association as an Approximation Using Continuous Channel Functions

What happens in the computing structure during training, is that the response channel vector $\mathbf{u}$ will associate with the feature channel signal vector $\mathbf{a}$. The association implies that the output response is approximated by a linear combination of the feature channel signals. This is illustrated for an actual case in Figure .

We can now compute the approximating function for a particular response node k over the sample points n:

$$u_{kn} = \sum_h c_{kh} a_{hn} \tag{22}$$

The association during the training implies finding the coefficients c_{kh} which implement this approximation. We can see that a particular response channel function is defined over some interval of samples, n, from the training set. We can vary u continuously, and as different channels ..., u_{k-1}, u_k, u_{k+1}, ... are activated,

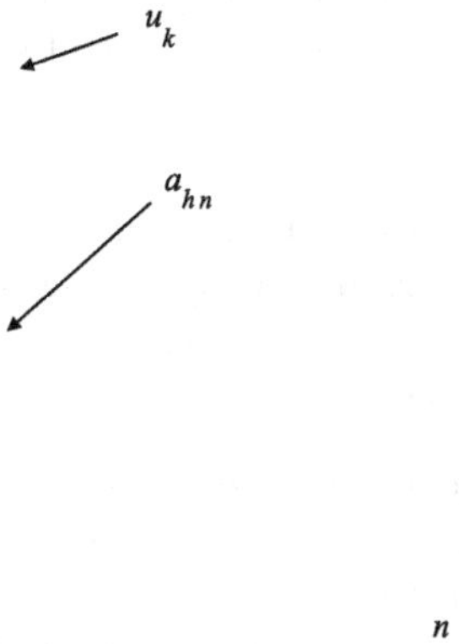

Fig. 9. Illustration of procedure to approximate a response channel function, u_k, with a set of feature channel functions, a_h, over an interval of sample points n

their approximation in terms of similarly activated features ..., a_{i-1}, a_i, a_{i+1} can be computed. The resulting optimization coefficients ..., c_{kh}, ... will constitute *quantitative links* in the linkage matrix $\mathbf{C}$ between the input feature side and the output response side.

Taken over all response nodes, k, this is written in matrix terms as:

$$\mathbf{u}_n = \mathbf{C}\mathbf{a}_n \tag{23}$$

For the entire training set of vectors $\mathbf{u}_n$ and $\mathbf{a}_n$ this is written in matrix form as before:

$$\mathbf{U} = \mathbf{C}\mathbf{A} \tag{24}$$

Having somehow completed a training procedure for the entire range of values of u, we can change the switch to output, from the training position $\mathtt{t}$, in Figure . After this we can present an unknown pattern with feature vector $\mathbf{a}$, within the definition range as input to the system, after which the system will use the linkage matrix derived, $\mathbf{C}$, to compute the actual value of $\mathbf{u}$.

$$\mathbf{u} = \mathbf{C}\mathbf{a} \tag{25}$$

When the training is completed, the computation of the preceding expression for an unknown vector $\mathbf{a}$ is extremely fast, due to the sparsity of the vectors and matrices involved.

9 Properties of the Linkage Matrix C

We have related the set of response states **U** and the corresponding percept vectors **A** with the matrix equation

$$\mathbf{U} = \mathbf{CA} \tag{26}$$

This matrix equation does not generally have a unique solution, but it can be underdetermined or overdetermined.

For the system to perform as desired, we require a solution with some particular properties:

1. The coefficients of matrix **C** shall be non-negative, as this gives a more sparse matrix and a more robust system. A traditional unrestricted least squares solution tends to give a full matrix with negative and positive coefficients, which do their best to push and pull the basis functions to minimize the error for the particular training set. A particular output may in this case be given by the difference between two large coefficients operating upon a small feature function, which leads to a high noise sensitivity.
2. The coefficients of matrix **C** shall be limited in magnitude, as this as well gives a more robust system. One of the ways to achieve this is to set elements of **A** below a certain threshold value to zero. This is related to the lower treshold part of the S curve, often assumed for the transfer function of real and artificial neurons.
3. Matrices **A** and **U** are sparse, and the entire system can be dealt with using fast and efficient procedures for solution of sparse systems of equations for values between two limits.
4. Coefficients of matrix **C** which are below a certain threshold shall be eliminated altogether, as this gives a matrix with lower density, which allows a faster processing using sparse matrix procedures. If desired, a re-optimization can be performed using this restricted set of coefficients.

After the linkage matrix **C** has been computed, we can obtain the response state **u** as a function of a particular feature vector **a** as

$$\mathbf{u} = \mathbf{Ca} \tag{27}$$

9.1 Multiple Response State Variables

So far we have only discussed the situation for a single response or state variable u. We will normally have a number of state variables $u, v, w, t, \ldots$. As we change the value of the additional variable v, however, the set of features which is involved for a particular value of u will vary, and we can suspect that different models would be required. This is true in general, but if feature vectors exhibit a sufficiently high degree of locality, the simple model structure proposed will still work. In such a case, the solution for three response variables in matrix terms can be expressed as

$$\begin{cases} \mathbf{U} = \mathbf{C}^u \mathbf{A} \\ \mathbf{V} = \mathbf{C}^v \mathbf{A} \mathbf{W} = \mathbf{C}^w \mathbf{A} \end{cases} \tag{28}$$

The reason why this works is again the extreme degree of locality of feature components. This means that as v varies, new feature components move into the mapping and old components move out transparently for the single model available. Due to the beauty of the channel representation, this means that channels which are not active will not disturb the matching process for those who are active.

10 Applications of Associative Structure

As the purpose of this paper is to give a description of the principles of the associative channel structure, we will in this context only give a few comments on results from applications.

The structure has with great success been used to estimate various properties in an image, ranging from description of line segments to structures containing corners or in general curvature. There are various approaches which can be used.

The structure has also been used for a view-centered object description procedure, which is able to recognize the object car from several different angles and also give an estimate of the view angle.

As in any other descriptive system, the mapping of certain properties are invariant, while others are not. In the associative procedure, the system will detect such properties by itself, or it can be guided in the choice of such properties.

11 Concluding Remarks

Learning in any robot system or biological system does not take place in parallel over a field of features and responses. Learning takes place along one-dimensional trajectories in a response state space. The reason for this is that a system, like a human or a robot, can only be at "one place at a time". As it moves from one place to another, which really implies from one state to another, it can only do so continuously due to its mass and limited power resources. Consequently, the system will move along a one-dimensional, continuous trajectory in a multidimensional space. This continuity is one of the few hard facts about its world, that the system has to its disposal to bring order into its perception of it, and it has to make the best possible use of it.

Acknowledgements

The author wants to acknowledge the financial support of the Swedish National Board of Technical Development, as well as of WITAS: The Wallenberg Laboratory for Information Technology and Autonomous Systems. Credits go to several people in the staff of the Computer Vision Laboratory of Linköping University, for discussions around ideas. Special thanks go to Per-Erik Forssén for his development of a MATLAB channel computation toolbox, something which has been invaluable in the development and tests of implementations.

References

1. M. Adlers. *Topics in Sparse Least Squares Problems.* PhD thesis, Linköping University, Linköping, Sweden, Dept. of Mathematics, 2000. Dissertation No. 634.
2. D. H. Ballard. Animate vision. Technical Report 329, Computer Science Department, University of Rochester, Feb. 1990.
3. M. F. Bear, B. W. Connors, and M. A. Paradiso. *Neuroscience. Exploring the Brain.* Williams & Wilkins, Baltimore, USA, 1996. ISBN 0–683–00488–3.
4. G. H. Granlund. The complexity of vision. *Signal Processing*, 74(1):101–126, April 1999. Invited paper.
5. G. H. Granlund and H. Knutsson. *Signal Processing for Computer Vision.* Kluwer Academic Publishers, 1995. ISBN 0-7923-9530-1.
6. Gösta Granlund. Does Vision Inevitably Have to be Active? In *Proceedings of SCIA99, Scandinavian Conference on Image Analysis*, Kangerlussuaq, Greenland, June 7–11 1999. Also as Technical Report LiTH-ISY-R-2247.
7. S. Haykin. *Neural Networks: A Comprehensive Foundation.* Macmillan College Publishing Company, 1994.
8. I. P. Howard and B. J. Rogers. *Binocular Vision and Stereopsis.* Number 29 in Oxford Psychology Series. Oxford University Press, New York, 1995. ISBN 0–19–508476–4.
9. K. Nordberg, G. Granlund, and H. Knutsson. Representation and Learning of Invariance. In *ICIP*, Austin, Texas, November 1994. IEEE.

The Structure of Colorimetry

Jan J. Koenderink and Andrea J. van Doorn

[1] Utrecht University
[2] Delft University of Technology

Abstract. We consider the structure of colorimetry, essentially of Graßmann's threedimensional linear space that summarized metamery (confusion of spectral distributions) for the human observer. We show that the definition of an orthonormal basis for this space requires a scalar product in both the space of physical beams, and the representation space on which Graßmann's manifold is mapped. The former of these scalar products has to be constructed on the basis of considerations of physics and physiology. The present standards (CIE) are very awkward. The latter of these scalar products can be choosen for reasons of convenience. After these choices "color space" becomes a "true image" of the space of physical beams, apart from the fact that all but three of the infinitely many dimensions are lost (the metamery). We show that the key operator of modern colorimetry, Cohen's "Matrix–R" (the projector on fundamental space that rejects the "metameric black" part of arbitrary physical beams) also requires these scalar products for its definition. In the literature such inner products are (implicitly) assumed with the unfortunate result that the awkward CIE definition is willy nilly accepted as the only possibility.

1 Introduction

Historically, "colorimetry" is one of the earliest success stories of visual psychophysics. Following Newton[ ], most of the structure was in place by the mid 19th c., the formal structure being due to Graßmann[ ], the empirical and methodological work to Maxwell[ ]. The field was polished off in the 1920's by Schrödinger[ ] after decisive work by Helmholtz[ ]. The only nameworthy development after this is due to Cohen[ ] by the 1970's.

As the field is presented in the standard texts it is somewhat of a chamber of horrors: Colorimetry proper is hardly distinguished from a large number of elaborations (involving the notion of "luminance" and of absolute color judgments for instance) and treatments are dominated by virtually ad hoc definitions (full of magical numbers and arbitrarily fitted functions). I know of no text where the essential structure is presented in a clean fashion. Perhaps the best textbook to obtain a notion of colorimetry is still Bouma's[ ] of the late 1940's, whereas the full impact of all sorts of vain ornamentation can be felt from a book like Wyszecki and Stiles[ ].

G. Sommer and Y. Y. Zeevi (Eds.): AFPAC 2000, LNCS 1888, pp. 69– , 2000.

2 The Colorimetric Paradigm

The basic facts are simple enough. If you look into a *beam of radiation* with radiant power in the range 400–700 nm you experience a *patch of light.* The apparent shape and size of the patch depend on beam geometry; the color on its spectrometric properties. I consider only beams of incoherent radiation. Such beams can be added and multiplied with non–negative factors through simple physical techniques. The "space of beams" $\mathbb{S}$ (say) is thus the non–negative part of a linear space. When two beams $\mathbf{a}$ and $\mathbf{b}$ yield patches that cannot be distinguished I will write the fact as $\mathbf{a} \Longleftrightarrow \mathbf{b}$. Such "colorimetric equivalence" can be objectively established. Notice that the observer is not even required to venture an opinion as to the "color of the patch". What makes this all interesting is that colorimetric equivalence does by no means imply radiometric identity. Of course two radiometrically identical beams are (indeed, trivially) colorimetrically equivalent. But *most* colorimetrically equivalent beams (drawn at random say) are unlikely to be radiometrically identical. This is the basic phenomenon of metamerism. The metamer of any beam $\mathbf{a}$ is the set of all its colorimetrically equivalent mates. *Colorimetry is the science of metamerism*, its aim is to parcellate the space of beams into distinct metamers.

The essential empirical facts are formalized as "Graßmann's Laws". These are idealizations of empirical observations: $\mathbf{a} \Longleftrightarrow \mathbf{b}$ implies $\mu\mathbf{a} \Longleftrightarrow \mu\mathbf{b}$ for any (non–negative) scalar μ, $\mathbf{a} \Longleftrightarrow \mathbf{b}$ implies $(\mathbf{a}+\mathbf{c}) \Longleftrightarrow (\mathbf{b}+\mathbf{c})$ for any beam $\mathbf{c}$, equivalent beams may be substituted for each other in any combination. A null beam exists (total darkness) which doesn't change any beam when you add it to it. Apart from these *generic* properties there is the *particular* fact that no more than three beams suffice to produce equivalent patches for all others. This is the human condition of *trichromacy.*

3 Geometrical Interpretation

3.1 Gauging the Spectrum

Maxwell was the first to "gauge the spectrum". The idea is that the Newtonian spectrum yields an exhaustive radiometric description of beams. Since Graßmann's Laws imply linearity it is sufficient to investigate the spectral components (monochromatic beams), then all other beams are treated as linear combinations of these.

Pick a set of three independent beams (the "primaries") $\{\mathbf{p_1}, \mathbf{p_2}, \mathbf{p_3}\}$ (no linear combination equivalent to the null beam). Almost any random triple will suffice. Denote the monochromatic beams of a given, fixed radiant power as $\mathbf{m}(\lambda)$ (λ the wavelength). Then "gauging the spectrum" consists of finding three "color matching functions" $a_i(\lambda)$ ($i = 1, 2, 3$) such that $(a_1(\lambda)\mathbf{p_1} + a_2(\lambda)\mathbf{p_2} + a_3(\lambda)\mathbf{p_3}) \Longleftrightarrow \mathbf{m}(\lambda)$. This is done in a wavelength by wavelength fashion. Notice that the color matching functions are not (necessarily) non–negative throughout: Graßmann's Laws enable you to handle that because formally $\mathbf{a} - \mathbf{b} \Longleftrightarrow \mathbf{c}$

implies $\mathbf{a} \iff \mathbf{b}+\mathbf{c}$. In practice one samples the spectral range at about a hundred locations.

Consider any beam $\mathbf{s}$ given by the radiant spectral density $s(\lambda)$. Then $(c_1\mathbf{p_1} + c_2\mathbf{p_2} + c_3\mathbf{p_3}) \iff \mathbf{s}$, where $c_i = \int a_i(\lambda)s(\lambda)d\lambda$. This is what "gauging the spectrum" buys you. The coefficients c_i are the "color coordinates" of the patch caused by the beam $\mathbf{s}$. They indicate a point $\mathbf{c}$ (the "color" of the beam $\mathbf{s}$) in three dimensional "color space" $\mathbb{C}$.

Of course the "color" changes when you swap primaries. You easily show that the new coordinates are a linear transformation of the old ones, the transformation being determined by the new pair of primaries and the (old) color matching functions (simply express the color of the new primaries in terms of the old system). This is the reason why textbooks tell you that "color space is only affine" and often make a show of plotting color coordinates on differently scaled oblique axes.

3.2 The Structure of Colorimetry

Notice that $c_i = \int a_i(\lambda)s(\lambda)d\lambda$ defines the color coordinates in terms of linear transformations $\langle\chi_i, \mathbf{s}\rangle$ of the radiant power spectrum of the beam. The χ_i are elements of the dual space $\mathbb{S}^*$ of the space of beams $\mathbb{S}$ ($\mathbb{S}^*$ is the space of linear functionals on $\mathbb{S}$) and $\langle\cdot,\cdot\rangle$ denotes the contraction of an element on a dual element. Thus $(\langle\chi_1, \mathbf{s}\rangle\mathbf{p_1} + \langle\chi_2, \mathbf{s}\rangle\mathbf{p_2} + \langle\chi_3, \mathbf{s}\rangle\mathbf{p_3}) \iff \mathbf{s}$. The relation $\{\mathbf{p_1}, \mathbf{p_2}, \mathbf{p_3}\} \to \{\chi_1, \chi_2, \chi_3\}$ is empirically determined through the gauging of the spectrum.

One fruitful way to understand the geometrical structure of colorimetry is due to Wyszecki and (especially) to Cohen[ ]. Think of the space of beams as the direct sum of a "fundamental space" $\mathbb{F}$ and a "black space" $\mathbb{B}$, thus $\mathbb{S} = \mathbb{F} + \mathbb{B}$. All elements of the black space are literally *invisible*, thus causally ineffective. Elements of fundamental space are causally maximally effective, that is to say, *colorimetric equivalence implies physical identity.* Any beam $\mathbf{s}$ (say) can be written as $\mathbf{s} = \mathbf{f} + \mathbf{b}$ and the metamer of $\mathbf{s}$ is $\mathbf{f} + \mathbb{B}$, *i.e.*, let the black component range over the full black space. Fundamental space has to be three dimensional and is isomorphic with color space $\mathbb{C}$.

Apply this to the primaries, *i.e.*, $\mathbf{p_i} = \mathbf{f_i} + \mathbf{b_i}$ (say). Clearly the black components are irrelevant: You may replace any primary with one of its metameric mates, it makes no difference. The fundamental components $\mathbf{f_i}$ are what matters, they form a basis for color space. Of course you *know* only the $\mathbf{p_i}$, not the $\mathbf{f_i}$ though. By picking the primaries you automatically select a basis for $\mathbb{F}$, only you don't know it!

3.3 Metric in the Space of Beams

In order to proceed you need additional structure, most urgently a metric or a scalar product in the space of beams. In many treatments (even Cohen's otherwise exemplary work) the existence of a scalar product on $\mathbb{S}$ is simply taken for granted, yet there are quite a few problems involved. Without a scalar product in both $\mathbb{S}$ and $\mathbb{C}$ you cannot introduce the transpose of a linear map (frequently done in the literature without so much as the drop of a hat) for instance.

A metric for the space of beams presupposes a decision on how to represent beams. The conventional way is through radiant power spectral density as a function of wavelength. Yet wavelength is *irrelevant* (should one take the vacuum wavelength or that in the acquous medium of the retina?) and photoreceptors count photon absorptions rather than integrate radiant power. For the interaction with photopigment it is the photon energy that is relevant. Once absorbed, the effect of any photon is like that of any other (*i.e.*, after absorption photon energy is irrelevant, the "Law of Univariance"). Thus only a spectral description in terms of photon number density as a function of photon energy makes any physiological sense.

Photon energy is still an awkward scale: Consider the notion of a "uniform spectrum". If you take constant spectral photon number flux distribution the result will depend on the particular unit (should you use electronVolts or ergs?). Clearly the notion of a "uniform spectrum" should not depend on such accidental choices! The only way to rid yourself of this problem is to take the uniform spectrum on a *logarithmic* photon energy scale. The spectral density $d\epsilon/\epsilon$ is invariant against changes of the units. The essential arguments are especially well laid out by Jaynes[ ].

Thus I will write the basic colorimetric equations as $c_i = \int_0^\infty s(\epsilon) g_i(\epsilon)\, d\epsilon/\epsilon$, where $s(\epsilon)$ denotes the spectral photon number flux density and the g_i are (new!) color matching functions.

With the representation in place you still need to define a scalar product. This again is a knotty problem. Various choices appear reasonable. The choice should be made according to the aspects of the relevant physics. Here I will settle on $\mathbf{a} \cdot \mathbf{b} = \int_0^\infty a(\epsilon) b(\epsilon)\, d\epsilon/\epsilon$.

3.4 Again: The Structure of Colorimetry

With the scalar product in the space of beams in place it is possible to venture some further advances. For instance, you can replace the basis of dual space $\mathbb{S}^*$ with a "dual basis" of $\mathbb{S}$ in the classical sense. Notice that to any dual vector σ (say) corresponds a unique vector $\mathbf{s}$ (say) when you require that $\langle \sigma, \mathbf{x} \rangle = \mathbf{s} \cdot \mathbf{x}$ for any vector $\mathbf{x}$. Let the vectors corresponding to the χ_i be denoted $\mathbf{g_i}$. Then $\{\mathbf{g_1}, \mathbf{g_2}, \mathbf{g_3}\}$ is the dual basis of $\{\mathbf{f_1}, \mathbf{f_2}, \mathbf{f_3}\}$ in fundamental space $\mathbb{F}$, that is to say, $\mathbf{s} = (\mathbf{g_1} \cdot \mathbf{s})\, \mathbf{f_1} + (\mathbf{g_2} \cdot \mathbf{s})\, \mathbf{f_2} + (\mathbf{g_3} \cdot \mathbf{s})\, \mathbf{f_3}$. (We also have the dual relation $\mathbf{s} = (\mathbf{f_1} \cdot \mathbf{s})\, \mathbf{g_1} + (\mathbf{f_2} \cdot \mathbf{s})\, \mathbf{g_2} + (\mathbf{f_3} \cdot \mathbf{s})\, \mathbf{g_3}$.)

Let me write the dual basis as $\mathbf{G} = \{\mathbf{g_1}, \mathbf{g_2}, \mathbf{g_3}\}$, *i.e.*, a matrix with columns equal to the color matching functions. Its Grammian (matrix with coefficients $G_{ij} = \mathbf{g_i} \cdot \mathbf{g_j}$) is $\mathbf{G^T G}$. It is a 3x3 symmetric, nonsingular matrix. You can use it to construct the basis of fundamental space: $\mathbf{F^T} = (\mathbf{G^T G})^{-1} \mathbf{G^T}$, where $\mathbf{F} = \{\mathbf{f_1}, \mathbf{f_2}, \mathbf{f_3}\}$. Thus the introduction of the scalar product has enabled us to find the basis of fundamental space induced by the (arbitrarily chosen!) primaries. This is real progress.

Notice that $\mathbf{g_i} \cdot \mathbf{s} = \int_0^\infty s(\epsilon) g_i(\epsilon)\, d\epsilon/\epsilon$, that is to say, *the coordinates of the dual basis vectors of fundamental space are the color matching functions.* "Duality" is expressed by the relations $(\mathbf{F^T F})^{-1} = \mathbf{G^T G}$ and $\mathbf{G^T F} = \mathbf{F^T G} = \mathbf{I_3}$.

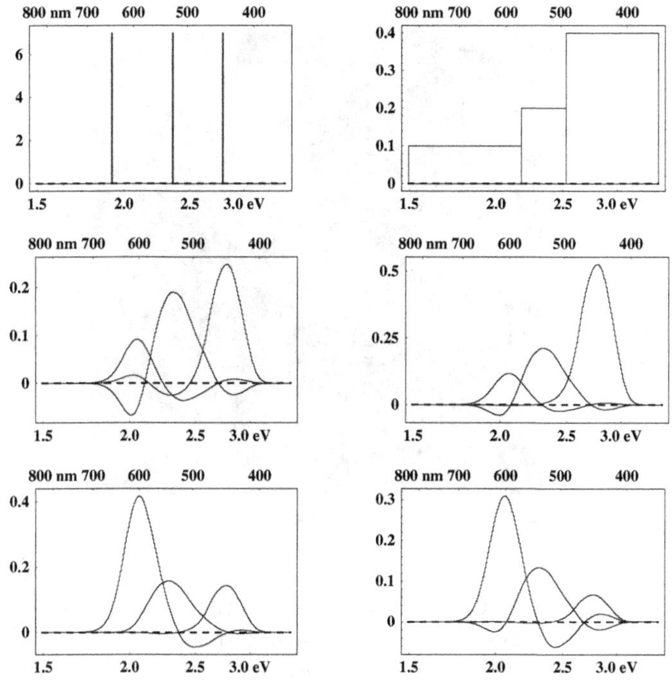

Fig. 1. In the upper row I show the spectra of two quite different sets of primaries. One set consists of monochromatic beams, the other of band pass spectra. In the middle row I plot the **F** basis, in the bottom row the corresponding **G** bases

Finally, $\mathbf{f} = \mathbf{F\,G^T s}$, for an arbitrary beam $\mathbf{s}$ with fundamental component $\mathbf{f}$. Notice that thus $\mathbf{f} = \mathbf{G(G^T G)^{-1} G^T s}$. The matrix $\mathbf{P_F} = \mathbf{G(G^T G)^{-1} G^T}$ is symmetrical, has rank 3, trace equal to 3 and satisfies $\mathbf{P_F} = \mathbf{P_F^2}$. It is the projection operator in $\mathbb{S}$ on fundamental space $\mathbb{F}$. Because of that it doesn't depend on the arbitrary choice of the primaries. You may easily check this explicitly by changing to another set of primaries. It is an invariant, complete description, the "holy grail" of colorimetry!

The projection operator is Cohen's "Matrix–R". From the present derivation its geometrical nature is immediately clear: $\mathbf{P_F} = \mathbf{FG^T} = \mathbf{GF^T}$. In plain words: *The fundamental spectra are simply linear combinations of the fundamental spectra of the primaries, the coefficients being the coordinates of the color.*

I show examples in figures and . In figure I consider two quite different sets of primaries and find the dual bases for each of them. Notice how these bases turn out to be quite different. However, when you calculate Cohen's matrix R from either set of bases you obtain the identical result (figure).

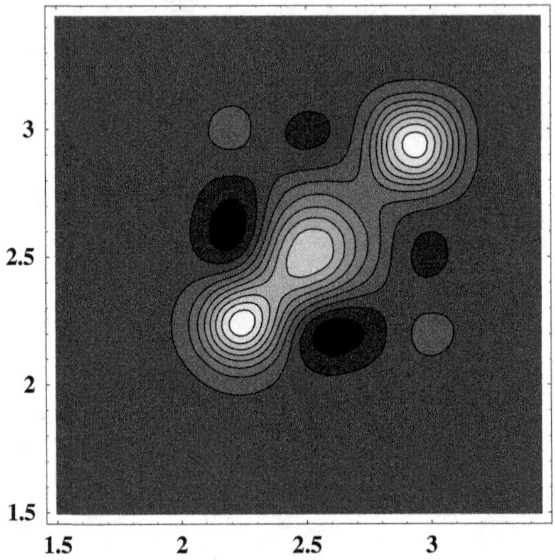

Fig. 2. A density plot of "Cohen's Matrix R". Either basis of figure yields the identical result. The projector on fundamental space is the main invariant of colorimetry. Matrix elements are labelled by photon energy in eV

4 True Images of Color Space

The usual textbooks tend to (over–)stress the point that color space is "affine", meaning that arbitrary linear transformations are irrelevant. Often they show figures in oblique, unequally divided axes to drive the point home. This is really counterproductive. "Color space" in the classical sense is an *image* of fundamental space. Why not try to construct an image that is as "true" as possible? Once you have a metric in the space of beams that question makes sense.

There are really two different issues here, one has to do with the structure of fundamental space, the other with "representation space". By the latter I mean that an "image" is usually drawn on some canvas. For instance, Cartesian graph paper is usually preferred over arbitrary, oblique, unequally divided axes and rightly so. It is not different for color space which is an image of fundamental space: The most convenient "canvas" is obviously three dimensional Cartesian space referred to an orthonormal basis. Notice that there is nothing "deep" going on here. Yet the point (though never mentioned) is important enough, because it means you already *have* a scalar product in color space, namely *by choice.* Since you already have a scalar product in fundamental space (induced by the scalar product in the space of beams), the notion of "true image" is well defined. An orthonormal basis in fundamental space should map on an orthonormal basis in representation space. In the conventional representation (the CIE convention[ ]) this is far from being the case though: the ratios of the lengths of the basis vectors are 1:0.97:0.43, whereas they subtend angles of 142°, 106° and 82°. Hardly a pleasant basis!

The standard way to construct a true image is to use the singular values decomposition (SVD) of the linear map, in this case of $\mathbf{G}$. Thus I write $\mathbf{G} = \mathbf{V}^{\mathbf{T}}\mathbf{W}\mathbf{U}$. The matrix $\mathbf{U}$ is the desired orthonormal basis of fundamental space, thus the problem is solved. The matrix $\mathbf{W}$ is a diagonal matrix containing the "singular values". There exist three nonvanishing singular values. Due to the arbitrary choice of primaries the singular values are unlikely to be equal, thus revealing the "distortion" of the original representation. (In the CIE convention the singular values are in the ratios 1:0.81:0.29.) The matrix $\mathbf{V}^{\mathbf{T}}$ is just an isometry in representation space, thus it doesn't spoil the "truthfulness" of the image.

In the "true image" a unit hypersphere in the space of beams maps upon a three dimensional unit sphere in color space. Infinitely many dimensions are simply lost in the image, but at least the ones that *are* preserved are represented truthfully. I illustrate this in figure . Images in the color spaces for the two sets of primaries introduced in figure of a (infinitely dimensional!) hypersphere in the space of beams turn out to be quite different triaxial ellipsoids. When I find canonical bases for either case (straight SVD) these turn into spheres that differ only by a rotation in (undeformed) color space. Of course arbitrary rotations don't deform color space, thus you may pick a convenient orientation according to some idiosyncratic criterion.

5 Conclusion

Although the linear structure was essentially understood by Maxwell[ ] and Graßmann[ ] around the 1850's, a geometrical interpretation had to wait till Cohen's[ ] work in the 1970's. Cohen's seminal work has yet to be absorbed into the textbooks. Current texts rarely approach the level of sophistication of Schrödinger[ ] who wrote in the 1920's.

What is lacking even in Cohen's treatment of colorimetry is a clear geometrical picture. For instance, the fundamental invariant of colorimetry "Cohen's Matrix–R" cannot be defined without a scalar product in the space of beams. Cohen—implicitly—used the Euclidian scalar product in the space of radiant power spectra on wavelength basis. The point of our discussion is that one has to make an explicit choice here. The choice makes a difference, because a space of spectral photon number density on photon energy basis (for example) is not linearly related to the conventional radiometric choice.

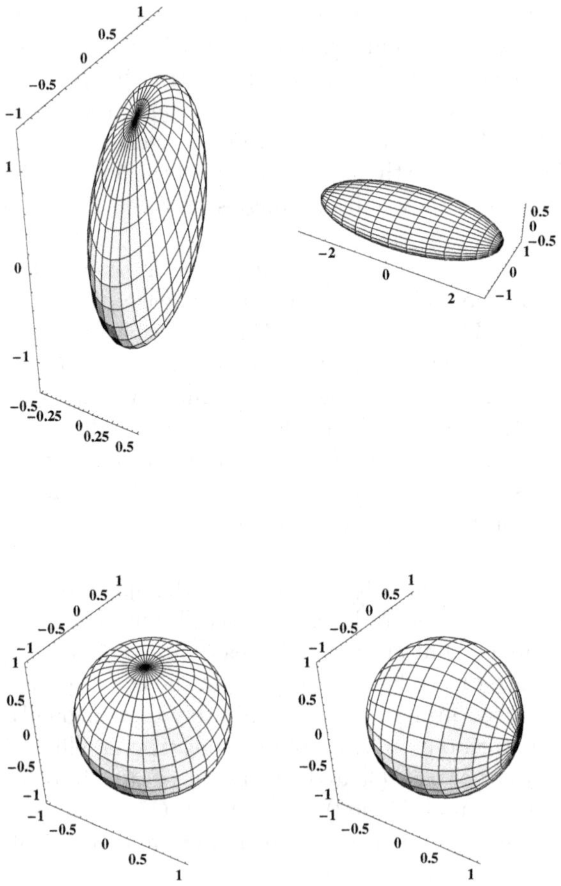

Fig. 3. Consider a hyperspere in the space of beams. In the top row I have plotted its image in the color spaces of the two bases introduced in figure . Notice that they are different and hardly spherical. Both color spaces yield deformed (thus misleading) images of fundamental space. In the bottom row I show the same configuration in the canonical color spaces obtained by straight SVD. These differ only by an isometry (rotation about the origin of color space). Both are "true images" of fundamental space, they merely show views from different directions

That "color space" also presupposes a "representation space" with a conventional scalar product (*e.g.*, the standard Euclidian one) is a never acknowledged, but necessary element of Cohen's analysis. For instance, the conventional "color matching matrix" $\mathbf{G}$ is a matrix with the color matching functions as columns. The color $\mathbf{c}$ of a beam $\mathbf{s}$ (say) is $\mathbf{c} = \mathbf{G}^{\mathbf{T}}\mathbf{s}$. Thus $\mathbf{G}^{\mathbf{T}}$ occurs as the map from $\mathbb{S}$ to $\mathbb{C}$. In Cohen's Matrix–R the transpose of this map ($\mathbf{G}$) also occurs. It is a map from $\mathbb{C}$ to $\mathbb{S}$. Although Cohen defines the transpose via swapping of rows and columns of the matrix associated with the map, the *geometrical* definition is via $\mathbf{x} \cdot \mathbf{G}^{\mathbf{T}}\mathbf{y} = \mathbf{G}\mathbf{x} \cdot \mathbf{y}$. Here $\mathbf{x} \in \mathbb{C}$ and $\mathbf{y} \in \mathbb{S}$, thus the first scalar product is taken in $\mathbb{C}$, the second in $\mathbb{S}$. Thus one needs to settle on scalar products in both spaces. One choice has to be decided on conventional grounds (essentially convenience), the other on conceptual grounds (physics and physiology). But the choices *have* to be made explicitly.

References

1. Bouma, P. J.: Physical Aspects of Colour. N.V. Philips Gloeilampenfabrieken, Eindhoven (1947)
2. CIE Proceedings. Cambridge University Press, Cambridge (1931)
3. Cohen, J. B., Kappauf, W. E.: Metameric Color Stimuli, Fundamental Metamers, and Wyszecki's Metameric Blacks. American Journal of Psychology **95** (1982) 537–564
4. Graßmann, H.: Zur Theorie der Farbenmischung. Annalen der Physik **89** (1853) 69–84
5. Helmholtz, H. von" Handbuch der Physiologischen Optik. Voss, Hamburg (1896)
6. Jaynes, E. T: Prior probabilities. IEEE **SSC–4** (1968) 227–241
7. Maxwell, J. C.: XVIII. Experiments on colour, as perceived by the eye, with remarks on colourblindness. Trans. of the Roy. Soc. Edinburgh (1855) 275–298
8. Newton, I.: Opticks. Dover Publications, Inc., New York (1952) (Based on the fourth edition London, 1730)
9. Schrödinger, E.: Grundlinien einer Theorie der Farbenmetrik im Tagessehen. Teil 1 und Teil 2. Annalen der Physik **63** (1920) 397–456
10. Wyszecki, G., Stiles, W. S.: Color Science. Concepts and Methods, Quantitative Data and Formulas. John Wiley & Sons, Inc., New York (1967)

Fast Calculation Algorithms of Invariants for Color and Multispectral Image Recognition

Valeri Labunets, Ekaterina Labunets-Rundblad, and Jaakko Astola

Tampere University of Technology, Signal Processing Laboratory,
Tampere, Finland

Abstract. We propose a novel method to calculate invariants of color and multicolor nD images. It employs an idea of multidimensional hypercomplex numbers and combines it with the idea of Fourier–Clifford–Galois Number Theoretical Transforms over hypercomplex algebras, which reduces the computational complexity of a global recognition algorithm from $\mathcal{O}(knN^{n+1})$ to $\mathcal{O}(kN^n \log N)$ for nD k–multispectral images. From this point of view the visual cortex of a primates brain can by considered as a "Fast Clifford algebra quantum computer".

1 Introduction

The moment invariants have found wide application in pattern recognition since they were proposed by HU [ ]. Traditionally, the moment invariants have been widely used in pattern recognition application to describe the geometrical shapes of the different objects. These invariants represent fundamental geometrical properties (*e.g.*, area, centroid, moment of inertia, skewness, kurtoses) of geometrical distortion images. Low–order moments are related to the global properties of the image and are commonly used to determine the position, orientation, and scale of the image. Higher–order moments contain information about image details. The image invariants are constructed in the two steps [ ]–[ ]. At the first step two–indices moments of the image $f(i_1, \ldots, i_n)$ are computed by the form:

$$m_{(p_1,p_2,\ldots,p_n)}\{f\} = \sum_{i_1=0}^{N-1} \cdots \sum_{i_n=0}^{N-1} f(i_1, i_2, \ldots, i_n) i_1^{p_1} i_2^{p_2} \cdots i_n^{p_n}, \tag{1}$$

where $p_1, \ldots, p_n = 0, 1, \ldots, N-1$. At the second step invariants $In_{(p_1,\ldots,p_n)}\{f\}$ are computed as spectral coefficients of the Kravtchook transform of moments [ ]. There are two drawbacks in this algorithm.

- First, fast calculation procedures of moments are not known today. The direct computation of the moments is too expensive in computation load. For example, for a nD window $W^n(N)$ containing N^n grey–level pixels the computational complexity is $O(nN^{n+1})$ of multiplications and additions. In many cases, however, especially in real-time industrial applications the computation speed is often the main limitation.

G. Sommer and Y. Y. Zeevi (Eds.): AFPAC 2000, LNCS 1888, pp. 78– , 2000.

In the literature, authors usually discuss fast computations for the moment invariants of binary images (see review [ ] and [ ]). The authors usually use Green's theorem to transform the double integrals in the 2–D domain $\mathbf{R}^2$ into the curve integral along the boundary of this binary domain. The computational complexity in this case is reduced to $\mathcal{O}(N)$. However, for the grey–level image intensity function such algorithms are not known.

- Secondly, it is not possible to estimate the moments of high order with fixed point arithmetics. These moments and their invariants have very large dynamic range. Such moments and invariants corresponding to them are said to be (in terms of quantum mechanics) un–observed. Hence, a computer with fixed point arithmetics can not observe small details of image. Moreover, these moments are sensitive to noise. The number of the observed moments and invariants can be increased using floating–point arithmetics. However, in this case approximation errors may become the value of the high–order moments. This means that the high–order moment invariants are not a "good" image features.

The second drawback is that it is impossible to remove for the computers with the finite capacity. This means that we have to use the other arithmetics to observe the high–order moments.

We propose to use modular arithmetic of the Galois field to develop a fast calculating algorithm for the low– and high–degree moments and invariants. Here a notion of modular invariant is introduced for the first time. The new moments exhibit some useful properties. First, dynamic range is the same for all moments. It can help us to overcome the diminishing problem of higher–order moments which occurs when other moment invariants are used. We can use the Fourier–Clifford–Galois Number Theoretical Transforms (FCG–NTTs) for the fast evaluation of low– and higher–order moment invariants which reduces the computational complexity of the nD grey–level, color and multicolor images recognition. Modular invariant hypercomplex–valued multicolor images can be calculated on the quantum computer working in modular $\mathbf{Z}/Q$–arithmetic.

2 Fast Calculation Algorithms of Real–Valued Invariants Based on Fourier–Galois NTTs

2.1 Modular Images and Moments

Moments and invariants are calculated based on the image models. The classical image model is a function defined on a window $W^n(N) = [0, N-1]^n$ with values either in the real numbers field or in the complex numbers field. When digital computers appeared it became perfectly clear that the result of any calculation can be only a rational number or an integer. Hence, it can be considered (up to constant multiplier 2^m, where m is the capacity of an analog–to–digital converter (ADC)) that an image has their values in the ring of integers: $f(i_1, i_2, \ldots, i_n) : W^n(N) \longrightarrow \mathbf{Z}$. If an image $f(i_1, i_2, \ldots, i_n)$ has 2^m gray–levels then there are no principal limits to consider an image mathematical model as a function having their values in the finite ring $\mathbf{Z}/Q$: i.e. as $f(i_1, i_2, \ldots, i_n) : W^n(N) \longrightarrow \mathbf{Z}/Q$, if $Q > 2^m$. This model can also give possibility to operate with pixels of image according to $\mathbf{Z}/Q$–arithmetic laws. The $\mathbf{Z}/Q$–valued function is called *modular image*. They form the space $L(W^n(N), \mathbf{Z}/Q)$.

If 2D images are processed on computer then intermediate and final results are expressed as integers on fixed point computers. This allows us to introduce new moments and invariants.

Definition 1 [ ]-[ ]. Functionals $\mathcal{M}_{(p_1,p_2,\ldots,p_n)} := m_{(p_1,p_2,\ldots,p_n)} \pmod{Q} =$

$$= \sum_{i_1=0}^{N-1} \cdots \sum_{i_n=0}^{N-1} f(i_1, i_2, \ldots, i_n) i_1^{p_1} i_2^{p_2} \cdots i_n^{p_n} \pmod{Q} \tag{2}$$

and $\mathcal{I}_{(p_1,p_2,\ldots,p_n)} := In_{(p_1,p_2,\ldots,p_n)} \pmod{Q}$ are called *modular moments* and *absolute modular* **G**–*invariants*, respectively.

Note that if Q is a prime then according to Euler's theorem, Eq. $i^{Q-1} = 1 \pmod{Q}$ holds for every element $i \in \mathbf{GF}(Q)$ [ ]. Hence $i_1^{p_1+r_1(Q-1)} = i_1^{p_1} \pmod{Q}$, $\ldots$, $i_n^{p_n+r_n(Q-1)} = i_n^{p_n} \pmod{Q}$ are true for all $p_1, \ldots, p_n = 0, 1, \ldots, Q-2$ and for all $r_1, \ldots, r_n = 0, 1, \ldots$. Therefore, the matrix $[\mathcal{M}_{(p_1,p_2,\ldots,p_n}] = [m_{p_1,p_2,\ldots,p_n}] \pmod{Q}$ is the periodical matrix: $\mathcal{M}_{(p_1+r_1(Q-1), p_2+r_2(Q-1), \ldots, p_n+r_n(Q-1)} = \mathcal{M}_{(p_1,p_2,\ldots,p_n)}$. The fundamental period is nD matrix $\mathcal{M}_{(p_1,p_2,\ldots,p_n)}$. For $n = 2$ we have

$$[\mathcal{M}_{(p,q)}] = \begin{bmatrix} \mathcal{M}_{0,0} & \mathcal{M}_{0,1} & \ldots & \mathcal{M}_{0,Q-2} \\ \mathcal{M}_{1,0} & \mathcal{M}_{1,1} & \ldots & \mathcal{M}_{1,Q-2} \\ \ldots & \ldots & \ldots & \ldots \\ \mathcal{M}_{Q-2,0} & \mathcal{M}_{Q-2,1} & \ldots & \mathcal{M}_{Q-2,Q-2} \end{bmatrix} = \left[\sum_{i=0}^{Q-1} \sum_{j=0}^{Q-1} f(i,j) i^p j^q \right] \pmod{Q}.$$

The matrix elements $\mathcal{M}_{(p_1,p_2,\ldots,p_n)}$ of the matrix $[\mathcal{M}_{(p_1,p_2,\ldots,p_n)}]$ are Fourier–Mellin–Galois spectral coefficients [ ]-[ ].

2.2 Fast Calculation Algorithms of Modular Invariants Based on NTTs over Galois Fields

As we have seen all calculations in Eq. ( ) can be realized according to the rules of $\mathbf{GF}(Q)$–arithmetics if Q is prime number. Let ε be a primitive root in the field $\mathbf{GF}(Q)$. Its different powers $1 = \varepsilon^{k_1}$, $2 = \varepsilon^{k_2}$, $3 = \varepsilon^{k_3}$, ..., $Q-1 = \varepsilon^{k_{Q-1}}$ cover the field $\mathbf{GF}(Q)$ for proper $k_0 = 0, k_1, k_2, \ldots, k_{Q-2}$. If $a = \varepsilon^k$, then k is called the index of a in the base ε and is denoted by $k = ind_\varepsilon a$. Indexes play the same role in the field $\mathbf{GF}(Q)$ as logarithms in the field of the real numbers.

Theorem 1 Additive (Ad) and multiplicative (Mu) computer complexities of modular moments are equal to

$$\text{Mu} = \text{Ad} = \begin{cases} \mathcal{O}(nN^n \log_2 N), & \text{if } N = Q-1, \\ \mathcal{O}(nN^n \log_2^2 N), & \text{if } N << Q, \end{cases} \quad \text{if } \varepsilon \neq 2, \tag{3}$$

$$\text{Mu} = 0, \quad \text{Ad} = \begin{cases} \mathcal{O}(nN^n \log_2 N), & \text{if} \quad N = Q - 1, \\ \mathcal{O}(nN^n \log_2^2 N), & \text{if} \quad N << Q, \end{cases} \quad \text{if } \varepsilon = 2. \tag{4}$$

Proof: 1. For $N = Q-1$ let us represent $i_1, i_1, \ldots, i_n$ in Eq. ( ) in the form $i_1 := \varepsilon^{\underline{i}_1}$, $i_2 := \varepsilon^{\underline{i}_2}, \ldots, i_n := \varepsilon^{\underline{i}_n}$, where $\underline{i}_1 := ind_\varepsilon i_1$, $\underline{i}_2 := ind_\varepsilon i_2$, $\ldots \underline{i}_n := ind_\varepsilon i_n$. Then

$$\mathcal{M}_{(p_1,p_2,\ldots,p_n)} = \sum_{\underline{i}_1=0}^{Q-2} \sum_{\underline{i}_2=0}^{Q-2} \cdots \sum_{\underline{i}_n=0}^{Q-2} f(\varepsilon^{\underline{i}_1}, \varepsilon^{\underline{i}_2}, \ldots, \varepsilon^{\underline{i}_n}) \varepsilon^{\underline{i}_1 p_1 + \underline{i}_2 p_2 + \ldots + \underline{i}_n p_n} =$$

$$= \sum_{\underline{\mathbf{i}} \in W^n(N)} \underline{f}(\underline{\mathbf{i}}) \varepsilon^{\langle \mathbf{p} | \underline{\mathbf{i}} \rangle} \pmod{Q}, \tag{5}$$

where $\underline{f}(\underline{\mathbf{i}}) = \underline{f}(\underline{i}_1, \underline{i}_2, \ldots, \underline{i}_n) := f(\varepsilon^{\underline{i}_1}, \varepsilon^{\underline{i}_2}, \ldots, \varepsilon^{\underline{i}_n})$, $\underline{\mathbf{i}} := (\underline{i}_1, \ldots, \underline{i}_n)^t = |\underline{\mathbf{i}}\rangle$, and $\mathbf{p} := (p_1, p_2, \ldots, p_n) = \langle \mathbf{p}|$, $\langle \mathbf{p} | \underline{\mathbf{i}} \rangle := \underline{i}_1 p_1 + \underline{i}_2 p_2 + \ldots + \underline{i}_n p_n$. We obtain new calculating algorithm for the modular moments $\mathcal{M}_{(p_1,p_2,\ldots,p_n)}$ as the nD *Fourier–Galois Number Theoretical Transform* (FG–NTT). Its computational complexity is defined complexity of the fast nD FG–NTT: $\mathcal{O}(nN^n \log_2 N)$ additions and multiplications [ ]. We will denote this algorithm as follows:

– $\mathbf{Alg}_1\Big(\text{FG–NTT}_n, \varepsilon, N = Q-1, \mathbf{GF}(Q), \mathcal{O}_{\text{AdMu}}(nN^n \log_2 N)\Big)$.

Computational complexity of the new algorithm can be reduced by special choice of the primitive root ε. Indeed, if $\varepsilon = 2$ then Eq. ( ) is reduced to the nD FG–NTT, which is fulfilled without multiplication. Computational complexity of such computational scheme is only $\mathcal{O}(nN^n \log_2 N)$ additions. We will denote this algorithm as follows:

– $\mathbf{Alg}_1^{'}\Big(\text{FG–NTT}_n, 2, N = Q-1, \mathbf{GF}(Q), \mathcal{O}_{\text{Ad}}(nN^n \log_2 N), 0\Big)$.

2. If $N << Q$ then Eq. ( ) is nD Vandermonde–Galois transform:

$$\mathcal{M}_{(p_1,p_2,\ldots,p_n)} = \sum_{\underline{i}_1=0}^{N-1} \sum_{\underline{i}_2=0}^{N-1} \cdots \sum_{\underline{i}_n=0}^{N-1} \underline{f}(\underline{i}_1, \underline{i}_2, \ldots, \underline{i}_n) \varepsilon^{\underline{i}_1 p_1 + \underline{i}_2 p_2 + \ldots + \underline{i}_n p_n} \pmod{Q}, \tag{6}$$

Computational complexity of the nD Vandermonde–Galois Number Theoretical Transform (VG–NTT) is $\mathcal{O}(nN^n \log_2^2 N)$ additions and multiplications [ ], if $\varepsilon \neq 2$ and $\mathcal{O}(nN^n \log_2^2 N)$ additions if $\varepsilon = 2$. We obtain now two versions of the second algorithm and will denote them as follows:

– $\mathbf{Alg}_2\Big(\text{VG–NTT}_n; \varepsilon; N << Q; \mathbf{GF}(Q); \mathcal{O}_{\text{AdMu}}(nN^n \log_2^2 N)\Big)$,
– $\mathbf{Alg}_2^{'}\Big(\text{VG–NTT}_n; 2; N << Q; \mathbf{GF}(Q); \mathcal{O}_{\text{Ad}}\ (nN^n \log_2^2 N);\ 0\Big)$. □

2.3 Fast Calculation Algorithms of Modular Invariants Based on the Discrete Radon Transform

Let $\{\mathbf{p}^\circ\} \in W^n(N)$ be minimal vector set such that all rays $a\mathbf{p}^\circ$, $a = 0, 1, ..., N - 1$ cover the whole window $W^n(N)$. Then we can write that

$$\mathcal{M}_{\mathbf{p}} = \mathcal{M}_{(a\mathbf{p}^0)} = \sum_{\underline{\mathbf{i}} \in W^n(N)} \underline{f}(\underline{\mathbf{i}})\varepsilon^{\langle \mathbf{p}^0 | \underline{\mathbf{i}} \rangle} = \sum_{p=0}^{q-1} \left(\sum_{<\mathbf{p}^\circ | \underline{\mathbf{i}}>=p} \underline{f}(\underline{\mathbf{i}}) \right) \varepsilon^{ap},$$

or

$$\mathcal{M}_{(a\mathbf{p}^0)} = \sum_{p=0}^{N-1} \widehat{f}(\mathbf{p}^\circ, p)\varepsilon^{ap}, \quad \text{where} \quad \widehat{f}(\mathbf{p}^\circ, p) := \mathcal{RD}_n\{\underline{f}(\underline{\mathbf{i}})\} = \sum_{<\mathbf{p}^\circ | \underline{\mathbf{i}}>=p} \underline{f}(\underline{\mathbf{i}}). \quad (7)$$

Definition 2 [ ]-[ ]. The function $\widehat{f}(\mathbf{p}^\circ, p)$ which is equal to the sum of values of the signal $\underline{f}(\underline{\mathbf{i}})$ on the discrete hyperplane $\langle \mathbf{p}^\circ | \underline{\mathbf{i}} \rangle = p$ is called *Discrete Radon Transform* (DRT) of $\underline{f}(\underline{\mathbf{i}})$.

The expression ( ) means that nD NTT is a composition of DRT $\mathcal{RD}_n$ and a set of 1D NTTs. The total number of 1D NTTs is equal to the power of the set $\{\mathbf{p}^\circ\}$. Every 1D NTT acts along the ray $a\mathbf{p}^\circ$. It is necessary to find such the set $\{\mathbf{p}^o\}$ that would give DRT with minimum computational complexity. Note that the classical "rown/column separable" n–D NTT is reduced to nN^{n-1} 1D NTT's.

Theorem 2 [ ]-[ ]. The total number of 1D NTTs in Eq. ( ) is equal to

$$\begin{cases} \frac{q^n - 1}{q - 1} \approx N^{n-1}, & \text{if } N = q \text{ is prime integer,} \\ \frac{q^{m(n-1)} - 1}{q^{n-1} - 1} \approx N^{n-1}, & \text{if } N = q^m, \\ \prod_{i=1}^{k} \frac{q_i^{m(n-1)} - 1}{q_i^{n-1} - 1} \approx N^{n-1}, & \text{if } N = q_1^{m_1} q_2^{m_2} \cdot q_k^{m_k}. \end{cases}$$

The total computational complexity of the proposed algorithm for nD FG–NNT$_n$ are $\mathcal{O}(\text{FG-NTT}_n) = \mathcal{O}(\mathcal{R}_n) + N^{n-1}\mathcal{O}(\text{FG-NTT}_1)$ instead of $nN^{n-1}\mathcal{O}(\text{FG-NTT}_1)$ for classical fast "rown/column separable" nD FG-NNT. As result we obtain the following total additive $\mathcal{O}_{\text{Ad}}(nN^n \log_2)$ and multiplicative $\mathcal{O}_{\text{Mu}}(N^n \log_2)$ complexities for nD FG–NNT. We will denote this algorithm as follows:

$$\mathbf{Alg}_3\Big(\text{DRT}_n, \text{FG–NTT}_1, \varepsilon, Q-1, \mathbf{GF}(Q), \mathcal{O}_{\text{Ad}}(nN^n \log_2 N), \mathcal{O}_{\text{Mu}}(N^n \log_2 N)\Big).$$

Therefore additive computer complexities of the present algorithm and algorithms $\mathbf{Alg}_1$ are equivalent, but multiplicative computer complexity of the new algorithm is in n times smaller.

2.4 Fast Calculation Algorithms of Modular Invariants Based on the NTTs over Direct Sum of Galois Fields

In algorithm, based on the FG–NTT on the modulo Q imposes limitations window size $W^n(N)$. In this case powers $i_1^{p_1} i_2^{p_2} \dots i_n^{p_n} \pmod Q$ cover the spatial window $W^n(N)$ (exclusively 0, null column and null line). Rigid dependence between N and Q ($N = Q-1 \simeq 2^m$) restricts the searching of N only to one value: $N = Q - 1$. Let us show that this limitation can be removed via the Chinese Reminder Theorem (CRT).

Let $Q_1, ..., Q_k$ be a set of k prime integers such that $\min(Q_1, ..., Q_k) > N$ and $Q_\Sigma = Q_1 \cdots Q_k$. Then we can imbed the $W^n(N)$ into k windows

$$W^n(N) \to W_1^n(Q_1) = [0, Q_1 - 1]^n, \quad \dots, \quad W^n(N) \to W_k^n(Q_k) = [0, Q_k - 1]^n, \quad (8)$$

and process the images separately in to windows by modulo $Q_1, \dots, Q_k$:

$$f_1(i_1, i_2, \dots, i_n) : W_1^n(Q_1) \longrightarrow \mathbf{GF}(Q_1),$$

$$\dots,$$

$$f_k(i_1, i_2, \dots, i_n) : W_k^n(Q_k) \longrightarrow \mathbf{GF}(Q_k).$$

This is equivalent to the image processing into one window by a "big" modulo $Q_\Sigma = Q_1 Q_2 \dots Q_k$, i.e. as $f(i_1, i_2, \dots, i_n) : W^n(N) \longrightarrow \mathbf{Z}/Q_\Sigma$.

According to the CRT the moments $\mathcal{M}_{(p_1, p_2, \dots, p_n)}$ can be calculated in these k windows using k $\mathbf{GF}(Q)$–arithmetics:

$$^1\mathcal{M}_{(p_{11}, p_{12}, \dots, p_{1n})} = \sum_{i_{11}=0}^{Q_1-2} \dots \sum_{i_{1n}=0}^{Q_1-2} i_{11}^{p_{11}} i_{12}^{p_{12}} \dots i_{1n}^{p_{1n}} f_1(i_{11}, i_{12}, \dots, i_{1n}) \pmod{Q_1},$$

$$\dots, \quad (9)$$

$$^k\mathcal{M}_{(p_{k1}, p_{k2}, \dots, p_{kn})} = \sum_{i_{k1}=0}^{Q_k-2} \dots \sum_{i_{kn}=0}^{Q_k-2} i_{k1}^{p_{k1}} i_{k2}^{p_{k2}} \dots i_{kn}^{p_{kn}} f_k(i_{k1}, i_{k2}, \dots, i_{kn}) \pmod{Q_k},$$

where $f_l(i_{l1}, i_{l2}, \dots, i_{ln}) = f(i_1, i_2, \dots, i_n) \pmod{Q_l}$, $\forall l = 1, 2, ..., k$ and

$$i_{l1}^{p_{l1}} \equiv i_1^{p_1} \pmod{Q_l}, \ i_{l2}^{p_{l2}} \equiv i_2^{p_2} \pmod{Q_l}, \ \dots, \ i_{ln}^{p_{ln}} \equiv i_n^{p_n} \pmod{Q_l}, \ \forall l = 1, 2, ..., k.$$

Let $\varepsilon_1, \dots, \varepsilon_k$ by primitive roots in the Galois fields $\mathbf{GF}(Q_1), ..., \mathbf{GF}(Q_k)$, respectively. If we substitute expressions

$$i_{11} := \varepsilon_1^{i_1}, \ i_{12} := \varepsilon_1^{i_2}, \dots, \ i_{1n} := \varepsilon_1^{i_n}, \quad i_{21} := \varepsilon_2^{i_1}, \ i_{22} := \varepsilon_2^{i_2}, \dots, \ i_{2n} := \varepsilon_2^{i_n},$$

$$i_{k1} := \varepsilon_k^{i_1}, \ i_{k2} := \varepsilon_k^{i_2}, \dots, \ i_{kn} := \varepsilon_k^{i_n}$$

into Eq. ( ) we obtain k NTTs:

$$
\begin{gathered}
{}^{1}\mathcal{M}_{(p_{11},\dots,p_{1n})}=\sum_{\underline{i}_{11}=0}^{Q_1-2}\cdots\sum_{\underline{i}_{1n}=0}^{Q_1-2}\varepsilon_1^{\underline{i}_{11}p_{11}+\ldots+\underline{i}_{1n}p_{1n}}\underline{f}_1(\underline{i}_{11},\dots,\underline{i}_{1n})\ (mod\,Q_1),\\
\dots,\\
{}^{k}\mathcal{M}_{(p_{k1},\dots,p_{kn})}=\sum_{\underline{i}_{k1}=0}^{Q_k-2}\cdots\sum_{\underline{i}_{kn}=0}^{Q_k-2}\varepsilon_k^{\underline{i}_{k1}p_{k1}+\ldots+\underline{i}_{kn}p_{kn}}\underline{f}_k(\underline{i}_{k1},\dots,\underline{i}_{kn})\ (mod\,Q_k),
\end{gathered}
\tag{10}
$$

acting in the k windows ( ).

It is not difficult to see that the computational complexity of such a scheme (using the FG–NTT [ ]–[ ]) is $\sum_{i=1}^{k} 2Q_i^2 \log_2 Q_i \approx 2kN^2 \log_2 N$ additions and multiplications. If $\varepsilon_i = 2$, $\forall i = 1, 2, \dots, k$ then computational complexity is reduced to $2kN^2 \log_2 N$ additions. We obtain two new algorithms:

- $\mathbf{Alg}_4\Big(\text{FG–NTT}_2; \{\varepsilon_i\}_{i=1}^k; N; \{\mathbf{GF}(Q_1)\}_{i=1}^k; \mathcal{O}_{\text{AdMu}}(2kN^2 \log_2 N)\Big)$,
- $\mathbf{Alg}_4^{'}\Big(\text{FG–NTT}_2; \{\varepsilon_i = 2\}_{i=1}^k; N; \{\mathbf{GF}(Q_i)\}_{i=1}^k; \mathcal{O}_{\text{Ad}}(2kN^2 \log_2 N);\ 0\Big)$.

2.5 Fast Calculation Algorithms of Invariant Correlation Function Based on NTTs over Galois Fields

Using the nD matrix of absolute invariants $\widehat{\mathbf{In}}\{f\} := [In_{(p_1,p_2,\dots,p_n)}\{f\}]$ we can construct generalized autocorrelation function $COR_f(x_1, x_2, ..., x_n)$ of the image $f(x_1, x_2, ..., x_n)$ as follows:

$$COR_f(x_1,x_2,...,x_n)=\sum_{p_1=0}^{\infty}...\sum_{p_n=0}^{\infty} In_{(p_1,p_2,\dots,p_n)}\{f\}e_{p_1}(x_1)e_{p_2}(x_2)...e_{p_n}(x_n),$$

or in the matrix form by

$$COR_f(x_1x_2,...,x_n)=[x_1^{p_1}]^{-1}\otimes[x_2^{p_2}]^{-1}\otimes\ldots\otimes[x_n^{p_n}]^{-1}[In_{(p_1,p_2,\dots,p_n)}\{f\}],$$

where $[e_{p_i}](x_i) := [x_i^{p_i}]^{-1}$, $i = 1, 2, \dots, n$ are the inverse Vandermonde Matrix Transforms. Note that all samples of autocorrelation functions have the same dynamic range whereas the moments (and the moment invariants) have different ranges. Thus information about the image represented in the moment invariants is not of equal value, but in the autocorrelation function it is represented by equal value.

For the modular autocorrelation function we have

$$\mathcal{COR}_f(i_1,i_2,\dots,i_n)=\sum_{p_1=0}^{N-1)}...\sum_{p_n=0}^{Q-1}\mathcal{I}_{(p_1,p_2,\dots,p_n)}i_1^{-p_1}i_2^{-p_2}\{f\}...i_1^{-p_2}\ (mod\,Q_\Sigma).$$

Let us represent $i_1, i_1, \ldots, i_n$ in the last equation in the form $i_1 := \varepsilon^{\underline{i}_1}$, $i_2 := \varepsilon^{\underline{i}_2}$, $\ldots$, $i_n := \varepsilon^{\underline{i}_n}$, where $\underline{i}_1 := ind_\varepsilon i_1$, $\underline{i}_2 := ind_\varepsilon i_2$, $\ldots$ $\underline{i}_n := ind_\varepsilon i_n$. Then

$$\mathcal{COR}_f(i_1, i_2, \ldots, i_n) = \sum_{p_1=0}^{N-1} \ldots \sum_{p_n=0}^{N-1} \mathcal{I}_{(p_1,p_2,\ldots,p_n)}\{f\}\varepsilon^{\underline{i}_1 p_1 + \underline{i}_2 p_2 + \ldots + \underline{i}_n p_n} \pmod{Q_\Sigma}.$$

We obtain two calculation algorithms of invariant correlation function

- $\mathbf{Alg}_5\Big(\text{FG–NTT}_n; \varepsilon; N << Q; \mathbf{Z}/Q_\Sigma; \mathcal{O}_{\text{AdMu}}(2nN^n \log_2^2 N)\Big)$,
- $\mathbf{Alg}_5^{'}\Big(\text{FG–NTT}_n; 2; N << Q; \mathbf{Z}/Q_\Sigma; \mathcal{O}_{\text{Ad}}\ (2nN^n \log_2^2 N);\ 0\Big)$.

  and
- $\mathbf{Alg}_6\Big(\text{VG–NTT}_n; \varepsilon; N << Q; \mathbf{Z}/_\Sigma; \mathcal{O}_{\text{AdMu}}(2nN^n \log_2^2 N)\Big)$,
- $\mathbf{Alg}_6^{'}\Big(\text{VG–NTT}_n; 2; N << Q; \mathbf{Z}/_\Sigma; \mathcal{O}_{\text{Ad}}\ (2nN^n \log_2^2 N);\ 0\Big)$.

Let us consider limitations which have to be imposed on $Q_\Sigma = Q_1 Q_2 \cdots Q_k$ and N if FG–NTT is used for the correlation function computation. If computations in modular arithmetic coincide with computations in the classical integer arithmetic then $\mathcal{COR}_f(i_1, i_2, ..., i_n) \equiv COR_f(i_1, i_2, ..., i_n) \leq Q_\Sigma$, $\forall i, j \in [0, N-1]^2$. If $\max\limits_{i_1,\ldots,i_n} f(i_1, i_2, ..., i_n) = A$, and

$$|COR(i_1, i_2, ..., i_n)| \leq \sum_{i_1=0}^{N-1} \ldots \sum_{i_n=0}^{N-1} |f(i_1, i_2, ..., i_n)|^2 \leq N^n A^2 \leq Q_\Sigma,$$

i.e. if $Q_\Sigma \geq N^n A^2$ or $A \leq \sqrt[n]{Q_\Sigma}/N$, then $\mathcal{COR}_f(i_1, i_2, ..., i_n) \equiv COR_f(i_1, i_2, ..., i_n)$.

For example, let $n = 2$, $N = 64$, and $Q_1 = 67$, $Q_2 = 71$, $Q_3 = 73$, $Q_4 = 79$. Then $A \leq \sqrt{Q_\Sigma}/N = \sqrt{67 \times 71 \times 73 \times 79}/64 \approx 81$. This means that $\mathcal{COR}_f(i_1, ..., i_n) \equiv COR_f(i_1, ..., i_n)$ if $0 \leq f(i_1, i_2, ..., i_n) \leq 81$.

3 Fast Calculation Algorithms of Complex–Valued Invariants Based on Fourier–Clifford–Gauss–Galois Transforms

In this section our aim is to reduce the complexity of the two previous stages in calculating algorithm applying continuous complex arithmetics. We propose new modular complex–valued moments. They are relative invariants with respect to the wide class of the geometrical distortions. The use of complex arithmetics makes it un necessary to complete the second step (Kravtchook transform) in the global recognition algorithm. Therefore, new invariants can be measured directly from an image without the calculation of moments. Further more, we use modular Clifford–Gauss–Galois arithmetics [ ]–[ ] for the fast computation of the low– and higher–order complex–valued invariants.

3.1 Complex Moments and Invariants of 2D Images

Let $f(x,y)$ be a 2D grey–level image, where $(x,y) \in \mathbf{R^2}$. This function can be considered on the generalized complex plane $\mathcal{GC}_2(\mathbf{R}|1,I) : f(\mathbf{z}) := f(x,y)$, where $\mathbf{z} = x + Iy \in \mathcal{GC}_2(\mathbf{R}|1,I)$, and I is the generalized imaginary units ($I^2 = -1, 0, +1$). These 2D generalized complex numbers form the 2D generalized spatial complex Cayley–Klein algebra $\mathcal{A}_2(\mathbf{R}|1,I)$ spanned on two main elements $1, I$, with $I^2 := \delta = -1, 0, 1$ [ ]. In the first case ($I^2 = i^2 = -1$) 2–D algebra forms the field of complex numbers, in second ($I^2 = \epsilon^2 = 0$) – one algebra of dual numbers and in the third ($I^2 = e^2 = 1$) case – algebra of double numbers that are denoted as $\mathcal{A}_2(\mathbf{R}|1,i) := \mathbf{R} + \mathbf{R}i$, $\mathcal{A}_2(\mathbf{R}|1,\epsilon) := \mathbf{R} + \mathbf{R}\epsilon$, $\mathcal{A}_2(\mathbf{R}|1,e) := \mathbf{R} + \mathbf{R}e$, respectively. When one speaks about all three algebras simultaneously then it concerns algebra of generalized complex numbers, that is denoted as $\mathcal{A}_2(\mathbf{R}|1,I) := \mathbf{R} + \mathbf{R}I$.

Let $\mathbf{c} \in \mathcal{GC}_2$ be the centroid of image $f(\mathbf{z})$.

Definition 3 [ ], [ ]. Functionals of the form

$$\mathbf{m}_p\{f\} = \int\limits_{\mathbf{z}\in\mathcal{GC}_2} (\mathbf{z}-\mathbf{c})^p f(\mathbf{z})\,d\mathbf{z}, \quad \mathbf{m}_{pq}\{f\} = \int\limits_{\mathbf{z}\in\mathcal{GC}_2} (\mathbf{z}-\mathbf{c})^p f(\mathbf{z})\,d\mathbf{z} \tag{11}$$

are called *one– and two–index $\mathcal{A}_2(\mathbf{R}|1,I)$–valued central fractional moments* of the 2D image $f(\mathbf{z})$, respectively, where $p,q \in \mathbf{Q}$ is the rational numbers.

If $f(\mathbf{z})$ is the initial image then $f_{\mathbf{v},\mathbf{w}}(\mathbf{z}) = f(\mathbf{v}(\mathbf{z}+\mathbf{w})) := f(\mathbf{z}^*)$ denotes its geometrical distortion copy. Here $\mathbf{v},\mathbf{w}$ are fixed complex numbers. Summing $\mathbf{w}$ with $\mathbf{z}$ brings us to image translation by the vector $\mathbf{w}$, multiplication $\mathbf{z}+\mathbf{w}$ by $\mathbf{v}$ equivalent to the rotation of the vector $\mathbf{z}+\mathbf{w}$ by angle φ (where $\varphi = \arg(\mathbf{v})$) and to the dilatation by factor $|\mathbf{v}|$.

Theorem 3 [ ]. Central moments of the image $f(\mathbf{z})$ are relative $\mathcal{A}_2(\mathbf{R}|1,I)$–valued invariants

$$\mathbf{m}\{f_{\mathbf{v},\mathbf{w}}\} = \mathbf{v}^p|\mathbf{v}|^2\mathbf{m}\{f\} = e^{pI\varphi}|\mathbf{v}|^{p+2}\mathbf{m}_p\{f\}, \tag{12}$$

with respect to the small affine group $\mathbf{aff}(\mathcal{GC}_2(\mathbf{R}|1,I))$ with $\mathcal{A}_2(\mathbf{R}|1,I)$–valued multiplicators $\mathbf{v}^p|\mathbf{v}|^2 = e^{Ip\varphi}|\mathbf{v}|^{p+2}$.

The following ratios $\eta_p := \mathbf{m}_p/\mathbf{m}_0^{\frac{p+2}{2}}$ are called *unary normalized moments.* These moments are *respective* $\mathcal{A}_2(\mathbf{R}|1,I)$–valued invariants with respect to the affine group $\mathbf{aff}(\mathcal{A}_2(\mathbf{R}|1,I))$ of the generalized complex plane $\mathcal{GC}_2(\mathbf{R}|1,I)$ with $\mathcal{A}_2(\mathbf{R}|1,I)$–valued multiplicators $e^{Ip\varphi}$, because $\eta_p\left\{f_{\varphi,|\mathbf{v}|,\mathbf{w}}\right\} := e^{Ip\varphi}\eta_p\left\{f\right\}$. Let us calculate module of the left and right parts of the last equation: $|\eta_p\left\{f_{\varphi,|\mathbf{v}|,\mathbf{w}}\right\}| = |\eta_p\left\{f\right\}|$. As result we obtain the following theorem.

Theorem 4 [ ]. Modules of unary moments $|\eta_p\{f\}|$ are absolute scalar–valued invariants $In_p\left\{\mathbf{aff}(\mathcal{A}_2(\mathbf{R}|1,I))|\ f\right\}$ of the small affine group $\mathbf{aff}(\mathcal{A}_2(\mathbf{R}|1,I))$.

3.2 Fast Calculation Algorithms of $\mathcal{A}_2(\mathrm{R}|1,I)$–Valued Modular Invariants Based on NTTs over $\mathbf{GF}(Q^2)$–Arithmetic

Let us concentrate on the problem of the fast calculation of $\mathcal{A}_2(\mathbf{R}|1,I)$–valued moments:

$$\mathbf{m}_p = \sum_{x=0}^{N-1}\sum_{y=0}^{N-1}(x+Iy)^p f(x,y), \quad \mathbf{m}_{pq} = \sum_{x=0}^{N-1}\sum_{y=0}^{N-1}(x+Iy)^p(x-Iy)^q f(x,y). \quad (13)$$

Here we can use modular complex arithmetics with different modules. We will consider two types of rings: $\mathbf{Z}I/Q$ and $\mathbf{Z}I/\mathbf{Q}$, where $Q \in \mathbf{Z}$ or $\mathbf{Q} \in \mathbf{Z}I$.

Let, for example, $I = i$. In the ring $\mathbf{Z}i/Q = \{a + ib \mid a, b \in \mathbf{GF}(Q)\} = \mathbf{GF}(Q) + i\mathbf{GF}(Q) = \mathbf{GF}(Q^2)$ $Q^2 - 1$ nonzero Clifford Gauss–integers (CG–integers) are contained. We have to work with the CG–integers from this ring as with ordinary complex integers and final results are calculated by modulo Q.

Case 1 (1–index moments). Let ε be the primitive root in the Galois field $\mathbf{GF}(Q^2)$, $t = ind_\varepsilon(x + iy)$, then $(x + iy) = \varepsilon^t$. Substituting the latter equation into Eq. ( ) we obtain

$$\mathcal{M}_p = \sum_{x=0}^{Q-1}\sum_{y=0}^{Q-1}(x + iy)^p f(x + iy) = \sum_{t=0}^{Q^2-1} \varepsilon^{tp} f(\varepsilon^t), \ (mod\ Q) \quad (14)$$

that is 1D $(Q^2 - 1)$–point Fourier–Clifford–Gauss–Galois NTT (FCGG–NTT) [ ],[ ]–[ ].

Case 2 (2–index moments). For two-indexes moments $\mathcal{M}_{pq}$ a slightly changed computational scheme is also valid. In the Galois field $\mathbf{GF}(Q^2)$ the following equation $(x + iy)^Q \equiv (x - iy)\ (mod\ Q)$ is holds. Hence, we can write

$$\mathcal{M}_{pq} = \sum_{x=0}^{Q-1}\sum_{y=0}^{Q-1}(x + iy)^{p+qQ} f(x + iy)\ (mod\ Q). \quad (15)$$

Let $r = p + Qq = (p, q)$ be a 2–bit number written in Q–radix number system. Then

$$\mathcal{M}_{(p,q)} = \mathcal{M}_r = \sum_{x=0}^{Q-1}\sum_{y=0}^{Q-1}(x + iy)^r f(x + iy)\ (mod\ Q). \quad (16)$$

Let ε be a primitive root in $\mathbf{GF}(Q^2)$. Then $\varepsilon^t = x + iy = (x, y)$ in $\mathbf{GF}(Q^2)$, where $t = ind_\varepsilon(x + iy)$. Substituting the later equation into Eq. ( ) we obtain

$$\mathcal{M}_{(p,q)} = \mathcal{M}_r = \sum_{t=0}^{Q^2-2} \varepsilon^{tr} f(\varepsilon^t)\ (mod\ Q),\ r = 0, 1, \ldots, Q^2 - 1. \quad (17)$$

As the result computations of complex moments are reduced to $(Q^2 - 1)$–point FCGG–NTT [ ],[ ]–[ ]. In both cases we get algorithms:

– $\mathbf{Alg}_7\Big(\text{FCGG–NTT}_1; \varepsilon; N = \sqrt{Q}; \mathbf{GF}(Q^2); \mathcal{O}_{\text{AdMu}}(2N^2 \log_2 N)\Big)$,
– $\mathbf{Alg}_7'\Big(\text{FCGG–NTT}_1; 2; N = \sqrt{Q}; \mathbf{GF}(Q^2); \mathcal{O}_{\text{Ad}}(2N^2 \log_2 N);\ 0\Big)$.

Computational complexity of this algorithm is equal to $\mathcal{O}_{\text{AdMu}}(2N^2 \log_2 N)$ (for $\mathbf{Alg}_7$) if $\varepsilon \neq 2$, and $\mathcal{O}_{\text{Ad}}(2N^2 \log_2 N)$ (for $\mathbf{Alg}_7'$) if $\varepsilon = 2$.

3.3 Fast Calculation Algorithms of $\mathcal{A}_2(\mathbf{R}|1, I)$–Valued Modular Invariants Based on NTTs over $\mathbf{Z}I/\mathbf{Q}$–Arithmetic

Let us investigate the case when module is a complex number $\mathbf{Q} = A + iB$, where $(A, B) = 1$.

Theorem 5 (The Gauss theorem [ ], [ ]). Let $\mathbf{Q} = A + iB$ and $(A, B) = 1$ then the ring $\mathbf{Z}i/(A+iB)$ is isomorphic to the ring $\mathbf{Z}/|\mathbf{Q}|$, where $|\mathbf{Q}| = A^2 + B^2$: $\mathbf{Z}i/(A+iB) \sim \mathbf{Z}/|\mathbf{Q}|$ and one–to–one correspondence is realized as a process of the complex CG–integers $a + ib$ *realification* $\mathcal{R}(a + ib) \longrightarrow h = a + \rho b$, where $\rho \equiv (-A/B) \,(mod\,|\mathbf{Q}|)$ and as a process of the real integers h *complexification*

$$\mathcal{C}(h) = x + iy = \frac{AhA_{|\mathbf{Q}|} + BhB_{|\mathbf{Q}|}}{|\mathbf{Q}|} + i\frac{BhA_{|\mathbf{Q}|} - AhB_{|\mathbf{Q}|}}{|\mathbf{Q}|}, \tag{18}$$

which gives the transition from h to $x + iy$, where $hA_{|\mathbf{Q}|} := hA \ (mod\,|\mathbf{Q}|)$ and $hB_{|\mathbf{Q}|} = hB \ (mod\,|\mathbf{Q}|)$.

The Gauss theorem means that the $\mathbf{Z}/\mathbf{Q}$– and $\mathbf{Z}/|\mathbf{Q}|$–arithmetic laws are a similar. For this case we have the following expression for modular moments

$$\mathcal{M}_p = \sum_{(x+iy)\in \mathbf{Z}i/\mathbf{Q}}\sum (x + iy)^p f(x + iy) \ (mod\ \mathbf{Q}), \tag{19}$$

instead of Eq. ( ). Let ρ be the Gauss coefficient. Realification of the left and right parts of Eq. ( ) gives

$$\mathcal{R}(\mathcal{M}_p) = \sum_{(x+\rho y)\in \mathbf{Z}/|\mathbf{Q}|} (x + \rho y)^p f(x + \rho y) = \sum_{h=0}^{|\mathbf{Q}|-1} h^p f(h) \ (mod\,|\mathbf{Q}|), \tag{20}$$

where $f(h) := f(x + \rho y)$ and $h = x + \rho y$. Let complex modulo $\mathbf{Q}$ be selected so that $|\mathbf{Q}|$ is a prime integer (for example, for $\mathbf{Q} = 1 + 16i$ we have $|\mathbf{Q}| = 1^2 + 16^2 = 257$). Then $\mathbf{Z}/|\mathbf{Q}| = \mathbf{GF}(|\mathbf{Q}|)$ is such Galois field. Let us select a primitive root ε in the Galois field. Let $h = \varepsilon^{\underline{h}}$. Then

$$\mathcal{R}(\mathcal{M}_p) = \sum_{\underline{h}=0}^{\mathbf{N}(|\mathbf{Q}|-2} \varepsilon^{p\underline{h}} f(\varepsilon^{\underline{h}}) \ (mod\,|\mathbf{Q}|) = \sum_{\underline{h}=0}^{|\mathbf{Q}|-2} \varepsilon^{p\underline{h}} \underline{f}(\underline{h}) \ (mod\,|\mathbf{Q}|) \tag{21}$$

is 1D $(|\mathbf{Q}| - 1)$–point FG–NTT, where $\underline{f}(\underline{h}) := f(\varepsilon^{\underline{h}})$. This case is reduced to the problem considered in the previous section and gives the following algorithms:

- $\mathbf{Alg}_8\Big(\text{FG–NTT}_1; \varepsilon; N = \sqrt{N(\mathbf{Q})}; \mathbf{Z}i/\mathbf{Q}; \mathcal{O}_{\text{AdMu}}(N^2 \log_2 N)\Big)$,
- $\mathbf{Alg}'_8\Big(\text{FG–NTT}_1; 2; N = \sqrt{N(\mathbf{Q})}; \mathbf{Z}i/\mathbf{Q}; \mathcal{O}_{\text{Ad}}(N^2 \log_2 N);\ 0\Big)$.

Complex–valued moments are easily reconstructed on calculated values: $\mathcal{M}_p = \mathcal{C}\{\mathcal{R}(\mathcal{M}_p)\}$ that describes the process of complexification of real–valued moments.

4 Fast Calculation Algorithms of Quaternion Invariants Based on Fourier–Clifford–Hamilton Transforms

There is currently a considerable interest in methods of invariant 3–D image recognition. Indeed, very often information about 3–D objects can be obtained by computer tomographic reconstruction, 3–D magnetic resonance imaging, passive 3–D sensors or active range finders. Due to that algorithms of systematic derivation of 3–D moment invariants should be developed for 3–D object recognition.

In this section we proposed an elegant theory which allows to describe many such invariants. Our theory is based on the quaternion theory. We propose quaternion–valued invariants, which are related to the descriptions of objects as the zero sets of implicit polynomials. These are global invariants which show great promise for recognition of complicated objects. Quaternion–valued invariants have good discriminating power for computer recognition of 3–D objects using statistical pattern recognition methods. For fast computation of low– and higher–order quaternion–valued invariants we will use modular arithmetic of Galois fields and rings, which maps calculation of quaternion-valued invariants to fast Fourier–Clifford–Hamilton-Galois NTT and which reduces the computation complexity of the global recognition algorithm from $\mathcal{Q}(3N^4)$ to $\mathcal{Q}(N^3)$ for 3D grey–level images.

The Hamilton quaternions (4–D hypercomplex numbers) of the form $\mathbf{q} = w + ix + jy + zk$, where $w, x, y, z \in \mathbf{R}$, form 4–D algebra $\mathcal{H}_4(\mathbf{R}|1, i, j, k) := \mathbf{R} + \mathbf{R}i + \mathbf{R}j + \mathbf{R}k$ [ ]. This noncommutative number system is therefore characterized by three imaginary units i, j, k which satisfy the following multiplication rules $i^2 = j^2 = k^2 = jk = -1$.

4.1 Generalized Quaternions

The 4–D Hamiltonian algebra $\mathcal{H}_4(\mathbf{R}|1, i, j, k) := \mathbf{R} + \mathbf{R}i + \mathbf{R}j + \mathbf{R}k$, can considered as 2–D algebra over the field of complex numbers:

$$\mathcal{H}_2(\mathbf{C}|1, j) := \mathbf{C} + \mathbf{C}j = \{\mathbf{a} + \mathbf{b}j \mid \mathbf{a}, \mathbf{b} \in \mathbf{C}\}. \tag{22}$$

Surely, in ( ) for hyperimaginary unit j we have $j^2 = -1$. But it can be set $j^2 = -1, 0, 1$ and take generalized complex numbers, then we obtain new quaternion algebra:

$$\mathcal{H}_2\Big(\mathcal{A}_2(\mathbf{R}|1,\imath)\,|\,1,\jmath\Big) = \mathcal{A}_2(\mathbf{R}|1,\imath) + \mathcal{A}_2(\mathbf{R}|1,\imath)\jmath$$
$$= \{\mathbf{a}+\mathbf{b}\jmath \mid \mathbf{a},\mathbf{b}\in\mathcal{A}_2(\mathbf{R}|1,\imath)\}, \tag{23}$$

where $\jmath^2 = -1, 0, 1$ and $\jmath^2 = -1, 0, 1$. This generalized quaternion algebra were proposed by Clifford [ ]. Introducing designation I, J, K for all three new hyperimaginary units we can represent this 2–D algebra as 4–D algebra over the real field

$$\mathcal{GHA}_4\Big(\mathbf{R}|1,I,J,K\Big) = \mathcal{A}_2(\mathbf{R}|1,I) + \mathcal{A}_2(\mathbf{R}|1,I)J = \mathbf{R} + \mathbf{R}I + \mathbf{R}J + \mathbf{R}K.$$

Every generalized quaternion $\mathbf{q}$ and its conjugate have the unique representation in the form $\mathbf{q} = t + xI + yJ + zK$, $\overline{\mathbf{q}} = t - xI - yJ - zK$, where t, x, y, z are real numbers, and product $\mathbf{q}\overline{\mathbf{q}}$ is equal to

$$\mathbf{q}\overline{\mathbf{q}} = ||a||_{\mathcal{A}_2(\mathbf{R})} - J^2||b||_{\mathcal{A}_2(\mathbf{R})} = (t^2 - I^2x^2) - J^2(y^2 - I^2z^2), \tag{24}$$

is called *pseudonorm* of generalized quaternion $\mathbf{q}$, where $||a||_{\mathcal{A}_2(\mathbf{R})}, ||b||_{\mathcal{A}_2(\mathbf{R})}$ are module of generalized complex numbers in the algebra $\mathcal{A}_2(\mathbf{R}|1,I)$. It can take both positive and negative values. If pseudodistance between two generalized quaternions $\mathbf{p}$ and $\mathbf{q}$ is defined as module of their difference $\rho(\mathbf{p},\mathbf{q}) = |\mathbf{p}-\mathbf{q}|$, then algebra $\mathcal{GHA}_4(\mathbf{R}|1,I,J,K)$ of generalized quaternions is transformed into 4–D pseudometric space designed as $\mathcal{GH}_4$. Surely, there are nine such spaces.

Subspace of pure vector generalized quaternions $xI + yJ + zK$ is 3–D space $\mathcal{GR}_3 := \mathbf{Vec}\{\mathcal{GH}_4\}$. Introduced in $\mathcal{GH}_4$ pseudometrics induce in $\mathcal{GR}_3$ corresponding pseudometrics, the expressions for which are obtained from $\rho(\mathbf{p},\mathbf{q})$ for $t = 0$. There are only three such non–trivial pseudometrics:

$$\rho(\mathbf{Vec}\{\mathbf{p}\},\mathbf{Vec}\{\mathbf{q}\}) = |(\mathbf{Vec}\{\mathbf{p}\} - \mathbf{Vec}\{\mathbf{q}\}| = |\mathbf{Vec}\{\mathbf{u}\}| =$$

$$= \sqrt{||xI + yJ + zK||_{\mathcal{GH}_4}} = \sqrt{||xI||^2_{\mathcal{GC}_2} - J^2||y + zI||^2_{\mathcal{GC}_2}} = \begin{cases} \sqrt{(x^2 + y^2 + z^2)}, \\ \sqrt{(x^2 - y^2 - z^2)}, \\ \sqrt{x^2} = |x|. \end{cases}$$

Corresponding 3–D metrical spaces will be denoted as ${}^{3,0,0}\mathbf{R}^3$, ${}^{1,0,2}\mathbf{R}^3$, ${}^{1,2,0}\mathbf{R}^3$. They form Euclidean, Minkovskean, Galilean 3–D pseudometric spaces. When we are dealing with all the three pseudometric spaces we will use symbol $\mathcal{GR}_3$.

As we known generalized complex number and classical quaternions of the unit modulo has the following form $\mathbf{z} = e^{I\phi} = \cos(\phi) + I\sin(\phi)$, $\mathbf{q} = e^{i_{\mathbf{u}}\phi} = \cos\phi + i_{\mathbf{u}}\sin\phi$, where $\cos\phi$, $\sin\phi$ are trigonometric functions in corresponding 2D $\mathcal{GC}_2$–geometries. Generalized quaternion of the unit modulus can be written in the such form [ ]: $\mathbf{q} = e^{I_{\mathbf{u}}\phi} = \cos(\phi) + I_{\mathbf{u}}\sin(\phi)$, where ϕ is a rotation angle around purely vector quaternion $\mathbf{u}$ of the unit modulus ($|\mathbf{u}| = 1$, $\mathbf{u} = -\overline{\mathbf{u}}_0$).

Definition 4 [ ]. Transformations $\mathbf{q}' = \mathbf{q} + \mathbf{p}$, $\quad \mathbf{q}' = \lambda\mathbf{q}$,

$$\mathbf{q}' = e^{I_{\mathbf{u}_1}\phi_1}\mathbf{q}, \quad \mathbf{q}'' = \mathbf{q}e^{-I_{\mathbf{u}_2}\phi_2}, \quad \mathbf{q}''' = e^{I_{\mathbf{u}_1}\phi_1/2}\mathbf{q}e^{-I_{\mathbf{u}_2}\phi_2/2}, \tag{25}$$

where $\mathbf{q},\mathbf{p} \in \mathcal{GH}_4$, $\lambda \in \mathbf{R}^+$, are called *translations*, *dilations* and *rotations* of 4–D space $\mathcal{GH}_4$, respectively.

They form the translation $\mathbf{Tr}(\mathcal{GH}_4)$, the dilation $\mathbf{M}(\mathcal{GH}_4)$, and rotation $\mathbf{SO}_L(\mathcal{GH}_4)$, $\mathbf{SO}_R(\mathcal{GH}_4)$, $\mathbf{SO}_{LR}(\mathcal{GH}_4)$ groups.

Theorem 6 [ ]. Transforms

$$\mathbf{q}' = e^{I_{\mathbf{u}_1}\phi_1}\mathbf{q} + \mathbf{p}, \quad \mathbf{q}'' = \mathbf{q}e^{-I_{\mathbf{u}_2}\phi_2} + \mathbf{p}, \quad \mathbf{q}''' = e^{I_{\mathbf{u}_1}\phi_1/2}\mathbf{q}e^{-I_{\mathbf{u}_2}\phi_2/2} + \mathbf{p}$$

form three groups of left $\mathbf{MOV}_L(\mathcal{GH}_4)$, right $\mathbf{MOV}_R(\mathcal{GH}_4)$, and double–side $\mathbf{MOV}_{LR}(\mathcal{GH}_4)$ motions of the space $\mathcal{GH}_4 = \mathbf{Sc}\{\mathcal{GH}_4\} \oplus \mathbf{Vec}\{\mathcal{GH}_4\} = \mathbf{R} \oplus \mathcal{GR}_3$.

If setting $\mathbf{u}_1 = \mathbf{u}_2$, $\phi_1 = \phi_2$, in LR–transform ( ) then it will maps real axis $\mathbf{R} = \mathbf{Sc}\{\mathcal{GH}_4\}$ and 3D vector subspace $\mathcal{GR}_3 = \mathbf{Vec}\{\mathcal{GH}_4\}$ into itself: $\mathbf{R} = \mathbf{q}^{-1}\mathbf{R}\mathbf{q}$, $\mathcal{GR}_3 = \mathbf{q}^{-1}\mathcal{GR}_3\mathbf{q}$. Hence, transforms $\mathbf{x}' = \mathbf{g}^{-1}\mathbf{x}\mathbf{q}$, $\mathbf{q} = e^{I_{\mathbf{u}}\phi/2}$, $\mathbf{x} \in \mathcal{GR}^3$ are rotations of $\mathcal{GR}_3$–space around point 0 and form the group of its rotations $\mathbf{SO}(\mathcal{GR}_3)$.

Theorem 7 [ ]. Every motion of 3D space $\mathcal{GR}_3$ is represented in the form

$$\mathbf{x}' = \mathcal{Q}^{-1}\mathbf{x}\mathcal{Q} + \mathbf{p}, \quad \mathcal{Q} := e^{I_{\mathbf{u}}\phi/2}, \ \mathbf{x} \in \mathcal{GR}_3. \tag{26}$$

4.2 Quaternion Moments and Invariants of 3–D Images

Let $f(\mathbf{q})$ be a 3D image depending on the pure vector generalized quaternion $\mathbf{q} \in \mathbf{Vec}\{\mathcal{GH}_4\} = \mathcal{GR}_3$. Let $\mathbf{c}$ be the centroid of image $f(\mathbf{q})$.

Definition 5 [ ],[ ]. Functionals of the form

$$\mathbb{M}_p\{f\} = \int_{\mathbf{q}\in\mathcal{GR}_3} (\mathbf{q}-\mathbf{c})^p f(\mathbf{q})d\mathbf{q} \tag{27}$$

are called *one–indexes $\mathcal{GHA}_4$–valued central fractional moments* of 3–D image $f(\mathbf{q})$, where $\mathcal{GHA}_4 := \mathcal{GHA}_4(\mathbf{R}|1, I, J, K)$, $d\mathbf{q} := dxdydz$, and $p \in \mathbf{Q}$ is a rational number.

Let us find rules of moments changing with respect to geometrical distortions of the initial image. These distortions will be caused by translation $\mathbf{q} \longrightarrow \mathbf{q} + \mathbf{a}$, rotation $\mathbf{q} \longrightarrow \mathcal{Q}\mathbf{q}\mathcal{Q}^{-1}$, where $\mathcal{Q} = e^{I_{\mathbf{u}}\phi/2}$ and dilatation: $\mathbf{x} \longrightarrow \lambda\mathbf{x}$, where $\lambda \in \mathbf{R}^+ \setminus 0$. If $f(\mathbf{q})$ is the initial image and $f_{\lambda\mathcal{Q}\mathbf{a}}(\mathbf{q})$ its distorted version then

$$f_{\lambda\mathcal{Q}\mathbf{a}}(\mathbf{q}) := f(\lambda\mathcal{Q}(\mathbf{q}+\mathbf{a})\mathcal{Q}^{-1}) = f(\mathbf{q}^*), \tag{28}$$

where $\mathbf{q}^* := \lambda\mathcal{Q}(\mathbf{q}+\mathbf{a})\mathcal{Q}^{-1}$.

Theorem 8 [ ]. Central moments $\mathbb{M}_p$ are the *respective* $\mathcal{GHA}_4$–valued invariants

$$\mathbb{M}_p\{f_{\lambda\mathcal{Q}\mathbf{a}}\} = \lambda^{p+3}\mathcal{Q}^p\mathbb{M}_p\{f\}\mathcal{Q}^{-p}, \tag{29}$$

of the small affine group $\mathbf{aff}(\mathcal{GR}_3)$ with left $\lambda^{p+3}\mathcal{Q}^p$ and right $\mathcal{Q}^{-p}$ multiplicators.

Being invariants, they will denoted as $\mathbb{J}_p\{\mathbf{aff}(\mathcal{GR}_3)|f\} := \mathbb{M}_p\{f,\mathbf{c}\} := \mathbb{M}_p\{f\}$. Obviously, the following ratios $\mathbb{N}_p := \mathbb{M}_p/\mathbb{M}_0^{\frac{p+3}{3}}$ can be call *normalized moments.* These moments are *respective* $\mathcal{GHA}_4$–valued invariants

$$\mathbb{N}_p\{f_{\lambda\mathcal{Q}\mathbf{a}}\} := \mathcal{Q}^p\mathbb{N}_p\{f\}\mathcal{Q}^{-p} \qquad (30)$$

of the small affine group $\mathbf{aff}(\mathcal{GR}_3)$ with left $\mathcal{Q}^p$ and right $\mathcal{Q}^{-p}$ multiplicators, respectively. As $\mathcal{GH}_4$–valued respective invariants have both left and right multiplicators common multiplication can not obtain absolute invariants. Let us show now that such invariants exist among module of unary moments. For this aim calculate modulus of the lefthand and righthand parts of Eq. ( ): $|\mathbb{N}_p\{f_{\lambda\mathcal{Q}\mathbf{a}}\}| = |\mathbb{N}_p\{f\}|$.

Theorem 9 [ ]. Module of moments

$$|\mathbb{N}_p\{f_{\lambda\mathcal{Q}\mathbf{a}}\}| = |\{\mathbb{N}_p\{f\}| = \mathbb{I}_p\{\mathbf{aff}(\mathcal{GR}_3)|\ f\}$$

are absolute scalar–valued invariants of the affine group.

4.3 Fast Calculation of Quaternion–Valued Invariants Based on Fourier–Clifford–Hamilton Transforms

For digital estimation one–index quaternion–valued moments we have

$$\mathbb{M}_p = \sum_{x=0}^{N-1}\sum_{y=0}^{N-1}\sum_{z=0}^{N-1}(xI + yJ + zK)^p f(x,y,z) =$$

$$= \sum_{x=0}^{N-1}\sum_{y=0}^{N-1}\sum_{z=0}^{N-1}\Big[(xI)^p + ((y+zI)J)^p\Big]f[xI + (y+zI)J] =$$

$$= \left[\sum_{x=0}^{N-1} x^p f_1(x)\right] I^p + \left[\sum_{y=0}^{N-1}\sum_{z=0}^{N-1}(y - zI)^{p-]\frac{p}{2}[}(y+zI)^{]\frac{p}{2}[} f_2(y,z)\right] J^p, \qquad (31)$$

where

$$f_1(x) := \sum_{y=0}^{N-1}\sum_{z=0}^{N-1} f[xI + (y+zI)J], \quad f_2(y,z) = \sum_{x=0}^{N-1} f[xI + (y+zI)J].$$

In Eq. ( ) we used the following equalities $[xI+(y+zI)J]^p = xI^p+[(y+zI)J]^p$ and $[(y+zI)J]^p = (y-zI)^{p-]\frac{p}{2}[}(y+zI)^{]\frac{p}{2}[}$ because $IJ = -JI$ and $(zJ)^2 = zJzJ = z\overline{z}JJ = \overline{z}J^2$.

In modular case 3D numbers of the form $[xI + (y+zI)J]$, where $x,y,z \in \mathbf{GF}(Q)$ form 3D modular space of the type

$$\mathbf{GF}(Q)I + \mathbf{GF}(Q)J + \mathbf{GF}(Q)K = \mathbf{GF}(Q)I + [\mathbf{GF}(Q) + \mathbf{GF}(Q)I]J.$$

This is the vector part of the modular 4D quaternion algebra $\mathbf{GF}(Q)+[\mathbf{GF}(Q)I+\mathbf{GF}(Q)J+\mathbf{GF}(Q)K]$. Let Q be such prime, that $[\mathbf{GF}(Q)+\mathbf{GF}(Q)I]$ is the Galois field $\mathbf{GF}(Q^2)$ then the vector part is represented as the following sum

$$\mathbf{GF}(Q)I + [\mathbf{GF}(Q) + \mathbf{GF}(Q)I]J = \mathbf{GF}(Q)I + \mathbf{GF}(Q^2)J.$$

Let ε an $\mathcal{E}$ be primitive root in Galois fields $\mathbf{GF}(Q)$, $\mathbf{GF}(Q^2)$, respectively. Obviously, $\varepsilon = \mathcal{E}^{Q+1}$. Then we can write $x = \varepsilon^t = \mathcal{E}^{(Q+1)t}$, $(y + Iz) = \mathcal{E}^s$, where $t = ind_\varepsilon(x)$, $x \in \mathbf{GF}(Q)$ and $s = ind_{\mathcal{E}}(y + Iz)$, $(y + Iz) \in \mathbf{GF}(Q^2)$. Substituting the latter equations into Eq. ( ) we obtain

$$\mathcal{M}_p = \mathbb{M}_p \,(mod\, Q) = \sum_{x=0}^{Q-1} \sum_{y=0}^{Q-1} \sum_{z=0}^{Q-1} [xI + (y + zI)J]^p f[xI + (y + zI)J](mod\, Q) =$$

$$= \left[\sum_{t=0}^{Q-2} \mathcal{E}^{(Q+1)pt} f_1(\mathcal{E}^t)\right] \cdot I^p + \left[\sum_{s=0}^{Q^2-1} \mathcal{E}^{\{(p-]\frac{p}{2}[)Q+]\frac{p}{2}[\}s} f_2(\mathcal{E}^s)\right] J^p \;(mod\, Q).$$

We obtain new algorithm for calculating modular quaternion–valued moments $\mathcal{M}_p$ as 2–D Fourier–Clifford–Hamilton–Galois NTT [ ],[ ] of the image $f(\varepsilon^t, \mathcal{E}^s)$. Its computational complexity is defined complexity of fast algorithm this transform [ ] : $[Q^2(Q-1) + Q\log_2 Q] + [Q(Q^2-1) + 2Q^2 \log_2 Q] \approx 2Q^3$ additions and $Q\log_2 Q + 2Q^2 \log_2 Q = (2Q+1)Q\log_2 Q$ multiplications. Computational complexity of this algorithm can be reduced by special choice of primitive roots. Indeed, if $\varepsilon = \pm 2$ and $\mathcal{E} = \pm 2, \pm 2(1 \pm I)$ computation complexity is reduced to only $2Q^3 \log_2 Q$ additions. We obtain now two versions of the new algorithm and will denote them as follows:

- $\mathbf{Alg}_9\Big(\text{FCHG–NTT}_n; \varepsilon; \mathcal{E}; Q^3; \mathbf{GF}(Q); \mathcal{O}_{\text{Ad}}(2Q^3), \mathcal{O}_{\text{Mu}}(2Q^2 \log_2 Q)\Big)$,
- $\mathbf{Alg}'_9\Big(\text{FCHG–NTT}_n;\ \pm 2;\ \pm 2(1 \pm I);\ Q^3; \mathbf{GF}(Q); \mathcal{O}_{\text{Ad}}(2Q^3);\ 0\Big)$.

5 Fast Calculation Algorithms of Invariants of Multicolor Images Based on the Multiplet–Fourier–Clifford–Galois NTTs

The concept of color and multispectral image recognition connects all the topics we have considered. In this work, the term "multicomponent (multispectral, multicolor) image" is defined for an image with more than one component. A RGB image is an example of a color image featuring three separate image components R(red), G(green), and B(blue).

Our main hypothesis is: the brain of primates calculates hypercomplex–valued invariants of an image during recognizing [ ]–[ ],[ ]–[ ]. Visual systems in primates and animals with different evolutionary history use different hypercomplex algebras. For example, the human brain uses 3D hypercomplex (triplet) numbers to recognize color (RGB)–images and mantis shrimps use 10D multiplet numbers to recognize multicolor images. From this point of view the visual cortex of a primate's brain can be considered as a "Clifford algebra quantum computer" [ ].

In this section we propose a novel method to calculate invariants of color and multicolor images. It employs an idea of multidimensional hypercomplex numbers and combines it with the idea of number theoretical transforms over hypercomplex algebras, which reduces the computational complexity of the global recognition algorithm from $\mathcal{O}(knN^{n+1})$ to $\mathcal{O}(kN^n \log N)$ for nD k–multispectral images.

5.1 Multiplet Numbers

The multicomponent color image of an object is measured as k–component vector

$$\mathbf{f}_{mcol}(\mathbf{x}) := \begin{bmatrix} f_1(x,y,z) \\ f_2(x,y,z) \\ \cdots \\ f_k(x,y,z) \end{bmatrix} = \begin{bmatrix} \int_\lambda S^{obj}(\lambda;x,y,z)H_1(\lambda)d\lambda \\ \int_\lambda S^{obj}(\lambda;x,y,z)H_2(\lambda)d\lambda \\ \cdots \\ \int_\lambda S^{obj}(\lambda;x,y,z)H_k(\lambda)d\lambda \end{bmatrix}, \tag{32}$$

where $H_1(\lambda), H_2(\lambda), \ldots, H_k(\lambda)$ are sensor sensitivity functions. For example, if $k = 3$ and $f_1(x,y,z) = f_R(x,y,z)$, $f_2(x,y,z) = f_G(x,y,z)$, $f_3(x,y,z) = f_B(x,y,z)$ then we have color (RGB) images. We will interpret such images as hypercomplex–valued signals

$$\mathbf{f}_{mcol}(x,y,z) = f_0 1 + f_1(x,y,z)\varepsilon^1_{mcol} + \ldots + f_{k-1}(x,y,z)\varepsilon^{k-1}_{mcol} \tag{33}$$

which takes values in the *multiplet algebra* $\mathcal{A}(\mathbf{R}|1, \varepsilon^1_{mcol}, \varepsilon^2_{mcol}, \ldots, \varepsilon^{k-1}_{mcol})$, where $\varepsilon^k_{mcol} = 1$. In particular, RGB–color images are represented as triplet–valued functions:

$$\mathbf{f}_{col}(x,y,z) = f_R(x,y,z)1_{col} + f_G(x,y,z)\varepsilon_{col} + f_B(x,y,z)\varepsilon^2_{col}.$$

Multiplet numbers are represented in its basic form by

$$\mathcal{C}_1 = c_0 1 + c_1\varepsilon_{mcol} + c_2\varepsilon^2_{mcol} + \ldots + c_{k-1}\varepsilon^{k-1}_{mcol},$$

where $1, \varepsilon^1_{mcol}, \varepsilon^2_{mcol}, \ldots, \varepsilon^{k-1}_{mcol}$ are hyperimaginary units and $\varepsilon^k_{mcol} = 1$. They form *multiplet algebra*:

$$\mathcal{A}_k(\mathbf{R}) = \mathcal{A}_k(\mathbf{R} \mid 1, \varepsilon^1_{mcol}, \ldots, \varepsilon^{k-1}_{mcol}) := \mathbf{R}1 + \mathbf{R}\varepsilon^1_{mcol} + \mathbf{R}\varepsilon^2_{mcol} + \ldots + \mathbf{R}\varepsilon^{k-1}_{mcol}$$

which we will call *multicolor algebras* and denote as $\mathcal{A}^{mcol}_k(\mathbf{R}|1, \varepsilon^1_{mcol}, \ldots, \varepsilon^{k-1}_{mcol})$, or for briefly as $\mathcal{A}^{mcol}_k$.

One can show (see [ ]) that multiplet algebra $\mathcal{A}^{mcol}_k$ is the direct sum of real and complex fields $\mathbf{R}$, $\mathbf{C}$:

$$\mathcal{A}^{mcol}_k = \begin{cases} \mathbf{R}\cdot\mathbf{e}^1_{lu} + \mathbf{R}\cdot\mathbf{e}^2_{lu} + \sum\limits_{j=1}^{\frac{k}{2}-1}[\mathbf{C}\cdot\mathbf{E}^j_{Ch}], & \text{if } k \text{ even}, \\ \mathbf{R}\cdot\mathbf{e}^1_{lu} + \sum\limits_{j=1}^{\frac{k-1}{2}}[\mathbf{C}\cdot\mathbf{E}^j_{Ch}], & \text{if } k \text{ odd}, \end{cases} \tag{34}$$

where $\mathbf{e}^i_{lu}$ and $\mathbf{E}^j_{Ch}$ are orthogonal idempotent "real" and "complex" hyperimaginary units, respectively, such that $(\mathbf{e}^1_{lu})^2 = \mathbf{e}^1_{mcol}$, $(\mathbf{e}^2_{lu})^2 = \mathbf{e}^2_{mcol}$, $(\mathbf{E}^j_{Ch})^2 = \mathbf{E}^j_{Ch}$ and $\mathbf{e}^i_{lu}\mathbf{E}^j_{Ch} = \mathbf{E}^j_{Ch}\mathbf{e}^i_{lu} = 0$, for all i, j. For example, triplet color algebra is the direct sum of real $\mathbf{R}$ and complex $\mathbf{C}$ fields:

$$\mathcal{A}_3^{col} = \mathcal{A}_3^{col}(\mathbf{R} \mid 1, \varepsilon_{col}, \varepsilon_{col}^2) := \mathbf{R}1_{col} + \mathbf{R}\varepsilon_{col} + \mathbf{R}\varepsilon_{col}^2 = \mathbf{R}\cdot\mathbf{e}_{lu} + \mathbf{C}\cdot\mathbf{E}_{Ch}. \quad (35)$$

where $\mathbf{e}_{lu} := (1 + \varepsilon_{col} + \varepsilon_{col}^2)/3$, $\mathbf{E}_{Ch} := (1 + \omega\varepsilon_{col} + \omega^2\varepsilon_{col}^2)/3$ are orthogonal idempotent "real" and "complex" hyperimaginary units, respectively and $\omega := e^{\frac{2\pi\sqrt{-1}}{3}}$, $\mathbf{e}_{lu}^2 = \mathbf{e}_{col}$, $\mathbf{E}_{Ch}^2 = \mathbf{E}_{Ch}$, $\mathbf{e}_{lu}\mathbf{E}_{Ch} = \mathbf{E}_{Ch}\mathbf{e}_{lu} = 0$. Here $\varepsilon_{col}^3 = 1$. Therefore, every triplet $\mathcal{C}$ is a linear combination $\mathcal{C} = a\cdot\mathbf{e}_{lu} + \mathbf{z}\cdot\mathbf{E}_{Ch}$ of the "scalar" and "complex" parts $a\mathbf{e}_{lu}$, $\mathbf{z}\mathbf{E}_{Ch}$, respectively. The real numbers $a \in \mathbf{R}$ we will call *intensity (lumiance) numbers* and complex numbers $\mathbf{z} = b + jc \in \mathbf{C}$ we will call *chromacity numbers.*

Let $k_{lu} = 0, 1, 2$ and $k_{Ch} = \frac{n}{2}, \frac{k}{2} - 1, \frac{n-1}{2}$. Every multiplet $\mathcal{C}$ can be represented as a linear combination of k_{lu} "scalar" parts and k_{Ch} "complex" parts:

$$\mathcal{C} = \sum_{i=1}^{k_{lu}}(a_i\cdot\mathbf{e}^i_{lu}) + \sum_{j=1}^{k_{Ch}}(\mathbf{z}_j\cdot\mathbf{E}^j_{Ch}).$$

The real numbers $a_i \in \mathbf{R}$ are called *intensity numbers* and complex numbers $\mathbf{z}_j = b + ic \in \mathbf{C}$ are called *multichromacity numbers.* Two main arithmetic operations have very simple form in such representation:

$$\mathcal{C}^1 + \mathcal{C}^2 = \left[\sum_{i=1}^{k_{lu}}(a_i^1\cdot\mathbf{e}^i_{lu}) + \sum_{j=1}^{k_{Ch}}(\mathbf{z}_j^1\cdot\mathbf{E}^j_{Ch})\right] \pm \left[\sum_{i=1}^{k_{lu}}(a_i^2\cdot\mathbf{e}^i_{lu}) + \sum_{j=1}^{k_{Ch}}(\mathbf{z}_j^2\cdot\mathbf{E}^j_{Ch})\right] =$$

$$= \sum_{i=1}^{k_{lu}}\left[a_i^1 \pm a_i^2\right]\cdot\mathbf{e}^i_{lu} + \sum_{j=1}^{k_{Ch}}\left[\mathbf{z}_j^1 \pm \mathbf{z}_j^2\right]\cdot\mathbf{E}^j_{Ch},$$

$$\mathcal{C}^1 \cdot \mathcal{C}^2 = \left[\sum_{i=1}^{k_{lu}}(a_i^1\cdot\mathbf{e}^i_{lu}) + \sum_{j=1}^{k_{Ch}}(\mathbf{z}_j^1\cdot\mathbf{E}^j_{Ch})\right] \cdot \left[\sum_{i=1}^{k_{lu}}(a_i^2\cdot\mathbf{e}^i_{lu}) + \sum_{j=1}^{k_{Ch}}(\mathbf{z}_j^2\cdot\mathbf{E}^j_{Ch})\right] =$$

$$= \sum_{i=1}^{k_{lu}}\left[a_i^1 a_i^2\right]\cdot\mathbf{e}^i_{lu} + \sum_{j=1}^{k_{Ch}}\left[\mathbf{z}_j^1\mathbf{z}_j^2\right]\cdot\mathbf{E}^j_{Ch}.$$

Multiplet algebras possess divisors of zero and form number rings.

5.2 Algebraic Model of Perceptual Multispectral Spaces

The multicomponent color 3D image ( )–( ) we will interpret as multiplet–valued 3D signal

$$\mathbf{f}_{mcol}(\mathbf{q}) = \sum_{i=1}^{k_{lu}}[f^i_{lu}(\mathbf{q})\cdot\mathbf{e}^i_{lu}] + \sum_{j=1}^{k_{Ch}}[f^j_{Ch}(\mathbf{q})\cdot\mathbf{E}^j_{Ch}],$$

$$k \qquad \mathbf{q} \in \mathcal{GR}_3^{Sp} = \mathcal{GR}(3|I, J, K), \quad (36)$$

which takes values in multiplet algebras $\mathcal{A}_k^{mcol}$. Such algebras generalize classical HSV–model perceptive color space on multispectral spaces.

We will use the generalized complex algebra $\mathcal{A}_2(\mathbf{R}|I)$ in ( ) instead of the complex field $\mathbf{C}$. As result we obtain new generalized multiplet algebras

$$\mathcal{GA}_k^{mcol} = \sum_{i=1}^{k_{lu}} [\mathbf{R} \cdot \mathbf{e}_{lu}^i] + \sum_{j=1}^{k_{Ch}} [\mathcal{A}_2^j(\mathbf{R}|I_j^{Ch}) \cdot \mathbf{E}_{Ch}^j]. \tag{37}$$

Definition 6 [ ],[ ],[ ]. A multispectral 3D image of the form

$$\mathbf{f}_{mcol}(\mathbf{q}) = \sum_{i=1}^{k_{lu}} [f_{lu}^i(\mathbf{q}) \cdot \mathbf{e}_{lu}^i] + \sum_{j=1}^{k_{Ch}} [f_{Ch}^j(\mathbf{q}) \cdot \mathbf{E}_{Ch}^j], \tag{38}$$

where $\mathbf{f}_{mcol}(\mathbf{q}) \in \mathcal{GA}_k^{mcol}$ is called *multiplet–valued image* and the generalized multiplet color algebra $\mathcal{GA}_k^{mcol}$ ( ) is called *generalized perceptive multicolor space.*

Definition 7 [ ],[ ]. If $\mathbf{c}$ is the centroid of the image $\mathbf{f}_{mcol}$ then functionals

$$\mathfrak{M}_p := \int\limits_{\mathbf{q} \in \mathcal{GR}_3^{Sp}} (\mathbf{q} - \mathbf{c})^p \mathbf{f}_{mcol}(\mathbf{q}) d\mathbf{q} =$$

$$= \sum_{i=1}^{k_{lu}} \left(\int\limits_{\mathbf{q} \in \mathcal{GR}_3^{Sp}} (\mathbf{q} - \mathbf{c})^p \mathbf{f}_{lu}^i(\mathbf{q}) d\mathbf{q} \right) \cdot \mathbf{e}_{lu} + \sum_{j=1}^{k_{Ch}} \left(\int\limits_{\mathbf{q} \in \mathcal{GR}_3^{Sp}} (\mathbf{q} - \mathbf{c})^q \mathbf{f}_{Ch}^j(\mathbf{q}) d\mathbf{q} \right) \cdot \mathbf{E}_{Ch}^j$$

are called *central* $\mathcal{GA}^{SpMcol}$*–valued moments* of multicolor 3D image $\mathbf{f}_{mcol}(\mathbf{q})$.

Here all products of the type $\mathbf{q}^p \mathbf{f}_{mcol}$, where $\mathbf{q} \in \mathcal{GR}_3^{Sp}$, $\mathbf{f}_{mcol} \in \mathcal{GA}_k^{mcol}$, are spatial–multicolor numbers belonging to the generalized spatial–color algebra [ ]

$$\mathcal{GA}^{SpMcol} = \begin{cases} \mathcal{GHA}_4^{Sp} \cdot \mathcal{GA}_k^{mcol} = \mathcal{GA}_{4(k_{lu}+k_{Ch})}^{SpMcol}, \text{ if } \mathcal{A}_2^{Sp}(\mathbf{R}|I^{Sp}) = \mathcal{A}_2^j(\mathbf{R}|I_j^{Ch}), \\ \qquad \forall j = 1, ..., k_{Ch}; \\ \\ \mathcal{GHA}_4^{Sp} \otimes \mathcal{GA}_k^{mcol} = \mathcal{GA}_{4(k_{lu}+2k_{Ch})}^{SpMcol}, \text{ if } \mathcal{A}_2^{Sp}(\mathbf{R}|I^{Sp}) \neq \mathcal{A}_2^j(\mathbf{R}|I_j^{Ch}), \\ \qquad \forall j = 1, ..., k_{Ch}. \end{cases}$$

For example, let $\mathbf{q} \in \mathcal{GHA}_4^{Sp}$ is a generalized quaternion and $\mathbf{f}_{mcol} = \mathbf{f}_{col} \in \mathcal{GA}_3^{Col}$, then product

$$\mathbf{q}\mathbf{f}_{col} \in \begin{cases} \mathcal{GHA}_4^{Sp} \cdot \mathcal{GA}_3^{Col} = \mathcal{GA}_8^{SpCol}, & \text{if } \mathcal{A}_2^{Sp}(\mathbf{R}|I^{Sp}) = \mathcal{A}_2(\mathbf{R}|I^{Ch}), \\ \mathcal{GHA}_4^{Sp} \otimes \mathcal{GA}_3^{Col} = \mathcal{GA}_{12}^{SpCol}, & \text{if } \mathcal{A}_2^{Sp}(\mathbf{R}|I^{Sp}) \neq \mathcal{A}_2(\mathbf{R}|I^{Ch}). \end{cases}$$

In the first case spatial–color numbers $\mathbf{qf}_{col}$ are 8D generalize biquaternions and in the second case they are 12D Hurwitz numbers belonging to the generalized Hurwitzion algebra [ ].

Changes in the surrounding world as such of intensity, multicolor or illuminations can be treated in the language of the multiplet algebra as action of the multiplicative group $[\mathbf{M}(\mathcal{GA}_k^{mcol}) \times \mathbf{SO}(\mathcal{GA}_k^{mcol})] := (\mathbf{M} \times \mathbf{SO})(\mathcal{GA}_k^{mcol})]$ in perceptual multicolor space $\mathcal{GA}_k^{mcol}$, where $\mathbf{M}(\mathcal{GA}_k^{mcol})$ and $\mathbf{SO}(\mathcal{GA}_k^{mcol})$ are the dilatation and the rotation groups of generalized perceptive multicolor space $\mathcal{GA}_k^{mcol}$, respectively.

Let $\mathcal{A} = \sum_{i=1}^{k_{lu}} [a_i^{lu} \cdot \mathbf{e}_{lu}^i] + \sum_{j=1}^{k_{Ch}} [\mathbf{A}^{Ch} \cdot \mathbf{E}_{Ch}^j] \in [(\mathbf{M} \times \mathbf{SO})(\mathcal{GA}_k^{mcol})]$. Let us clarify the rules of moments transformation with respect to $[(\mathbf{M} \times \mathbf{SO})(\mathcal{GA}_k^{mcol})]$–multicolor and $[\mathbf{aff}(\mathcal{GR}_3^{Sp})]$–geometrical distortions of initial images. If

$$\mathbf{f}_{mcol}(\mathbf{z}) = \sum_{i=1}^{k_{lu}} [f_{lu}^i(\mathbf{z}) \cdot \mathbf{e}_{lu}^i] + \sum_{j=1}^{k_{Ch}} [f_{Ch}^j(\mathbf{z}) \cdot \mathbf{E}_{Ch}^j],$$

$$\mathbf{f}_{mcol}(\mathbf{q}) = \sum_{i=1}^{k_{lu}} [f_{lu}^i(\mathbf{q}) \cdot \mathbf{e}_{lu}^i] + \sum_{j=1}^{k_{Ch}} [f_{Ch}^j(\mathbf{q}) \cdot \mathbf{E}_{Ch}^j]$$

are initial 2D and 3D multicolor images then

$$_{\mathbf{v},\mathbf{a}}f_{mcol}^{\mathcal{A}}(z) = \mathcal{A}\mathbf{f}_{mcol}(\mathbf{v}(\mathbf{z}+\mathbf{a})) =$$

$$= \sum_{i=1}^{k_{lu}} [a_i^{lu} f_{lu}^i(\mathbf{v}(\mathbf{z}+\mathbf{a}))\mathbf{e}_{lu}^i] + \sum_{j=1}^{k_{Ch}} [A_j^{Ch} f_{Ch}^j(\mathbf{v}(\mathbf{z}+\mathbf{a}))\mathbf{E}_{Ch}^j],$$

$$_{\lambda\mathcal{Q}\mathbf{a}}f_{mcol}^{\mathcal{A}}(q) = \mathcal{A}\mathbf{f}_{mcol}(\lambda\mathcal{Q}(\mathbf{q}+\mathbf{a})\mathcal{Q}^{-1}) =$$

$$= \sum_{i=1}^{k_{lu}} [a_i^{lu} f_{lu}^i(\lambda\mathcal{Q}(\mathbf{q}+\mathbf{a})\mathcal{Q}^{-1})\mathbf{e}_{lu}^i] + \sum_{j=1}^{k_{Ch}} [A_j^{Ch} f_{Ch}^j(\lambda\mathcal{Q}(\mathbf{q}+\mathbf{a})\mathcal{Q}^{-1})\mathbf{E}_{Ch}^j]$$

denote their multicolor and geometric distorted copies, respectively.

Theorem 10 [ ]. The central moments $\mathfrak{M}_p$ of the color images $\mathbf{f}_{mcol}(\mathbf{z})$, $\mathbf{f}_{mcol}(\mathbf{q})$ are relative $\mathcal{GA}^{SpMcol}$–valued invariants

$$\mathfrak{J}_p\{_{\mathbf{v},\mathbf{a}}\mathbf{f}_{mcol}^{\mathcal{A}}\} := \mathfrak{M}_p\{_{\mathbf{v},\mathbf{a}}\mathbf{f}_{mcol}^{\mathcal{A}}\} = \mathcal{A}\mathbf{v}^p|\mathbf{v}|^2[\mathfrak{M}_p\{\mathbf{f}_{mcol}\}], \tag{39}$$

$$\mathfrak{J}_p\{_{\lambda\mathcal{Q}\mathbf{a}}\mathbf{f}_{mcol}^{\mathcal{A}}\} := \mathfrak{M}_p\{_{\lambda\mathcal{Q}\mathbf{a}}\mathbf{f}_{mcol}^{\mathcal{A}}\} = \lambda^{p+3}\mathcal{A}\mathcal{Q}^p[\mathfrak{M}_p\{\mathbf{f}_{mcol}\}]\mathcal{Q}^{-p} \tag{40}$$

with respect to the spatial–multicolor groups $[\mathbf{aff}(\mathcal{GC}_2^{Sp})] \times [(\mathbf{M} \times \mathbf{SO})(\mathcal{GA}^{Mcol})]$ and $[\mathbf{aff}(\mathcal{GR}_3^{Sp})] \times [(\mathbf{M} \times \mathbf{SO})(\mathcal{GA}^{Mcol})]$, respectively.

Let us consider 2D multicolor images.

Definition 8 [ ]. The products $\mathfrak{M}_{p_1}^{k_1}\mathfrak{M}_{p_2}^{k_2}\cdots\mathfrak{M}_{p_s}^{k_s}$ are called *s–ary* $\mathcal{GA}^{SpMcol}$*–valued central moments,* where $k_i \in \mathbf{Q}$.

Theorem 11 [ ]. The *s*–ary central moments of the 2D image $f(\mathbf{z})$ are relative $\mathcal{GA}^{SpMcol}$–valued invariants

$$\mathfrak{J}_{p_1,\dots,p_s}^{k_1,\dots,k_s}\{\ {}_{\mathbf{v},\mathbf{a}}\mathbf{f}_{mcol}^{\mathcal{A}}\} = \mathfrak{M}_{p_1}^{k_1}\mathfrak{M}_{p_2}^{k_2}\cdots\mathfrak{M}_{p_s}^{k_s}\{{}_{\mathbf{v},\mathbf{a}}\mathbf{f}_{mcol}^{\mathbf{A}}\} =$$

$$= \mathbf{v}^{(p_1k_1+\dots+p_sk_s)}(\mathcal{A}|\mathbf{v}|^2)^{(k_1+\dots+k_s)}\mathfrak{M}_{p_1}^{k_1}\mathfrak{M}_{p_2}^{k_2}\cdots\mathfrak{M}_{p_s}^{k_s}\{\mathbf{f}_{mcol}\}$$

with respect to the spatial–multicolor group $[\mathbf{Aff}(\mathcal{GC}_2^{Sp})]\times[(\mathbf{M}\times\mathbf{SO})(\mathcal{GA}^{Mcol})]$ with $\mathcal{GA}^{SpMcol}$–valued multiplicators $\mathbf{v}^{(p_1k_1+\dots+p_sk_s)}(\mathcal{A}|\mathbf{v}|^2)^{(k_1+\dots+k_s)}$, which have *s* free parameters $k_1,\dots,k_s$.

The *s*–ary moments of the form $\mathfrak{M}_{p_1}^{k_1}\mathfrak{M}_{p_2}^{k_2}\cdots\mathfrak{M}_{p_s}^{k_s}$, where $k_1p_1+\dots+k_sp_s=0$, and $k_1+\dots k_s=0$, are called *normalized central one–index moments.* Normalized central one–index moments are by definition absolute complex–valued invariants with respect to the spatial–multicolor group $[\mathbf{aff}(\mathcal{GC}_2^{Sp})]\times[(\mathbf{M}\times\mathbf{SO})(\mathcal{GA}^{Mcol})]$. Being invariants, they be denoted as $\mathfrak{J}_{p_1,\dots,p_s}^{k_1,\dots,k_s}\{\mathbf{f}_{mcol}\}$.

For 3D multicolor images we obtain the following results. Obviously, the following ratios $\mathfrak{N}_p := \mathfrak{M}_p/\mathfrak{M}_0^{\frac{p+3}{3}}$ are normalized moments. They are *respective* $\mathcal{GA}^{SpMcol}$–valued invariants

$$\mathfrak{N}_p\left\{{}_{\lambda\mathcal{Q}\mathbf{a}}\mathbf{f}_{mcol}^{\mathcal{A}}\right\} := \mathcal{Q}^p\mathfrak{N}_p\left\{\mathbf{f}_{mcol}\right\}\mathcal{Q}^{-p}. \tag{41}$$

with respect to the spatial–multicolor group $[\mathbf{aff}(\mathcal{GC}_2^{Sp})]\times[(\mathbf{M}\times\mathbf{SO})(\mathcal{GA}^{Mcol})]$ with left $\mathcal{Q}^p$ and right $\mathcal{Q}^{-p}$ multiplicators, respectively.

Theorem 12 Module of unary moments $|\mathfrak{N}_p\{{}_{\lambda\mathcal{Q}\mathbf{a}}\mathbf{f}_{mcol}^{\mathcal{A}}\}| = |\mathfrak{N}_p\{\mathbf{f}_{mcol}\}|$ are absolute scalar–valued invariants with respect to the spatial–multicolor group $[\mathbf{aff}(\mathcal{GC}_3^{Sp})]\times[(\mathbf{M}\times\mathbf{SO})(\mathcal{GA}^{Mcol})]$.

5.3 Fast Calculation Algorithms of Multiplet Invariants of Multispectral Images Based on Multiplet–Fourier–Clifford Transforms

As every term of a discrete multicolor image

$$\mathbf{f}_{mcol}(m,n) = f_0(m,n)1 + f_1(m,n)\varepsilon_{mcol}^1 + \dots + f_{k-1}(m,n)\varepsilon_{mcol}^{k-1}$$

has 2^n gray–levels that there are no principal limits for considering the mathematical model of every term as function which has their values in the Galois field $\mathbf{GF}(Q)$:

$$f_i(m,n) : [0,N-1]^2 \longrightarrow \mathbf{GF}(Q), \quad i = 0,1,\dots,k-1,$$

if $Q > 2^n$. In this case numbers of the form $a_0 + a_1\varepsilon^1_{mcol} + ... + a_{k-1}\varepsilon^{k-1}_{mcol}$, where $a_i \in \mathbf{GF}(Q)$, are called *modular multiplets*. They form the *modular multiplet algebra*:

$$\mathcal{GA}^{Mcol}(\mathbf{GF}(Q) \mid 1, \varepsilon^1_{mcol}, \ldots, \varepsilon^{k-1}_{mcol}) = \mathbf{GF}(Q) + \mathbf{GF}(Q)\varepsilon^1_{mcol} + \ldots + \mathbf{GF}(Q)\varepsilon^{k-1}_{mcol}.$$

One can show that for special cases Q modular multiplet algebra is the direct sum of the Galois fields $\mathbf{GF}(Q)$ and $\mathbf{GF}(Q^2)$:

$$\mathcal{GA}^{Mcol}(\mathbf{GF}(Q)) := \sum_{i=1}^{k_{lu}} \mathbf{GF}(Q) \cdot \mathbf{e}_{lu} + \sum_{j=1}^{k_{Ch}} \mathbf{GF}(Q^2) \cdot \mathbf{E}^j_{Ch}.$$

Definition 9 A multispectral discrete image of the form

$$\mathbf{f}_{mcol}(m,n) : \mathcal{GC}^{Sp}_2 \to \mathcal{GA}^{Mcol}(\mathbf{GF}(Q)) := \sum_{i=1}^{k_{lu}} \mathbf{GF}(Q)\mathbf{e}^i_{lu} + \sum_{j=1}^{k_{Ch}} \mathbf{GF}(Q^2)\mathbf{E}^j_{Ch}$$

is called *a modular multiplet–valued image*.

This model can also does computers to process values of image according to $\mathbf{GF}(Q)$–arithmetic laws.

Definition 10 Functionals $\mathcal{M}_p\{\mathbf{f}_{mcol}\} := \mathfrak{M}_p\{\mathbf{f}_{mcol}\} \,(\mathrm{mod}\, Q) :=$

$$= \sum_{m=0}^{Q-1}\sum_{n=0}^{Q-1} (m+In)^p \mathbf{f}_{mcol}(m+In) = \sum_{i=1}^{k_{lu}} \left(\sum_{m=0}^{Q-1}\sum_{n=0}^{Q-1} (m+In)^p f^i_I(m+In) \right) \cdot \mathbf{e}^i_{lu} +$$

$$+ \sum_{j=1}^{k_{Sh}} \left(\sum_{m=0}^{Q-1}\sum_{n=0}^{Q-1} (m+In)^q f^j_{Ch}(m+In) \right) \cdot \mathbf{E}^j_{Ch} \,(\mathrm{mod}\, Q)$$

are called *modular $\mathcal{GA}^{SpCol}(\mathbf{GF}(Q))$–valued moments* of multicolor image $\mathbf{f}^k_{mcol}(\mathbf{z})$.

Let $\mathcal{E}$ be a primitive root in the Galois field $\mathbf{GF}(Q^2)$ then $m + In = \mathcal{E}^k$, and

$$\mathcal{M}_p\{\mathbf{f}_{mcol}\} = \sum_{i=1}^{k_{lu}} \left(\sum_{i=0}^{Q^2-1} \mathcal{E}^{pk} \mathbf{f}^i_{lu}(\mathcal{E}^k) \right) \mathbf{e}^i_{lu} + \sum_{j=1}^{k_{Sh}} \left(\sum_{k=0}^{Q^2-1} \mathcal{E}^{pk} \mathbf{f}^j_{Ch}(\mathcal{E}^k) \right) \mathbf{E}^j_{Ch} \,(\mathrm{mod}\, Q).$$

We obtain new algorithm for calculating modular moments $\mathcal{M}_p$ as the $(k_{lu} + k_{Ch})$ *Multiplet–Fourier–Glifford–Gauss–Galois NTTs*:

$$\mathbf{Alg}_{10}\,(\text{MFCGG–NTT}_1,\ \mathcal{E}\ , N = Q, \mathcal{A}_3(\mathbf{GF}(Q)), \\ \mathcal{O}_{\text{AdMu}}((k_{lu} + k_{Ch})N^2 \log_2 N))\,.$$

Its computational complexity is defined complexity of $(k_{lu} + k_{Ch})$ fast Multiplet–Fourier– Clifford Transforms: $[k_{lu} + k_{Ch}][N^2 \log_2 N]$ additions and multiplications.

Computational complexity of new algorithm can be reduced by special choice of primitive root $\mathcal{E}$. Indeed, if $\mathcal{E} = \pm 2, \pm 2I, \pm(1 \pm I)$ or $2(1 \pm I)$ then Multiplet–Fourier–Glifford transform is reduced to the computation of 2–D fast Rader–transform, which can be done without multiplication. Computational complexity of such computational scheme is only $[k_{lu} + k_{Ch}]N^2 \log_2 N$ additions.

6 Conclusions

Higher speed of computation is the most important property of the pattern recognition algorithms. Unfortunately, the direct method suffers from high complexity. This algorithm needs $\mathcal{O}(nN^{n+1})$ operations to evaluate N^n moment invariants. Using modular arithmetics of the Galois fields and fast number theoretical transforms we reduced the computer complexity of the first stage of the global recognition algorithm from $\mathcal{O}(nN^{n+1})$ to $\mathcal{O}(nN^n \log_2 N)$ for the nD grey–level images. We developed six algorithms with low complexities $\mathbf{Alg}_1$–$\mathbf{Alg}_6$.

We have shown that it is possible to use complex Clifford–Gauss arithmetics and different Fourier–Clifford–Gauss–Galois NTT for direct and fast calculation of absolute complex–valued invariants. We constructed two new fast algorithms $\mathbf{Alg}_7$, $\mathbf{Alg}_8$. Computational complexity of these algorithms is equal to $\mathcal{O}(2N^2 \log_2 N)$. Naturally, the method is subjected to several conditions and assumptions. Foremost among them is spatial window limitation. However, this limit can be removed using the multiwindow technique in combination with the Chinese Remainder Theorem.

We have presented a novel algebraic tool for the integration of data from multiple sensors into a uniform representation. We have provided an explicit expressions for relative and absolute quaternion–valued invariants of color and k–multispectral 2D and 3D images with respect to geometrical and color distortions. The behavior of relative invariants with respect to the more important subgroups of the spatial–color groups is studied in detail. Our technique uses high–dimensional hypercomplex algebras and reduces the computational complexity of global recognition algorithm from $\mathcal{O}(kN^3)$ to $\mathcal{O}(kN^2 \log N)$ for 2–D k–multicolor images.

References

1. Hu, M. K.(1962): Visual pattern recognition by moment invariants. IEEE Trans. on Information Theory **IT–8**, 179–187
2. Abu–Mustafa, Y., Psaltis, D. (1984): Recognitive aspects of the moment invariants. IEEE Trans. Pattern Anal. Mach. Intell. **6**, 1698–1706
3. Budricis, Z., Haymian, M. (1984): Moment calculations by digital filtres. ATT Bell Labs Tech. J. **63**, 217– 229
4. Dudani, S., Breeding, K., McChee, R. (1977): Aircraft identification by moment nvariants. IEEE Trans. Comput. **26**, 39–45
5. Lucas, D. (1983): Moment techniques in picture analysis. Int. Conf. on Comput Vision and Patt. Recogn. 138–143

6. Reeves, A. P. (1981): The general theory of moments and the parallel implementation of moment operations. Technical Report TR–EE, **10**, 81–91
7. Sadjadi, F. A., Hall, E. L. (1980): Three–dimensional moment invariants. IEEE Trans. Pattern Anal. Mach. Intell. **2**, 127–136
8. Teague, M. (1980): Image analysis via the general theory of moments. J. Opt. Soc. Am. **2**, 70–80, 697–699, 920–930
9. Teh, C., Chin, R. (1988): On image analysis by the methods of moments. IEEE Trans. Pattern Anal. Mach. Intell. **10**, 496–512
10. Labunets–Rundblad E. V., Labunets V. G. (2000): Spatial–Colour Clifford Algebra for Invariant Image Recognition. (Geometric Computing with Clifford Algebra), Springer, Berlin Heideberg, be published
11. Bing–Cheng, Li Jun Shen (1991): Fast computation of moment invariants. Pattern Recognition, **24**(8), 807–813
12. Yang, L., Albergsten, F. (1984): Fast computation of invariant geometric moments: a new method giving correct results. Proc. IEEE Int. Conf. on Image Proc., 201–204
13. Labunets E. V. (1996): Group–Theoretical Methods in Image Recognition. Part 1. Report No LiTH–ISY–R–1827. Linköping University, 1–84
14. Labunets E. V. (1996): Group–Theoretical Methods in Image Recognition. Part 2. Report No LiTH–ISY–R–1840, Linköping University, 85–141
15. Labunets E. V. (1996): Group–Theoretical Methods in Image Recognition. Part 3. Report No LiTH–ISY–R–1852, Linköping University, 142–226
16. Labunets E. V. (1996): Group–Theoretical Methods in Image Recognition. Part 4. Report No LiTH–ISY–R–1854, Linkoping University, 227–281
17. Labunets E. V. (1996): Group–Theoretical Methods in Image Recognition. Report No LiTH–ISY–R–1855. Linköping University, 1–281
18. Labunets E. V., Labunets V. G. (1997): Towards an "Erlangen program" for pattern recognition. Part 1. Geometry and invariants. (Automatic and Information Technologies) UGTU–UPI Scientific Schools **1**, (Russian) Urals State Technical University, Ekaterinburg, 15–28
19. Labunets E. V., Labunets V. G. (1997): Towards an "Erlangen program" for pattern recognition. Part 2. 2D and nD image recognition. (Automatic and Information Technologies) UGTU–UPI Scientific Schools **1**, (Russian) Urals State Technical University, Ekaterinburg, 29–40
20. Labunets E. V., Uteschev J. V. (1997): Towards an "Erlangen program" for pattern recognition. Part 3. Projective model of invariant screen. (Automatic and Information Technologies) UGTU–UPI Scientific Schools **1**, (Russian) Urals State Technical University, Ekaterinburg, 41–52
21. Labunets E. V., Labunets V. G. (1999): Algebraic–geometry theory of pattern recognition. (Automatic and Information Technologies). UGTU–UPI Scientific Schools **5**, (Russian) Urals State Technical University, Ekaterinburg, 245–256
22. Lidl, R. (1983): Finite field. (Encyclopedia of Mathematics and Its Applications), London
23. Labunets V. G. (1984): Algebraic Theory of Signals and Systems (Russian). Krasnoyarsk State University Press, Krasnoyarsk
24. Varichenko L. V., Labunets, V. G., Rakov, M. A.(1986): Abstract Algebraical Structures and Digital Signal Processing (Russian). Naukova Dumka Press, Kiev 1986

25. Driscoll J. R., Healy D. M., Rockmore D. N. (1997): Fast discrete polynomial transforms with applications to data analysis for distance transitive graphs. SIAM J. Comput. **26**(4), 1066-1099
26. Labunets E. V., Labunets V. G. (1998): New fast algorithms of multidimensional Fourier and Radon discrete transforms. First Int. Workshop on Transforms and Filter Banks, Tampere, Finland, TICSP Series, **1**, 147–178
27. Labunets E. V., Labunets V. G., Egiazarian K., Astola, J. (1999): New fast algorithms of multidimensional Fourier and Radon discrete transforms". IEEE Int. Conf. on ASSP, Arizona, USA, 3193–3196
28. Rundblad–Labunets E. V., Labunets V. G., Astola, J., Egiazarian, K. (1999): Fast fractional Fourier–Clifford transforms. Second Int. Workshop on Transforms and Filter Banks. Tampere, Finland, TICSP Series, **4**, 376–405
29. Rundblad–Labunets E. V., Labunets V. G., Astola, L., Astola, J., Egiazarian, K.: (1999): Fast algorithms of multidimensional discrete non–separable $\mathcal{K}$–wave transformations and Volterra filtering". Second Int. Workshop on Transforms and Filter Banks. Tampere, Finland, TICSP Series **4**, 337–375
30. Rundblad–Labunets E. V., Labunets V.G, Egiazarian, K., Astola, J. (1999): A superfast convolutions technique for Volterra filtering. *Proc. of IEEE–EURASIP Workshop on Nonlinear Signal and Image Processing.* Antalya, Turkey, 399–403
31. Labunets E. V., Labunets V. G., Egiazarian, K., Astola, J. (1999): New Fast Algorithms of Multidimensional Fourier and Radon Discrete Transforms. IEEE Int. Conf. on ASSP, Arizona, USA, 3193–3196
32. Rundblad–Labunets E. V., Labunets V. G., Egiazarian, K., Astola, J. (1999): A superfast convolutions technique for Volterra filtering". Proc. of IEEE–EURASIP Workshop on Nonlinear Signal and Image Processing. Antalya, Turkey, 399–403
33. Labunets, V. G. (1990): Fast spectral algorithms of invariant pattern recognition and image matching based on modular invariants. 1st Int. Conf. on Informat. Techn. for Image Analysis and Pattern Recognition. Lviv, USSR, 70–89 ,
34. Assonov M. B., Labunets E. V., Labunets V. G., Lenz, R. (1996): Fast spectral algorithms for invariant pattern recognition and image matching based on modular invariants". ICIP'96, Switzerland, 284–288
35. Assonov M. B., Labunets E. V., Labunets V. G., Lenz, R. (1996): Fast spectral algorithms for invariant pattern recognition and image matching based on modular invariants. Report No LiTh–ISY–R–1850, Linköping University
36. Labunets E. V., Labunets V. G., Egiazarian, K., Astola, J. (1999): Fast spectral algorithms of invariants calculation. Proc. 10th Inter. Conf. on Image Analysis and Processing ICIAP'99, Venice, Italy, 203–208 ,
37. Labunets V.G, Labunets E. V., Egiazarian, K., Astola, J. (1988): Hypercomplex moments application in invariant image recognition. Proc. of IEEE Int. Conf. on Image Processing, Chicago, Illinois, **2**, 257–261 ,
38. Labunets E. V., Labunets V. G. (1996): Hypercomplex moments using in pattern invariant recognition. (New Information Methods In Research of Discrete Structures) (Russian). IMM UD RAS, Ekaterinburg, 58–63
39. Labunets E. V., Labunets V. G. (1995): Hypercomplex moments application in pattern invariant recognition. Part 1. Generalized complex moments and invariants. (The Century of Radio) (Russian). IMM UD RAS, Ekaterinburg, 125–148
40. Labunets E. V., Labunets V. G. (1995): Hypercomplex moments application in pattern invariant recognition. Part 2. Quaternion moments and invariants. (The Century of Radio) (Russian). IMM UD RAS, Ekaterinburg, 149–163

41. Assonov M. V., Labunets E. V., Labunets V. G., Lenz, R. (1995): Hypercomplex moments application in pattern invariant recognition. Part 3. Fast algorithms of image recognition and matching based on modular invariants. (The Century of Radio) (Russian). IMM UD RAS, Ekaterinburg, 164–189 ,
42. Gauss, C. F. (1870–1927): Werke. Bd. 1–12. Göttingen
43. Hamilton, W. R. (1853): Lectures on quaternions. Dublin
44. Clifford, N. K. (1968): Mathematical papers. N. Y. ,
45. Labunets E. V., Labunets V. G. (1996): Quaternion moments using in pattern invariant recognition. (New Information Methods in Research of Discrete Structures) (Russian). IMM UD RAS, Ekaterinburg, 63–69 ,
46. Cronin, T. W., Marschal, N. J. (1989): A retina with at least ten spectral types of photoreceptors in a mantis shrimp. Nature, **339**, 137–140
47. Van der Varden B. L. (1968): Algebra. Springer–Verlag, Berlin, New York
48. Rundblad–Labunets E. V., Labunets V. G., Astola, J., Egiazarian, K., Polovnev S. V. (1999): Fast invariant recognition of colour images based on Triplet–Fourier–Gauss transform". Proc. of Int. Conf. Computer Science and Information Technologies, Yerevan, Armenia, 265–268
49. Rundblad–Labunets E. V., Labunets V. G. (1999): Fast invariant recognition of multicolour images based on Triplet–Fourier–Gauss transform. Second Int. Workshop on Transforms and Filter Banks, Tampere, Finland, TICSP Series, **4**, 405–438
50. Hurwitz A. (1896): Über die Zahlntheorie der Quaternion. Nach. Geselschaft wiss. Göttingen, Math–Phys. Klasse, 313–340.

Modelling Motion: Tracking, Analysis and Inverse Kinematics

Joan Lasenby, Sahan Gamage, and Maurice Ringer

Department of Engineering, University of Cambridge
Trumpington Street, Cambridge CB2 1PZ, UK
{jl,ssh23,mar39}@eng.cam.ac.uk
http://www-sigproc.eng.cam.ac.uk/vision/

Abstract. In this paper we will use the mathematical framework of *geometric algebra (GA)* to illustrate the construction of articulated motion models. The advantages of solving the *forward kinematics* of a given problem in this way are that the equations can be constructed in a coordinate-free fashion using the GA representations of rotation, *rotors*. One can then use the rotor parameters as our variables in a tracking scheme where we use Kalman filters to track real motion data from moving subjects – this has various advantages over the standard Euler angle approach. The paper then looks at the advantages of this system for solving the *inverse kinematics* – estimating the model given the positions/observations. It will be shown that the often complicated inversion procedures can be simplified by a combination of incidence geometry and rotor inversion.

Keywords: Rotations, geometric algebra, articulated motion, motion estimation, motion modelling, tracking, Kalman filters, forward and inverse kinematics, conformal geometry.

1 Introduction

The main driving force behind the development of the modelling techniques we will describe in subsequent sections has been the need to provide fast and efficient algorithms for optical motion capture. Optical motion capture is a relatively cheap method of producing 3D reconstructions of a subject's motion over time, the results of which can be used in a variety of applications; biomechanics, robotics, medicine, animation etc. Using a system with few cameras (3 or 4) we find that in order to reliably match and track the data (consisting of bright markers placed at strategic points on the subject) we must use realistic models of the possible motion. Once the data has been tracked using such models, we are in a position to analyse the motion in terms of the rotors we have recovered.

The mathematical language we will use throughout will be that of geometric algebra (GA). This language is based on the algebras of Clifford and Grassmann and the form we follow here is that formalised by David Hestenes [ ]. There are now many texts and useful introductions to GA, [, , ,], so we do no more here than outline why it is so useful for the problems we will discuss.

G. Sommer and Y. Y. Zeevi (Eds.): AFPAC 2000, LNCS 1888, pp. 104– , 2000.

In a geometric algebra of n-dimensions, we have the standard *inner* product which takes two vectors and produces a scalar, plus an outer or *wedge* product that takes two vectors and produces a new quantity we call a *bivector* or oriented area. Similarly, the outer product between three vectors produces a *trivector* or oriented volume etc. Thus the algebra has basic elements which are oriented geometric objects of different orders. The highest order object in a given space is called the *pseudoscalar* with the unit pseudoscalar denoted by I, e.g. in 3D I is the unit trivector $e_1 \wedge e_2 \wedge e_3$ for basis vectors $\{e_i\}$. Multivectors are quantities which are made up of linear combinations of these different geometric objects. More fundamental than the inner or wedge products is the *geometric product* which can be defined between any multivectors – the geometric product, unlike the inner or outer products, is invertible. For vectors the inner and outer products are the symmetric and antisymmetric parts of the geometric product;

$$ab = a \cdot b + a \wedge b \tag{1}$$

In effect the manipulations within geometric algebra are keeping track of the objects of different grades that we are dealing with (much as complex number arithmetic does). For a general multivector X, we will use the notation $\langle X \rangle_r$ to denote the rth grade part of X.
In what follows we shall use the convention that vectors will be represented by non-bold lower case roman letters, while we use non-bold, upper case roman letters for multivectors – exceptions to this are stated in the text. Unless otherwise stated, repeated indices will be summed over.

2 Rotations

If, in 3D, we consider a rotation to be made up of two consectutive reflections, one in the plane perpendicular to a unit vector m and the next in the plane perpendicular to a unit vector n, it can easily be shown [ ] that we can represent this rotation by a quantity R we call a **rotor** which is given by

$$R = nm$$

Thus a rotor in 3D is made up of a scalar plus a bivector and can be written in one of the following forms

$$R = e^{-B/2} = \exp\left(-I\frac{\theta}{2}n\right) = \cos\frac{\theta}{2} - In\sin\frac{\theta}{2}, \tag{2}$$

which represents a rotation of θ radians about an axis parallel to the unit vector n in a right-handed screw sense. Here the bivector B represents the plane of rotation. Rotors act two-sidedly, ie. if the rotor R takes the vector a to the vector b then

$$b = Ra\tilde{R}.$$

where $\tilde{R} = mn$ is the reversion of R (i.e the order of multiplication of vectors in any part of the multivector is reversed). We have that rotors must therefore satisfy the constraint that $R\tilde{R} = 1$. One huge advantage of this formulation is that rotors take the same form, i.e. $R = \pm\exp(B)$ in any dimension (we can define hyperplanes or bivectors in any space) and can rotate any objects, not just vectors; e.g.

$$\begin{aligned} R(a \wedge b)\tilde{R} &= \langle Rab\tilde{R}\rangle_2 = \langle Ra\tilde{R}Rb\tilde{R}\rangle_2 \\ &= Ra\tilde{R} \wedge Rb\tilde{R} \end{aligned} \tag{3}$$

gives the formula for rotating a bivector.

Before we leave the topic of rotations, we will outline one property of rotors which will turn out to be familiar to us from classical Euler angle descriptions of 3D rotations. Consider an orthonormal basis for 3-space, $\{e_1, e_2, e_3\}$; suppose we perform a rotation R_1, where $R_1 = \mathrm{e}^{-I\theta_1 e_1}$, i.e. we first rotate an angle θ_1 about an axis e_1. We then follow this by a rotation of θ_2 about *the rotated* e_2 axis – this second rotor, R_2, is given by

$$R_2 = \mathrm{e}^{-I\theta_2 R_1 e_2 \tilde{R}_1}$$

The combined rotation is therefore given by $R_T = R_2 R_1$ – this can be written as follows:

$$\begin{aligned} R_T &= \{\cos\theta_2/2 - IR_1 e_2 \tilde{R}_1 \sin\theta_2/2\} R_1 \\ &= R_1\{\cos\theta_2/2 - Ie_2 \sin\theta_2/2\}\tilde{R}_1 R_1 \\ &= R_1 R_2' \end{aligned} \tag{4}$$

since $R_1\tilde{R}_1 = 1$ and $R_1 \alpha \tilde{R}_1 = \alpha$ for α a scalar.. Thus if R_2' is the rotation of θ_2 about the *non-rotated* axis (i.e. just e_2 in this case), we see that the compound rotation can be written in two ways

$$R_2 R_1 = R_1 R_2' \tag{5}$$

Now recall the classical Euler angle formulation: any general rotation can be expressed as follows: a rotation of ϕ about the $\boldsymbol{e}_3$ axis, followed by a rotation of θ about the **rotated** $\boldsymbol{e}_1$ axis, followed by a rotation of ψ about the **rotated** $\boldsymbol{e}_3$ axis [ ], as shown in figure

Something we always want to do is to apply such a rotation to a vector x. In GA terms we have 3 rotors representing the 3 rotations:

$$R_1 = \exp\{-I\frac{\phi}{2}e_3\},\ R_2 = \exp\{-I\frac{\theta}{2}e_1'\},\ R_3 = \exp\{-I\frac{\psi}{2}e_3''\}$$

where $\boldsymbol{e}_1' = R_1\boldsymbol{e}_1\tilde{R}_1$ and $\boldsymbol{e}_3'' = R_2R_1\boldsymbol{e}_3\tilde{R}_1\tilde{R}_2$. The combined rotor is

$$R_T = R_3R_2R_1 \ \text{ so that } \ x' = R_T x \tilde{R}_T$$

This is all very straightforward, mainly because we are dealing with *active* transformations.

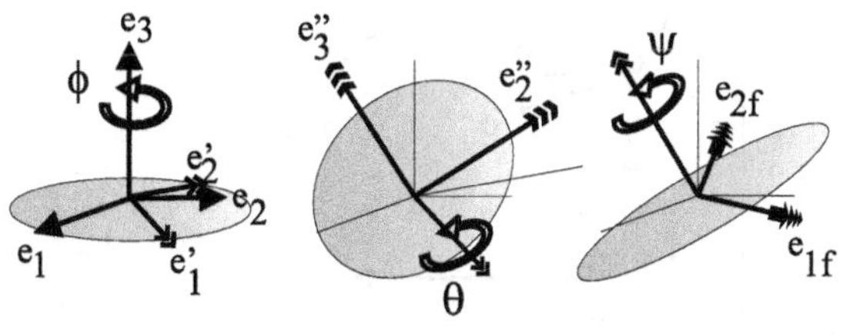

Fig. 1. Sketch of the three elementary rotations in the Euler angle formulation – in which initial axes (e_1, e_2, e_3) are rotated to final axes (e_{1f}, e_{2f}, e_{3f})

Now, if we implement our Euler angle formulation via rotation matrices, [ ], we see that we have 3 rotations matrices:

$$A_1 = \begin{pmatrix} \cos\phi & \sin\phi & 0 \\ -\sin\phi & \cos\phi & 0 \\ 0 & 0 & 1 \end{pmatrix} \quad A_2 = \begin{pmatrix} 1 & 0 & 0 \\ 0 & \cos\theta & \sin\theta \\ 0 & -\sin\theta & \cos\theta \end{pmatrix}$$

$$A_3 = \begin{pmatrix} \cos\psi & \sin\psi & 0 \\ -\sin\psi & \cos\psi & 0 \\ 0 & 0 & 1 \end{pmatrix}$$

which represent the rotations about the non-rotated axes and we apply these matrices in *reverse* order to form

$$A_T = A_1 A_2 A_3 \quad \text{so that} \quad x' = A_T x$$

If R_1', R_2', R_3' are the rotors representing the rotations encoded in A_1, A_2, A_3 (i.e. rotations about the non-rotated axes), then we therefore see that (noting $R_1' = R_1$)

$$R_1' R_2' R_3' = R_3 R_2 R_1$$

which is precisely the formula that we know relates rotations about rotated and non-rotated axes given in equation . Confusion often arises due to the *passive* nature of the Euler angle formulation as given in standard textbooks – there is no such confusion possible if we work totally with *active* transformations, as one is forced to do with the rotor formulation.

3 Articulated Motion Models: Forward Kinematics and Tracking

We begin by considering a simple model of a leg as two linked rigid rods shown in figure . Let us assume that the first rod, AB, can rotate with all degrees of freedom about point A but that the second rod, BC, can only rotate in the plane formed by the two rods (i.e. about an axis which is perpendicular to both rods and initially aligned with the e_2 axis). In reality more complex constraints can be considered. e_1, e_2, e_3 form a fixed orthonormal basis oriented

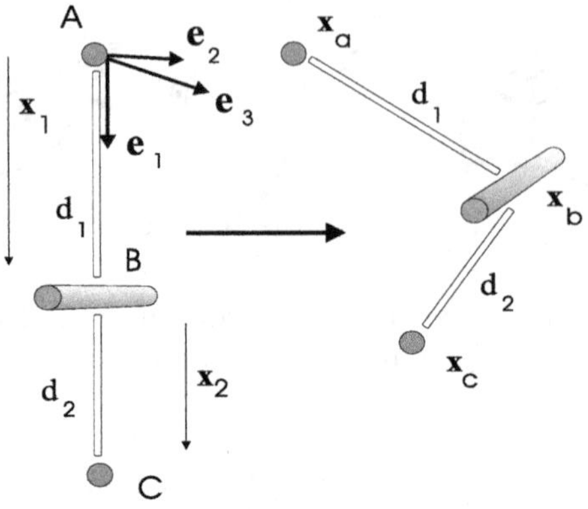

Fig. 2. Two linked rigid rods used to simulate the leg

as shown. x_a, x_b, x_c are the vectors representing the 3D positions of A, B, C respectively and $x_1 = x_b - x_a$ and $x_2 = x_c - x_b$, which, initially, take the values $d_1 e_1$ and $d_2 e_1$. We can immediately write down the position of points B and C as

$$x_b = x_a + d_1 R_1 e_1 \tilde{R}_1 \tag{6}$$

$$x_c = x_b + d_2 R_2 R_1 e_1 \tilde{R}_1 \tilde{R}_2 \tag{7}$$

where we have $R_1 = \exp\{-I\frac{\theta_1}{2} n_1\}$ and we allow for the fact that the point A may move in space (note here that we allow any rotation of rod AB about A, although we may want to only have 2 dof rather than 3 if we are not interested in the orientation of the axes at A). We also have that $R_2 = \exp\{-I\frac{\theta_2}{2} n_2'\}$ with $n_2' = R_1 e_2 \tilde{R}_1$. Using the fact that $R_2 R_1 = R_1 R_2'$ with $R_2' = \exp\{-I\frac{\theta_2}{2} e_2\}$ we are able to give the position of the ankle, x_c as

$$x_c = x_a + d_1 R_1 e_1 \tilde{R}_1 + d_2 R_1 R_2' e_1 \tilde{R}_2' \tilde{R}_1 \equiv x_a + R_1 \{d_1 e_1 + d_2 R_2' e_1 \tilde{R}_2'\} \tilde{R}_1 \tag{8}$$

Thus we are able to write down, in a manner which deals only with active transformations, such forward kinematics equations for arbitrarily complex mechanisms. But this is not the only advantage of this approach; we can now have the elements of our *state* as *rotors* – it is well known that singularities can occur using Euler angles (i.e. when an angle goes to zero, 90° or other specific ranges) and we can avoid many of these singularities using the rotor components as our variables. The use of such models in optical tracking scenarios is briefly discussed here.

In a typical multi-camera tracking problem where we place markers on joints, the measurements (2D points in the camera planes) will be related to the state via a *measurement equation*:

$$\boldsymbol{y}(k) = H_k(\boldsymbol{x}(k)) + \boldsymbol{w}(k) \tag{9}$$

where the $\boldsymbol{y}(k)$ is our set of measurements (observations) at time $t = k$, $\boldsymbol{x}(k)$ is the state at time $t = k$ (parameters describing our model(s)), and $\boldsymbol{w}(k)$ is a zero-mean random vector representing noise at the detection points. The function H_k relates the model parameters to the observations. In this case we take our model parameters to be the coefficients of the bivectors representing the rotors ($B = b_1 I e_1 + b_2 I e_2 + b_3 I e_3$) and then use expressions such as equation to relate these to our observations.

The *process* equation

$$\boldsymbol{x}(k+1) = F_k(\boldsymbol{x}(k)) + \boldsymbol{v}(k)$$

tells us how our system (model) evolves in time; here $\boldsymbol{v}(k)$ represents the process noise. In the case described, F_k tells us how we believe the bivectors to be evolving – one might argue that the variation of the bivectors will be smoother than the evolution of separate Euler-angles.

In general, H_k will be extremely non-linear and so the above problem can be solved by applying an extended Kalman filter (EKF) to update our model estimates and predicted observations at each time step.

A detailed comparison of the difference between using Euler-angles and using bivector coefficients as the scalar model parameters in such tracking problems will be given elsewhere.

4 Inverse Kinematics (IK)

Inverse kinematics is the procedure of recovering the model or state parameters given the measurements – in particular, when incomplete sets of measurements are given (i.e. not all the joint coordinates) we can, in certain cases, recover a unique model or a specified family of solutions. In this section we shall outline the use of GA in solving IK problems by consideration of a particular, fairly simple, example. The example we choose is the following (it is one which often appears in standard texts); a system consisting of three linked rigid rods representing a typical insect leg – such a setup is commonly used in walking robots and is illustrated in figure . Here we fix a set of axes represented by unit vectors (e_1, e_2, e_3) (note that in the figure, $-e_1, -e_2, e_3$ are shown) at the basal joint, so that the angle of the first link, or coxa, is given by the Euler angles (θ, λ, μ), and the rotor representing this rotation is

$$R_A = \mathrm{e}^{-I\frac{\mu}{2}e_3}\mathrm{e}^{-I\frac{\lambda}{2}e_1}\mathrm{e}^{-I\frac{\theta}{2}e_3} \equiv R_\mu R_\lambda R_\theta \tag{10}$$

Generally the angles (λ, μ) are taken as known, so that θ alone describes the position of the first link. The second (femur) and third (tibia) links are such that only rotation in the plane of the three links is allowed, so that the positions of the leg are fully described by two further angles, ϕ and ψ as shown in figure . If we take our initial configuration to be that in which the leg is fully extended with all links lying along the rotated (by R_A) e_2 direction (i.e. ϕ and $\psi = 0$),

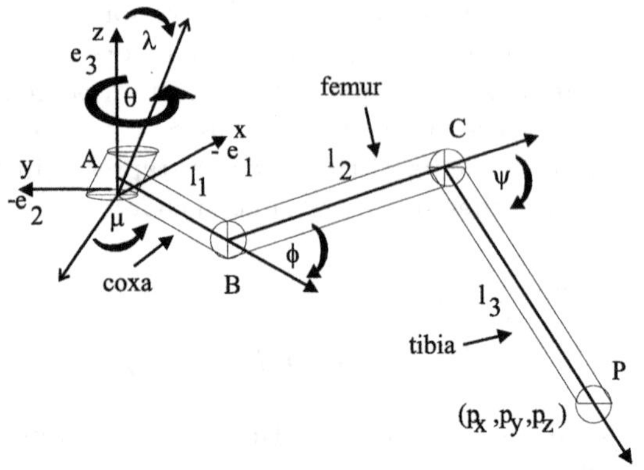

Fig. 3. Three linked rigid rods representing the leg of an insect

then the rotations at joints B and C are given by

$$R_B = \mathrm{e}^{I\frac{\phi}{2}e_1'} \quad \text{and} \quad R_C = \mathrm{e}^{I\frac{\psi}{2}e_1''} \tag{11}$$

where $e_1' = R_A e_1 \tilde{R}_A$, $e_1'' = R_B R_A e_1 \tilde{R}_A \tilde{R}_B$. Note here that we are rotating about the $-e_1$ direction in order to give the sense of ϕ and ψ shown in figure . We are thus able to write down the postion vectors of all joints and finally of the foot position x_P as follows

$$x_B = x_A + R_A(-l_1 e_2)\tilde{R}_A \equiv R_\mu R_\lambda R_\theta(-l_1 e_2)\tilde{R}_\theta \tilde{R}_\lambda \tilde{R}_\mu \tag{12}$$

$$x_C = x_B + R_B R_A(-l_2 e_2)\tilde{R}_A \tilde{R}_B = x_B + R_A R_B'(-l_2 e_2)\tilde{R}_B' \tilde{R}_A \tag{13}$$

$$\begin{aligned} x_P &= x_C + R_C R_B R_A(-l_3 e_2)\tilde{R}_A \tilde{R}_B \tilde{R}_C = \\ &\quad x_C + R_A R_B' R_C'(-l_3 e_2)\tilde{R}_C' \tilde{R}_B' \tilde{R}_A \end{aligned} \tag{14}$$

where $R_B' = \mathrm{e}^{I\frac{\phi}{2}e_1}$ and $R_C' = \mathrm{e}^{I\frac{\psi}{2}e_1}$. We can therefore write x_P as

$$x_P = x_A + R_A\{-l_1 e_2 + R_B'\{-l_2 e_2 + R_C'(-l_3 e_2)\tilde{R}_C'\}\tilde{R}_B'\}\tilde{R}_A \tag{15}$$

This uniquely gives the forward kinematic equations in terms of the three rotors R_A, R_B, R_C; if one was to convert this to angles one gets the following equations (which are conventionally obtained when one uses transformation matrices to denote position of one joint relative to the previous joint [ ]):

$$\begin{aligned} p_x &= (\cos\mu\cos\tilde{\theta} - \cos\lambda\sin\mu\sin\tilde{\theta})[l_2\cos\phi + l_3\cos(\phi+\psi) + l_1] \\ &\quad + \sin\lambda\sin\mu[l_2\sin\phi + l_3\sin(\phi+\psi)] \\ p_y &= (\sin\mu\cos\tilde{\theta} + \cos\lambda\cos\mu\sin\tilde{\theta})[l_2\cos\phi + l_3\cos(\phi+\psi) + l_1] \\ &\quad - \sin\lambda\cos\mu[l_2\sin\phi + l_3\sin(\phi+\psi)] \\ p_z &= \sin\lambda\sin\tilde{\theta}[l_2\cos\phi + l_3\cos(\phi+\psi) + l_1] + \cos\lambda[l_2\sin\phi + l_3\sin(\phi+\psi)] \end{aligned} \tag{16}$$

In the above, $\tilde{\theta} = \theta - \pi/2$, since the convention (following Denavit-Hartenberg) is to measure this basal rotation angle from the e_2 axis rather than from the e_1 axis as our rotor formulation has done. Now, the inverse kinematics comes in when we try to recover the joint angles $(\tilde{\theta}, \phi, \psi)$ given (p_x, p_y, p_z) (and the origin of coordinates). Conventionally the solution is obtained by a series of fairly involved matrix manipulations to give the following expressions for the joint angles:

$$\tilde{\theta} = \arctan\left(\frac{-p_x \cos\lambda \sin\mu + p_y \cos\lambda \cos\mu + p_z \sin\lambda}{p_x \cos\mu + p_y \sin\mu}\right)$$

$$\psi = \arctan\left(-\sqrt{1 - \left(\frac{z^2 + x^2 + y^2 - l_2^2 - l_3^2}{2l_2 l_3}\right)^2} \Big/ \frac{z^2 + x^2 + y^2 - l_2^2 - l_3^2}{2l_2 l_3}\right)$$

$$\phi = \arctan\left(\frac{z}{x^2 + y^2}\right) - \arctan\left(\frac{l_3 \sin\psi}{l_2 + l_3 \cos\psi}\right) \tag{17}$$

where

$$x = p_x \cos\mu + p_y \sin\mu - l_1 \cos\tilde{\theta} \tag{18}$$

$$y = -p_x \cos\lambda \sin\mu + p_y \cos\lambda \cos\mu + p_z \sin\lambda - l_1 \sin\tilde{\theta} \tag{19}$$

$$z = p_x \sin\lambda \sin\mu - p_y \sin\lambda \cos\mu + p_z \cos\lambda \tag{20}$$

In standard texts it is often noted that it is better to express joint angles in terms of arctangent functions to avoid quadrant polarities – we will return to this point later when discussing problems with this Euler angle formulation. Suppose that we have the points x_a, x_b, x_c, x_p, we will now show that it is straightforward, from equations - , to recover each of the rotors, R_A, R_B, R_C. In order to do this we shall use a simple result from GA (see [ ] for more details). Suppose that a set of three (non-coplanar and not necessarily orthonormal) unit vectors e_1, e_2, e_3 is rotated by a rotor R into a set of three other (necessarily non-coplanar) unit vectors f_1, f_2, f_3 – then the unique rotor which performs this job is given by

$$R \propto 1 + e^i f_i \tag{21}$$

where the proportionality factor is easily found by ensuring $R\tilde{R} = 1$ and $\{e^i\}$ denotes the reciprocal frame of $\{e_i\}$. The reciprocal frame $\{e^i\}$ is such that $e^i \cdot e_j = \delta^i_j$ and can be formed (for 3D) as follows

$$e^1 = \frac{1}{\alpha} I e_2 \wedge e_3$$

$$e^2 = \frac{1}{\alpha} I e_3 \wedge e_1 \tag{22}$$

$$e^3 = \frac{1}{\alpha} I e_1 \wedge e_2, \tag{23}$$

where $I\alpha = e_3 \wedge e_2 \wedge e_1$.

This provides us with a remarkably easy way of extracting rotors if we know the joint coordinates. Let us first consider equation for R_A. We can rewrite this equation as

$$\tilde{R}_\lambda \tilde{R}_\mu (x_B - x_A) R_\mu R_\lambda = R_\theta(-l_1 e_2)\tilde{R}_\theta \tag{24}$$

From this we can see that the vector $f_1 = -l_1 e_2$ is rotated into the vector $g_1 = \tilde{R}_\lambda \tilde{R}_\mu (x_B - x_A) R_\mu R_\lambda$ and also that, since $R_\theta = \mathrm{e}^{-I\frac{1}{2}\theta e_3}$, the vector $f_2 = e_3$ is rotated into itself, i.e. $g_2 = e_3$. From this it follows that $f_3 = If_1 \wedge f_2$ must be rotated into $g_3 = Ig_1 \wedge g_2$. Thus, using equations we can form $\{f^i\}$ and the rotor R_θ as follows

$$\begin{aligned} R_\theta &\propto 1 + f^i g_i \\ \text{where} \quad [f_1, f_2, f_3] &= [-l_1 e_2, e_3, If_1 \wedge f_2] \\ \text{and} \quad [g_1, g_2, g_3] &= [\tilde{R}_\lambda \tilde{R}_\mu (x_B - x_A) R_\mu R_\lambda, e_3, Ig_1 \wedge g_2] \end{aligned} \tag{25}$$

Thus R_A is then recovered from equation . Using this we can now look at equation which can be rewritten as

$$\tilde{R}_A (x_C - x_B) R_A = R'_B (-l_2 e_2) \tilde{R}'_B \tag{26}$$

We can then invert as above to give

$$\begin{aligned} R'_B &\propto 1 + f^i g_i \\ \text{where} \quad [f_1, f_2, f_3] &= [-l_2 e_2, e_1, If_1 \wedge f_2] \\ \text{and} \quad [g_1, g_2, g_3] &= [\tilde{R}_A (x_C - x_B) R_A, e_1, Ig_1 \wedge g_2] \end{aligned} \tag{27}$$

Finally, R'_C can be recovered by precisely the same means using

$$\begin{aligned} R'_C &\propto 1 + f^i g_i \\ \text{where} \quad [f_1, f_2, f_3] &= [-l_3 e_2, e_1, If_1 \wedge f_2] \\ \text{and} \quad [g_1, g_2, g_3] &= [\tilde{R}'_B \tilde{R}_A (x_P - x_C) R_A R'_B, e_1, Ig_1 \wedge g_2] \end{aligned} \tag{28}$$

Thus, we see that we are able to invert our forward kinematic equations trivially if we have the coordinates of the joints. Of course, the IK problem as we described it involved being given only x_A and x_P. The plan we advocate is therefore to find x_B and x_C by purely geometric means as an initial stage, followed by the rotor inversion process described above. To illustrate this, consider how we would find x_B, x_C for the given example.

Taking x_A at the origin, we know that e'_3 and x_P must define the plane in which all the links must lie, call this plane Φ – see figure . We can form x_B via

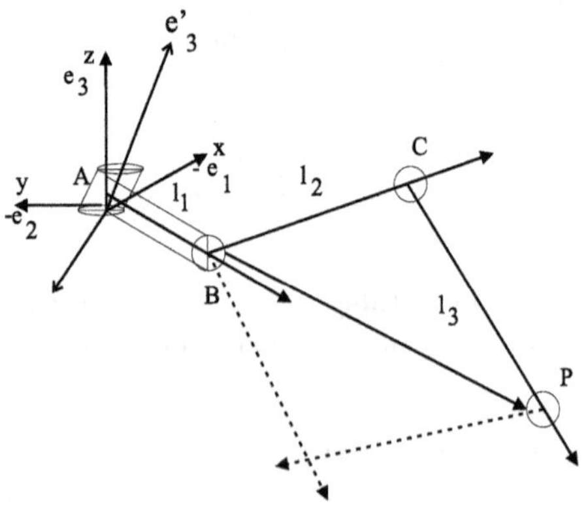

Fig. 4. Figure illustrating setup used to determine joint positions from geometry

$$x_B = l_1 \frac{(x_p - (x_P{\cdot}e_3')e_3')}{|x_p - (x_P{\cdot}e_3')e_3'|} \tag{29}$$

There are clearly two possibilities for x_C, given by the intersections of the circles lying in the plane Φ having centres and radii given by (x_B, l_2) and (x_P, l_3). If we then define $e_\| = (x_P - x_B)/(|x_P - x_B)|$ and $e_\perp$ a vector perpendicular to $e_\|$ lying in Φ, it is not hard to show that x_C is given by

$$x_C = x_\| e_\| + x_\perp e_\perp$$

where

$$x_\| = \frac{l_2^2 - l_3^2 + (x_P - x_B)^2}{2|x_P - x_B|} \tag{30}$$

$$x_\perp^2 = -\left\{ \frac{[(l_2 - l_3)^2 - (x_P - x_B)^2][(l_2 + l_3)^2 - (x_P - x_B)^2]}{4|x_P - x_B|} \right\} \tag{31}$$

When the geometry is more complex than given in this example (indeed, things will get more complicated if we also have prismatic joints rather than simple revolute joints) the joint positions, or family of joint positions are found by intersecting circles, spheres, planes and lines (with possible dilations) in 3D. The system that we are currently working on performs this initial geometric stage in the 5D conformal geometric algebra [,]. This framework provides a very elegant means of dealing with incidence geometry and extends the functionality of projective geometry to include circles and spheres. A feature of the conformal setting is that rotations, translations, dilations and inversions in 3D all become rotors in 5D.

We now return to the question of whether we gain any advantages from doing our IK problems in geometric algebra. In the simple case illustrated, simulations have shown that we can recover the rotors (there always exist two sets of solutions) exactly for any combination of angles – there is no need to restrict any of the angles to particular ranges. However, when the equations in are used to recover angles, we find that the whole process is plagued with conditionals, i.e. the correct solutions are obtained only if signs of various terms are checked for various angles in various ranges. From a computing point of view this is expensive and may ultimately lead to hard-to-track-down errors.

5 Conclusions

In this paper we have illustrated how the geometric algebra, and particularly the rotor formulation within the algebra, can be used as a mathematical system in which forward kinematics, motion modelling and inverse kinematics can be elegantly expressed. The formulations given have been put to use in tracking problems in which optical motion capture data is tracked via constrained articulated models and in inverse kinematics of simple leg structures. We believe that the system as outlined here has enormous potential in more complex inverse kinematics problems where we would like to define families of possible solutions – the key here would be to do the initial geometry stage via a 5D conformal geometric algebra. Work in progress also includes the analysis of human motion data via our articulated models in an attempt to understand how motions are described using the rotor formulation.

References

1. D. Hestenes and G. Sobczyk, *Clifford Algebra to Geometric Calculus: A Unified Language for Mathematics and Physics.* D. Reidel: Dordrecht, 1984.
2. D. Hestenes, *New Foundations for Classical Mechanics.* Kluwer Academic Publishers, 2nd edition., 1999.
3. C. Doran and A. Lasenby, "Physical applications of geometric algebra," Available at `http://www.mrao.cam.ac.uk/ clifford/ptIIIcourse/`. 1999.
4. J. Lasenby, W. Fitzgerald, A. Lasenby, and C. Doran, "New geometric methods for computer vision: An application to structure and motion estimation," *International Journal of Computer Vision*, vol. 26, no. 3, pp. 191–213, 1998. , ,
5. L. Dorst, S. Mann, and T. Bouma, "GABLE: A MATLAB tutorial for geometric algebra," Available at `http://www.carol.wins.uva.nl/~leo/clifford/gablebeta.html`.2000.
6. H. Goldstein, "Classical Mechanics," Second Edition. Addison Wesley. 1980. ,
7. J.M. McCarthy, "An Introduction to Theoretical Kinematics". MIT Press, Cambridge, MA. 1990.
8. H. Li, D. Hestenes, A. Rockwood. " Generalized Homogeneous Coordinates for Computational Geometry". G. Sommer, editor, *Geometric Computing with Clifford Algebras.* Springer. To appear Summer 2000.
9. D. Hestenes, D. "Old wine in new bottles: a new algebraic framework for computational geometry". Advances in Geometric Algebra with Applications in Science and Engineering, Eds. Bayro-Corrochano, E. D. and Sobcyzk, G. Birkhauser Boston. To appear Summer 2000.

The Lie Model for Euclidean Geometry*

Hongbo Li

Academy of Mathematics and Systems Science, Chinese Academy of Sciences
Beijing 100080, P. R. China

Abstract. In this paper we investigate the Lie model of Lie sphere geometry using Clifford algebra. We employ it to Euclidean geometric problems involving oriented contact to simplify algebraic description and computation.

Keywords: Euclidean geometry, Lie sphere geometry, Clifford algebra.

1 Introduction

According to Cecil (1992), Lie (1872) introduced his sphere geometry in his dissertation to study contact transformations. The subject was actively pursued through the early part of the twentieth century, culminating with the publication of the third volume of Blaschke's *Vorlesungen über Differentialgeometrie* (1929), which is devoted entirely to Lie sphere geometry and its subgeometries, particularly in dimensions two and three. After this, the subject fell out of favor until 1981, when Pinkall used it as the principal tool to classify Dupin hypersurfaces in $\mathcal{R}^4$ in his dissertation. Since then, it has been employed by several differential geometers to study Dupin, isoparametric and taut submanifolds (*eg.* Cecil and Chern, 1987). It has also been used by Wu (1984/1994) in automated geometry theorem proving.

Despite its important role played in differential geometry, Lie sphere geometry has limited applicability in classical geometry. This is because a Lie sphere transformation has classical geometric interpretation only when it is a Möbius transformation or the orientation-reversing transformation. In other words, a general Lie sphere transformation does NOT have classical geometric interpretation.

For classical geometry, Lie sphere geometry can contribute to simplifying description and computation of tangencies of spheres and hyperplanes. Because of this, we would like to "attach" Lie sphere geometry to the homogeneous model of Euclidean geometry in (Li, Hestenes and Rockwood, 2000a) as a supplementary tool. This goal is achieved in this paper.

The tool, called the Lie model, is essentially a coordinate-free reformulation of Lie sphere geometry for the purpose of applying it to Euclidean geometry. It

* This paper is supported partially by the Grant NKBRSF of China, the Hundred People Program of the Chinese Academy of Sciences, and the Qiu Shi Science and Technology Foundations of Hong Kong.

G. Sommer and Y. Y. Zeevi (Eds.): AFPAC 2000, LNCS 1888, pp. 115– , 2000.

is used to solve geometric problems involving oriented contact of spheres and hyperplanes, and can help obtaining simplifications. Unfortunately, this may be as much as it can contribute to classical geometry. The model can also be extended to spherical and hyperbolic geometries via their homogeneous models in (Li, Hestenes and Rockwood, 2000b, c).

This paper is arranged as follows: in section we introduce the Lie model using the language of Clifford algebra (Hestenes and Sobczyk, 1984); in section we investigate further basic properties of this model; in section we provide examples to illustrate how to apply it to Euclidean geometry.

2 The Lie Model

The Lie model will be established upon the homogeneous model. So first let us review the homogeneous model for Euclidean geometry (Li, Hestenes and Rockwood, 2000a; Li, 1998).

2.1 The Homogeneous Model

Let $\{\mathbf{e}_1, \ldots, \mathbf{e}_n\}$ be an orthonormal basis of $\mathcal{R}^n$. A point $\mathbf{c}$ of $\mathcal{R}^n$ corresponds to the vector from the origin to the point, denoted by $\mathbf{c}$ as well. The origin corresponds to the zero vector.

We embed $\mathcal{R}^n$ into a Minkowski space of $n+2$ dimensions as a subspace. Denote the Minkowski space by $\mathcal{R}^{n+1,1}$. Let $\{e_{-2}, e_{-1}, \mathbf{e}_1, \ldots, \mathbf{e}_n\}$ be an an orthonormal basis of $\mathcal{R}^{n+1,1}$, where $-e_{-2}^2 = e_{-1}^2 = 1$. The 2-space spanned by e_{-2}, e_{-1} is Minkowski, and has two null 1-subspaces. Let e, e_0 be null vectors in the two 1-subspaces respectively. Rescale them to make $e \cdot e_0 = -1$.

Now we map $\mathcal{R}^n$ in a one-to-one manner into the null cone of $\mathcal{R}^{n+1,1}$ as follows:

$$\mathbf{c} \mapsto \acute{\mathbf{c}} = e_0 + \mathbf{c} + \frac{\mathbf{c}^2}{2} e, \text{ for } \mathbf{c} \in \mathcal{R}^n. \tag{2.1}$$

The range of the mapping is

$$\{x \in \mathcal{R}^{n+1,1} | x^2 = 0, x \cdot e = -1\}. \tag{2.2}$$

By this mapping, a point $\mathbf{c}$ of $\mathcal{R}^n$ can be represented by the null vector $\acute{\mathbf{c}}$. In particular, the origin of $\mathcal{R}^n$ can be represented by e_0. Vector e corresponds to the unique point at infinity for the compactification of $\mathcal{R}^n$. This representation is is called the *homogeneous model* for Euclidean geometry.

The homogeneous model can also be described as follows. Any null vector x of $\mathcal{R}^{n+1,1}$ represents a point or the point at infinity of $\mathcal{R}^n$. It represents the point at infinity if and only if $x \cdot e = 0$. Two null vectors represent the same point or the point at infinity if and only if they differ by a nonzero scale.

The following is a fundamental property of the homogeneous model:

Theorem 2.1. *Let $B_{r-1,1}$ be a Minkowski r-blade in $\mathcal{G}_{n+1,1}$, $2 \leq r \leq n+1$. Then $B_{r-1,1}$ represents an $(r-2)$-dimensional sphere or plane in the sense that a point represented by a null vector a is on it if and only if $a \wedge B_{r-1,1} = 0$. It represents a plane if and only if $e \wedge B_{r-1,1} = 0$. The representation is unique up to a nonzero scale.*

The $(r-2)$-dimensional sphere passing through r affinely independent points $\mathbf{a}_1, \ldots, \mathbf{a}_r$ of $\mathcal{R}^n$ can be represented by $\acute{\mathbf{a}}_1 \wedge \cdots \wedge \acute{\mathbf{a}}_r$; the $(r-2)$-dimensional plane passing through $r-1$ affinely independent points $\mathbf{a}_1, \ldots, \mathbf{a}_{r-1}$ of $\mathcal{R}^n$ can be represented by $e \wedge \acute{\mathbf{a}}_1 \wedge \cdots \wedge \acute{\mathbf{a}}_{r-1}$.

When $r = n+1$, the dual form of the above theorem is

Theorem 2.2. *Let s be a vector of $\mathcal{R}^{n+1,1}$ satisfying $s^2 > 0$. Then it represents a sphere or hyperplane in the sense that a point represented by a null vector a is on it if and only if $a \cdot s = 0$. It represents a hyperplane if and only if $e \cdot s = 0$. The representation is unique up to a nonzero scale.*

The following are *standard representations* in the homogeneous model.

1. A sphere with center $\mathbf{c}$ and radius $\rho > 0$ is represented by $\acute{\mathbf{c}} - \frac{\rho^2}{2}e$.
2. A hyperplane with unit normal $\mathbf{n}$ and distance $\delta > 0$ away from the origin in the direction of $\mathbf{n}$ is represented by $\mathbf{n} + \delta e$.
3. A hyperspace with unit normal $\mathbf{n}$ is represented by either of $\pm\mathbf{n}$.
4. A sphere with center $\mathbf{c}$ and passing through point $\mathbf{a}$ is represented by $\acute{\mathbf{a}} \cdot (e \wedge \acute{\mathbf{c}})$.
5. A hyperplane with normal $\mathbf{n}$ and passing through point $\mathbf{a}$ is represented by $\acute{\mathbf{a}} \cdot (e \wedge \mathbf{n})$.

The following are formulas and explanations for some inner products in $\mathcal{R}^{n+1,1}$:

- For two points $\acute{\mathbf{c}}_1$ and $\acute{\mathbf{c}}_2$,

$$\acute{\mathbf{c}}_1 \cdot \acute{\mathbf{c}}_2 = -\frac{(\acute{\mathbf{c}}_1 - \acute{\mathbf{c}}_2)^2}{2} = -\frac{|\mathbf{c}_1 - \mathbf{c}_2|^2}{2}. \tag{2.3}$$

- For point $\acute{\mathbf{c}}$ and hyperplane $\mathbf{n} + \delta e$,

$$\acute{\mathbf{c}} \cdot (\mathbf{n} + \delta e) = \mathbf{c} \cdot \mathbf{n} - \delta. \tag{2.4}$$

  It is positive, zero or negative if the vector from the hyperplane to the point is along $\mathbf{n}$, zero or along $-\mathbf{n}$ respectively. Its absolute value equals the distance between the point and the hyperplane.

- For point $\acute{\mathbf{c}}_1$ and sphere $\acute{\mathbf{c}}_2 - \frac{\rho^2}{2}e$,

$$\acute{\mathbf{c}}_1 \cdot (\acute{\mathbf{c}}_2 - \frac{\rho^2}{2}e) = \frac{\rho^2 - |\mathbf{c}_1 - \mathbf{c}_2|^2}{2}. \tag{2.5}$$

  It is positive, zero or negative if the point is inside, on or outside the sphere respectively. Its absolute value equals half the distance between the point and the sphere.

- For two hyperplanes $\mathbf{n}_1 + \delta_1 e$ and $\mathbf{n}_2 + \delta_2 e$,

$$(\mathbf{n}_1 + \delta_1 e) \cdot (\mathbf{n}_2 + \delta_2 e) = \mathbf{n}_1 \cdot \mathbf{n}_2. \tag{2.6}$$

- For hyperplane $\mathbf{n} + \delta e$ and sphere $\acute{\mathbf{c}} - \frac{\rho^2}{2} e$,

$$(\mathbf{n} + \delta e) \cdot (\acute{\mathbf{c}} - \frac{\rho^2}{2} e) = (\mathbf{n} + \delta e) \cdot \acute{\mathbf{c}}. \tag{2.7}$$

  It is positive, zero or negative if the vector from the hyperplane to the center $\mathbf{c}$ is along $\mathbf{n}$, zero or along $-\mathbf{n}$ respectively. Its absolute value equals the distance between the center and the hyperplane.
- For two spheres $\acute{\mathbf{c}}_1 - \frac{\rho_1^2}{2} e$ and $\acute{\mathbf{c}}_2 - \frac{\rho_2^2}{2} e$,

$$(\acute{\mathbf{c}}_1 - \frac{\rho_1^2}{2} e) \cdot (\acute{\mathbf{c}}_2 - \frac{\rho_2^2}{2} e) = \frac{\rho_1^2 + \rho_2^2 - |\mathbf{c}_1 - \mathbf{c}_2|^2}{2}. \tag{2.8}$$

  It is zero if the two spheres are perpendicular to each other. When the two spheres intersect, ( ) equals cosine the angle of intersection multiplied by $\rho_1 \rho_2$.

The following are geometric interpretations of the outer product of s_1, s_2, which are vectors of nonnegative square in $\mathcal{R}^{n+1,1}$:

1. If $s_1 \wedge s_2 = 0$, s_1, s_2 represent the same the geometric object. If $s_1 \wedge s_2 \neq 0$ but $(s_1 \wedge s_2)^2 = 0$, then
   - if s_1 represents a point or the point at infinity, it must be on the sphere or hyperplane represented by s_2;
   - if s_1, s_2 represent two spheres or a sphere and a hyperplane, they must be tangent to each other;
   - if s_1, s_2 represent two hyperplanes, they must be parallel to each other.

   In all these cases we say the geometric objects represented by s_1 and s_2 are in *contact*.
   Let $s_1 \wedge s_2 \neq 0$. The blade represents the pencil of spheres and hyperplanes that contact both s_1 and s_2, together with the point of contact, in the sense that a point, the point at infinity, a sphere or a hyperplane represented by a vector s of nonnegative square is in contact with both s_1, s_2 if and only if $s \wedge s_1 \wedge s_2 = 0$. The pencil is called a *contact pencil.*
2. If $(s_1 \wedge s_2)^2 < 0$, then s_1, s_2 must represent two intersecting spheres or hyperplanes. The blade $s_1 \wedge s_2$ represents the pencil ofaa spheres and hyperplanes that pass through the intersection of s_1 and s_2, together with the points of intersection. The pencil is called a *concurrent pencil.*
3. If $(s_1 \wedge s_2)^2 > 0$, then s_1, s_2 are separate from each other. $s_1 \wedge s_2$ represents x, y which are two points or a point and the point at infinity, together with the pencil of spheres and hyperplanes with respect to which x, y are inversive. The pencil is called a *Poncelet pencil.*
   Let $\lambda = s_1 \cdot s_2\, s_1 \cdot e\, s_2 \cdot e$.

- If s_1, s_2 represent a point and a sphere, the point is inside the sphere if $\lambda > 0$, outside if $\lambda < 0$.
- If s_1, s_2 represent two spheres, they are *inclusive*, i.e., one is inside the other, if $\lambda > 0$; they are *exclusive*, i.e., any sphere is outside the other, if $\lambda < 0$.

Conformal geometry is the geometry on Möbius transformations. *Möbius transformations* of $\mathcal{R}^n$ are orthogonal transformations of $\mathcal{R}^{n+1,1}$ with $\pm Id$ identified, where Id denotes the identity transformation. Möbius transformations can be studied by means of spinors in $\mathcal{G}_{n+1,1}$.

2.2 Lie Spheres

A *Lie sphere* of $\mathcal{R}^n$ refers to an oriented sphere, or an oriented hyperplane, or a point, or the point at infinity. First let us discuss the orientations of hyperplanes and spheres.

A hyperplane has two orientations. Let $\mathbf{a}_1, \ldots, \mathbf{a}_n$ be n affinely independent points of $\mathcal{R}^n$, i.e., $J_{n-1} = \partial(\mathbf{a}_1 \wedge \cdots \wedge \mathbf{a}_n) \neq 0$, where

$$\partial(\mathbf{a}_1 \wedge \cdots \wedge \mathbf{a}_n) = \sum_{i=1}^{n} (-1)^{i+1} \mathbf{a}_1 \wedge \cdots \wedge \check{\mathbf{a}}_i \wedge \cdots \wedge \mathbf{a}_n. \tag{2.9}$$

$\check{\mathbf{a}}_i$ denotes that $\mathbf{a}_i$ does not occur in the outer product. The n points generate a hyperplane of $\mathcal{R}^n$, and J_{n-1} determines an orientation of the hyperplane. Alternatively, the vector

$$\mathbf{n} = (-1)^{n-1} J_{n-1}^{\sim} \tag{2.10}$$

is normal to the hyperplane, and satisfies $(J_{n-1} \wedge \mathbf{n})^{\sim} > 0$. It can be used to indicate the same orientation. *The two normal directions indicate the two orientations of the hyperplane.*

A sphere also has two orientations. Let $\mathbf{a}_1, \ldots, \mathbf{a}_{n+1}$ be $n+1$ affinely independent points of $\mathcal{R}^n$. Then $J_n = \partial(\mathbf{a}_1 \wedge \cdots \wedge \mathbf{a}_{n+1}) \neq 0$ and determines an orientation of the sphere. If $J_n^{\sim} > 0$, the sphere is said to have *positive* orientation; otherwise it is said to have *negative* orientation.

Let $\mathbf{a}$ be a point on the sphere, then the blade $J_{n-1} = \mathbf{a} \cdot J_n$ determines an orientation of the hyperplane tangent to the sphere at $\mathbf{a}$, called the *induced* orientation of the tangent hyperplane. The normal direction of the hyperplane with the induced orientation, called the *induced radial direction* of the sphere at $\mathbf{a}$, is

$$\mathbf{n} = (-1)^{n-1} J_n^{\sim} \mathbf{a}. \tag{2.11}$$

$\mathbf{n}$ is in the direction of $(-1)^{n-1}\mathbf{a}$ if and only if the sphere has positive orientation. *The two radial directions (inward and outward directions) indicate the two orientations of the sphere.* For even dimensional spaces, the positive orientation of a sphere is inward, while for odd dimensional spaces it is outward.

An oriented hyperplane and an oriented sphere are said to be in *oriented contact* if they are tangent to each other and at the point of tangency the normal direction of the oriented hyperplane is the induced radial direction of the oriented sphere. Two oriented spheres are said to be in *oriented contact* if they are tangent to each other and at the point of tangency they have the same induced radial directions.

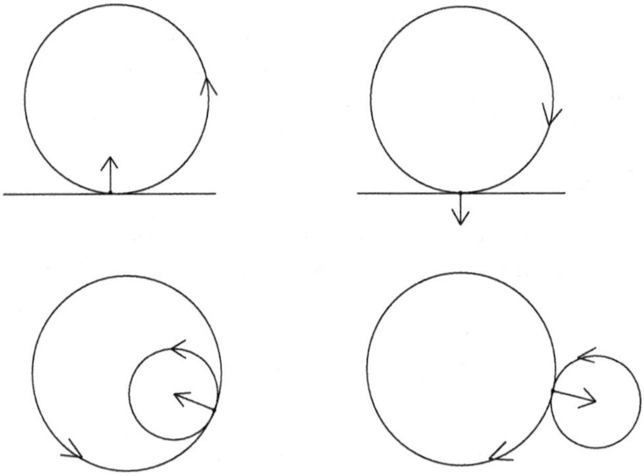

Fig. 1. Oriented contact

In the previous subsection, we have seen that a sphere or hyperplane can be represented by a Minkowski $(n+1)$-blade (or dually by a vector of positive square) in $\mathcal{R}^{n+1,1}$. Two such blades (or vectors) represent the same sphere or hyperplane if and only if they differ by a nonzero scale. Since a blade $B_{n,1}$ (or vector s) also represents an oriented vector space, we can use $\pm B_{n,1}$ (or $\pm s$) to represent the same sphere or hyperplane with different orientations, i.e., two Minkowski $(n+1)$-blades (or two vectors of positive square) represent the same oriented sphere or hyperplane if and only if they differ by a positive scale.

A better representation is provided by Lie (1872), where an oriented sphere or hyperplane is represented by a null vector, and two null vectors represent the same oriented sphere or hyperplane if and only if they differ by a NONZERO scale. Lie's construction can be described as follows. $\mathcal{R}^{n+1,1}$ is embedded into $\mathcal{R}^{n+1,2}$ as a hyperspace. Let $\{e_{-2}, e_{-1}, \mathbf{e}_1, \ldots, \mathbf{e}_n\}$ be an orthonormal basis of $\mathcal{R}^{n+1,1}$. An orthonormal basis of $\mathcal{R}^{n+1,2}$ is $\{e_{-2}, e_{-1}, \mathbf{e}_1, \ldots, \mathbf{e}_n, e_{n+1}\}$, where $e_{n+1}^2 = -1$. Null vectors of $\mathcal{R}^{n+1,1}$ are also null vectors of $\mathcal{R}^{n+1,2}$. They are the set

$$\mathcal{N}_0 = \{x \in \mathcal{R}^{n+1,2} | x^2 = 0, x \cdot e_{n+1} = 0\}. \tag{2.12}$$

A vector in $\mathcal{N}_0$ represents a point or the point at infinity of $\mathcal{R}^n$. Two such vectors represent the same geometric object if and only if they differ by a nonzero scale. Vector x represents the point at infinity if and only if $x \cdot e = 0$.

In $\mathcal{R}^{n+1,1}$, a sphere or hyperplane is represented by a vector s of positive square. The mapping

$$s \mapsto \acute{s} = s + |s|e_{n+1} \tag{2.13}$$

maps all such vectors from $\mathcal{R}^{n+1,1}$ to the set

$$\mathcal{N}_* = \{x \in \mathcal{R}^{n+1,2} | x^2 = 0, x \cdot e_{n+1} \neq 0\}. \tag{2.14}$$

In particular, $\pm s$ are mapped to different null vectors. Let τ be the transformation of $\mathcal{R}^{n+1,2}$ which changes e_{n+1} to $-e_{n+1}$ while keeping $e_{-2}, e_{-1}, \mathbf{e}_1, \ldots, \mathbf{e}_n$ invariant. Then $(-s)' = -\tau\acute{s}$. A point or the point at infinity of $\mathcal{R}^n$ represented by a null vector x is on the sphere or hyperplane represented by vector s of $\mathcal{R}^{n+1,1}$ if and only if $x \cdot \acute{s} = 0$, or equivalently, $x \cdot (\tau\acute{s}) = 0$. The equalities are invariant when $\acute{s}$ and $\tau\acute{s}$ are rescaled.

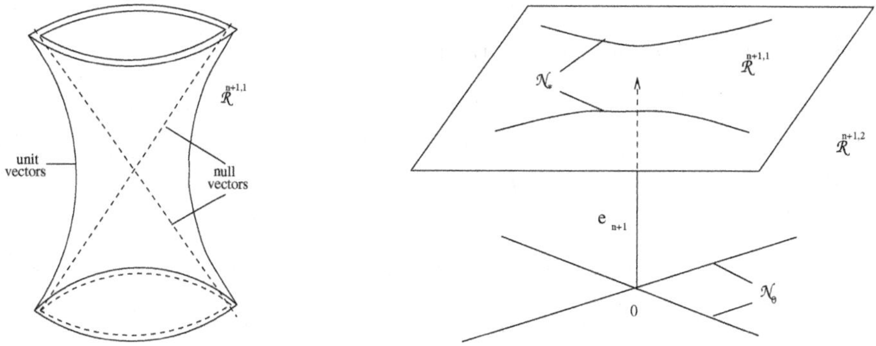

Fig. 2. Lie's construction

Based on the mapping ( ) and the division of null vectors into $\mathcal{N}_0, \mathcal{N}_*$, the Lie model of Lie spheres can be defined by the following theorem:

Theorem 2.3. [Lie] *Let E_3 be a Minkowski 3-blade of $\mathcal{R}^{n+1,2}$. Let e_{n+1} be a unit vector in the blade, e, e_0 be null vectors orthogonal to e_{n+1} in the blade. Then any null vector s of $\mathcal{R}^{n+1,2}$ represents a Lie sphere in the sense that a point represented by a null vector x is on it if and only if $x \cdot s = 0$. Two null vectors represent the same Lie sphere if and only if then differ by a nonzero scale. A null vector s represents the point at infinity if $s \cdot e = s \cdot e_{n+1} = 0$, a point if $s \cdot e_{n+1} = 0$, $s \cdot e \neq 0$, an oriented hyperplane if $s \cdot e = 0$, $s \cdot e_{n+1} \neq 0$, and an oriented sphere otherwise. This algebraic representation of Lie spheres is called the Lie model.*

The following are *standard representations* of Lie spheres:

1. The point at infinity is represented by e.
2. Point $\mathbf{c}$ of $\mathcal{R}^n$ is represented by $\acute{\mathbf{c}}$.

3. The hyperplane with unit normal $\mathbf{n}$ and distance $\delta \geq 0$ away from the origin in the direction of $\mathbf{n}$ has two orientations. The oriented hyperplane with normal $\mathbf{n}$ is represented by $\mathbf{n} + \delta e + e_{n+1}$; the oriented hyperplane with normal $-\mathbf{n}$ is represented by $\mathbf{n} + \delta e - e_{n+1}$.
4. The sphere with center $\mathbf{c}$ and radius $\rho > 0$ has two orientations. The oriented sphere with inward orientation is represented by $\acute{\mathbf{c}} - \frac{\rho^2}{2} e + \rho e_{n+1}$; the one with outward orientation is represented by $\acute{\mathbf{c}} - \frac{\rho^2}{2} e - \rho e_{n+1}$.

Let $\acute{s}_1, \acute{s}_2$ be null vectors of $\mathcal{R}^{n+1,2}$, and let $\epsilon, \epsilon_1, \epsilon_2 = \pm 1$. The following are formulas on the inner product $\acute{s}_1 \cdot \acute{s}_2$.

- If $\acute{s}_1 = \acute{\mathbf{c}}_1$, $\acute{s}_2 = \acute{\mathbf{c}}_2$, then

$$\acute{s}_1 \cdot \acute{s}_2 = -\frac{|\acute{\mathbf{c}}_1 - \acute{\mathbf{c}}_2|^2}{2}. \tag{2.15}$$

- If $\acute{s}_1 = \acute{\mathbf{c}}$, $\acute{s}_2 = \mathbf{n} + \delta e + \epsilon e_{n+1}$, then

$$\acute{s}_1 \cdot \acute{s}_2 = \mathbf{c} \cdot \mathbf{n} - \delta. \tag{2.16}$$

- If $\acute{s}_1 = \acute{\mathbf{c}}_1$, $\acute{s}_2 = \acute{\mathbf{c}}_2 - \frac{\rho^2}{2} e + \epsilon \rho e_{n+1}$, then

$$\acute{s}_1 \cdot \acute{s}_2 = \frac{\rho^2 - |\acute{\mathbf{c}}_1 - \acute{\mathbf{c}}_2|^2}{2}. \tag{2.17}$$

- If for $i = 1, 2$, $\acute{s}_i = \mathbf{n}_i + \delta_i e + \epsilon_i e_{n+1}$, then

$$\acute{s}_1 \cdot \acute{s}_2 = \epsilon_1 \epsilon_2 \mathbf{n}_1 \cdot \mathbf{n}_2 - 1. \tag{2.18}$$

- If $\acute{s}_1 = \mathbf{n} + \delta e + \epsilon_1 e_{n+1}$, $\acute{s}_2 = \acute{\mathbf{c}} - \frac{\rho^2}{2} e + \epsilon_2 \rho e_{n+1}$, then

$$\acute{s}_1 \cdot \acute{s}_2 = \mathbf{c} \cdot \mathbf{n} - \delta - \epsilon_1 \epsilon_2 \rho. \tag{2.19}$$

- If for $i = 1, 2$, $\acute{s}_i = \acute{\mathbf{c}}_i - \frac{\rho_i^2}{2} e + \epsilon_i \rho_i e_{n+1}$, then

$$\acute{s}_1 \cdot \acute{s}_2 = \frac{(\rho_1 - \epsilon_1 \epsilon_2 \rho_2)^2 - |\acute{\mathbf{c}}_1 - \acute{\mathbf{c}}_2|^2}{2}. \tag{2.20}$$

The *oriented contact distance* between two Lie spheres $\acute{s}_1, \acute{s}_2$ is defined by $|\acute{s}_1 - \acute{s}_2| = \sqrt{2|\acute{s}_1 \cdot \acute{s}_2|}$, where the null vectors take the forms in the above formulas. *Two Lie spheres are in oriented contact if and only if their oriented contact distance is zero.*

- When $\acute{s}_1, \acute{s}_2$ are both points, $|\acute{s}_1 - \acute{s}_2|$ equals the distance between them.
- When one is a point and the other is a hyperplane, $|\acute{s}_1 - \acute{s}_2|$ equals $\sqrt{2}$ times the distance between them.

– When one is a point and the other is a sphere, $|\acute{s}_1 - \acute{s}_2|$ equals the distance between them.
– When both are hyperplanes, $|\acute{s}_1 - \acute{s}_2|$ equals $2\sin\frac{\theta}{2}$, where θ is the angle between vectors $\epsilon_1\mathbf{n}_1, \epsilon_2\mathbf{n}_2$.
– When one is a hyperplane and the other is a sphere, the set of signed distances from the hyperplane to the points on the sphere has a unique maximum and a unique minimum, denoted by $d_{\max}$ and $d_{\min}$ respectively. Let ϵ be the sign of $\mathbf{c}\cdot\mathbf{n} - \delta$ when it is nonzero.
  - If $\mathbf{c}\cdot\mathbf{n} - \delta = 0$, then $|\acute{s}_1 - \acute{s}_2|$ equals the radius of the sphere.
  - If $\epsilon\epsilon_1\epsilon_2 = 1$, then $|\acute{s}_1 - \acute{s}_2| = \sqrt{2}|d_{\max}|$. In particular, when $|\acute{s}_1 - \acute{s}_2| = 0$, the hyperplane and the sphere are in oriented contact.
  - If $\epsilon\epsilon_1\epsilon_2 = -1$, then $|\acute{s}_1 - \acute{s}_2| = \sqrt{2}|d_{\min}|$. In particular, when $|\acute{s}_1 - \acute{s}_2| = 0$, the hyperplane and the sphere are in oriented contact.
– When both are spheres, then
  - if they have the same orientation and are not inclusive, $|\acute{s}_1 - \acute{s}_2|$ equals the *outer tangential distance* between the two spheres, i.e., the distance between the two points of tangency in the common tangent hyperplane of which the spheres are on the same side; in particular, if $|\acute{s}_1 - \acute{s}_2| = 0$, the two spheres are inner tangent to each other;
  - if they have different orientations and are outer tangent to each other, $|\acute{s}_1 - \acute{s}_2| = 0$;
  - if they have different orientations and are exclusive, $|\acute{s}_1 - \acute{s}_2|$ equals the *inner tangential distance* between the two spheres, i.e., the distance between the two points of tangency in the common tangent hyperplane of which the spheres are on different sides.

2.3 Lie Sphere Geometry

Lie sphere transformations are orthogonal transformations of $\mathcal{R}^{n+1,2}$ with $\pm Id$ identified. Geometrically, Lie sphere transformations are transformations in the set of Lie spheres preserving oriented contact. *Lie sphere geometry* is the geometry on Lie sphere transformations.

Laguerre transformations are Lie transformations keeping the 1-subspace spanned by e invariant. Geometrically, Laguerre transformations are Lie sphere transformations keeping the set of hyperplanes invariant. *Laguerre geometry* is the geometry on Laguerre transformations.

Möbius transformations are Lie sphere transformations keeping vector e_{n+1} invariant.

From the definition of Lie sphere transformations, it is clear that spinors can play an important role in the study of Lie sphere transformations and Laguerre transformations. These are not to be discussed in this paper.

3 Further Properties of the Lie Model

In this section we further investigate basic properties of the Lie model for the purpose of applying it to Euclidean geometry. All the results hold for n dimensions by obvious revisions. We let $n = 2$ here only to make the material more easily understood.

The Lie model for the plane is in the space $\mathcal{R}^{3,2}$. Let I_2 be a unit 2-blade determining the orientation of $\mathcal{R}^2$. The orientation of $\mathcal{R}^{3,2}$ is determined by the unit pseudoscalar $I_{3,2} = e \wedge e_0 \wedge I_2 \wedge e_3$. We have

$$I_{3,2}^{-1} = I_{3,2}^{\dagger} = I_{3,2}. \tag{3.1}$$

3.1 One Lie Circle

Since $n = 2$, the positive orientation of a circle is inward. Let $\acute{s}$ be a null vector in $\mathcal{R}^{3,2}$. Then $\acute{s} \cdot e\, \acute{s} \cdot e_3 > 0$ if $\acute{s}$ represents a positive circle; $\acute{s} \cdot e\, \acute{s} \cdot e_3 < 0$ if it represents a negative circle.

- Let $\mathbf{c}_1, \mathbf{c}_2, \mathbf{c}_3$ be three non-collinear points. The oriented circle passing through them and whose orientation is from $\mathbf{c}_1$ to $\mathbf{c}_2$ to $\mathbf{c}_3$ can be represented by
$$\acute{s} = (\acute{\mathbf{c}}_1 \wedge \acute{\mathbf{c}}_2 \wedge \acute{\mathbf{c}}_3)e_3^{\sim} - |\acute{\mathbf{c}}_1 \wedge \acute{\mathbf{c}}_2 \wedge \acute{\mathbf{c}}_3| e_3. \tag{3.2}$$
Notice the negative sign. One can verify that $\acute{s} \cdot e > 0$ if the orientation from $\mathbf{c}_1$ to $\mathbf{c}_2$ to $\mathbf{c}_3$ is positive. For n dimensions the sign is $(-1)^{n-1}$.
- Let $\mathbf{c}_1, \mathbf{c}_2$ be two distinct points. The directed line passing through them and whose direction is from $\mathbf{c}_1$ to $\mathbf{c}_2$ can be represented by
$$(e \wedge \acute{\mathbf{c}}_1 \wedge \acute{\mathbf{c}}_2)e_3^{\sim} + |e \wedge \acute{\mathbf{c}}_1 \wedge \acute{\mathbf{c}}_2| e_3 = (e \wedge \acute{\mathbf{c}}_1 \wedge \acute{\mathbf{c}}_2)e_3^{\sim} + |\mathbf{c}_1 - \mathbf{c}_2| e_3. \tag{3.3}$$
- Let $\mathbf{c}$ be a point, $\mathbf{a}$ be a unit vector of $\mathcal{R}^2$. The directed line passing through $\mathbf{c}$ and with direction $\mathbf{a}$ can be represented by
$$(e \wedge \acute{\mathbf{c}} \wedge \mathbf{a})e_3^{\sim} + |e \wedge \acute{\mathbf{c}} \wedge \mathbf{a}| e_3 = (e \wedge \acute{\mathbf{c}} \wedge \mathbf{a})e_3^{\sim} + e_3. \tag{3.4}$$
- The inward circle with center $\mathbf{c}$ and passing through point $\mathbf{a}$ can be represented by
$$\acute{\mathbf{a}} \cdot (e \wedge \acute{\mathbf{c}}) - |(e \wedge \acute{\mathbf{c}}) \cdot \acute{\mathbf{a}}| e_3 = \acute{\mathbf{a}} \cdot (e \wedge \acute{\mathbf{c}}) - |\mathbf{c} - \mathbf{a}| e_3. \tag{3.5}$$

3.2 Two Lie Circles

Let $\acute{s}_1, \acute{s}_2$ represent two Lie distinct circles. They represent the same circle or line with opposite orientations if and only if

$$e_3 \wedge \acute{s}_1 \wedge \acute{s}_2 = 0. \tag{3.6}$$

The blade $\acute{s}_1^{\sim}$ can represent the set of Lie circles that are in oriented contact with the Lie circle $\acute{s}_1$ in the sense that, a Lie circle $\acute{s}$ is in oriented contact with $\acute{s}_1$ if and only if $\acute{s} \wedge \acute{s}_1^{\sim} = 0$. For two Lie circles, the blade $(\acute{s}_1 \wedge \acute{s}_2)^{\sim} = \acute{s}_1^{\sim} \vee \acute{s}_2^{\sim}$ represents the set of Lie circles that are in oriented contact with both Lie circle. For example, if we use "$\simeq$" to denote that the two sides of the symbol are equal up to a nonzero scale, then for an oriented circle or line $\acute{s}$, the blade $(\acute{s} \wedge (\tau \acute{s}))^{\sim} \simeq (e_3 \wedge s)^{\sim}$ represents points on the circle, or points on the line and the point at infinity.

The blade $T_3 = (\acute{s}_1 \wedge \acute{s}_2)^{\sim}$ has two possibilities:

1. If $\acute{s}_1 \cdot \acute{s}_2 = 0$, then $\acute{s}_1$ and $\acute{s}_2$ are in oriented contact, T_3 represents a *parabolic pencil* of Lie circles, i.e., the set of Lie circles that are in oriented contact with both Lie circles at the point or point at infinity where the two Lie circles are in oriented contact. It is a contact pencil of circles and lines together with the common point of contact, with the circles and lines assigned compatible orientations.
2. If $\acute{s}_1 \cdot \acute{s}_2 \neq 0$, then T_3 is Minkowski. The set of common oriented contact Lie circles is topologically a circle.

Let $\acute{s}_1, \acute{s}_2$ be two circles. They are inclusive if and only if $(e_3 \wedge \acute{s}_1 \wedge \acute{s}_2)^2 < 0$ and $(e \wedge \acute{s}_1 \wedge \acute{s}_2)^2 < 0$; they are exclusive if and only if $(e_3 \wedge \acute{s}_1 \wedge \acute{s}_2)^2 < 0$ and $(e \wedge \acute{s}_1 \wedge \acute{s}_2)^2 > 0$.

This can be proved as follows. Since $(e_3 \wedge \acute{s}_1 \wedge \acute{s}_2)^2 = -(s_1 \wedge s_2)^2$, the two circles are separate if and only if $(e_3 \wedge \acute{s}_1 \wedge \acute{s}_2)^2 < 0$. If they are exclusive, then they have four common tangent lines, which means that the two oriented circles and the point at infinity have two common oriented contact Lie circles. By Theorem in the next subsection, this is equivalent to $(e \wedge \acute{s}_1 \wedge \acute{s}_2)^2 > 0$. If they are inclusive, they do not have any common tangent line, by the same theorem, $(e \wedge \acute{s}_1 \wedge \acute{s}_2)^2 < 0$.

3.3 Three Lie Circles

Let $\acute{s}_1, \acute{s}_2, \acute{s}_3$ be three distinct Lie circles. $\acute{s}_1 \wedge \acute{s}_2 \wedge \acute{s}_3 = 0$ if and only if they belong to a parabolic pencil. Assume that $T_3 = \acute{s}_1 \wedge \acute{s}_2 \wedge \acute{s}_3 \neq 0$. Consider the following problem: when do they have a common oriented contact Lie circle, and what kind of common oriented contact Lie circles do they have?

Theorem 3.1. *When $T_3^2 > 0$, $= 0$ or < 0, the number of common oriented contact Lie circles is 2, 1 or 0 respectively.*

Proof. Since $\acute{s}_1, \acute{s}_2, \acute{s}_3$ are all null vectors, T_3 has only three possible signatures: $(2,1,0), (1,1,1), (1,2,0)$. In the three cases, $T_3^2 > 0$, $= 0$, < 0 respectively. The corresponding 2-blade $T_3^{\sim}$ has the following signatures respectively: $(1,1,0), (1,0,1), (2,0,0)$. The number of null 1-subspaces in $T_3^{\sim}$, which equals the number of common oriented contact Lie circles, is 2, 1, 0 respectively.

If $\acute{s}_1, \acute{s}_2, \acute{s}_3$ are three points, there is a circle or line passing through them with two possible orientations. If they are three pairwise intersecting oriented lines, then besides the point at infinity, there exists another common oriented contact Lie circle, which is either the inscribed circle or an escribed circle of the triangle formed by the lines, depending on the orientations of the lines.

Theorem 3.2. *Let $\acute{s}_1, \acute{s}_2, \acute{s}_3$ be three oriented Lie circles having two common oriented contact Lie circles $\acute{s}_4, \acute{s}_5$.*

1. *$\acute{s}_4, \acute{s}_5$ are both points if and only if the Lie circles belong to a concurrent pencil together with the points of intersection.*

2. *If $\acute{s}_1, \acute{s}_2, \acute{s}_3$ are all circles, then $\acute{s}_4, \acute{s}_5$ are both lines if and only if*

$$\frac{\mathbf{c}_1 - \mathbf{c}_2}{\mathbf{c}_1 - \mathbf{c}_3} = \frac{\epsilon_1\rho_1 - \epsilon_2\rho_2}{\epsilon_1\rho_1 - \epsilon_3\rho_3}. \tag{3.7}$$

3. *If at least one of $\acute{s}_1, \acute{s}_2, \acute{s}_3$ is a circle or point, then $\acute{s}_4, \acute{s}_5$ are two circles of different orientations if and only if $e \wedge T_3 \neq 0$, $e_3 \wedge T_3 \neq 0$, but $(e \wedge T_3) \cdot (e_3 \wedge T_3) = 0$.*

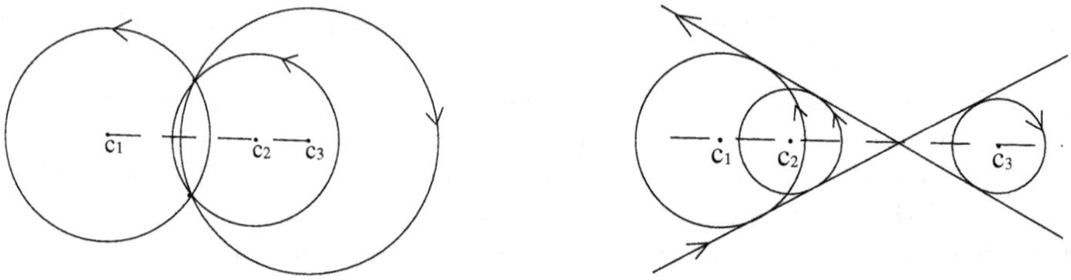

Fig. 3. Theorem , 1 and 2

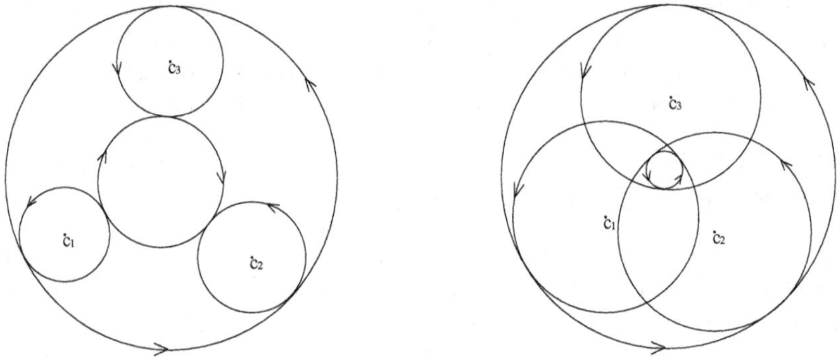

Fig. 4. Theorem , 3

Proof. 1. This can be obtained from the fact that a concurrent pencil of circles and lines which is not a parabolic pencil must have two points as the intersection.

2. Let $\acute{s}_i = \acute{\mathbf{c}}_i + \epsilon_i\rho_i e_3$ for $i = 1, 2, 3$. $T_3^{\sim}$ corresponds to two lines if and only if $e \wedge T_3 = 0$. Expanding this equality, we get

$$\begin{cases} e \wedge \acute{\mathbf{c}}_1 \wedge \acute{\mathbf{c}}_2 \wedge \acute{\mathbf{c}}_3 = 0 \\ e \wedge e_3 \wedge (\epsilon_1\rho_1\acute{\mathbf{c}}_2 \wedge \acute{\mathbf{c}}_3 + \epsilon_2\rho_2\acute{\mathbf{c}}_3 \wedge \acute{\mathbf{c}}_1 + \epsilon_3\rho_3\acute{\mathbf{c}}_1 \wedge \acute{\mathbf{c}}_2) = 0 \end{cases}$$

which can be written as ( ).

3. Let $B_2 = \acute{s}_4 \wedge \acute{s}_5$. Then $\acute{s}_4 \cdot \acute{s}_5 \neq 0$. If at least one of $\acute{s}_1, \acute{s}_2, \acute{s}_3$ is a circle or point, then neither $\acute{s}_4$ nor $\acute{s}_5$ is collinear with e, i. e., $e \cdot \acute{s}_4$ and $e_3 \cdot \acute{s}_4$ cannot be both zero, and the same is true for $\acute{s}_5$.
   So B_2 corresponds to two points if and only if $e_3 \cdot \acute{s}_4 = e_3 \cdot \acute{s}_5 = 0$, and corresponds to two lines if and only if $e \cdot \acute{s}_4 = e \cdot \acute{s}_5 = 0$. When B_2 corresponds to neither two points nor two lines, then $e_3 \cdot \acute{s}_4$ and $e_3 \cdot \acute{s}_5$ cannot be both zero, and the same is true when e_3 is replaced by e. In this case, from

$$(e \wedge T_3) \cdot (e_3 \wedge T_3) = (e \cdot B_2) \cdot (e_3 \cdot B_2) = -\acute{s}_4 \cdot \acute{s}_5 (e \cdot \acute{s}_4 \, e_3 \cdot \acute{s}_5 + e \cdot \acute{s}_5 \, e_3 \cdot \acute{s}_4) \quad (3.8)$$

   we get that ( ) equals zero if and only if $e \cdot \acute{s}_4 / e_3 \cdot \acute{s}_4 = -e \cdot \acute{s}_5 / e_3 \cdot \acute{s}_5$, which is neither zero nor infinity. So $\acute{s}_4, \acute{s}_5$ must represent two circles with different orientations.

When $\acute{s}_1, \acute{s}_2, \acute{s}_3$ have a unique common oriented contact Lie circle, then at least one of the $\acute{s}_i \cdot \acute{s}_j$, $1 \leq i < j \leq 3$, equals zero, but not all of them are zero. Assume that $\acute{s}_1 \cdot \acute{s}_2 \neq 0$. Let

$$t = (\acute{s}_1 \wedge \acute{s}_2) \cdot (\acute{s}_1 \wedge \acute{s}_2 \wedge \acute{s}_3). \quad (3.9)$$

Then t is a null vector representing the common oriented contact Lie circle.

The unique common oriented contact Lie circle of three Lie circles $\acute{s}_1$, $\acute{s}_2$, $\acute{s}_3$ is a point if and only if

$$(e_3 \wedge \acute{s}_i \wedge \acute{s}_j) \cdot (\acute{s}_1 \wedge \acute{s}_2 \wedge \acute{s}_3) = 0 \quad (3.10)$$

for any $1 \leq i < j \leq 3$; it is a line if and only if

$$(e \wedge \acute{s}_i \wedge \acute{s}_j) \cdot (\acute{s}_1 \wedge \acute{s}_2 \wedge \acute{s}_3) = 0. \quad (3.11)$$

3.4 Four Lie Circles

If four Lie circles $\acute{s}_i$, $i = 1, \ldots, 4$ have a common contact Lie circle, the vector $x = (\acute{s}_1 \wedge \acute{s}_2 \wedge \acute{s}_3 \wedge \acute{s}_4)^{\sim}$ is either zero or null. In both cases we have

$$(\acute{s}_1 \wedge \acute{s}_2 \wedge \acute{s}_3 \wedge \acute{s}_4)^2 = 0. \quad (3.12)$$

Conversely, if $x \neq 0$, it must represent the unique common oriented contact Lie circle. If $x = 0$, then if $\acute{s}_i \wedge \acute{s}_j \wedge \acute{s}_k = 0$ for any $1 \leq i < j < k \leq 3$, the four Lie circles belong to a parabolic pencil, and have infinitely many common oriented contact Lie circles; if $\acute{s}_1 \wedge \acute{s}_2 \wedge \acute{s}_3 \neq 0$, then any Lie circle that is in common oriented contact with $\acute{s}_1, \acute{s}_2, \acute{s}_3$ must be in oriented contact with $\acute{s}_4$.

When the $\acute{s}_i$ are points, ( ) can be written as

$$(\acute{s}_1 \wedge \acute{s}_2 \wedge \acute{s}_3 \wedge \acute{s}_4)^2 = \det(\acute{s}_i \cdot \acute{s}_j)_{i,j=1..4} = \frac{1}{16} \det(|\mathbf{c}_i - \mathbf{c}_j|^2)_{i,j=1..4} = 0.$$

After factorization, we get

Theorem 3.3. [Ptolemy Theorem] *If four points* $\mathbf{c}_i$, $i = 1, \ldots, 4$ *are on the same circle, then*

$$d_{12}d_{34} \pm d_{14}d_{23} \pm d_{13}d_{24} = 0, \tag{3.13}$$

where d_{ij} *is the distance between point* $\mathbf{c}_i$ *and point* $\mathbf{c}_j$.

When the $\acute{s}_i$ are circles, ( ) can be written as

$$(\acute{s}_1 \wedge \acute{s}_2 \wedge \acute{s}_3 \wedge \acute{s}_4)^2 = \det(\acute{s}_i \cdot \acute{s}_j)_{i,j=1..4} = \frac{1}{16}\det(|\acute{s}_i - \acute{s}_j|^2)_{i,j=1..4} = 0.$$

After factorization, we get

Theorem 3.4. [Casey Theorem] *If four circles* $\mathbf{c}_i$, $i = 1, \ldots, 4$ *are tangent to the same circle, then*

$$T_{12}T_{34} \pm T_{14}T_{23} \pm T_{13}T_{24} = 0, \tag{3.14}$$

where T_{ij} *is the tangential distance between circle* $\mathbf{c}_i$ *and circle* $\mathbf{c}_j$.

When the $\acute{s}_i$ are lines, ( ) is always true because the point at infinity is on every line. There exists another common contact Lie circle if and only if

$$\acute{s}_1 \wedge \acute{s}_2 \wedge \acute{s}_3 \wedge \acute{s}_4 = 0, \tag{3.15}$$

and either the four lines pass through a common point, or at least three of them have a common oriented contact Lie circle other than the point at infinity.

4 Illustrative Examples

Example 1. Let ABC be a triangle in the plane. Let $\mathbf{a}, \mathbf{b}, \mathbf{c}$ be unit vectors along sides AB, BC, CA respectively, and let $|AB| = l$. Represent the inscribed circle of the triangle with $A, l, \mathbf{a}, \mathbf{b}, \mathbf{c}$.

Below we use four different Clifford algebraic models to solve this problem.

Approach 1. The Clifford model $\mathcal{G}_2$.

Let I be the center of the inscribed circle. Line IA bisects $\angle BAC$, and line IB bisects $\angle ABC$. In the language of vectors, vector $I - A$ is parallel to vector $\mathbf{c} - \mathbf{a}$, and vector $I - B$ is parallel to vector $\mathbf{a} - \mathbf{b}$. These constraints can be represented by

$$\begin{cases} (I - A) \wedge (\mathbf{c} - \mathbf{a}) = 0 \\ (I - B) \wedge (\mathbf{a} - \mathbf{b}) = 0 \end{cases}.$$

From $B - A = l\mathbf{a}$, we get $I - B = I - A + A - B = I - A - l\mathbf{a}$. The second equation can be written as

$$(I - A) \wedge (\mathbf{a} - \mathbf{b}) = -l\mathbf{a} \wedge \mathbf{b}.$$

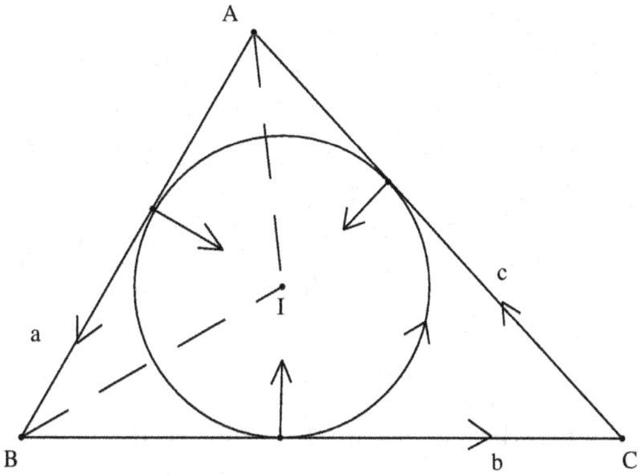

Fig. 5. Example 1

So

$$I - A = -l \frac{\mathbf{a} \wedge \mathbf{b}}{(\mathbf{c} - \mathbf{a}) \wedge (\mathbf{a} - \mathbf{b})} (\mathbf{c} - \mathbf{a}). \tag{4.1}$$

The radius ρ of the circle equals the distance from the center to line AB:

$$\rho = |P_{\mathbf{a}}^{\perp}(I - A)| = |\mathbf{a} \wedge (I - A)| = l \frac{|\mathbf{a} \wedge \mathbf{b}||\mathbf{c} \wedge \mathbf{a}|}{|(\mathbf{c} - \mathbf{a}) \wedge (\mathbf{a} - \mathbf{b})|}. \tag{4.2}$$

Approach 2. The Grassmann model $\mathcal{G}_3$.

In this model, the plane is embedded in $\mathcal{R}^3$ as an affine plane away from the origin. Since vector $I - A$ is parallel to vector $\mathbf{c} - \mathbf{a}$, and vector $I - B$ is parallel to vector $\mathbf{a} - \mathbf{b}$, line IA can be represented by $A \wedge (\mathbf{c} - \mathbf{a})$, and line IB can be represented by $B \wedge (\mathbf{a} - \mathbf{b})$. The intersection of the two lines is

$$\begin{aligned} I &\simeq (B \wedge (\mathbf{a} - \mathbf{b})) \vee (A \wedge (\mathbf{c} - \mathbf{a})) \\ &= (A \wedge B \wedge (\mathbf{a} - \mathbf{b}))^{\sim} (\mathbf{c} - \mathbf{a}) + (B \wedge (\mathbf{c} - \mathbf{a}) \wedge (\mathbf{a} - \mathbf{b}))^{\sim} A \\ &= (B \wedge (\mathbf{c} - \mathbf{a}) \wedge (\mathbf{a} - \mathbf{b}))^{\sim} \left(A + \frac{(A \wedge (B - A) \wedge (\mathbf{a} - \mathbf{b}))^{\sim}}{(B \wedge (\mathbf{c} - \mathbf{a}) \wedge (\mathbf{a} - \mathbf{b}))^{\sim}} (\mathbf{c} - \mathbf{a}) \right). \end{aligned}$$

So $I - A = -l \dfrac{\mathbf{a} \wedge \mathbf{b}}{(\mathbf{c} - \mathbf{a}) \wedge (\mathbf{a} - \mathbf{b})} (\mathbf{c} - \mathbf{a})$.

The radius equals the distance from I to line AB:

$$\rho = |I \wedge A \wedge \mathbf{a}| = |\mathbf{a} \wedge (I - A)| = l \frac{|\mathbf{a} \wedge \mathbf{b}||\mathbf{c} \wedge \mathbf{a}|}{|(\mathbf{c} - \mathbf{a}) \wedge (\mathbf{a} - \mathbf{b})|}.$$

Approach 3. The homogeneous model $\mathcal{G}_{3,1}$.

Similar to the Grassmann model, line IA can be represented by $e \wedge \acute{A} \wedge (\mathbf{c} - \mathbf{a})$, and line IB can be represented by $e \wedge \acute{B} \wedge (\mathbf{a} - \mathbf{b})$. The intersection of the two lines is

$$\begin{aligned} e\wedge \acute{I} &\simeq (e\wedge \acute{B}\wedge (\mathbf{a}-\mathbf{b}))\vee (e\wedge \acute{A}\wedge (\mathbf{c}-\mathbf{a}))\\ &= (e\wedge \acute{A}\wedge \acute{B}\wedge (\mathbf{a}-\mathbf{b}))^{\sim} e\wedge (\mathbf{c}-\mathbf{a})\\ &\qquad +(e\wedge \acute{B}\wedge (\mathbf{c}-\mathbf{a})\wedge (\mathbf{a}-\mathbf{b}))^{\sim} e\wedge \acute{A}\\ &= (e\wedge \acute{B}\wedge (\mathbf{c}-\mathbf{a})\wedge (\mathbf{a}-\mathbf{b}))^{\sim}\ e\wedge\\ &\qquad \left(\acute{A}+\frac{(e\wedge \acute{A}\wedge (\acute{B}-\acute{A})\wedge (\mathbf{a}-\mathbf{b}))^{\sim}}{(e\wedge \acute{B}\wedge (\mathbf{c}-\mathbf{a})\wedge (\mathbf{a}-\mathbf{b}))^{\sim}}(\mathbf{c}-\mathbf{a})\right). \end{aligned}$$

So $I-A=-l\dfrac{\mathbf{a}\wedge\mathbf{b}}{(\mathbf{c}-\mathbf{a})\wedge(\mathbf{a}-\mathbf{b})}(\mathbf{c}-\mathbf{a})$.

The radius equals the distance from I to line AB:

$$\rho = |e\wedge \acute{I}\wedge \acute{A}\wedge \mathbf{a}| = |\mathbf{a}\wedge (I-A)| = l\frac{|\mathbf{a}\wedge\mathbf{b}||\mathbf{c}\wedge\mathbf{a}|}{|(\mathbf{c}-\mathbf{a})\wedge(\mathbf{a}-\mathbf{b})|}.$$

Approach 4. The Lie model $\mathcal{G}_{3,2}$.

Directed lines AB, BC, CA are represented by null vectors $(e\wedge\acute{A}\wedge\mathbf{a}\wedge e_3)^{\sim}+e_3$, $(e\wedge\acute{B}\wedge\mathbf{b}\wedge e_3)^{\sim}+e_3$, $(e\wedge\acute{A}\wedge\mathbf{c}\wedge e_3)^{\sim}+e_3$ respectively. The inscribed oriented circle corresponds to the null 1-subspace other than the one generated by e in the 2-blade

$$\begin{aligned} B_2 &= ((e\wedge \acute{A}\wedge \mathbf{a}\wedge e_3)^{\sim}+e_3)\wedge ((e\wedge \acute{B}\wedge \mathbf{b}\wedge e_3)^{\sim}+e_3)\\ &\qquad \wedge((e\wedge \acute{A}\wedge \mathbf{c}\wedge e_3)^{\sim}+e_3))^{\sim}\\ &= (e\wedge \acute{A}\wedge \mathbf{a}\wedge e_3+\tilde{e_3})\vee (e\wedge \acute{B}\wedge \mathbf{b}\wedge e_3+\tilde{e_3})\vee (e\wedge \acute{A}\wedge \mathbf{c}\wedge e_3+\tilde{e_3})\\ &= e\wedge e_3(e\wedge e_3\wedge \acute{A}\wedge \mathbf{a}\wedge \mathbf{c})^{\sim}(e\wedge e_3\wedge \acute{B}\wedge \mathbf{b}\wedge \acute{A})^{\sim}\\ &\qquad +e\wedge (\mathbf{c}-\mathbf{a})(e\wedge e_3\wedge \acute{B}\wedge \mathbf{b}\wedge \acute{A})^{\sim}\\ &\qquad\quad -e\wedge \acute{A}(e\wedge e_3\wedge \acute{A}\wedge (\mathbf{a}\wedge\mathbf{b}+\mathbf{b}\wedge\mathbf{c}+\mathbf{c}\wedge\mathbf{a}))^{\sim}\\ &= -(e\wedge e_3\wedge \acute{A}\wedge (\mathbf{a}\wedge\mathbf{b}+\mathbf{b}\wedge\mathbf{c}+\mathbf{c}\wedge\mathbf{a}))^{\sim}\\ &\qquad e\wedge\left(\acute{A}-\frac{(e\wedge e_3\wedge \acute{A}\wedge (\acute{B}-\acute{A})\wedge \mathbf{b})^{\sim}}{(e\wedge e_3\wedge \acute{A}\wedge (\mathbf{a}\wedge\mathbf{b}+\mathbf{b}\wedge\mathbf{c}+\mathbf{c}\wedge\mathbf{a}))^{\sim}}(\mathbf{c}-\mathbf{a})\right.\\ &\qquad\qquad \left.+\frac{(e\wedge e_3\wedge \acute{A}\wedge \mathbf{c}\wedge \mathbf{a})^{\sim}(e\wedge e_3\wedge \acute{A}\wedge (\acute{B}-\acute{A})\wedge \mathbf{b})^{\sim}}{(e\wedge e_3\wedge \acute{A}\wedge (\mathbf{a}\wedge\mathbf{b}+\mathbf{b}\wedge\mathbf{c}+\mathbf{c}\wedge\mathbf{a}))^{\sim}}e_3\right). \end{aligned}$$

So the center of the circle is $I=A-l\dfrac{\mathbf{a}\wedge\mathbf{b}}{(\mathbf{c}-\mathbf{a})\wedge(\mathbf{a}-\mathbf{b})}(\mathbf{c}-\mathbf{a})$, the radius is $l\dfrac{|\mathbf{a}\wedge\mathbf{b}||\mathbf{c}\wedge\mathbf{a}|}{|(\mathbf{c}-\mathbf{a})\wedge(\mathbf{a}-\mathbf{b})|}$.

A comparison of the four approaches shows that, the computation based on the Lie model is not necessarily the simplest, considering the additional three dimensions it requires. However, in the Lie model it is the original definition of the inscribed circle of a triangle that is used in algebraic description, the center and the radius are directly computed at the same time, instead of the center being computed first and being used to compute the radius. The Lie model behaves more "dummy-proof" in algebraic description and computation.

Example 2. Let there be a convex polyhedron of five faces in the space. Find the condition for the existence of an inscribed sphere of the polyhedron.

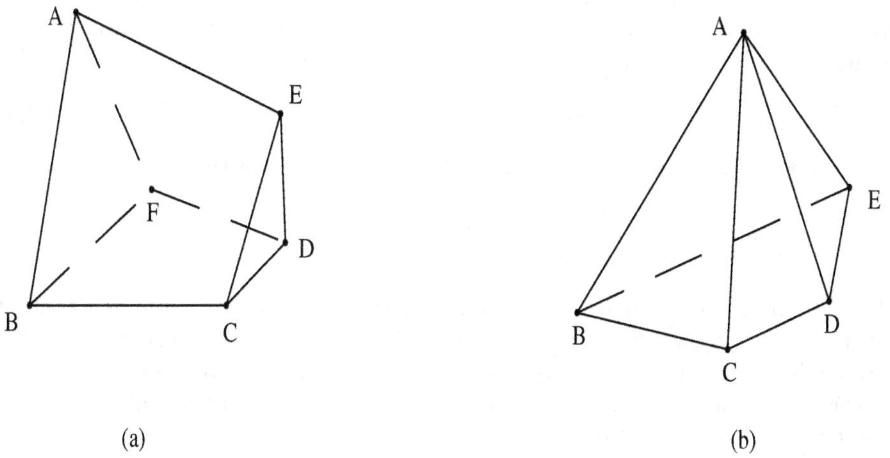

Fig. 6. Example 2

Let the unit outer normals of the five faces be $\mathbf{n}_i$, $i = 1, \ldots, 5$ respectively. Let A be the intersection of the three faces with normals $\mathbf{n}_1$, $\mathbf{n}_2$, $\mathbf{n}_3$, and let δ_4, δ_5 be the distances from A to the faces with normals $\mathbf{n}_4$, $\mathbf{n}_5$, respectively.

Choose A to be the origin of the space. The five faces can be represented by

$$\begin{cases} \acute{s}_1 = \mathbf{n}_1 + e_4 \\ \acute{s}_2 = \mathbf{n}_2 + e_4 \\ \acute{s}_3 = \mathbf{n}_3 + e_4 \\ \acute{s}_4 = \mathbf{n}_4 + \delta_4 e + e_4 \\ \acute{s}_5 = \mathbf{n}_5 + \delta_5 e + e_4 \end{cases}.$$

For a convex 5-faced polyhedron, the five faces do not possess a common point and at least four of them have a common tangent outward sphere. So the existence of an inscribed sphere is equivalent to

$$\begin{aligned} &\acute{s}_1 \wedge \acute{s}_2 \wedge \acute{s}_3 \wedge \acute{s}_4 \wedge \acute{s}_5 \\ &= e_4 \wedge e \wedge (\delta_4(\mathbf{n}_2 \wedge \mathbf{n}_3 \wedge \mathbf{n}_5 - \mathbf{n}_1 \wedge \mathbf{n}_3 \wedge \mathbf{n}_5 + \mathbf{n}_1 \wedge \mathbf{n}_2 \wedge \mathbf{n}_5 - \mathbf{n}_1 \wedge \mathbf{n}_2 \wedge \mathbf{n}_3) \\ &\qquad -\delta_5(\mathbf{n}_2 \wedge \mathbf{n}_3 \wedge \mathbf{n}_4 - \mathbf{n}_1 \wedge \mathbf{n}_3 \wedge \mathbf{n}_4 + \mathbf{n}_1 \wedge \mathbf{n}_2 \wedge \mathbf{n}_4 - \mathbf{n}_1 \wedge \mathbf{n}_2 \wedge \mathbf{n}_3)) \\ &= 0, \end{aligned}$$

i.e.,

$$\frac{\delta_4}{\delta_5} = \frac{\partial(\mathbf{n}_1 \wedge \mathbf{n}_2 \wedge \mathbf{n}_3 \wedge \mathbf{n}_4)}{\partial(\mathbf{n}_1 \wedge \mathbf{n}_2 \wedge \mathbf{n}_3 \wedge \mathbf{n}_5)}. \tag{4.3}$$

A 5-faced convex polyhedron has an inscribed sphere if and only if for any vertex A, the relation ( ) holds. The right-hand side of ( ) equals the ratio of the signed volumes of two tetrahedra whose vertices are respectively points $\mathbf{n}_1, \mathbf{n}_2, \mathbf{n}_3, \mathbf{n}_4$ and points $\mathbf{n}_1, \mathbf{n}_2, \mathbf{n}_3, \mathbf{n}_5$ on the unit sphere of the space. In particular, if for some vertex A, $\delta_4 = 0$, then ( ) becomes

$$\partial(\mathbf{n}_1 \wedge \mathbf{n}_2 \wedge \mathbf{n}_3 \wedge \mathbf{n}_4) = 0, \tag{4.4}$$

i.e., the four points $\mathbf{n}_1, \mathbf{n}_2, \mathbf{n}_3, \mathbf{n}_4$ are on an affine plane of the space.

5 Conclusion

The Lie model is principally for geometric problems involving oriented contact of spheres and hyperplanes. This model cannot deal with problems on conformal properties without resorting to the homogeneous model. It cannot represent lower dimensional spheres and planes. It can serve as a supplementary tool for the homogeneous model in classical geometry.

References

1. W. Blaschke (1929). *Vorlesungen über Differentialgeometrie und geometrische Grundlagen von Einsteins Relativitätstheorie.* Vol. 3, Springer, Berlin.
2. T. E. Cecil (1992). *Lie Sphere Geometry.* Springer, New York.
3. T. E. Cecil and S.-S. Chern (1987). Tautness and Lie sphere geometry. Math. Ann **278**: 381–399.
4. T. Havel and A. Dress (1993). Distance geometry and geometric algebra, Found. Phys. **23**: 1357–1374.
5. D. Hestenes and G. Sobczyk (1984). *Clifford Algebra to Geometric Calculus*, D. Reidel, Dordrecht, Boston.
6. D. Hestenes (1991). The design of linear algebra and geometry, Acta Appl. Math. **23**: 65–93.
7. H. Li (1998). Some applications of Clifford algebra to geometries. *Automated Deduction in Geometries*, LNAI **1669**, X.-S. Gao, D. Wang, L. Yang (eds.), pp. 156-179.
8. H. Li, D. Hestenes and A. Rockwood (2000a). Generalized homogeneous coordinates for computational geometry. In: *Geometric Computing with Clifford Algebra*, G. Sommer (ed.), Springer, Heidelberg.
9. H. Li, D. Hestenes and A. Rockwood (2000b). Spherical conformal geometry with geometric algebra. In: *Geometric Computing with Clifford Algebra*, G. Sommer (ed.), Springer, Heidelberg.
10. H. Li, D. Hestenes and A. Rockwood (2000c). A universal model for conformal geometries of Euclidean, spherical and double-hyperbolic spaces. In: *Geometric Computing with Clifford Algebra*, G. Sommer (ed.), Springer, Heidelberg.
11. S. Lie (1872). Über Komplexe, inbesondere Linien- und Kugelkomplexe, mit Anwendung auf der Theorie der partieller Differentialgleichungen. Math. Ann. **5**: 145–208, 209–256. (Ges. Abh. **2**: 1–121)

12. U. Pinkall (1981). Dupin'sche Hyperflächen. Dissertation, Univ. Freiburg.
13. W.-T. Wu (1994). *Mechanical Theorem Proving in Geometries: Basic Principle* (translated from Chinese edition 1984), Springer, Wien.

On the Geometric Structure of Spatio-temporal Patterns

Erhardt Barth[1] and Mario Ferraro[2]

[1] Institute for Signal Processing, Medical University of Lübeck
Ratzeburger Allee 160, 23538 Lübeck
barth@isip.mu-luebeck.de
http://www.isip.mu-luebeck.de
[2] Dipartimento di Fisica Sperimentale and INFM
via Giuria 1, Torino
ferraro@to.infn.it

Abstract. The structure of hypersurfaces corresponding to different spatio-temporal patterns is considered, and in particular representations based on geometrical invariants, such as the Riemann and Einstein tensors and the scalar curvature are analyzed. The spatio-temporal patterns result from translations, Lie-group transformations, accelerated and discontinuous motions and modulations. Novel methods are obtained for the computation of motion parameters and the optical flow. Moreover, results obtained for accelerated and discontinuous motions are useful for the detection of motion boundaries.

Keywords: dynamic features, motion, flow field, differential geometry, curvature tensor, Lie transformation groups.

1 Introduction

The input to the human and most technical vision systems is light intensity f as a function of space and time. This function defines a hypersurface

$$S = \{x, y, t, f(x, y, t)\} \tag{1}$$

which has the form of a 3-dimensional Monge patch. From a geometric point of view the curvature is the most important property of the surface in that it determines the intrinsic structure of the manifold [ ], so it is of interest to investigate how different types of visual inputs are represented by the curvature tensor of ( ). Further, two other geometric invariants, namely the scalar curvature and the Einstein tensor, will be also considered. The goal is, to gain a better understanding of multidimensional signals and visual processing. In vision-science terms, nonlinear representations of dynamic visual inputs are considered. Such representations are generic but of interest to the perception-action cycle. For example, the points on ( ) with significant curvature can track moving patterns and the curvature tensor can be used to compute motion parameters [, ,]. In

G. Sommer and Y. Y. Zeevi (Eds.): AFPAC 2000, LNCS 1888, pp. 134– , 2000.

this paper, however, we consider the theoretical aspects only. Applications have been presented elsewhere, including models of biological visual processing [, ,].

Geometric methods in computer vision most often deal with the extrinsic geometry of objects in 3D space and how these objects and their motions project on the image plane. However, the geometry of the hypersurface ( ) has been used for motion detection [ ] with an algorithm based on the gradient of ( ). It has also been shown that the Gaussian curvature of ( ) can be used to detect motion discontinuities [ ]. Our approach is related to the so-called structure-tensor method - see [ ] for a review - and this relationship will be discussed in a forthcoming paper [ ].

2 Translation with Constant Velocity

If the image sequence $f(x, y, t)$ results from any spatial pattern moving with constant velocity $\boldsymbol{v} = \{v_x, v_y\}$, f is assumed to satisfy the constraint [ ]

$$f(x, y, t) = f(x + dx, y + dy, t + dt), \tag{2}$$

that leads to [ ]

$$-\frac{\partial f}{\partial t} = \boldsymbol{\nabla} f \cdot \boldsymbol{v}, \tag{3}$$

with $\boldsymbol{\nabla} f$ being the spatial gradient of f. Finally the solution of ( ) is

$$f(x, y, t) = f(x - v_x t, y - v_y t), \tag{4}$$

showing that the image can be thought of as a "solitary wave" which moves, without distortion, with constant velocity along a straight line and whose shape is determined at any given time t by $f(\cdot, t)$.

2.1 Riemann Curvature Tensor

In this section we first summarize results that have been obtained previously [,] and that will be compared to the results in the following sections.

If we compute the components of the curvature tensor (see Eq. in the Appendix) for the specific function f in Eq. , and then simplify all possible ratios of components, we obtain the following results:

$$\begin{aligned} \boldsymbol{v}_1 &= \{R_{3221}, -R_{3121}\}/R_{2121} \\ \boldsymbol{v}_2 &= \{R_{3231}, -R_{3131}\}/R_{3121} \\ \boldsymbol{v}_3 &= \{R_{3232}, -R_{3231}\}/R_{3221}. \end{aligned} \tag{5}$$

Here indices simply denote the fact that we obtain different expressions for $\boldsymbol{v}$. All representations $\boldsymbol{v}_i$ were obtained by assuming the constant brightness constraint.

[1] These and the following simplifications have been performed with the aid of the software *Mathematica* [ ].

Note that $\boldsymbol{v}_1$ is the classical solution obtained for the optical flow under the assumption of constant spatial gradient [ ] (this is not surprising since this assumption is more general and includes the constraint in Eq.).

From Eqs. we can obtain a further motion vector $\boldsymbol{v}_4 = \{v_{4x}, v_{4y}\}$ with

$$v_{4x} = \text{sign}(v_{1x})\sqrt{R_{3232}/R_{2121}}, v_{4y} = \text{sign}(v_{1y})\sqrt{R_{3131}/R_{2121}}. \tag{6}$$

It seems an interesting result that the sectional curvatures (cf. Eq.) determine the direction of motion (but for the sign which is here taken from the vector $\boldsymbol{v}_1$).

To summarize, we found four different combinations of $\boldsymbol{R}$ components that are equal and equal to the motion vector in case that Eq. holds ($\boldsymbol{v} = \boldsymbol{v}_1 = \boldsymbol{v}_2 = \boldsymbol{v}_3 = \boldsymbol{v}_4$). We shall see in later sections, how these expressions might differ for patterns other than ( ).

2.2 Einstein Tensor

As for the curvature tensor, we can obtain four expressions for the motion vector by simplifying the components of the Einstein tensor $\mathbf{G}$ that is obtained from the Riemann tensor through a contraction of the indices (see [ ] for definition and properties):

$$\begin{aligned} \boldsymbol{v}_1 &= \{G_{11}, G_{21}\}/G_{31} \\ \boldsymbol{v}_2 &= \{G_{21}, G_{22}\}/G_{32} \\ \boldsymbol{v}_3 &= \{G_{31}, G_{32}\}/G_{33}. \end{aligned} \tag{7}$$

The expressions for the components of $\boldsymbol{G}$ contain first and second order derivatives. Unfortunately, these expressions are too large to be printed here but are available on this paper's website [ ].

As in Section we can obtain a further motion vector $\boldsymbol{v}_4$ from the relation $\{v_{4x}^2, v_{4y}^2\} = \{G_{11}, G_{22}\}/G_{33}$.

2.3 Scalar Curvature

So far we have considered tensor-based representations of spatio-temporal patterns. It can be useful, however, to consider also scalar quantities that can be derived from S. The scalar curvature C is a contraction of $\mathbf{R}$ [,]. Under the constraint ( ) C simplifies to

$$C = \frac{2\,(1 + \boldsymbol{v}\cdot\boldsymbol{v})\,\left(f_{\xi\xi} f_{\chi\chi} - {f_{\chi\xi}}^2\right)}{\left(1 + \boldsymbol{\nabla} f \cdot \boldsymbol{\nabla} f + (\boldsymbol{\nabla} f \cdot \boldsymbol{v})^2\right)^2} \tag{8}$$

[2] Note that if we simplify the indices in Eqs. and , i.e., we just set $3221/2121 = 3/1$, $3121/2121 = 3/2$, ..., we obtain $\{3/1, 3/2\}$ for the first three vectors and $\{33/11, 33/22\}$ for $\{v_{4x}^2, v_{4y}^2\}$.

with f being a function of (χ, ξ) where $\chi = x - v_x t$, $\xi = y - v_y t$, and $\boldsymbol{\nabla} f = \{f_\chi, f_\xi\}$. The dot "·" denotes the scalar product and indices in f_χ and $f_{\chi\chi}$ denote first- and second order partial derivatives respectively. Note that for zero velocity, the 3D scalar curvature is just the 2D Gaussian curvature in (x, y), as should be expected.

3 Lie Transformation Groups

So far we have considered spatio-temporal patterns that arise from a translation, however, spatio-temporal patterns can result from a variety of transformations. To investigate how the constant brightness constraint is modified in this case we shall make use of the theory of Lie transformation groups [ ].

If the image is transformed by the action of a linear one-parameter Lie transformation group, whose infinitesimal operator $X_\lambda = a_1(x,y)\partial/\partial x + a_2(x,y)\partial/\partial y$, λ being the parameter of the transformation, then the fundamental flow constraint can be written as

$$f(\boldsymbol{r}, t) = f(\boldsymbol{r}', t + dt), \tag{9}$$

where $\boldsymbol{r} = \{x, y\}$ and $\boldsymbol{r}' = \boldsymbol{r} + d\boldsymbol{r}$. The transformation $\boldsymbol{r} \to \boldsymbol{r}'$ results in

$$dx = x' - x = a_1(x,y)d\lambda \qquad dy = y' - y = a_2(x,y)d\lambda. \tag{10}$$

A straightforward application (omitted here for brevity) of Lie group theory shows that Eq. leads to

$$-\frac{\partial f}{\partial t} = X_\lambda f \frac{d\lambda}{dt} = \boldsymbol{\nabla} f \cdot \boldsymbol{a} \frac{d\lambda}{dt}, \tag{11}$$

$\boldsymbol{a}(x,y) = \{a_1(x,y), a_2(x,y)\}^T$ and here λ has been considered a function of t, as it must be in case of motion.

If several transformation groups are considered, with differential operators X_{λ_j} then Eq. becomes

$$-\frac{\partial f}{\partial t} = \sum_j X_{\lambda_j} f \frac{d\lambda_j}{dt} = \boldsymbol{\nabla} f \cdot \left(\sum_j \boldsymbol{a}_j \frac{d\lambda_j}{dt} \right). \tag{12}$$

Suppose $d\lambda_j/dt = \nu_j$ to be constant and that $a_{j1} = a_{j1}(y)$, $a_{j2} = a_{j1}(x)$, then the solution of Eq. is

$$f(x,y,t) = f\left(x - \sum_j a_{j1}(y)\nu_j t, y - \sum_j a_{j2}(x)\nu_j t \right). \tag{13}$$

For instance consider the general rigid motion in $2D$, that is given by two translations along the coordinate axis and by a rotation; in this case $\boldsymbol{a}_1 = \{1, 0\}, \nu_1 = v_{Ox}$, $\boldsymbol{a}_2 = \{0, 1\}, \nu_2 = v_{Oy}$, where O is the center of rotation. Then the velocity of the center of rotation is just $\boldsymbol{v}_O = \nu_1 \boldsymbol{a}_1 + \nu_2 \boldsymbol{a}_2 = \{v_{Ox}, v_{Oy}\}$ that can also be obtained by usual kinematics. The rotation around O is given by $\boldsymbol{a}_3 = \{-y, x\}$, $\nu_3 = \omega$, where ω is the angular velocity. Suppose $\boldsymbol{v}_O$ and ω constant; Eq. becomes

$$f(x,y,t) = f\left(x - v_{Ox}t + \omega y t, y - v_{Oy}t - \omega x t\right), \tag{14}$$

where x, y are coordinates with respect to O.

For this general case, however, it seems difficult to analyze the effect of such patterns on the spatio-temporal curvature without additional assumptions.

From Eq. a rotation constraint is defined by

$$f(x,y,t) = f\left(x + \omega y t, y - \omega x t\right). \tag{15}$$

As Eq. itself, the results for this transformation can be obtained by simply setting $v_{Ox} = 0$ and $v_{Oy} = 0$ in Eq. and in the equations obtained below for the transformation ( ).

3.1 Riemann Curvature Tensor

For this type of input the vectors v_i differ, and they depend on $x, y, t, v_{Ox}, v_{Oy}, \omega$ and the first and second order derivatives of $f(\chi, \xi)$ with $\chi = x - v_{Ox}t + \omega y t$ and $\xi = y - v_{Oy}t - \omega x t$.

However, we obtain interesting results if we further assume that the gradient of f vanishes. In this case the components of $\mathbf{R}$ are:

$$\begin{aligned}
R_{2121} &= D\big(1 + t^2\omega^2\big)^2 \\
R_{3131} &= D\big(v_{Oy} + t v_{Ox}\omega + x\omega - t y \omega^2\big)^2 \\
R_{3232} &= D\big(v_{Ox} - t v_{Oy}\omega - y\omega - t x \omega^2\big)^2 \\
R_{3121} &= D\left(1 + t^2\omega^2\right)\left(-v_{Oy} - t v_{Ox}\omega - x\omega + t y \omega^2\right) \\
R_{3221} &= -D\left(1 + t^2\omega^2\right)\left(-v_{Ox} + t v_{Oy}\omega + y\omega + t x \omega^2\right) \\
R_{3231} &= -D\left(v_{Ox} - t v_{Oy}\omega - y\omega - t x \omega^2\right)\left(v_{Oy} + t v_{Ox}\omega + x\omega - t y \omega^2\right) \\
&\mathit{with:} \\
D &= f_{\chi\chi} f_{\xi\xi} - {f_{\chi\xi}}^2
\end{aligned} \tag{16}$$

and f as a function of (χ, ξ) defined as above. It is straightforward to check that in this case all the vectors v_i (Eqs. and) point in the same direction, which is the direction of the vector:

$$\{v_{Ox} - t v_{Oy}\omega - y\omega - t x \omega^2, -(v_{Oy} + t v_{Ox}\omega + x\omega - t y \omega^2)\} \tag{17}$$

3.2 Einstein Tensor

For the Einstein tensor, again, we could not obtain useful simplifications but for the case of zero gradient. Surprisingly, the independent components of $\mathbf{G}$ are equal to those of $\mathbf{R}$ in this case (but for the signs):

$$\begin{aligned} G_{33} &= -\,R_{2121} \\ G_{22} &= -\,R_{3131} \\ G_{11} &= -\,R_{3232} \\ G_{32} &= \quad R_{3121} \\ G_{31} &= -\,R_{3221} \\ G_{21} &= \quad R_{3231} \end{aligned} \tag{18}$$

3.3 Scalar Curvature

For zero gradient the scalar curvature simplifies to

$$\begin{aligned} C = &-2D\left(1+t^2\omega^2\right) \\ &\left(1+{v_{Ox}}^2+{v_{Oy}}^2+2v_{Oy}x\omega-2v_{Ox}y\omega+\left(t^2+x^2+y^2\right)\omega^2\right) \end{aligned} \tag{19}$$

Note that for zero rotation and velocity, C is, in coordinates (χ, ξ), the 2D Gaussian curvature (with zero gradient).

4 Translation with Time-Dependent Velocity

We now consider the more general case where the image shift contains higher-order terms, i.e., the motion can be accelerated, i.e.,

$$f(x,y,t) = f(x-d_1(t), y-d_2(t)). \tag{20}$$

4.1 Riemann Curvature Tensor

With the constraint in Eq. , we still obtain for the curvature tensor

$$\{R_{3221}, -R_{3121}\}/R_{2121} = \{d_1'(t), d_2'(t)\}, \tag{21}$$

but the other three expressions $\{R_{3231}, -R_{3131}\}/R_{3121}$, $\{R_{3232}, -R_{3231}\}/R_{3221}$, and $\{R_{3232}, R_{3131}\}/R_{2121}$ do not simplify to yield the velocity components.

However, if we assume that the gradient of $f(\chi,\xi)$ vanishes (${f_\chi}^2 + f_\xi^2 = 0$), we obtain the following relations:

$$\begin{aligned} \{R_{3231}, -R_{3131}\}/R_{3121} &= \{d_1'(t), d_2'(t)\} \\ \{R_{3232}, -R_{3231}\}/R_{3221} &= \{d_1'(t), d_2'(t)\} \\ \{R_{3232}, R_{3131}\}/R_{2121} &= \{d_1'(t)^2, d_2'(t)^2\} \end{aligned} \tag{22}$$

i.e., the motion vectors $\boldsymbol{v}_2$, $\boldsymbol{v}_3$, and $\boldsymbol{v}_4$ are obtained only for local extrema of $f(\chi,\xi)$ (that are extrema of $f(x,y)$ also).

4.2 Einstein Tensor

For the Einstein tensor under the constraint ( ) we could not obtain any simplifications. However, under the additional constraint of zero gradient (see above) we obtain:

$$\begin{aligned} \{G_{11}, G_{21}\}/G_{31} &= \{d_1'(t), d_2'(t)\} \\ \{G_{21}, G_{22}\}/G_{32} &= \{d_1'(t), d_2'(t)\} \\ \{G_{31}, G_{32}\}/G_{33} &= \{d_1'(t), d_2'(t)\} \\ \{G_{11}, G_{22}\}/G_{33} &= \{d_1'(t)^2, d_2'(t)^2\} \end{aligned} \tag{23}$$

4.3 Scalar Curvature

In case of the additional assumption of zero gradient (see above), the scalar curvature simplifies to:

$$C = 2(1 + d_1'(t)^2 + d_2'(t)^2)(f_{\xi\xi} f_{\chi\chi} - f_{\chi\xi}{}^2) \tag{24}$$

with f being a function of (χ, ξ) where $\chi = x - d_1(t)$, $\xi = y - d_2(t)$.

5 Discontinuous Motion

In this section we consider different types of motion discontinuities and how they are represented by the curvature tensor. In particular, we will show that the expressions for the vectors $\boldsymbol{v}_i$ in Eqs. and differ. Therefore the differences can be used as indicators of discontinuous motions [,]. An exception are the locations where the gradient of f vanishes (local extrema).

5.1 Velocity Step

We first consider the case where the velocity vector changes suddenly from zero to $\{v_x, v_y\}$, i.e. the image-sequence intensity $f(x, y, t)$ is defined by

$$f(x, y, t) = f(x - v_x \gamma(t), y - v_y \gamma(t)) \tag{25}$$

where $\gamma(t)$ is the unit step function. We obtain

$$\{R_{3221}, -R_{3121}\}/R_{2121} = \{\delta(t) v_x, -\delta(t) v_y\} \tag{26}$$

where $\delta(t)$ is the Dirac-Delta distribution.

Note that this vector is different from zero only at $t = 0$ when it points in the direction of the motion vector $\{v_x, v_y\}$, i.e., $-R_{3121}/R_{3221} = v_y/v_x$. This is not the case for the other three vectors in Eqs. and . For example, for $\boldsymbol{v}_2$ we obtain:

$$-R_{3131}/R_{3231} = \frac{v_y^2 \delta(t)^2 f_{\chi\xi}{}^2 + v_y \delta'(t) f_\xi f_{\chi\chi} - v_y{}^2 \delta(t)^2 f_{\xi\xi} f_{\chi\chi} + v_x \delta'(t) f_\chi f_{\chi\chi}}{\delta(t)^2 (v_x v_y f_{\xi\xi} f_{\chi\chi} - v_x v_y f_{\chi\xi}{}^2)} \tag{27}$$

with f as a function of (χ, ξ) and $\chi = x - v_x\gamma(t)$, $\xi = y - v_y\gamma(t)$. Similar but different expressions are obtained for R_{3231}/R_{3232} and R_{3131}/R_{3232}. For the extrema of $f(\chi, \xi)$ (assumption of zero gradient as above), however, all the four vectors (Eq. and) point in the direction of $\{v_x, v_y\}$.

5.2 Onset of a Spatial Pattern

Here we consider the case:

$$f(x, y, t) \to f(x, y)\gamma(t) \tag{28}$$

i.e., the spatial pattern $f(x, y)$ is turned on at time $t = 0$.

We obtain the following results:

$$\begin{aligned}
\frac{\{R_{3221}, -R_{3121}\}}{R_{2121}} &= \delta(t)\{\frac{f_{yy}f_x - f_y f_{xy}}{{f_{xy}}^2 - f_{yy}f_{xx}}, \frac{f_x f_{xy} - f_y f_{xx}}{{f_{xy}}^2 - f_{yy}f_{xx}}\} \\
\frac{\{R_{3231}, -R_{3131}\}}{R_{3121}} &= \{\frac{\delta(t)^2 f_y f_x - f(x,y)\gamma(t)\delta'(t)f_{xy}}{\delta(t)(\gamma(t)f_x f_{xy} - \gamma(t)f_y f_{xx})}, \frac{-(\delta(t)^2 {f_x}^2) + f(x,y)\gamma(t)\delta'(t)f_{xx}}{-(\delta(t)\gamma(t)f_x f_{xy}) + \delta(t)\gamma(t)f_y f_{xx}}\} \\
\frac{\{R_{3232}, -R_{3231}\}}{R_{3221}} &= \{\frac{-\delta(t)^2 {f_y}^2 + f(x,y)\gamma(t)\delta'(t)f_{yy}}{-\delta(t)\gamma(t)f_{yy}f_x + \delta(t)\gamma(t)f_y f_{xy}}, \frac{-\delta(t)^2 f_y f_x + f(x,y)\gamma(t)\delta'(t)f_{xy}}{-\delta(t)\gamma(t)f_{yy}f_x + \delta(t)\gamma(t)f_y f_{xy}}\} \\
\frac{\{R_{3232}, R_{3131}\}}{R_{2121}} &= \{\frac{-\delta(t)^2 {f_y}^2 + f(x,y)\gamma(t)\delta'(t)f_{yy}}{-\gamma(t){f_{xy}}^2 + \gamma(t)f_{yy}f_{xx}}, \frac{-\delta(t)^2 {f_x}^2 + f(x,y)\gamma(t)\delta'(t)f_{xx}}{-\gamma(t){f_{xy}}^2 + \gamma(t)f_{yy}f_{xx}}\}
\end{aligned} \tag{29}$$

Note that the four expressions, which are equal for translations, differ for this specific dynamic pattern. For this type of input (Eq.) it is interesting to look at the components of $\boldsymbol{R}$ for the case of zero spatial gradient. We obtain the following results:

$$\begin{aligned}
R_{2121} &= \gamma(t)(-{f_{xy}}^2 + f_{yy}f_{xx})/N \\
R_{3131} &= (f(x,y)\gamma(t)\delta'(t)f_{xx})/N \\
R_{3232} &= (f(x,y)\gamma(t)\delta'(t)f_{yy})/N \\
R_{3121} &= 0 \\
R_{3221} &= 0 \\
R_{3231} &= (f(x,y)\gamma(t)\delta'(t)f_{xy})/N \\
\text{with:} & \\
N &= 1 + \delta(t)^2 f(x,y)^2
\end{aligned} \tag{30}$$

Note that two of the components are zero, such the the vector $\boldsymbol{v}_1$ is zero and the vectors $\boldsymbol{v}_2$ and $\boldsymbol{v}_3$ are undefined due to a zero denominator.

By substituting δ for γ, δ' for δ, and δ'' for δ' we obtain the results for flashing pattern, i.e., $f(x, y, t) \to f(x, y)\delta(t)$. The above results are a special case of modulation, i.e. $f(x, y, t) \to f(x, y)a(t)$ with $a(t) = \gamma(t)$ and $a'(t) = \delta(t)$.

6 Discussion

Differential geometry provides powerful tools for analyzing the geometric structure of multidimensional manifolds. With these tools it is possible to construct invariants that capture the structure of the manifold [,]. We have considered the visual input as a manifold with a specific metric that is defined by image intensity $f(x, y, t)$ (it is the metric of the hypersurface in Eq.), and we have looked at the curvature tensor $\mathbf{R}$ of that manifold as the most prominent geometric invariant and at two specific contractions of $\mathbf{R}$ (we had also looked at the Ricci tensor but had not obtained any meaningful result). In particular, we have shown how selected constraints on f, that are related to motion, affect these geometric invariants. By doing so we have found novel methods for the computation of motion parameters.

Thus, the reported results show that relevant information about spatio-temporal patterns can be gained by analyzing the above-mentioned curvature measures. We have first considered translations and have obtained new expressions for the flow fields in terms of the components of the Einstein tensor. We have then generalized the usual constraint Eqs. and to the more general case of transformations that form Lie transformation groups. For these transformations we have shown how the transformations giving rise to spatio-temporal patterns are encoded by the curvature measures. Meaningful results, however, have been obtained only for zero gradient, i.e. the local extrema of $f(\chi, \xi)$ (that are extrema of $f(x, y)$ as well) with coordinates (χ, ξ) depending on the transformation. Finally, we have also considered discontinuous motions that have been described by step functions and Dirac-Delta distributions. These functions have been analyzed analytically as global patterns, but, of course, the scope is to detect local discontinuities and motion boundaries. In practical applications, the size of the local neighborhood will be determined by the filters used to compute the derivatives, and these filters can be implemented on multiple scales.

Methods based on the four motion vectors derived from $\mathbf{R}$, i.e. Eqs. and , have already been applied, both to obtain robust motion estimations and to model biological motion sensitivity [, ,]. The authors had assumed that the four motion vectors will differ in case of discontinuous motions and have used these differences as indicators of occlusions and noise. Here we have shown that the vectors do indeed differ for such patterns. It seems an important result that the vector $\boldsymbol{v}_1$ still yields the correct motion in case of accelerated motions (Eq.) but the other three vectors do not (except for zero gradient). For discontinuous motions the vector $\boldsymbol{v}_1$ again plays a distinct role and thus supports the idea of confidence measures based on the differences among the vectors $\boldsymbol{v}_i$. The question of how many vectors to use in which combinations still needs further investigation, but applications show that the use of all four vectors improves the results compared to using only two or three vectors.

In conclusion, we have shown that the intrinsic geometry of spatio-temporal patterns, generated by specific transformations, provides useful information on the parameters of the transformations and new insights for the coding of motion and dynamic features.

Acknowledgment

We thank Cicero Mota and the reviewers for comments on the manuscript.

References

1. http://www.isip.mu-luebeck/~barth/papers/afpac2000.
2. D. Ballard and C. Brown. *Computer Vision.* Prentice Hall, Englewood Cliffs, New Jersey, 1982.
3. E. Barth. Bewegung als intrinsische Geometric von Bildfolgen. In W. Forster, J. M. Buhmann, A. Faber, and P. Faber, editors, *Mustererkennung 99,* pages 301-308, Bonn, 1999. Springer, Berlin.
4. E. Barth. The minors of the structure tensor. In G. Sommer, editor, *Mustererkennung 2000,* Kiel, 2000. Springer, Berlin, in print.
5. E. Barth. Spatio-temporal curvature and the visual coding of motion. In *Neural Computation,* Berlin, 2000. Proceedings of the International ICSC Congress.
6. E. Barth and A. B. Watson. Nonlinear spatio-temporal model based on the geometry of the visual input. *Invest. Ophthalm. Vis. Sci.,* 39-4 (Supplement) :S-2110, 1998.
7. B. Jahne, H. HauBecker, and P. GeiBler, editors. *Handbook of Computer Vision and Applications,* volume 2, chapter 13 by HauBecker and Spies. Academic Press, Boston, 1999.
8. S.-P. Liou and R. C. Jain. Motion detection in spatio-temporal space. *Computer Vision, Graphics, and Image Processing,* 45:227-50, 1989.
9. P. J. Olver. *Applications of Lie Groups to Differential Equations.* Springer, New York, Berlin, 1986.
10. B. Schutz. *Geometrical methods of mathematical physics.* Cambridge University Press, Cambridge, 1980.
11. B. Schutz. *A first course in general relativity.* Cambridge University Press, Cambridge, 1985.
12. O. Tretiak and L. Pastor. Velocity estimation from image sequences with second order differential operators. In *Proc. 7th Int. Conf. Pattern Recognition,* pages 16-19, Montreal, Canada, 1984. IEEE Computer Society Press.
13. S. Wolfram. *Mathematica: a system for doing mathematics by computer.* Adison-Wesley Publishing Co., Redwood City, CA, 2 edition, 1991.
14. C. Zetzsche and E. Barth. Direct detection of flow discontinuities by 3D curvature operators. *Pattern Recognition Letters,* 12:771-9, 1991.

A Components of the Riemann Curvature Tensor

$$
\begin{aligned}
R_{2121} &= (f_{yy}f_{xx} - {f_{xy}}^2)/(1+\nabla f) \\
R_{3131} &= (f_{tt}f_{xx} - {f_{xt}}^2)/(1+\nabla f) \\
R_{3232} &= (f_{tt}f_{yy} - {f_{yt}}^2)/(1+\nabla f) \\
R_{3121} &= (f_{yt}f_{xx} - f_{xt}f_{xy})/(1+\nabla f) \\
R_{3221} &= (f_{yt}f_{xy} - f_{yy}f_{xt})/(1+\nabla f) \\
R_{3231} &= (f_{tt}f_{xy} - f_{xt}f_{yt})/(1+\nabla f) \\
&\text{with:} \\
1+\nabla f &= 1+{f_x}^2+{f_y}^2+{f_t}^2
\end{aligned}
\tag{31}
$$

Learning Geometric Transformations with Clifford Neurons

Sven Buchholz and Gerald Sommer

Department of Computer Science, University of Kiel
Preusserstr. 1-9, 24105 Kiel, Germany
{sbh,gs}@ks.informatik.uni-kiel.de

Abstract. In this paper we propose a new type of neuron developed in the framework of Clifford algebra. It is shown how this novel Clifford neuron covers complex and quaternionic neurons and that it can compute orthogonal transformations very efficiently. The introduced framework can also be used for neural computation of non–linear geometric transformations which makes it very promising for applications. As an example we develop a Clifford neuron that computes the cross-ratio via the corresponding Möbius transformation. Experimental results for the proposed novel neural models are reported.

1 Introduction

The aim of neural computation is to understand neural and cognitive processes in biological systems, in particular in the human brain, in computational terms and to design and analyze models and algorithms towards these goals. The connectionist approaches among these are known as *Neural Networks* (NNs).
Neural networks are interconnections of artificial "neurons" that are greatly simplified versions of biological neurons [14].
The first model of an artificial neuron was proposed already more than 50 years ago by McCulloch and Pitts in 1943 [10]. However, the concept reaches wide public in the Artifical Intelligence community not before the end of the 80s, when suitable learning algorithms for training of multi–layer perceptrons (MLPs) were (re–)discovered in [13]. Nowadays artificial neural networks are very common and used in many application fields ranging from classical ones like pattern recognition or computer linguistics to modern ones like financial forecasting, medical diagnosis systems and energy management systems. Therefore, some kind of paradigm shift from symbolic approaches (programming) to neural networks (learning) could be noticed. A change that has also been pushed by many strong mathematical properties of neural networks that could be proven, e.g. the universal approximation property of sigmoidal MLPs [4].
On the other hand, NNs (like any other existing approach) are still far away from brains in any aspect. Besides that distance to the ultimate goal, there are computational problems too. One is the famous bias/variance trade-off formulated exactly in terms of statistical learning theory in [6], which arises from the

G. Sommer and Y. Y. Zeevi (Eds.): AFPAC 2000, LNCS 1888, pp. 144-153, 2000.

fact that NNs can only learn (in practice) from a finite set of examples. Roughly spoken, the only solution is to integrate a priori knowledge. Actually, many other problems that one faces in neural computation reduce in a broader sense to this type of problem.

For example, it is impossible to predict whether a real valued MLP or a complex valued MLP will better perform a task of learning a complex valued function [9]. Both networks are universal approximators. The therefore required high non–linearities can create networks with high "variance", which also makes rule extraction from such networks difficult. In that kind of monolithic structured networks the integration of domain or task specific knowledge ("bias") is not easy to do.

One solution to overcome that "black–box" behavior of NNs is the design of small specific subnetworks that perform simpler tasks and build up more complicated networks from such a functional base. In terms of supervised learning this so called "model–based" approach [3] also has the advantage that optimization is simplified due to constrained lower dimensional weight spaces. In a recent paper [5] the approach is extended to the neural computation of algebraic and geometric structures like eigenvalues of matrices or surface representations.

In this paper we will follow similar intentions, but will focus exclusively on neural computation of geometric transformations. We will do this because of the fundamental importance of geometric transformations in all aspects of the perception–action cycle related to both pattern generation and recognition. Thus, by following Felix Klein's Erlangen program, we will demonstrate how to design NNs as building blocks of geometric expert knowledge by algebraic constraining neural computation. Linear transformations of real vector spaces can be computed easily by a Single Layer Neural Network (SL–NN). This will be our starting point in the next section. There we will also see, that SL–NNs have some drawbacks. The first and fundamental one is obviously the limitation to real linear transformations, which excludes many interesting transformation like projective or Möbius transformations. The second one results from the fact that only certain linear transformations are needed in most applications, like affine transformations in object recognition or rigid motions in robotics. Both drawbacks result from the fact that no additional geometrical a priori knowledge can be integrated into SL–NNs.

Therefore, we propose instead a new type of neuron — the Clifford Neuron, which is formulated in terms of Clifford algebra. Since a Clifford algebra is constructed from a real vector space with a quadratic form on it, concepts of metric geometry can be used easily. It is shown that Clifford neurons compute orthogonal transformations very efficiently. Moreover, they can be used to compute non–linear geometric transformations which allows many interesting applications. As an example, we develop a Clifford neuron that computes the cross-ratio via the corresponding Möbius transformation. Experimental results to confirm the proposed novel neural models are presented.

2 Single Layer Neural Networks

A formal neuron is a computational unit of the form shown in Figure 1. First,

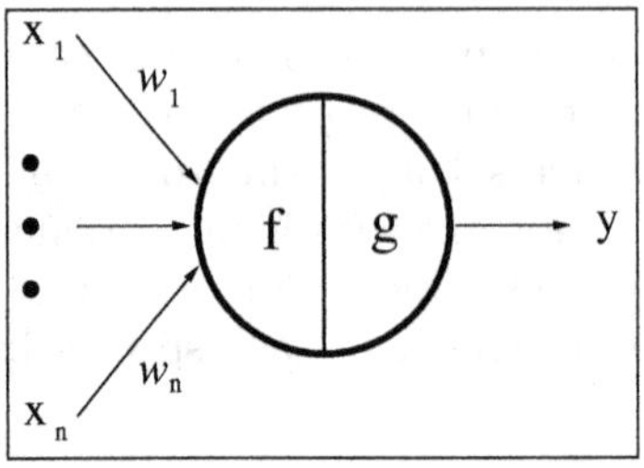

Fig. 1. Formal neuron

a propagation function f associates the input vector x with the weight vector w, which comprises the learnable parameters of the neuron. Then, an activation function g is applied. Using the identity as activation function and the scalar product for weight association gives a linear neuron

$$y = \sum_{j=1}^{n} w_j x_j + \theta \;\; =< (w, \theta), (x, 1) > . \tag{1}$$

The goal of learning is to minimize a given error function E on a training set $T := \{(x^1, t^1), \ldots, (x^m, t^m) \mid x^i \in \mathbb{R}^n, t^i \in \mathbb{R}\}$. Minimization is done by using gradient descent resulting in the update rule

$$\Delta w_j = -\eta \frac{\partial E}{\partial w_j} = \eta \sum_{i=1}^{m} (t^i - \sum_{j=1}^{n} w_j x_j^i) x_j^i \tag{2}$$

for the common sum–of–squared error (SSE). Composing linear neurons as shown in Figure 2 gives a single layer NN (SL–NN).

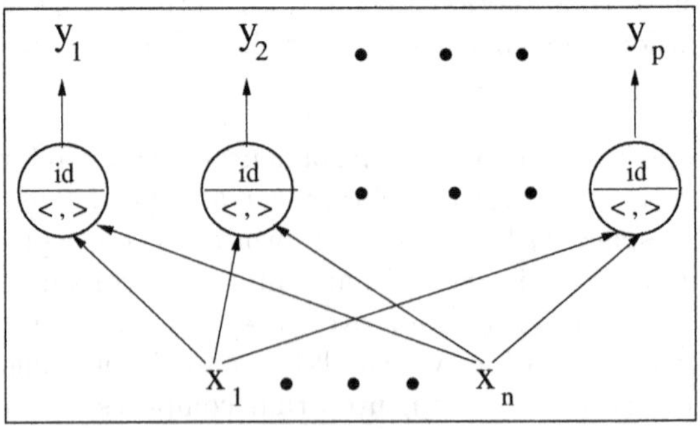

Fig. 2. Single Layer Neural Network (SL–NN)

A SL–NN computes $y = Wx$, whereby the matrix W is composed of the single weight vectors. In particular, if W is square a general linear transformation is computed. In addition to that fundamental limitation, SL–NNs have other drawbacks too.
Suppose, we want to learn an unknown transformation of an object like the one shown in Figure 3.

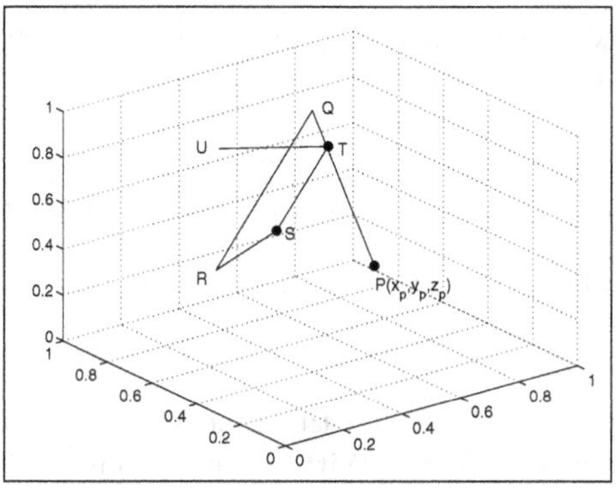

Fig. 3. Wire frame object

Thus, the input consists of 6 points in space. We may also have additional knowledge about corresponding points in transformed output views. However, the only thing we can put in a SL–NN is one big 18–dimensional input vector. Therefore, also the transformation is searched by the SL–NN in 18–dimensional space. Reflecting the above, we can come up with the following observations.
SL–NN fail to incorporate any geometrical a priori knowledge, since there is no representation of geometric entities of objects (e.g. points and lines) available.
Actually, learning a geometric transformation means to learn the action of the corresponding transformation group on geometric entities.
In the next section we will introduce the framework of Clifford algebra to develop neurons that are designed in such a way and will allow powerful integration of geometrical a priori knowledge.

3 Clifford Algebra and Clifford Neurons

The drawbacks of SL–NN in computing geometric transformations seen before come from the underlying concept. The structural concept of a vector space is too weak for such purposes. An additional structure of a vector space is that implied by a quadratic form Q on it. From this an algebra with the desired properties can be constructed in the following way.
For all $v = (v_1, \ldots, v_n) \in \mathbb{R}^n$ there exists a basis of $\mathbb{R}^n$ such that

$$Q(v) = -v_1^2 - v_2^2 - \ldots - v_p^2 + v_{p+1}^2 \cdots + v_{p+q}^2 , \tag{3}$$

with $p, q \in \mathbb{N}$ and $p + q = n$. Thus, Q is already determined by (p, q) and so is a scalar product on $\mathbb{R}^n$. Together we call this the quadratic space $\mathbb{R}^{p,q}$. In particular, we obtain from (3) for an orthonormal basis $\{e_1, \ldots, e_n\}$

$$-Q(e_i) = +1 \quad \text{if} \quad i \leq p \tag{4}$$

$$-Q(e_i) = -1 \quad \text{if} \quad i > p\,. \tag{5}$$

This allows the following definition of a Clifford algebra [8].

Definition 1 *An associative algebra with unity 1 over $\mathbb{R}^{p,q}$ containing $\mathbb{R}^{p,q}$ and $\mathbb{R}$ as distinct subspaces is called the Clifford algebra $\mathcal{C}_{p,q}$ of $\mathbb{R}^{p,q}$, iff*

(a) $v \otimes_{p,q} v = -Q(v)\,, \quad v \in \mathbb{R}^{p,q}$

(b) $\mathcal{C}_{p,q}$ is generated as an algebra by $\mathbb{R}^{p,q}$

(c) $\mathcal{C}_{p,q}$ is not generated by any proper subspace of $\mathbb{R}^{p,q}$.

Condition (a) implies for all $i, j \in \{1, \ldots, p+q\}$

$$e_i \otimes_{p,q} e_j = -e_j \otimes_{p,q} e_i\,. \tag{6}$$

Together with (b) and (c) follows that the dimension of $\mathcal{C}_{p,q}$ is 2^{p+q} [12] and $\mathcal{C}_{p,q}$ is commutative only if $p + q <= 1$. With the notations

$$\mathcal{A} := \{\{a_1, \ldots, a_r\} \in \mathcal{P}(\{1, \ldots, n\}) \mid 1 \leq a_1 \leq \ldots \leq a_r \leq n\}\,, \tag{7}$$

where $\mathcal{P}(\{1, \ldots, n\})$ denotes the power set and defining for all $A \in \mathcal{A}$

$$e_A := e_{a_1} \ldots e_{a_r}\,, \tag{8}$$

we obtain a basis $\{e_A \mid A \in \mathcal{A}\}$ of $\mathcal{C}_{p,q}$. Then the following involutions of $\mathcal{C}_{p,q}$ can be defined

$$\hat{x} = \sum_{A \in \mathcal{A}} (-1)^{|A|}\, x_A e_A \tag{9}$$

$$\tilde{x} = \sum_{A \in \mathcal{A}} (-1)^{\frac{|A|\,(|A|-1)}{2}} x_A e_A \tag{10}$$

$$\bar{x} = \sum_{A \in \mathcal{A}} (-1)^{\frac{|A|\,(|A|+1)}{2}} x_A e_A\,. \tag{11}$$

After all these preliminaries we shall look at some Clifford algebra concretely. Any Clifford algebra is isomorphic to a matrix algebra. Table 1 gives an overview in low dimensions, containing the real numbers ($\mathcal{C}_{0,0}$), the complex numbers ($\mathcal{C}_{0,1}$) and the quaternions ($\mathcal{C}_{0,2}$). In this cases we can directly define meaningful Clifford neurons via the algebra multiplication. Complex NNs are well known (see e.g. [7]), whereas the study of quaternionic NNs just started recently [1]. However, our focus here is on the neural computation of geometric transformations. So far, we have only embedded quadratic spaces in Clifford algebras. Our next goal is to determine elements of $\mathcal{C}_{p,q}$ that act as geometric transformations. Therefore, we have to study transformation groups that act on a special subspace $\lambda\mathbb{R}^{p,q} := \mathbb{R} \oplus \mathbb{R}^{p,q}$ of $\mathcal{C}_{p,q}$.

$p\backslash q$	0	1	2	3	4
0	$\mathbb{R}$	$\mathbb{C}$	$\mathbb{H}$	$^2\mathbb{H}$	$\mathbb{H}(2)$
1	$^2\mathbb{R}$	$\mathbb{R}(2)$	$\mathbb{C}(2)$	$\mathbb{H}(2)$	$^2\mathbb{H}(2)$
2	$\mathbb{R}(2)$	$^2\mathbb{R}(2)$	$\mathbb{R}(4)$	$\mathbb{C}(4)$	$\mathbb{H}(4)$
3	$\mathbb{C}(2)$	$\mathbb{R}(4)$	$^2\mathbb{R}(4)$	$\mathbb{R}(8)$	$\mathbb{C}(8)$
4	$\mathbb{H}(2)$	$\mathbb{C}(4)$	$\mathbb{R}(8)$	$^2\mathbb{R}(8)$	$\mathbb{R}(16)$

Table 1. Low dimensional Clifford algebras
(^{2} denotes direct product; R(x) denotes x×x matrices)

Definition 2 *The Clifford group* $\Gamma_{\mathrm{p,q}}$ *of a Clifford algebra* $\mathcal{C}_{p,q}$ *is defined by its action on* $\lambda\mathbb{R}^{p,q}$ *as*

$$\Gamma_{\mathrm{p,q}} := \{s \in \mathcal{C}_{p,q}, s \otimes_{p,q} \tilde{s} = \pm 1 \mid \forall x \in \lambda\mathbb{R}^{p,q} : s \otimes_{p,q} x \otimes_{p,q} \hat{s}^{-1} \in \lambda\mathbb{R}^{p,q}\}. \quad (12)$$

This action is in fact an orthogonal transformation of $\mathbb{R}^{p,q+1}$.

Theorem 1 *The map* $\Psi_s : \Gamma_{\mathrm{p,q}} \to O(p, q+1)$; $s \mapsto \psi_s$ *is a group epimorphism, whereby* ψ_s *denotes the function* $x \mapsto s \otimes_{p,q} x \otimes_{p,q} \hat{s}^{-1}$.

Thus, ψ_s implemented as shown in Figure 4 gives a neuron with only one weight s that computes an orthogonal transformation according to Theorem 1.

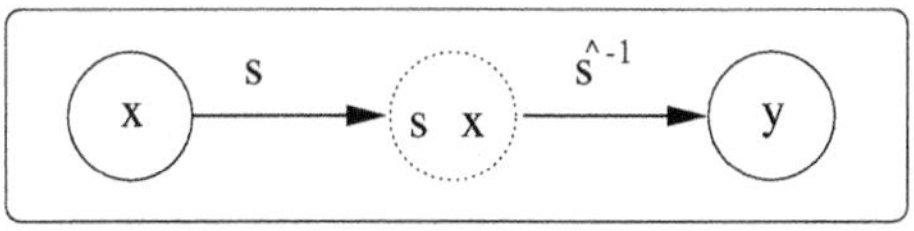

Fig. 4. Clifford neuron (CN)

Let us briefly review the first non trivial case of quaternions $\mathcal{C}_{0,2}$. For a more detailed study the interested reader is refered to [2], which also discusses the complex case that reduces to one multiplication.
Let q_0 denote the scalar part of a quaternion q and $\boldsymbol{q}$ its vector part, respectively. The update rule for the weight s of a quaternionic Clifford neuron is given by

$$\Delta s = -\eta \frac{\partial E}{\partial s} = \eta \sum_{i=1}^{m} (t^i - y^i) \otimes_{0,2} \delta, \quad (13)$$

$\delta = \nabla(s_0^2 q_0 + (\boldsymbol{s} \cdot \boldsymbol{s}) q_0,\ s_0^2 \boldsymbol{q} + (\boldsymbol{s} \cdot \boldsymbol{q})\boldsymbol{s} + 2 s_0 (\boldsymbol{s} \times \boldsymbol{q}) - \boldsymbol{s} \times \boldsymbol{q} \times \boldsymbol{s})$. For illustration purposes we made the following simulation with a quaternionic CN and a 3×3 SL-NN. The given task was to learn the transformation consisting of rotation of -60° about the axis [0.5, $\sqrt{0.5}$, 0.5] and translation about [0.2,-0.2,0.3] from 5 randomly chosen points with added normal noise of 20%. Figure 5 gives the obtained results from the trained networks of a test object.

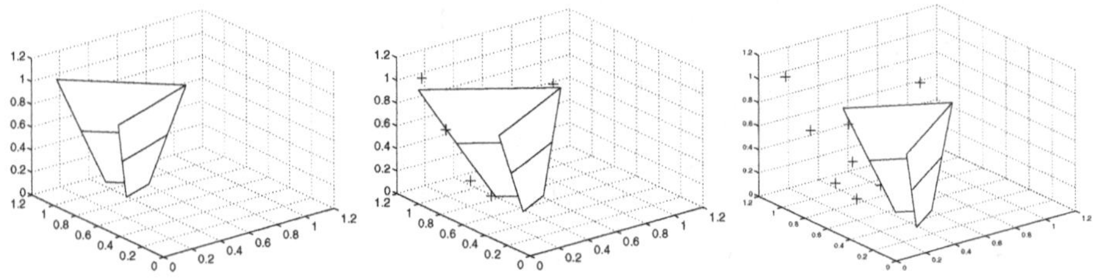

Fig. 5. Expected output test data (left), output quaternionic CN (middle) and output SL-NN (right)

Obviously, the SL–NN just learned the noise in the data which then resulted in a complete wrong pose of the test object. This means that no generalization was possible. On the other hand, the generalization performance of the quaternionic CN was quiet satisfactorily. Figure 6 showing numerically training and generalization error by different noise levels confirms the drawn conclusions.

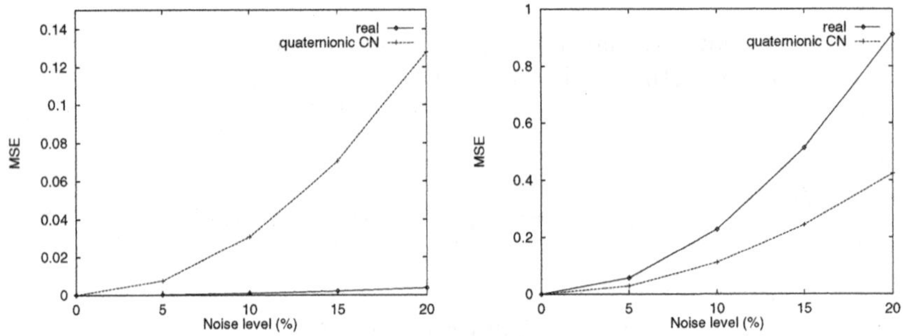

Fig. 6. Training errors (left) and generalization errors (right) by different noise levels

In the next section we will see exemplarily how the introduced Clifford neurons can be used to compute non–linear geometric transformations.

4 Computing the Cross–Ratio with Clifford Neurons

The computation of invariants of geometric transformations is fundamental in many areas of pattern recognition and computer vision. In the latter, a particular interesting class is those of projective invariants. The basic projective invariant is the cross–ratio from which many further invariants can be derived [15].
The cross–ratio of 4 points $z, q, r, s \in \mathbb{C}$ is given by

$$[z, q, r, s] = \frac{(z-q)(r-s)}{(z-s)(r-q)} . \tag{14}$$

The cross–ratio depends on the order of the points. The given definition is very common, but other definitions are common too [11]. To our knowledge, there are no neural algorithms for the computation of cross–ratios proposed so far.
In this section we will show how this attractive non–linear task can be computed by a single Clifford neuron. Thereby, we will use the fact that the cross–ratio is closely related to Möbius transformations. A Möbius (or linear–fractional) transformation is a conformal mapping of the extended complex plane of the form

$$w = M(z) = \frac{a + b\,i}{c + d\,i}, \tag{15}$$

whereby a, b, c, d are complex constants such that $ad - bc \neq 0$. The group of that transformations will be denoted by $\mathcal{M}$. A Möbius transformation is uniquely determined by 3 given points. Moreover, the cross–ratio is invariant under Möbius transformations and two given cross–ratios are related by a unique Möbius transformation vice versa. Thus, one obtains the following known theorem (see e.g. [11]).

Theorem 2 *The cross–ratio* [z,q,r,s] *is the image of z under that Möbius transformation that maps q to 0, r to 1 and s to ∞, respectively.*

Applying the group isomorphisms $\mathcal{M} \cong O(1,2) \cong \Gamma_{1,2}$ [12] we can study Möbius transformations in the 8–dimensional Clifford algebra $\mathcal{C}_{1,2}$. According to Table 1, $\mathcal{C}_{1,2}$ is isomorphic to the matrix algebra $\mathbb{C}(2)$. First, we have to find an appropriate coding of complex numbers in $\mathcal{C}_{1,2}$. This coding is given by [2]

$$co(z) = (Re(z), \frac{1}{2}(1 + z\bar{z}), Im(z), \frac{1}{2}(1 - z\bar{z}), 0, 0, 0, 0) \tag{16}$$

or in matrix notation by

$$Co(z) = \begin{pmatrix} z & z\bar{z} \\ 1 & \bar{z} \end{pmatrix}. \tag{17}$$

Finally, the Clifford neuron that computes the cross–ratio via the corresponding Möbius transformation has the form

$$y = s \otimes_{1,2} c(z) \otimes_{1,2} \tilde{s}, \tag{18}$$

from which we get

$$Co(y) = \lambda \begin{pmatrix} z' & z'\bar{z'} \\ 1 & \bar{z'} \end{pmatrix}, \tag{19}$$

with $\lambda = |cz + d|^2$ and $z' = (az + b)/(cz + d)$ [2].
Due to the limited space we have to omit the derivation of the learning algorithm here. In order to test our proposed Clifford neuron we made the following little experiment.
The task was to compute the cross–ratio [z,q,r,s] with $q = 0.2 + 0.3\,i, r = 0.4 - 0.7\,i$, and $s = 0.6 - 0.2\,i$. The numerical problem we had to face was of course the coding of ∞ to which the point s should be mapped. After some trials we have found out, that 10^{15} was a reasonable value with respect to both stability

of learning algorithm and accuracy of the results. However, we still had to use a very low learning rate $\eta = 0.00001$ in order to guarantee stability. With this rate the convergence of learning in terms of epochs was only slow, as reported in Figure 7.

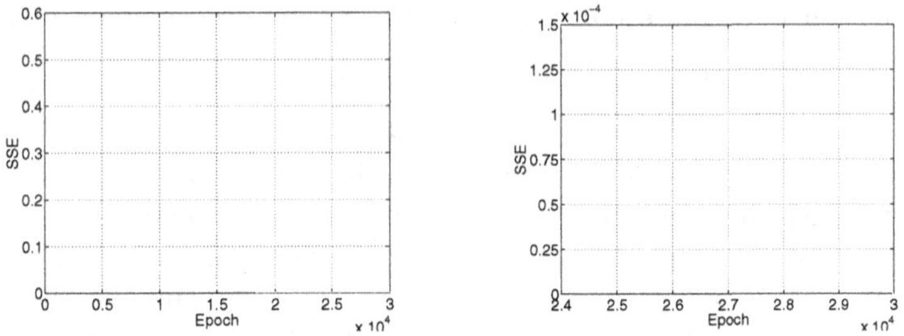

Fig. 7. Convergence of learning: all epochs (left) and last epochs only (right)

But, computational time was less than 10 seconds. The SSE dropped under 0.00001 after 30000 epochs, which gave a weight vector corresponding to the transformation parameters in Table 2. The difference of the learned transformation parameters to the real values was quiet acceptable. Tabel 3 gives as confirmation the results obtained for some test points.

Parameter	Value	Learned parameters
a	0.2 + 0.5i	0.20019 + 0.50010i
b	0.11-0.16i	0.11001 - 0.15992i
c	-0.2+i	-0.20079 + 0.99939i
d	-0.08-0.64i	-0.07930 - 0.64007i

Table 2. Transformation parameters

z	Value	Clifford neuron output
2+3i	0.3578-0.3357i	0.3577-0.3364i
4-7i	0.4838-0.3044i	0.4838-0.3051i
0.3+0.1i	0.0077-0.6884i	-0.0082+0.6884i

Table 3. Cross–Ratios of test points (rounded)

The achieved accuracy could still be improved by longer learning times. The Clifford neuron behaved very well in the experiment, so limitation is machine accuracy only.

5 Conclusion

In this paper we presented novel artificial neurons in the framework of Clifford algebra. We showed how they cover complex and quaternionic neurons and how they are related to orthogonal transformations. The potential power of the proposed Clifford neurons was illustrated by the example of computing the cross-ratio with a single Clifford neuron, which is a non–linear task. Future work will concentrate on enlarging the functional basis of Clifford neurons (projective transformations) and then build up networks of it for applications with the emphasis on object recognition.

References

1. P. Arena, L. Fortuna, G. Muscato, and M. G. Xibilia. *Neural Networks in Multidimensional Domains.* Number 234 in LNCIS. Springer–Verlag, 1998.
2. S. Buchholz and G. Sommer. *Introduction to Neural Computation in Clifford Algebra,* chapter 13. G. Sommer (Ed.), Geometric Computing with Clifford Algebra. Springer–Verlag, 2000 (to appear).
3. T. Caelli, D. Squire, and T. Wild. Model–Based Neural Networks. *Neural Networks,* 6:613–625, 1993.
4. G. Cybenko. Approximation by superposition of a sigmoidal function. *Mathematics of Control, Signals and Systems,* 2:303–314, 1989.
5. M. Ferraro and T. Caelli. Neural computation of algebraic and geometrical structures. *Neural Networks,* 11:669–707, 1998.
6. S. Geman, E. Bienenstock, and R. Doursat. Neural networks and the bias/variance dilemma. *Neural Computation,* 4:1–58, 1992.
7. G. Georgiou and C. Koutsougeras. Complex domain backpropagation. *IEEE Trans. Circ. and Syst. II,* 39:330–334, 1992.
8. P. Lounesto. *Clifford Algebras and Spinors.* Cambridge University Press, 1997.
9. T. Masters. *Signal and Image Processing with Neural Networks.* John Willey and Sons, 1994.
10. W. S. McCulloch and W. Pitts. A Logical Calculus of Ideas Immanent in Nervous Activity. *Bulletin of Mathematical Biophysics,* 5:115–133, 1943.
11. T. Needham. *Visual Complex Analysis.* Clarendon Press, Oxford, 1997.
12. I. R. Porteous. *Clifford Algebras and the Classical Groups.* Cambridge University Press, Cambridge, 1995.
13. D.E. Rumelhart, G.E. Hinton, and R.J. Williams. Learning internal representations by error propagation. In *Parallel Distributed Processing,* volume 1. MIT Press, Cambridge MA, 1986.
14. M. Vidyasagar. *A Theory of Learning and Generalization.* Springer–Verlag, 1997.
15. K. Voss and H. Süsse. *Adaptive Modelle und Invarianten für zweidimensionale Bilder.* Verlag Shaker, Aachen, 1995.

Hurwitzion Algebra and its Application to the FFT Synthesis *

Vladimir M. Chernov
Image Processing Systems Institute (IPSI RAS),
151 Molodogvardejskaya St, 443001, Samara, Russia
e-mail: vche@smr.ru

Abstract

The main idea of the paper is that fast algorithms, like FFT, can be made more efficient in the context of an algebra, rather than in the more singular quaternion or complex algebras structure. However, the complex algebra structure can then be recovered as a projection from the larger algebra in which it is embedded. Namely, the 12-dimensional algebra (hurwitzion algebra) having the basis elements associated with the integer Hurwitz quaternions is introduced. The computational aspects of the hurwitzion arithmetic are considered. The overlapped fast algorithms of two-dimensional discrete Fourier transform of an RGB image are also developed.

Keywords: quaternion algebra, hurwitzion, FFT

1 Introduction

The basis for the well-known "overlapped" one-dimensional FFT [1] is the possibility of obtaining additional computational advantages at the expense of the redundancy of representation of complex basis functions with respect to a real input signal $x(n)$. Putting it more exactly, the possibility of constructing overlapped algorithms exists due to the presence of a non-trivial automorphism (the complex conjugation) of complex field $\mathbf{C}$, acting identically upon $\mathbf{R}$.

Actually, let

$$\hat{x}(m) = \sum_{n=0}^{N-1} x(n) \exp\{2\pi i \frac{mn}{N}\}, \quad m = 0, \ldots, N-1, \quad N = 2^r, \quad x(n) \in \mathbf{R}. \tag{1}$$

*This work was performed with financial support from the Russian Foundation for Basic Research (Grant 00-01-00600).

G. Sommer and Y. Y. Zeevi (Eds.): AFPAC 2000, LNCS 1888, pp. 154-163, 2000.

Let us form an auxiliary sequence

$$z(n) = x(2n) + ix(2n+1) = x_0(n) + ix_1(n). \tag{2}$$

$$\hat{z}(m) = \sum_{n=0}^{\frac{N}{2}-1} z(n) \exp\{2\pi i \frac{2mn}{N}\}, \quad m = 0, \ldots, \frac{N}{2} - 1.$$

Then, "partial spectra"

$$\hat{x}_0(m) = \sum_{n=0}^{\frac{N}{2}-1} x_0(n) \exp\{2\pi i \frac{2mn}{N}\}, \quad \hat{x}_1(m) = \sum_{n=0}^{\frac{N}{2}-1} x_1(n) \exp\{2\pi i \frac{2mn}{N}\} \tag{3}$$

can be found from the relations

$$2\hat{x}_0(m) = \hat{z}(m) + \overline{\hat{z}(-m)}, \quad 2i\hat{x}_1(m) = \hat{z}(m) - \overline{\hat{z}(-m)}. \tag{4}$$

The full spectrum reconstruction is realized by relation

$$\hat{x}(m) = \hat{x}_0(m) + \hat{x}_1(m) \exp\{2\pi i \frac{2m}{N}\}. \tag{5}$$

Note that the computer-aided transition to the complex-conjugate number does not result in additional arithmetic operations.

The majority of fast algorithms (of Cooley-Tuckey type) of the discrete Fourier transform (DFT) have the complexity

$$W(N) = \lambda N \log_2 N + O(N),$$

where the constant λ characterizes a particular scheme of the algorithm [5]. Thus, the complexity of the "ovelapped" FFT is

$$W(N) = \frac{1}{2} \lambda N \log_2 N + O(N).$$

If the technique discussed above is used for multidimensional DFTs (in particular 2D transforms) then certain difficulties may be encountered: the field **C** has "too few" automorphisms admitting repeated overlaps for each argument with the possibility of the subsequent separation of spectra. This reasoning leads to the necessity of immersion of the field **R** into algebraic structures possessing a sufficiently large number of trivially implemented automorphisms over **R**.

In some papers (e.g. [7], [8], [6]) I used this approach to solve the problem of DFT fast algorithms synthesis. The idea of this approach is as follows: we explicitly construct the immersion of the complex field into the algebraic structures having sufficient number of trivially implemented automorphisms. Somewhat increased computational complexity of the main arithmetic operations in this case can be compensated by the possibility of the "overlapped" calculation of spectrum portions. This is guaranteed by the existence of the above automorphisms. The interpretation of the obtained results of an auxiliary discrete transform (ADT) with values in a chosen algebraic structure is a homomorphism. In a number of cases, the calculational complexity of such an interpretation is asymptotically negligible.

In particular, the FFTs with the overlapped two-dimensional DFT-spectrum calculation in the four-dimensional quaternion algebra are considered in [7], [8].

The tasks similar to those of DFT fast algorithm complexity reduction (the task of simultaneous "overlapped" calculation of (several) DFT-spectra for multichannel signal) are solved using similar methods.

Actually, $x_0(n)$ and $x_1(n)$ can be treated as two independent sequences. Thus, using (3) and (4) and omitting (5) we can reconstruct the full spectrum.

In order to calculate three complex spectra simultaneously the algebra of at least three dimensions is needed. Thus, to simultaneously calculate two-dimensional spectrum of a real numbers array the 12-dimensional $\mathbf{R}$-algebra with the sufficient number of trivially implemented automorthisms is needed. The algebra having the basis associated with the integer Hurwitz quaternions can be used for this purpose.

2 Algebra of hurwitzions: definition and some properties

According to [2], [3], [4] we shall give the following definition.

Definition 1 *The* ***algebra of hurwitzions*** *is a twelve-dimensional* $\mathbf{R}$*-algebra* **Hu** *with the basis* $\mathcal{E}$ *given by*

$$\mathcal{E} = \{\mathbf{e}, \mathbf{i}, \mathbf{j}, \mathbf{k}, \mathbf{w}, \mathbf{w}^{i}, \mathbf{w}^{j}, \mathbf{w}^{k}, \bar{\mathbf{w}}, \bar{\mathbf{w}}^{i}, \bar{\mathbf{w}}^{j}, \bar{\mathbf{w}}^{k}\} \tag{6}$$

and the multiplication rules induced by multiplication rules for the elements of set $\mathcal{E}$ *(see Table 1). The elements of hurwitzion algebra are called the* ***hurwitzions.***

TABLE 1

e	i	j	k	$\mathbf{w}$	$\mathbf{w}^{i}$	$\mathbf{w}^{j}$	$\mathbf{w}^{k}$	$\bar{\mathbf{w}}$	$\bar{\mathbf{w}}^{i}$	$\bar{\mathbf{w}}^{j}$	$\bar{\mathbf{w}}^{k}$
i	-e	k	-j	$\mathbf{w}^{k}$	$\mathbf{w}^{j}$	$-\mathbf{w}^{i}$	$-\mathbf{w}$	$-\bar{\mathbf{w}}^{j}$	$-\bar{\mathbf{w}}^{k}$	$\bar{\mathbf{w}}$	$\bar{\mathbf{w}}^{i}$
j	-k	-e	i	$\mathbf{w}^{i}$	$-\mathbf{w}$	$\mathbf{w}^{k}$	$-\mathbf{w}^{j}$	$-\bar{\mathbf{w}}^{k}$	$\bar{\mathbf{w}}^{j}$	$-\bar{\mathbf{w}}^{i}$	$\bar{\mathbf{w}}$
k	j	-i	-e	$\mathbf{w}^{j}$	$-\mathbf{w}^{k}$	$-\mathbf{w}$	$\mathbf{w}^{i}$	$-\bar{\mathbf{w}}^{i}$	$\bar{\mathbf{w}}$	$\bar{\mathbf{w}}^{k}$	$-\bar{\mathbf{w}}^{j}$
$\mathbf{w}$	$\mathbf{w}^{j}$	$\mathbf{w}^{k}$	$\mathbf{w}^{i}$	$\bar{\mathbf{w}}$	$-\bar{\mathbf{w}}^{j}$	$-\bar{\mathbf{w}}^{k}$	$-\bar{\mathbf{w}}^{i}$	e	-j	-k	-i
$\mathbf{w}^{i}$	$\mathbf{w}^{k}$	$-\mathbf{w}^{j}$	$-\mathbf{w}$	$-\bar{\mathbf{w}}^{k}$	$\bar{\mathbf{w}}^{i}$	$-\bar{\mathbf{w}}$	$-\bar{\mathbf{w}}^{j}$	j	e	-i	k
$\mathbf{w}^{j}$	$-\mathbf{w}$	$\mathbf{w}^{i}$	$-\mathbf{w}^{k}$	$-\bar{\mathbf{w}}^{i}$	$-\bar{\mathbf{w}}^{k}$	$\bar{\mathbf{w}}^{j}$	$-\bar{\mathbf{w}}$	k	i	e	-j
$\mathbf{w}^{k}$	$-\mathbf{w}^{i}$	$-\mathbf{w}$	$\mathbf{w}^{j}$	$-\bar{\mathbf{w}}^{j}$	$-\bar{\mathbf{w}}$	$-\bar{\mathbf{w}}^{i}$	$\bar{\mathbf{w}}^{k}$	i	-k	j	e
$\bar{\mathbf{w}}$	$-\bar{\mathbf{w}}^{k}$	$-\bar{\mathbf{w}}^{i}$	$-\bar{\mathbf{w}}^{j}$	e	k	i	j	$\mathbf{w}$	$-\mathbf{w}^{k}$	$-\mathbf{w}^{i}$	$-\mathbf{w}^{j}$
$\bar{\mathbf{w}}^{i}$	$-\bar{\mathbf{w}}^{j}$	$\bar{\mathbf{w}}$	$\bar{\mathbf{w}}^{k}$	-k	e	-j	i	$-\mathbf{w}^{j}$	$\mathbf{w}^{i}$	$-\mathbf{w}^{k}$	$-\mathbf{w}$
$\bar{\mathbf{w}}^{j}$	$\bar{\mathbf{w}}^{i}$	$-\bar{\mathbf{w}}^{k}$	$\bar{\mathbf{w}}$	-i	j	e	-k	$-\mathbf{w}^{k}$	$-\mathbf{w}$	$\mathbf{w}^{j}$	$-\mathbf{w}^{i}$
$\bar{\mathbf{w}}^{k}$	$\bar{\mathbf{w}}$	$\bar{\mathbf{w}}^{j}$	$-\bar{\mathbf{w}}^{i}$	-j	-i	k	e	$-\mathbf{w}^{i}$	$-\mathbf{w}^{j}$	$-\mathbf{w}$	$\mathbf{w}^{k}$

In other words, the element $\mathbf{i}$ (and elements $\mathbf{j}, \mathbf{k}$) "behaves " as quaternion i while it is not such.

The proofs of Propositions 1-5 and Proposition 7 given below can be obtained by direct calculations.

Proposition 1 *The mapping* $\Psi : \mathbf{Hu} \to \mathbf{H}$ *such that*

$$\Psi : \mathbf{e} \longmapsto 1, \quad \Psi : \mathbf{i} \longmapsto i, \quad \Psi : \mathbf{j} \longmapsto j, \quad \Psi : \mathbf{k} \longmapsto k,$$

$$\begin{array}{ll} \mathbf{w} = \mathbf{e}^{-1}\mathbf{we} \longmapsto \frac{1}{2}(-1+i+j+k), & \bar{\mathbf{w}} = \mathbf{e}^{-1}\bar{\mathbf{w}}\mathbf{e} \longmapsto \frac{1}{2}(-1-i-j-k), \\ \mathbf{w}^i = \mathbf{i}^{-1}\mathbf{wi} \longmapsto \frac{1}{2}(-1+i-j-k), & \bar{\mathbf{w}} = \mathbf{i}^{-1}\bar{\mathbf{w}}\mathbf{i} \longmapsto \frac{1}{2}(-1-i+j+k), \\ \mathbf{w}^i = \mathbf{j}^{-1}\mathbf{wj} \longmapsto \frac{1}{2}(-1-i+j-k), & \bar{\mathbf{w}} = \mathbf{j}^{-1}\bar{\mathbf{w}}\mathbf{j} \longmapsto \frac{1}{2}(-1+i-j+k), \\ \mathbf{w}^i = \mathbf{k}^{-1}\mathbf{wk} \longmapsto \frac{1}{2}(-1-i-j+k), & \bar{\mathbf{w}} = \mathbf{k}^{-1}\bar{\mathbf{w}}\mathbf{k} \longmapsto \frac{1}{2}(-1+i+j-k), \end{array}$$

can be linearly extended up to homomorphism of algebra $\mathbf{Hu}$ *into algebra* $\mathbf{H}$.

The homomorphism (projection) Ψ of hurwitzion algebra $\mathbf{Hu}$ into quaternion algebra $\mathbf{H}$ transforms the elements $\mathbf{i}, \mathbf{j}, \mathbf{k}$ into the "true" quaternions i, j, k.

Proposition 2 *The set*

$$\mathbf{H}^*\mathbf{u} = \{x\mathbf{e}+y(\mathbf{i}+\mathbf{j}+\mathbf{k}), \quad x, y \in \mathbf{R}\}$$

is a subalgebra of the algebra $\mathbf{Hu}$.

Proposition 3 *The hurwitzion* $\mathbf{W} \in \mathbf{H}^*\mathbf{u}$:

$$\mathbf{W} = \mathbf{e}\cos\frac{2\pi}{K} + \frac{1}{\sqrt{3}}(\mathbf{i}+\mathbf{j}+\mathbf{k})\sin\frac{2\pi}{K} \tag{7}$$

is the Kth primitive root of unit $\mathbf{e} \in \mathbf{Hu}$. *The quaternion*

$$w = \Psi(W) = \cos\frac{2\pi}{K} + \frac{1}{\sqrt{3}}(i+j+k)\sin\frac{2\pi}{K} \tag{8}$$

is the Kth primitive root of unit $1 \in \mathbf{H}$.

Proposition 4 *Let*

$$\mathbf{V} = \mathbf{w}\left(\frac{1}{\sqrt{3}}\sin\frac{2\pi}{K} - \cos\frac{2\pi}{K}\right) - \bar{\mathbf{w}}\left(\frac{1}{\sqrt{3}}\sin\frac{2\pi}{K} + \cos\frac{2\pi}{K}\right) \tag{9}$$

then $\Psi(\mathbf{W}) = \Psi(\mathbf{V}) = w$.

Proposition 5 *There exists such an element* $h \in \mathbf{H}$ *that*

$$h^{-1}wh = \cos\frac{2\pi}{K} + i\sin\frac{2\pi}{K} = \omega. \tag{10}$$

Proposition 6 *Let* $\mathbf{X}, \mathbf{V} \in \mathbf{Hu}$ *be such that*

$$\mathbf{X} = A\mathbf{e}+B\mathbf{i}+C\mathbf{j}+D\mathbf{k}+E\mathbf{w}+F\mathbf{w}^i+G\mathbf{w}^j+H\mathbf{w}^k+P\bar{\mathbf{w}}+Q\bar{\mathbf{w}}^i+R\bar{\mathbf{w}}^j+S\bar{\mathbf{w}}^k, \tag{11}$$

$$\mathbf{V} = x\mathbf{w}+y\bar{\mathbf{w}}. \tag{12}$$

Then, the calculation of product $\mathbf{XV}$ *requires only 16 real multiplications.*

Proof. Using Table 1, we get

$$\begin{aligned}\mathbf{XV} = & \left[(Px+Ey)\,\mathbf{e}+(Ax+Py)\,\mathbf{w}+(Ex+Ay)\,\bar{\mathbf{w}}\right] \\ & +\left[(-Rx+Hy)\,\mathbf{i}+(-Hx-By)\,\bar{\mathbf{w}}^i+(Bx-Ry)\,\mathbf{w}^k\right] \\ & +\left[(-Sx+Fy)\,\mathbf{j}+(Cx-Sy)\,\mathbf{w}^i+(-Fx-Cy)\,\bar{\mathbf{w}}^k\right] \\ & +\left[(-Qx+Gy)\,\mathbf{k}+(Dx-Qy)\,\mathbf{w}^j+(-Gx-Dy)\,\bar{\mathbf{w}}^i\right].\end{aligned} \tag{13}$$

The elements ζ,η,ξ :

$$\zeta=\beta x+\gamma y, \quad \eta=\alpha y+\gamma x, \quad \xi=\alpha x+\beta y.$$

are components of a three-point convolution

$$\begin{pmatrix}\zeta\\ \eta\\ \xi\end{pmatrix}=\begin{pmatrix}0 & x & y\\ y & 0 & x\\ x & y & 0\end{pmatrix}\begin{pmatrix}\alpha\\ \beta\\ \gamma\end{pmatrix}.$$

According to [5], [10], in order to calculate ζ,η,ξ four real multiplications are required. ■

By $\mathrm{Rot}_{\mathbf{a}}(\mathbf{x})$ denote the mapping (an automorphism)

$$\mathrm{Rot}_{\mathbf{a}}(\mathbf{x})=\mathbf{a}^{-1}\mathbf{x}\mathbf{a}, \quad \mathbf{x},\mathbf{a}\in\mathbf{Hu}.$$

Table 2 defines the $\mathrm{Rot}_{\mathbf{a}}(\mathbf{x})$ mapping action for $\mathbf{x},\mathbf{a}\in\mathcal{E}$. Columns of Table 1 are the permutations of the element (11) components induced by the $\mathrm{Rot}_{\mathbf{a}}(\mathbf{x})$ actions.

TABLE 2

a \ x	e	i	j	k	w	$\mathbf{w}^i$	$\mathbf{w}^j$	$\mathbf{w}^k$	$\bar{\mathbf{w}}$	$\bar{\mathbf{w}}^i$	$\bar{\mathbf{w}}^j$	$\bar{\mathbf{w}}^k$
e	A	B	C	D	E	F	G	H	P	Q	R	S
i	A	B	-C	-D	F	E	H	-G	Q	P	S	-R
j	A	-B	C	-D	G	-H	E	F	R	-S	P	Q
k	A	-B	-C	D	H	G	-F	E	S	R	-Q	P
w	A	C	D	B	E	G	H	F	P	R	S	Q
$\mathbf{w}^i$	A	-C	D	-B	H	F	E	G	S	Q	P	R
$\mathbf{w}^j$	A	-C	-D	B	F	H	G	E	Q	S	R	P
$\mathbf{w}^k$	A	C	-D	-B	G	E	F	H	R	P	Q	S
$\bar{\mathbf{w}}$	A	D	B	C	E	H	F	G	P	S	Q	R
$\bar{\mathbf{w}}^i$	A	D	-B	-C	G	F	H	E	R	Q	S	P
$\bar{\mathbf{w}}^j$	A	-D	B	-C	H	E	G	F	S	P	R	Q
$\bar{\mathbf{w}}^k$	A	-D	-B	C	F	G	E	H	Q	R	P	S

Remark 1 *The group of* $\mathrm{Rot}_{\mathbf{a}}(\mathbf{x})$ *mapping actions on algebra* **Hu** *is easy to interpret geometrically. Let*

$$\mathfrak{R}=\{\pm\mathrm{Rot}_{\mathbf{a}}(\bullet), \quad \mathbf{a}\in\mathcal{E}\},$$

then the group $\mathfrak{R}$ is isomorphic to the group of self-coincidences of "quaternion cube"

$$\Delta_{\pm 1} = \{\delta : \delta = \pm i \pm j \pm k\} \subset \mathbf{H}.$$

Proposition 7 *Let the rows of matrix* $\mathbf{T}$ *be equal to the rows of Table 2. Then* Rank $\mathbf{T} = 12$.

3 Fast algorithm for calculating three DFT- spectra with overlapping in the hurwitzion algebra

Let $x^{(\alpha)}(n_1, n_2)$ be $RGB-$components of a color image, $\alpha = 0, 1, 2$; $N = 2K = 2^r$. Let $\{\hat{x}^{(\alpha)}(m_1, m_2)\}$ be three DFT-spectra:

$$\hat{x}^{(\alpha)}(m_1, m_2) = \sum_{n_1=0}^{N-1} \sum_{n_2=0}^{N-1} x^{(\alpha)}(n_1, n_2) e^{2\pi i \frac{n_1 m_1 + n_2 m_2}{N}}; \; m_1, m_2 = 0, ..., N-1.$$

We claim that the following Theorem 1 is true.

Theorem 8 *There is an algorithm for calculating of three discrete two-dimensional Fourier spectra, which requires no more than*

$$M^*(N^2) = N^2 \log_2 N + O(N^2) \tag{14}$$

real multiplications for each spectrum.

An analysis of the structure of the described DFT fast algorithms allows us to mark out six steps of this algorithm.

Step 1. Creation of an auxiliary hurwitzion-valued sequence.

By $x_{\beta\gamma}^{(\alpha)}(n_1, n_2), \mathbf{X}^{(1)}(n_1, n_2)$ denote

$$x_{\beta\gamma}^{(\alpha)}(n_1, n_2) = x_{\beta\gamma}^{(\alpha)}(2n_1 + \alpha, 2n_2 + \beta) \quad \alpha = 0, 1, 2; \; \beta, \gamma = 0, 1;$$

$$\begin{aligned} \mathbf{X}^{(1)}(n_1, n_2) &= \left(x_{00}^{(1)}\mathbf{e} + x_{10}^{(1)}\mathbf{i} + x_{01}^{(1)}\mathbf{j} + x_{11}^{(1)}\mathbf{k}\right)(n_1, n_2), \\ \mathbf{X}^{(2)}(n_1, n_2) &= \left(x_{00}^{(1)}\mathbf{w} + x_{10}^{(1)}\mathbf{w}^i + x_{01}^{(1)}\mathbf{w}^j + x_{11}^{(1)}\right)(n_1, n_2), \\ \mathbf{X}^{(3)}(n_1, n_2) &= \left(x_{00}^{(1)}\bar{\mathbf{w}} + x_{10}^{(1)}\bar{\mathbf{w}}^i + x_{01}^{(1)}\bar{\mathbf{w}}^j + x_{11}^{(1)}\bar{\mathbf{w}}^k\right)(n_1, n_2), \end{aligned}$$

Consider an auxiliary hurwitzion-valued sequence such that

$$\mathbf{X}(n_1, n_2) = \mathbf{X}^{(1)}(n_1, n_2) + \mathbf{X}^{(2)}(n_1, n_2) + \mathbf{X}^{(3)}(n_1, n_2).$$

Step 2. Calculation of an auxiliary discrete transform (ADT) .

Let $\mathbf{V}$ be denoted by (9). Consider an auxiliary hurwitzion-valued discrete transform (ADT) such that

$$\hat{\mathbf{X}}(m_1, m_2) = \sum_{n_1=0}^{K-1} \sum_{n_2=0}^{K-1} \mathbf{X}(n_1, n_2) \mathbf{V}^{n_1 m_1 + n_2 m_2}. \tag{15}$$

Using Radix-2 DFT decomposition scheme [5]

$$\hat{\mathbf{X}}(m_1,m_2)=\sum_{a,b=0}^{1}\mathbf{V}^{am_1+bm_2}\sum_{n_1=0}^{K-1}\sum_{n_2=0}^{K-1}\mathbf{X}(2n_1+a,2n_2+b)\mathbf{V}^{2(n_1m_1+n_2m_2)},$$

we get the estimate of multiplicative complexity

$$\mu(K^2)=12K^2\log_2 K+O(K^2).$$

Step 3. Separation of the "partial ADT-spectra".

It follows from Proposition 7 that "partial ADT-spectra"

$$\hat{\mathbf{X}}^{(\alpha)}_{\beta\gamma}(m_1,m_2)=\sum_{n_1=0}^{K-1}\sum_{n_2=0}^{K-1}x^{(\alpha)}_{\beta\gamma}(n_1,n_2)\mathbf{V}^{n_1m_1+n_2m_2},\quad \alpha=0,1,2;\ \beta,\gamma=0,1$$

are uniquely determined by the system of equations

$$\mathrm{Rot}_{\mathbf{a}}\left(\hat{\mathbf{X}}(m_1,m_2)\right)=\mathrm{Rot}_{\mathbf{a}}\left(\sum_{n_1=0}^{K-1}\sum_{n_2=0}^{K-1}\mathbf{X}(n_1,n_2)\mathbf{V}^{n_1m_1+n_2m_2}\right),\quad \mathbf{a}\in\mathcal{E}.$$

For calculation of $\hat{\mathbf{X}}^{(\alpha)}_{\beta\gamma}(m_1,m_2)$ only $O(N^2)$ real arithmetical operations are required.

Step 4. Separation of the "partial quaternion spectra".

Under the conditions of Proposition 4 the mapping Ψ transforms $\hat{\mathbf{X}}^{(\alpha)}_{\beta\gamma}(m_1,m_2)$ into

$$\hat{X}^{(\alpha)}_{\beta\gamma}(m_1,m_2)=\sum_{n_1=0}^{K-1}\sum_{n_2=0}^{K-1}x^{(\alpha)}_{\beta\gamma}(n_1,n_2)w^{n_1m_1+n_2m_2},\quad \alpha=0,1,2;\ \beta,\gamma=0,1$$

The mapping requires no real multiplications.

Step5. Separation of the "partial complex spectra".

Under the conditions of Proposition 5 the mapping

$$\hat{X}^{(\alpha)}_{\beta\gamma}(m_1,m_2)\longmapsto h^{-1}\hat{X}^{(\alpha)}_{\beta\gamma}(m_1,m_2)h$$

transforms $\hat{X}^{(\alpha)}_{\beta\gamma}(m_1,m_2)$ into

$$\hat{x}^{(\alpha)}_{\beta\gamma}(m_1,m_2)=\sum_{n_1=0}^{K-1}\sum_{n_2=0}^{K-1}x^{(\alpha)}_{\beta\gamma}(n_1,n_2)\omega^{n_1m_1+n_2m_2},\quad \alpha=0,1,2;\ \beta,\gamma=0,1.$$

Only $O(N^2)$ real arithmetical operations are required to calculate $\hat{x}^{(\alpha)}_{\beta\gamma}(m_1,m_2)$.

Step 6. Full DFT-spectra reconstruction.

The full DFT-spectra $\hat{x}^{(\alpha)}(m_1, m_2)$ can be reconstructed by

$$\hat{x}^{(\alpha)}(m_1, m_2) = \sum_{\beta,\gamma=0}^{1} e^{\frac{2\pi}{N}(\beta m_1 + \gamma m_2)} \hat{x}^{(\alpha)}_{\beta\gamma}(m_1, m_2).$$

This reconstruction requires $O(N^2)$ real arithmetical operations.

Finally, summing the complexity estimates for Steps 1-6, we obtain (14).

In [9], the similar task was solved using quaternion DFT (QDFT):

$$\hat{Q}(m_1, m_2) = \sum_{n_1,n_2=0}^{N-1} e^{2\pi i \frac{n_1 m_1}{N}} Q(n_1, n_2) e^{2\pi j \frac{n_2 m_2}{N}}, \tag{16}$$

where

$$Q(n_1, n_2) = (iR(n_1, n_2) + jG(n_1, n_2) + kB(m_1, m_2)).$$

The transform (16) is introduced in [7] as an auxiliary transform, and is analyzed as a self-sufficient transform in [13], [12]. The applications of QDFT to "anisotropic" tasks are discussed in [11]. The $RGB-$spectra calculation method, proposed in [9] requires

$$M^*(N^2) = \frac{8}{3} N^2 \log_2 N + O(N^2)$$

real multiplications. This is almost three-times worse relative to the method considered above.

Remark 2 *Note that there exist the 2D FFTs with the slightly better computational complexity characteristics. However these algorithms can hardly be used by an ordinary user because of the complex structure. Despite its non - triviality, the algorithm introduced above uses only standard computational means (Radix-2 or row-column FFT, solution of the system of 12 linear equations etc.). The main distinction is in the non-complex* **Hu** *algebra arithmetic.*

4 Conclusion

In the author's opinion, the capabilities of the approach described in this paper are not limited to the applications considered above.

In [14] the author used similar approach to develop the following DFT algorithms.

- Fast algorithm for calculating three two-dimensional Fourier spectra with overlapping in the eight-dimensional group algebra $\mathbf{A}(\mathbf{R}, \mathbf{D}_4)$ with

  $$M^*(N^2) = \frac{5}{4} N^2 \log_2 N + O(N^2),$$

  where $\mathbf{D}_4$ is the eight-element dihedral group.

- Fast algorithms for calculating five complex Fourier spectra of complex data with overlapping in the six-dimensional group algebra $\mathbf{A}(\mathbf{R}, \mathbf{S}_3)$ with

$$M^*(N) = \frac{4}{5} N \log_2 N + O(N),$$

where $\mathbf{S}_3$ is the six-element permutation group.

It seems to be quite interesting to consider the 120-dimensional icosian algebra which allow to one calculate a DFT with the 60-fold overlapping. The basis of this algebra is associated with the so-called icosians (integer quaternions over $\mathbf{Q}(\sqrt{5})$ (see [4], [15])).

$$\frac{1}{2}(\pm 1, 0, 0, 0)^\alpha, \quad \frac{1}{2}(\pm 1, \pm 1, \pm 1, \pm 1)^\alpha, \quad \frac{1}{2}(0, \pm 1, \pm \sigma, \pm \tau)^\alpha,$$

where by (x, y, z, t) we denote the quaternion $x + iy + jz + tk$,

$$\sigma = \frac{1}{2}\left(1 - \sqrt{5}\right), \quad \tau = \frac{1}{2}\left(1 + \sqrt{5}\right);$$

(α is the permutation action, $\alpha \in \mathbf{A}_4 \subset \mathbf{S}_4$,and $\mathbf{A}_4 \subset \mathbf{S}_4$ is the subgroup of even permutations).

Certainly, significant emphasis should be laid on the development of multiplication rules like those of Proposition 6 and the development of an appropriate mathematical formalism.

One of the main goals of this article is to attract the reasearchers' attention to the development of such a formalism.

References

[1] Elliot D.F., Rao K.R. *Fast Transforms.* New York.: Academic, 1982.

[2] Hurwitz A. Über die Zahlentheorie der Quaternionen. *Nach. Gesellschaft Wiss. Göttingen,* Math.-Phys. Klasse, 313-340, 1896.

[3] Coxeter H.S. *Twelve Geometric Essays.* Southern Illinois Press, Carbondale IL, 1968.

[4] Conway J.H., Sloan N.J.A. *Sphere Packing, Lattices and Groups,* Springer Verlag, 1988.

[5] Blahut R. E. *Fast Algorithms for Digital Signal Processing.* Addison-Wesley, 1985.

[6] Chernov V.M. Parametrization of Some Classes of Fast Algorithms for Discrete Orthogonal Transforms (1). *Pattern Recognition and Image Analysis,* 5 (2), pp.238-245, 1995.

[7] Chernov V.M. Arithmetic Methods in the Theory of Discrete Orthogonal Transforms. *Proc. SPIE.* V. 2363, pp. 134-141, 1994.

[8] Chernov V.M. Discrete Orthogonal Transforms with Data Representation in Composition Algebras. *Proceedings of The 9th Scandinavian Conference on Image Analysis.* Uppsala, Sweden, 1995. V.1. pp.357-364.

[9] Sangwine, S,J.: Fourier Transforms of Color Images Using Quaternion, or Hypercomplex, Numbers. *Electronics Letters.* 32 (21), pp.1979-1980, 1996.

[10] Nussbaumer H. J., *Fast Fourier Transform and Convolution Algorithms.* Springer-Verlag, Berlin, 1982.

[11] T. Buelow and G. Sommer. Multi-Dimensional Signal Processing Using an Algebraically Extended Signal Representation. In: *G. Sommer and J.J. Koenderink (Eds.). Algebraic Frames for Perception-Action Cycle,* Springer Verlag, LNCS 1395, pp. 148-163, 1997.

[12] V.M. Chernov, T. Buelow and M. Felsberg. Synthesis of Fast Algorithms for Discrete Fourier-Clifford Transform. *Pattern Recognition and Image Analysis,* 8(2), pp. 274-275, 1998.

[13] M.A. Chichyeva and M.V. Pershina. On Various Schemes of 2D-DFT Decomposition with Data Representation in the Quaternion Algebra. *Image Processing and Communications,* 2(1) pp. 13-20, 1996.

[14] Chernov V.M. Clifford Algebras as Projections of Group Algebras: How Can We Profit From It? In: *E.Bayro-Corrochano, G.Sobczyk (Eds), Advanches in Geometric Algebra with Applications in Science and Engineering.* Birkhauser, Boston, pp.467-482, 2000.

[15] Tits J. Quaternions over $\mathbf{Q}(\sqrt{5})$, Leech's Lattice and the Sporadic Group of Hall-Janko, *J.Algebra,* 63, pp.56-75, 1980.

Diffusion–Snakes Using Statistical Shape Knowledge

Daniel Cremers, Christoph Schnörr,
Joachim Weickert, and Christian Schellewald

Computer Vision, Graphics, and Pattern Recognition Group
Department of Mathematics and Computer Science
University of Mannheim, 68131 Mannheim, Germany
cremers@uni-mannheim.de
http://www.cvgpr.uni-mannheim.de/

Abstract. We present a novel extension of the Mumford–Shah functional that allows to incorporate statistical shape knowledge at the computational level of image segmentation. Our approach exhibits various favorable properties: non–local convergence, robustness against noise, and the ability to take into consideration both shape evidence in given image data and knowledge about learned shapes. In particular, the latter property distinguishes our approach from previous work on contour–evolution based image segmentation. Experimental results confirm these properties.

Keywords: image segmentation, shape recognition, statistical learning, variational methods, diffusion, active contour models, diffusion–snake

1 Introduction

Robust shape recognition is a fundamental task in a perception–action cycle. Two fundamental issues in shape recognition are image segmentation and the representation and incorporation of previously learned shape knowledge in order to cope with missing and imperfect data.

Variational methods provide a conceptually clear approach to image segmentation. They have been studied by Mumford and Shah [ ] and others [, , ,]. The basic idea is to approximate a given grey–value image with a piecewise smooth function by minimizing a suitable functional. There has been recent interest in knowledge acquisition through learning by examples, and appropriate modifications of the Mumford–Shah functional were proposed [,].

In this paper, we present a novel extension of the Mumford–Shah functional in order to combine powerful bottom–up diffusion filtering with top–down incorporation of statistical knowledge about previously learned shapes. As a first step, shape knowledge is represented in terms of the principal components of samples in a linear vector space of contours. The principal component analysis (PCA) was shown to be quite effective in modeling shape variation [, , ,]. Its combination with variational segmentation, however, is new.

G. Sommer and Y. Y. Zeevi (Eds.): AFPAC 2000, LNCS 1888, pp. 164– , 2000.

Our paper is organized as follows: in Section we extend the Mumford–Shah functional by an energy term representing statistical shape knowledge. The underlying curve representation and statistical learning process is explained in Section . Section describes the curve evolution scheme governing the snake and its interaction with diffusion filtering. The key properties of our approach are demonstrated by numerical results in Section .

2 Extending the Mumford–Shah Functional

The well–known approach to image segmentation suggested by Mumford and Shah [ ] is to compute a segmented version $u(x)$ of an input image $f(x)$ by minimizing the following energy functional:

$$E_{MS}(u, C) = \frac{1}{2} \int_{\Omega} (f - u)^2 \, d^2x \; + \; \lambda^2 \frac{1}{2} \int_{\Omega - C} |\nabla u|^2 \, d^2x \; + \; \nu |C| \,, \tag{1}$$

where $\Omega \subset \mathbb{R}^2$ denotes the image plane. This functional is simultaneously minimized with respect to the segmented image u and a contour C, across which u may be discontinuous.

The three energy terms in ( ) express the following constraints: The segmented image u should approximate the input image f; it should be smooth but jumps are allowed at the contour C, and the contour length $|C|$ should be minimal. λ and $\nu \geq 0$ are weighting factors.

In this paper, we propose to extend this energy functional to

$$E(u, C) = E_{MS}(u, C) \; + \; \alpha \, E_c(C) \,, \tag{2}$$

where the contour energy term $E_c(C)$ accounts for previously learned shape information. Minimization with respect to C should then favor contours which the segmentation process is " familiar" with due to the acquired shape knowledge.

How we describe shape, how we acquire shape statistics, and how we incorporate these in the contour energy E_c, will be explained next.

3 Modeling Shape Knowledge

3.1 Shape Representation

In this paper we represent the shape of an object as a quadratic B-spline curve. We chose this representation for the following reasons: a quadratic spline curve gives a simple mathematical description of an object contour. Moreover, it is differentiable such that curvature and normal vectors can easily be determined. The number of control points can be adapted to obtain a detailed contour description in terms of a moderate amount of data per shape. In particular, the shape statistics of various objects can conveniently be learned by examining the distribution of control points in a finite–dimensional vector space.

Thus, in this paper a contour will be a closed planar curve of the form

$$C : [0, 1] \to \Omega\,, \qquad C(s) = \sum_{n=1}^{N} \begin{pmatrix} x_n \\ y_n \end{pmatrix} B_n(s)\,, \tag{3}$$

where (x_n, y_n) are the control points and $B_n(s)$ are quadratic periodic B-spline basis functions [,].

Given the shape of an object as a binary image, the spline curve is automatically fitted around the object contours. As a result, the objects shape is represented by the set of control points $\{(x_n, y_n)\}_{n=1..N}$ for which we use the compact notation:

$$\boldsymbol{z} = (x_1, y_1, \ldots, x_N, y_N)^t \tag{4}$$

Each shape therefore corresponds to a vector $\boldsymbol{z} \in \mathbb{R}^{2N}$.

3.2 Shape Statistics

We now assume that numerous examples of the appearance of an object are given by a set of shapes $\mathcal{X} = \{\boldsymbol{z}_1, \boldsymbol{z}_2, \ldots\}$ with $\boldsymbol{z}_i \in \mathbb{R}^{2N}$ for all i. To analyze the variation within this set of shapes, we perform a principal component analysis (PCA) of the points $\boldsymbol{z}_i$ (cf. [,]). That is, we determine the mean $\boldsymbol{z}_0 = E[\boldsymbol{z}_i]$, and the covariance matrix

$$\Sigma = E\left[(\boldsymbol{z}_i - \boldsymbol{z}_0)(\boldsymbol{z}_i - \boldsymbol{z}_0)^t\right], \tag{5}$$

where $E[\cdot]$ denotes the sample average over the shapes $\boldsymbol{z}_i$ in our learning set $\mathcal{X}$. This is equivalent to modeling the distribution of shapes in $\mathbb{R}^{2N}$ as a Gaussian distribution. The eigenvectors associated with the largest eigenvalues σ_i^2 of the covariance matrix Σ are the (significant) principal components and indicate the directions of largest shape variation. The square root σ_i of the eigenvalue is a measure for the amount of shape variation in a given direction.

The shape variation can be visualized by sampling along an eigenvector in both directions from the mean. As an example, we present in Figure an arbitrary set of six different ellipses (shown on the left) and a visualization of the first two principal components (middle and right images). In this example, the visualization allows for a very intuitive interpretation of the principal components: the first one describes a variation in the size and the second one a variation in the aspect ratio of the ellipses. Thus, the principal component analysis imposes an order on the learned shapes, a ranking by importance, and simultaneously induces a compact description of shape variation. Moreover, having learned a number of shapes in this way, the covariance matrix allows us to define a probability measure on the shape space. If the covariance matrix Σ is of full rank, then the inverse Σ^{-1} exists and the Gaussian probability distribution of shapes $\boldsymbol{z}$ is

$$\mathcal{P}(\boldsymbol{z}) \propto \exp\left(-\frac{1}{2}(\boldsymbol{z} - \boldsymbol{z}_0)^t\, \Sigma^{-1}\, (\boldsymbol{z} - \boldsymbol{z}_0)\right). \tag{6}$$

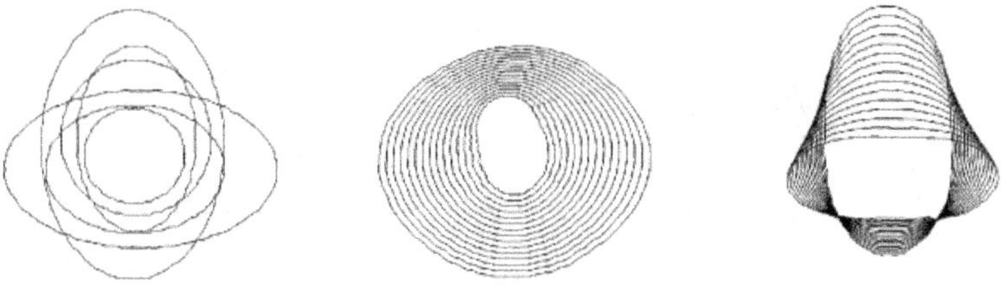

Fig. 1. On the left are shown six arbitrary ellipses. The middle and right images show the first and second principal component: sampling was done around the mean contour by two standard deviations in both directions

In general, however, the covariance matrix Σ may not have full rank, such that an inverse is not defined. Therefore, we replace it by a modified pseudoinverse Σ^*, which we construct as follows: assume the square roots of the eigenvalues of Σ to be ordered such that $\sigma_1 > \sigma_2 > \ldots > \sigma_r$, where r is the rank of Σ. We define the modified pseudoinverse as

$$\Sigma^* := T \begin{pmatrix} \sigma_1^{-2} & & & \\ & \sigma_2^{-2} & & \\ & & \ddots & \\ & & & \sigma_r^{-2} \end{pmatrix} T^t + \sigma_\perp^{-2} \left(I - T T^t \right) , \tag{7}$$

where T is the $2N \times r$–matrix, the columns of which are the first r eigenvectors of Σ. The constant $\sigma_\perp$ was introduced to account for all directions orthogonal to the eigendirections. It should be of the order of σ_r.

The geometric meaning of this modified pseudoinverse is the following: if the rank of Σ is less than the dimension $2N$ of the parameter space, then the assumed Gaussian distribution does not extend into the space orthogonal to the eigendirections. This means that shapes lying outside the eigenspace are not admissible. By adding the second term, the Gaussian distribution is artificially expanded into the orthogonal space. This allows to also take into consideration evidence of given image data that does not correspond to the learned shape knowledge. Enforcing $\sigma_\perp \leq \sigma_r$ would, for example, mean that the probability of a shape variation outside the learned shape space is not necessarily zero, but smaller than or equal to any shape variation encountered within the set of learned shapes. Decreasing $\sigma_\perp$ will continually suppress all unknown shape variations.

3.3 Shape Energy Functional

From the previously constructed shape distribution (cf. equation ( )) and the Gibbs identity

$$\mathcal{P}(\boldsymbol{z}) \propto \exp\left(-E(\boldsymbol{z})\right) , \tag{8}$$

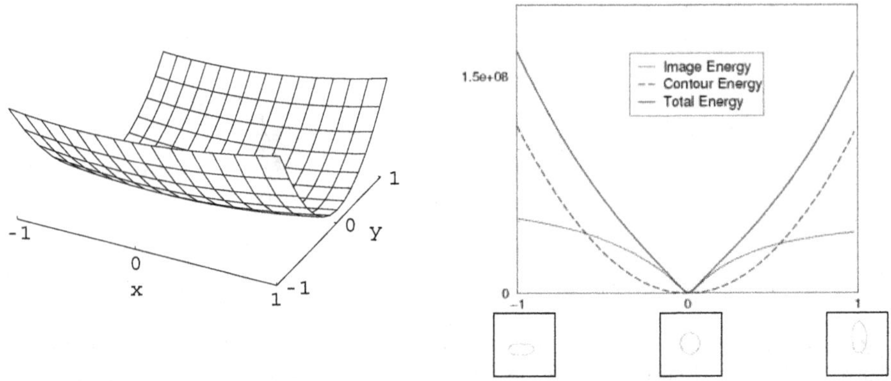

Fig. 2. Energy plots. The left side shows a schematic plot of the contour energy ( ): learned directions (eigenvectors of the covariance matrix Σ) are represented by x and orthogonal directions by y. The right image shows the image energy E_{MS}, the contour energy E_c and the sum of both for a fixed input image as a function of different contours. For the parameter values $-1, 0, 1$, the three respective contours are shown below

we derive the following shape energy functional:

$$E_c(\boldsymbol{z}) = \frac{1}{2}(\boldsymbol{z} - \boldsymbol{z}_0)^t \, \Sigma^* \, (\boldsymbol{z} - \boldsymbol{z}_0) \,. \tag{9}$$

To visualize the meaning of the orthogonal correction $\sigma_\perp$ in the definition of Σ^*, Figure , left side, shows a schematic plot of the shape energy. For the purpose of clarity the shape space is reduced to two dimensions — a learned direction x and an orthogonal direction y. The increase of shape energy in the learned directions is smaller than the one in orthogonal directions, such that the entire space of shape variations is basically reduced to an *elastic tunnel of familiar shapes*. Restricting the shape variability to this tunnel amounts to a drastic reduction in the effective dimensionality of the search space.

Adding the contour energy ( ) to the Mumford–Shah functional ( ) we obtain the total energy

$$E(u, C(\boldsymbol{z})) = E_{MS}(u, C(\boldsymbol{z})) \; + \alpha \, \frac{1}{2}(\boldsymbol{z} - \boldsymbol{z}_0)^t \, \Sigma^* \, (\boldsymbol{z} - \boldsymbol{z}_0) \,, \tag{10}$$

which is now a function of both the image u and the contour control points $\boldsymbol{z}$. Increasing the parameter $\alpha > 0$ allows to continuously shift from an image–based segmentation to a knowledge–based segmentation. On the other hand, the limit $\alpha \to 0$ results in a pure Mumford–Shah segmentation for closed spline contours. The effect of the contour energy upon the total energy can be seen in Figure , right side: image energy E_{MS}, contour energy E_c and total energy are plotted for various contours. The middle contour is the one which optimally describes the input image.

Several favorable properties are obtained by adding the contour energy: the basin of attraction is greatly increased. For sufficiently large learning strength

α, the total energy will be convex. Moreover, the position of the minimum can be shifted: this leads to a minimizing contour which displays a compromise between knowledge–based and data–driven segmentation. This effect will be demonstrated by examples in Section .

4 Implementation

The total energy in ( ) needs to be minimized with respect to both the segmented image u and the spline control points $\boldsymbol{z}$ defining the shape of the segmenting contour in the image plane. This is done by iterating two fractional steps: the first minimizes ( ) with respect to the contour C (Subsection), the second minimizes ( ) with respect to the image u (Subsection).

4.1 Curve Evolution

As shown in [ ], minimizing $E_{MS}(u, C)$ with respect to the contour C leads to the equation

$$\frac{\delta E_{MS}}{\delta C} = e^+(s) - e^-(s) - \kappa(s) = 0 \quad \forall s \,. \tag{11}$$

Here, s is the curve parameter, κ is the curvature of the contour, e^+ and e^- refer to the integrand in $E_{MS}(u, C)$ on the outside and the inside of the contour C, respectively. Minimization can be implemented by a curve evolution equation with an artificial time parameter t:

$$\frac{dC(s,t)}{dt} = \left(e^+(s,t) - e^-(s,t) - \kappa(s,t)\right) \boldsymbol{n}(s,t) \quad \forall s \,, \tag{12}$$

such that the contour is modified along its outer normal vector $\boldsymbol{n}(s,t)$ only.

Inserting the contour definition ( ) results in an evolution equation for the control points $\boldsymbol{z}$. Without loss of generality, this will be described for the x–components only:

$$\sum_{m=1}^{N} B_m(s) \, \frac{dx_m(t)}{dt} = \left(e^+(s,t) - e^-(s,t) - \kappa(s,t)\right) \boldsymbol{n}_x(s,t) \quad \forall s \,, \tag{13}$$

where $B_m(s)$ are the spline basis functions and $\boldsymbol{n}_x$ denotes the x–component of the normal vector.

This equation has to be satisfied for every point along the curve. The description of the contour as a spline curve induces a discretization of ( ) along the parameter s with the nodes $s_i = (i + 1/2)/N$, $i = 1, \ldots, N$, which are the maxima of the respective basis functions $B_i(s)$. We end up with a set of N coupled linear differential equations:

$$\sum_{m=1}^{N} B_{mi} \, \frac{dx_m(t)}{dt} = \left(e^+(s_i,t) - e^-(s_i,t) - \kappa(s_i,t)\right) \boldsymbol{n}_x(s_i,t) \,. \tag{14}$$

The coefficients $B_{mi} = B_m(s_i)$ form an $N \times N$ tridiagonal matrix $\boldsymbol{B}$. Inversion leads to the evolution equation for the control points. Adding the term induced by the statistical shape information ( )

$$-\frac{dE_c}{d\boldsymbol{z}} = \Sigma^* \left(\boldsymbol{z} - \boldsymbol{z}_0\right), \tag{15}$$

we obtain the evolution equation of the control points for the total energy ( ):

$$\begin{aligned}\frac{dx_m(t)}{dt} = \left(\boldsymbol{B}^{-1}\right)_{mi} \left(e^+(s_i,t) - e^-(s_i,t) - \kappa(s_i,t)\right) \boldsymbol{n}_x(s_i,t) \\ - \left[\Sigma^* \left(\boldsymbol{z} - \boldsymbol{z}_0\right)\right]_{2m-1}.\end{aligned} \tag{16}$$

In the corresponding equation for the y-components, $\boldsymbol{n}_x$ has to be replaced by $\boldsymbol{n}_y$ and $\left[\Sigma^* \left(\boldsymbol{z} - \boldsymbol{z}_0\right)\right]_{2m-1}$ by $\left[\Sigma^* \left(\boldsymbol{z} - \boldsymbol{z}_0\right)\right]_{2m}$.

4.2 Inhomogeneous Linear Diffusion

To minimize the total energy $E(u, C)$ in ( ) with respect to the image u, we rewrite the Mumford–Shah functional:

$$E_{MS}(u, C) = \frac{1}{2}\int_\Omega (f-u)^2\, d^2x \,+\, \lambda^2 \frac{1}{2}\int_\Omega w_c(x) |\nabla u|^2\, d^2x \,+\, \nu |C|. \tag{17}$$

The contour dependence is now implicitly represented by an indicator function

$$w_c : \Omega \to \{0, 1\}, \qquad w_c(x) = \begin{cases} 0 & \text{if } x \in C \\ 1 & \text{otherwise} \end{cases}. \tag{18}$$

For fixed C, the corresponding inhomogeneous linear diffusion equation

$$\frac{du}{dt} = -\frac{dE}{du} = (f-u) \,-\, \lambda^2\, \nabla \cdot \left(w_c(x)\, \nabla u\right) \tag{19}$$

converges to the minimum of ( ) for $t \to \infty$.

We used a modified explicit scheme to solve ( ), and iterated the contour evolution equation ( ) alternatingly.

5 Results

Implementation Details. For the set of six ellipses, shown in Figure on the left, mean contour and covariance matrix were determined. The modified pseudoinverse Σ^* (equation ( )) was determined, the parameter for orthogonal correction was chosen $\sigma_\perp = 1.2\, \sigma_r$. The contour was initialized to the mean contour and the following steps were iterated: calculate the indicator function $w(x)$ (equation ( )) from the current contour C, diffuse the image u with the inhomogeneous diffusivity $w(x)$ (equation ( )), update the contour control points (equation ( )) and determine the new contour C (equation ( )). This procedure

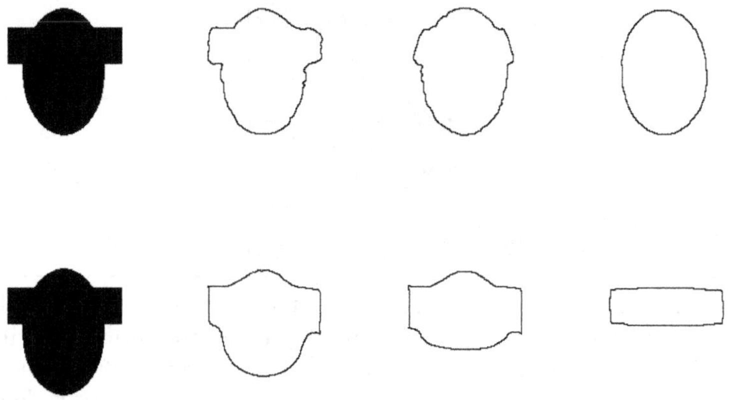

Fig. 3. The input image is shown on the left. The information energy was constructed upon the 6 ellipses of Figure for the top row and a set of bars for the bottom row. For three strength values $\alpha = 2000, 22000, 840000$ (top row) and $\alpha = 300, 2000, 50000$ (bottom row) the final contour was determined. Increasing the knowledge energy the overlapping part (bar or ellipse) is continuously removed

was done for various values of the parameter α, which determines the relative strength of the knowledge term in the energy functional (equation ( )).

Image–based vs. knowledge–based segmentation. The input image – a rectangle overlapping an ellipse – and the resulting contours C for the three different values of α are shown in Figure , top row. To elaborate the effect of the acquired shape statistics, we analyzed the same input image – Figure left – but this time we constructed the contour energy based on a different set of learned images, namely a set of bars.

The results show that for increasing knowledge strength parameter α, the segmentation process continuously ignores shapes that do not correspond to learned shape statistics. Note, however, that the resulting contour does *not* correspond simply to the mean shape, i.e. the most probable shape. The segmentation process still incorporates evidence given by the input image data.

Similar results are obtained (for fixed $\alpha > 0$) by continuously decreasing the parameter $\sigma_\perp$ which suppresses all contour motion in directions orthogonal to the learned ones.

Non–local convergence and noise robustness. In a second example, we show the robustness of the method against noise and its independence of the initial contour. We trained the system with the set of ellipses. The input image is an ellipse (which, again, is not the mean contour of the training set) with 75% noise, that is three out of four pixels were replaced by an arbitrary grey value.

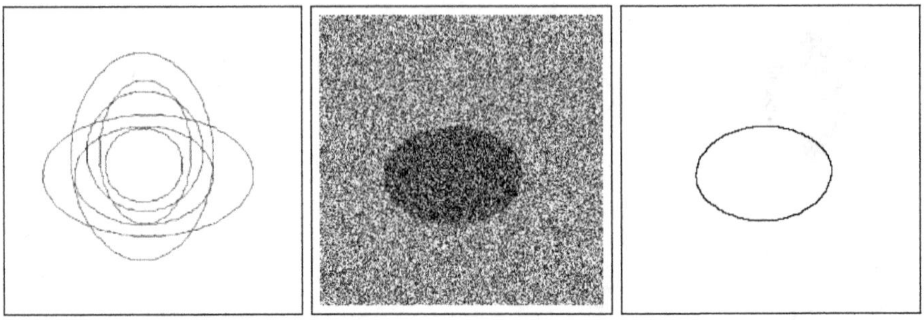

Fig. 4. Robustness: The contour energy is constructed upon the set of ellipses on the left. The input image (middle) is an ellipse with uniformly distributed noise where 75% of the pixels were replaced by arbitrary grey values. The right side shows the final contour. The initialization and iteration process is depicted in Figure

Figure shows that the correct segmentation is obtained despite the large amount of noise. Figure depicts three samples of the dynamic segmentation process: the diffused images u and the associated contours are plotted at various times. Note, that the initial shape (left) is “far away” from the correct shape (right). Nevertheless, the process converges. This behavior corresponds to the shape of the energy functional depicted in Figure , right: shape knowledge increases the basin of attraction considerably.

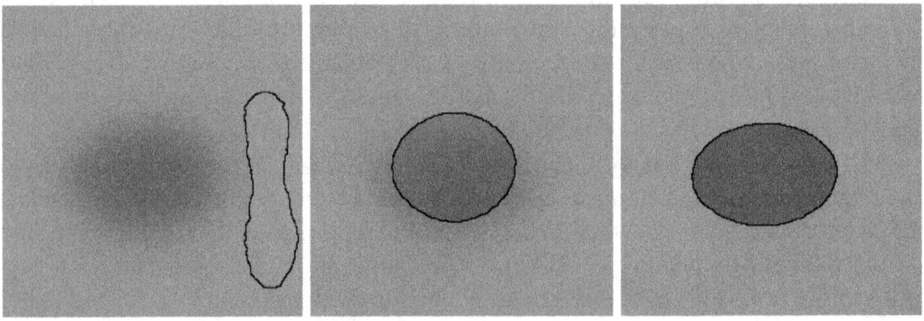

Fig. 5. From left to right: a series of steps in the minimization process showing the diffused image and the current contour (in black) for the noisy input image in Figure . The parameters were chose $\lambda^2 = 400, \nu = 800, \alpha = 100000, \sigma_\perp = 0.4\,\sigma_r$. Note, that the final contour (bottom right) does not correspond to the mean shape of the learning set

6 Conclusion

We presented a novel extension of the Mumford–Shah functional which allows to integrate statistical shape knowledge already at the computational level of image segmentation. As a result, we obtained a snake–like segmentation approach which exhibits the following favorable properties over classical snake approaches [, ,]:

(i) Non–local convergence and noise robustness: these properties are due to the diffusion filtering, which is a global smoothing process interacting over large distances within the image plane. By contrast, the region of convergence of traditional snake approaches is restricted to narrow valleys typically defined by locally estimated gradients of noisy image data.

(ii) The forces driving the snake do not solely depend on signal transitions of given image data, but also take into consideration previously experienced appearances of known objects.

Our method allows to continuously shift from an image–based to a knowledge–based segmentation. First experiments indicate its high robustness against noise.

Our further work will focus on more sophisticated representations of statistical shape information going beyond standard PCA, and on deriving meaningful criteria to determine the "knowledge strength" parameter α automatically.

Acknowledgements

We thank Laurenz Wiskott and Gabriele Steidl for some helpful discussions.

References

1. L. Ambrosio. Variational problems in SBV and image segmentation. *Acta Appl. Math.*, 17:1–40, 1989.
2. A. Blake and M. Isard. *Active Contours.* Springer, London, 1998. , ,
3. T. F. Cootes, C. J. Taylor, D. M. Cooper, and J. Graham. Active shape models – their training and application. *Comp. Vision Image Underst.*, 61(1):38–59, 1995. ,
4. L. Dryden and K. V. Mardia. *Statistical Shape Analysis.* Wiley, Chichester, 1998. ,
5. G. Farin. *Curves and Surfaces for Computer–Aided Geometric Design.* Academic Press, San Diego, 1997.
6. M. Kass, A. Witkin, and D. Terzopoulos. Snakes: Active contour models. *Int. J. of Comp. Vision*, 1(4):321–331, 1988.
7. C. Kervrann and F. Heitz. Statistical deformable model-based segmentation of image motion. *Neural Computation*, 8:583–588, 1999.
8. F. Leymarie and M. D. Levine. Tracking deformable objects in the plane using an active contour model. *IEEE Trans. Patt. Anal. Mach. Intell.*, 15(6):617–634, 1993.
9. J.-M. Morel and S. Solimini. *Variational Methods in Image Segmentation.* Birkhäuser, Boston, 1995.

10. D. Mumford and J. Shah. Optimal approximations by piecewise smooth functions and associated variational problems. *Comm. Pure Appl. Math.*, 42:577–685, 1989.
11. N. Nordström. Biased anisotropic diffusion - a unified regularization and diffusion approach to edge detection. *Image and Vis. Comp.*, 8(4):318–327, 1990.
12. C. Schnörr. Unique reconstruction of piecewise smooth images by minimizing strictly convex non-quadratic functionals. *J. Math. Imag. Vision*, 4:189–198, 1994.
13. L. Younes. Calibrating parameters of cost functionals. In *ECCV*, Dublin, 2000. Springer Lecture Notes in Computer Science. To appear.
14. S. C. Zhu and D. Mumford. Prior learning and Gibbs reaction–diffusion. *IEEE Trans. Patt. Anal. Mach. Intell.*, 19(11):1236–1250, 1997.

The Multidimensional Isotropic Generalization of Quadrature Filters in Geometric Algebra*

Michael Felsberg and Gerald Sommer

Institute of Computer Science and Applied Mathematics, Cognitive Systems
Christian-Albrechts-University of Kiel, Preußerstraße 1-9, 24105 Kiel, Germany
Tel: +49 431 560433, Fax: +49 431 560481
{mfe,gs}@ks.informatik.uni-kiel.de

Abstract. In signal processing, the approach of the analytic signal is a capable and often used method. For signals of finite length, quadrature filters yield a bandpass filtered approximation of the analytic signal. In the case of multidimensional signals, the quadrature filters can only be applied with respect to a preference direction. Therefore, the orientation has to be sampled, steered or orientation adaptive filters have to be used. Up to now, there has been no linear approach to obtain an isotropic analytic signal which means that the amplitude is independent of the local orientation. In this paper, we present such an approach using the framework of geometric algebra. Our result is closely related to the Riesz transform and the structure tensor. It is seamless embedded in the framework of Clifford analysis. In a suitable coordinate system, the filter response contains information about local amplitude, local phase and local orientation of intrinsically one-dimensional signals. We have tested our filters on two- and three-dimensional signals.

Keywords: quadrature filter, analytic signal, Riesz transform

1 Introduction

In image and image sequence processing, different paradigms of interpreting the signals exist. Regardless of they are following a constructive or an appearance based strategy, they all need a capable low-level preprocessing scheme.

For one-dimensional signals, the analytic signal and the quadrature filters are capable theoretical and practical methods, respectively. The analytic signal codes the local properties of structure in an optimal way. Using quadrature filters, it is simple to detect steps and spikes in the signal.

Accordingly, in image processing, the detection of edges and lines is a frequently discussed topic, which suffers from the fact that there has been no odd filter with isotropic energy up to now (e.g. [ ]). The corresponding problem in

* This work has been supported by German National Merit Foundation and by DFG Graduiertenkolleg No. 357 (M. Felsberg) and by DFG So-320-2-2 (G. Sommer).

G. Sommer and Y. Y. Zeevi (Eds.): AFPAC 2000, LNCS 1888, pp. 175– , 2000.

the frequency domain is that one cannot define positive and negative frequencies (see [ ]), such that it is not possible to create a 'real' 2D quadrature filter.

To overcome this problem, several approaches has been developed in the past, all using the quadrature filters with respect to a preference direction:

1. orientation adaptive filtering using the structure tensor, e.g. [,]
2. sampling of the orientation, e.g. [,]
3. steerable filters, e.g. [,]

The first two approaches are non-linear and the corresponding algorithms have high complexities (compared to convolutions). The steerable filters are linear and fast, but they are not related to a generalized analytic signal and only yield approximative quadrature filters with steered preference direction. In our opinion, the structure tensor is the method which is closest to a generalized quadrature filter. It is isotropic but not linear because the phase is neglected. Actually, the phase contains all information about the characteristic of structure [ ]!

Therefore, one should keep the phase, which is automatically fulfilled if a linear approach is developed. Since the preprocessing is only the first link in a long chain of operations, it is also useful to have a linear approach, because otherwise it would be nearly impossible to design the higher-level processing steps. If the preprocessing is linear, one can consider simple cases because the effect in a more complex signal is simply the sum of the parts.

On the other hand, we need a rich representation if we want to treat as much as possible in the preprocessing stage. Furthermore, the representation of the signal during the different operations should be complete, in order to prevent a loss of information. These constraints enforce us to leave the approach of complex analysis and to use the framework of geometric algebra instead which is also advantageous if we combine image processing with neural computing and robotics (see [ ]).

In this paper, we introduce a new approach for the 2D analytic signal which enables us to substitute the structure tensor by an entity which is linear, preserves the split of the identity and has a geometrically meaningful representation. We have overcome the problem of odd filters in higher dimensions, the resulting method is of low complexity and is naturally embedded as a generalized analytic signal in the field of Clifford analysis.

2 Fundamentals

Since we work on signals in Euclidean space ($\mathbb{R}^n$), we have to use the geometric algebra $\mathbb{R}_{0,n}$. That is, for 1D signals we use $\mathbb{R}_{0,1}$ (isomorphic to the algebra of complex numbers), for image processing we use $\mathbb{R}_{0,2}$ (isomorphic to the algebra of quaternions $\mathbb{H}$), and for image sequences we use $\mathbb{R}_{0,3}$. The classical complex signal theory naturally embeds in these algebras, since the algebra of complex numbers can be considered as a subalgebra.

The base vectors of $\mathbb{R}^n$ are denoted

$$\mathbf{e}_1, \mathbf{e}_2, \dots, \mathbf{e}_n \qquad \text{where } \mathbf{e}_k\mathbf{e}_k = -1,\ k \in \{1, \dots, n\}$$

and the base elements of $\mathbb{R}_{0,n}$ are denoted

$$\mathbf{e}_0, \mathbf{e}_1, \mathbf{e}_2, \ldots, \mathbf{e}_n, \mathbf{e}_{12}, \mathbf{e}_{13}, \ldots, \mathbf{e}_{(n-1)n}, \mathbf{e}_{123}, \ldots, \mathbf{e}_{1\ldots n}$$

where $\mathbf{e}_0$ is the base element for the scalar part, i.e. it commutes with all base elements and squares to 1. The subspace of $\mathbb{R}_{0,n}$ consisting of k-vectors is denoted $\mathbb{R}^k_{0,n}$. The conjugation (inversion and reversion) of $a \in \mathbb{R}_{0,n}$ is denoted by $\bar{a}$. For a complete introduction to geometric algebra see e.g. [ ].

All signals are considered to be defined on vector spaces, hence a real 1D signal is not any longer a function $\mathbb{R} \to \mathbb{R}$ but a function $\mathbf{e}_1\mathbb{R} \to \mathbf{e}_0\mathbb{R}$ or a curve in $\mathbb{R}^1_{0,1} \oplus \mathbb{R}^0_{0,1}$. Accordingly, an image is a surface in $\mathbb{R}^1_{0,2} \oplus \mathbb{R}^0_{0,2}$ (i.e. in a 3D space), and an image sequence is a 3D subspace in $\mathbb{R}^1_{0,3} \oplus \mathbb{R}^0_{0,3}$. Vectors (elements of $\mathbb{R}^1_{0,n} \cong \mathbb{R}^n$) are denoted in bold face: $\boldsymbol{x} = \sum_{k=1}^n x_k \mathbf{e}_k$. Elements of $\mathbb{R}^1_{0,n} \oplus \mathbb{R}^0_{0,n}$ (*paravectors*, [ ]) are denoted in normal face: $x = \sum_{k=0}^n x_k \mathbf{e}_k = x_0\mathbf{e}_0 + \boldsymbol{x}$.

The Fourier transform of the nD signal $f(\boldsymbol{x})$ is denoted

$$f(\boldsymbol{x}) \circ\!\!-\!\!\bullet\, F(\boldsymbol{u}) = \int_{\mathbb{R}^1_{0,n}} f(\boldsymbol{x}) \exp(\mathbf{e}_1 2\pi \boldsymbol{u} \cdot \boldsymbol{x})\, d\boldsymbol{x} \ .$$

Since we want to extend the analytic signal, we briefly introduce the 1D approach (see e.g. [ ]). The Hilbert transform of a 1D signal f is denoted f_H and it is obtained by the transfer function $H(\boldsymbol{u}) = \mathbf{e}_1 \operatorname{sign}(\boldsymbol{u})$ in the frequency domain or by the convolution kernel $-\frac{1}{\pi x_1} = \frac{\boldsymbol{x}\mathbf{e}_1}{\pi|\boldsymbol{x}|^2}$. The analytic signal is obtained by the sum $f_A = f - f_H \mathbf{e}_1$.

The typical property of the analytic signal is the *split of identity* which means that it contains local phase and local amplitude. While the local phase represents a *qualitative* measure of a structure, the local amplitude represents a *quantitative* measure of a structure (see e.g. [,]). For higher dimensions, a consequent generalization of the analytic signal should keep the property of splitting the identity.

Local structures in multi-dimensional signals can be classified in different categories according to the intrinsic dimensionality (see [ ]). In our approach, we keep with a single structure phase. Therefore, we can only classify intrinsically 1D signals. What is left for a complete description of structure is the local orientation. Obviously, local orientation is not independent of the phase because the local direction of a signal fixes both properties at the same time (see [ ]). The resulting constraint is fulfilled by our approach which we will introduce in the next section.

3 The Monogenic Signal

The constraint which must be fulfilled by the multidimensional extension of the analytic signal is the following: If the signal is rotated such that it is reflected wrt. the origin (e.g. 2D: rotation of π), the change of the orientation phase must

[1] Note that for most dimensions, it is not possible to find a rotation which is identical to a reflection through the origin for arbitrary objects. In our case, we only consider

yield a negation of the structure phase. This is fulfilled if we use the standard spherical coordinates and assign the first angle to the structure phase:

$$\begin{aligned} x_0 &= r\cos\theta_1 \\ x_1 &= r\sin\theta_1\cos\theta_2 \\ &\vdots \\ x_{n-1} &= r\sin\theta_1\ldots\sin\theta_{n-1}\cos\theta_n \\ x_n &= r\sin\theta_1\ldots\sin\theta_{n-1}\sin\theta_n \end{aligned}$$

where $r = \sqrt{x\bar{x}}$, $\theta_2, \ldots, \theta_n \in [0;\pi)$ and $\theta_1 \in [0;2\pi)$.
For the 2D case this coordinate system is illustrated in Fig. . The reflection of

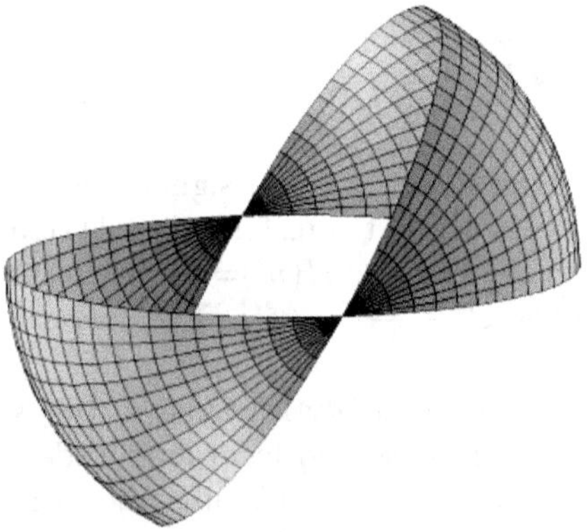

Fig. 1. Spherical coordinate system for 2D

a point wrt. the origin in vector space corresponds to a rotation of the angular coordinate θ_2 by π. This yields a negation of $\boldsymbol{x}$ (identical to the conjugation of $\boldsymbol{x}$) and therefore, the structure phase is negated as well. In other words, we have to find an operator $\mathcal{O}$ such that

$$\mathcal{O}\{f(\boldsymbol{x})\} \in \mathbb{R}^1_{0,n} \oplus \mathbb{R}^0_{0,n} \qquad \text{and} \tag{1}$$

$$\mathcal{O}\{f(-\boldsymbol{x})\} = \overline{\mathcal{O}\{f(\boldsymbol{x})\}} \tag{2}$$

which is a multidimensional generalization of the Hermite symmetry of the analytic signal. Consequently, the $\mathbf{e}_0$-part (real part) of $\mathcal{O}\{f(\boldsymbol{x})\}$ must be even and the vector part of $\mathcal{O}\{f(\boldsymbol{x})\}$ must be odd. Up to now, it was a common opinion that no odd operator with isotropic energy exists (e.g. [ ]). Actually, this statement is true if the operator is required to consist of only one scalar

intrinsically 1D signals which means that it is sufficient to reflect wrt. the direction in which the signal changes (normal vector). This is always done by a rotation around an axis orthogonal to the normal vector through the origin.

valued component. If we extend our operator space to vector valued operators, the statement is not true any more.

The operator $\mathcal{O}$ that fulfills the constraints ( ) and ( ) would be the multidimensional generalization of the analytic signal we are looking for. The even part of $\mathcal{O}$ can be adopted from the 1D analytic signal, since the Dirac impulse $\delta_0(\boldsymbol{x})$ is even and isotropic. The harder task is to find the odd part of $\mathcal{O}$. If we have an intrinsically 1D signal with normal vector $\boldsymbol{n}$ (i.e. $f(\boldsymbol{x}) = g((\boldsymbol{x} \cdot \boldsymbol{n})\mathbf{e}_1)$), a good choice for the odd part of $\mathcal{O}$ would transform $f(\boldsymbol{x})$ such that

$$\mathcal{O}_{\text{odd}}\{f(\boldsymbol{x})\} = \pm \boldsymbol{n} g_H((\boldsymbol{x} \cdot \boldsymbol{n})\mathbf{e}_1) \tag{3}$$

where g_H is the 1D Hilbert transform of g. The factor $\boldsymbol{n}$ yields the odd symmetry we need.
In order to obtain $\mathcal{O}_{\text{odd}}$ we have to look at the Fourier transform of $f(\boldsymbol{x})$:

$$f(\boldsymbol{x}) \circ\!\!-\!\!\bullet\, F(\boldsymbol{u}) = \delta_0(\boldsymbol{u} \wedge \boldsymbol{n}) G((\boldsymbol{u} \cdot \boldsymbol{n})\mathbf{e}_1) \ ,$$

$(\delta_0(\boldsymbol{u} \wedge \boldsymbol{n}) = \delta_0(-(\boldsymbol{u} \wedge \boldsymbol{n}) \cdot \mathbf{e}_{12}) \ldots \delta_0(-(\boldsymbol{u} \wedge \boldsymbol{n}) \cdot \mathbf{e}_{(n-1)n}))$ and therefore,

$$\mathcal{O}_{\text{odd}}\{f(\boldsymbol{x})\} \circ\!\!-\!\!\bullet\, \pm \boldsymbol{n} \delta_0(\boldsymbol{u} \wedge \boldsymbol{n}) \operatorname{sign}(-\boldsymbol{u} \cdot \boldsymbol{n}) G((\boldsymbol{u} \cdot \boldsymbol{n})\mathbf{e}_1)\mathbf{e}_1 \ .$$

Since $\boldsymbol{n}$ is equal to $\pm \frac{\boldsymbol{u}}{|\boldsymbol{u}|}$, we obtain

$$\mathcal{O}_{\text{odd}}\{f(\boldsymbol{x})\} \circ\!\!-\!\!\bullet\, \pm \frac{\boldsymbol{u}}{|\boldsymbol{u}|} \delta_0(\boldsymbol{u} \wedge \boldsymbol{n}) G((\boldsymbol{u} \cdot \boldsymbol{n})\mathbf{e}_1)\mathbf{e}_1 \ .$$

The transfer function of the generalized Hilbert transform reads

$$H(\boldsymbol{u}) = \frac{\boldsymbol{u}}{|\boldsymbol{u}|} \ , \tag{4}$$

$(\mathcal{O}_{\text{odd}}\{f(\boldsymbol{x})\} \circ\!\!-\!\!\bullet\, -H(\boldsymbol{u})F(\boldsymbol{u})\mathbf{e}_1)$ where we chose the sign according to the 1D case. The transfer function of the Hilbert transform reads $\frac{u}{|u|} = \operatorname{sign}(u)\mathbf{e}_1$.

The spatial representation of ( ) which can be obtained using the Hankel transform (see [,])

$$h(\boldsymbol{x}) = \frac{\Gamma((n+1)/2)}{\pi^{(n+1)/2}} \frac{\bar{\boldsymbol{x}}}{|\boldsymbol{x}|^{n+1}} \mathbf{e}_1 \ , \tag{5}$$

is the kernel of the *Riesz transform* which is the multidimensional generalization of the Hilbert transform from a mathematician's point of view (see e.g. [ ] and also [ ]). The factor in ( ) is two times one over the surface area of the n-dimensional unit-sphere (Γ: Gamma function). Actually, the kernel of the Riesz transform is closely related to the Cauchy kernel

$$E(x) = \frac{\Gamma((n+1)/2)}{2\pi^{(n+1)/2}} \frac{\bar{x}}{|x|^{n+1}} = \frac{\Gamma((n+1)/2)}{2\pi^{(n+1)/2}} \frac{x_0\mathbf{e}_0 - \boldsymbol{x}}{|x_0\mathbf{e}_0 + \boldsymbol{x}|^{n+1}} \tag{6}$$

(we have $E(x)|_{x_0=0} = -\frac{1}{2}h(\boldsymbol{x})\mathbf{e}_1$) which is the fundamental solution of the generalized Cauchy-Riemann differential equations:

$$\sum_{k=0}^{n} \mathbf{e}_k \frac{\partial}{\partial x_k} f(x) = 0 \ . \tag{7}$$

If a (Clifford valued) function $f(x)$ solves this system of differential equations, it is called a (left) *monogenic* function. Therefore, the monogenic function is the multidimensional generalization of the analytic function.

The analytic signal got its name from the analytic function because the Hilbert transform is identical to a convolution with the 1D Cauchy kernel for $x_0 = 0$ (up to a factor of two) and therefore, the Hilbert transform of an analytic signal reads

$$f_A(\boldsymbol{x}) = -\frac{1}{\pi} \int_{\mathbb{R}} \frac{f_A(\boldsymbol{t})}{\boldsymbol{x} - \boldsymbol{t}} \, dt \tag{8}$$

which is quite similar to the Cauchy formula for analytic functions (for details see [ ]).

Since the Riesz transform is a convolution with the nD Cauchy kernel for $x_0 = 0$ (up to a factor of two) and the Riesz transform of the signal

$$f_M(\boldsymbol{x}) = f(\boldsymbol{x}) - f_H(\boldsymbol{x})\mathbf{e}_1 \tag{9}$$

fulfills

$$f_M(\boldsymbol{x}) = \frac{\Gamma((n+1)/2)}{\pi^{(n+1)/2}} \int_{\mathbb{R}^n} \frac{\overline{(\boldsymbol{x} - \boldsymbol{t})} f_M(\boldsymbol{t})}{|\boldsymbol{x} - \boldsymbol{t}|^{n+1}} \, dt \tag{10}$$

which can be obtained from the generalized Cauchy formula for monogenic functions, the signal $f_M(\boldsymbol{x})$ is called the *monogenic signal.*

We conclude this section with a last remark. The monogenic signal can be obtained in three ways:

1. by the transfer function $1 - \frac{\boldsymbol{u}}{|\boldsymbol{u}|}\mathbf{e}_1$
2. by the convolution kernel $\delta_0 + \frac{\Gamma((n+1)/2)}{\pi^{(n+1)/2}} \frac{\bar{\boldsymbol{x}}}{|\boldsymbol{x}|^{n+1}}\mathbf{e}_1$, and
3. by a modified inverse Fourier transform (see [ ] for the 2D case)

Finally, the proofs of the relations from Clifford analysis can be found in [ ] and some proofs of the relations of the monogenic signal can be found in [ ].

4 Spherical Quadrature Filters

In practical cases of signal processing, signals are of finite length. Therefore, the Hilbert transform is calculated for a bandpass filtered version of the signal. The Hilbert transform of the bandpass filter and the bandpass filter itself form a pair of quadrature filters. This approach can also be applied to the Riesz transform in order to obtain the multidimensional generalization of quadrature filters: the *spherical quadrature filters* (SQF).

The SQF are an $(n+1)$-tuple of filters which are created by a radial bandpass filter and the convolution of the Riesz kernel (n components) with this bandpass filter. The energy of the filter is isotropic (if the effect of the cubic filter mask can be neglected, see e.g. [,]) and it estimates the local amplitude, local phase and local orientation with only $n+1$ convolutions. Hence, it is quite fast and should be real time capable.

The SQF are somehow related to the steerable quadrature filters (e.g. [ ]), since the vector part is steerable. But there are some differences: firstly, the steered quadrature pair is an approximation to a *classical* quadrature filter with arbitrary preference orientation.

Secondly, the orientation parameter is directly obtained from the vector part of the SQF in contrast to the steered quadrature filters. The reason for this is the degree of the polynomials which has to be at least two in order to obtain a sufficient approximation of the Hilbert transform. The SQF correspond to polynomials of degree zero and one, such that the filter responses are constant and linear with the orientation vector.

Thirdly, the radial bandpass is derived from a Gaussian function in [ ], whereas for the spherical quadrature filters any bandpass can be chosen. In our approach, we use the lognormal bandpass because it has some fundamental advantages wrt. the Gaussian function (see also [ ]): it allows arbitrary large bandwidth while always being DC-free.

Therefore, our bandpass filter is represented in the frequency domain by

$$B(\boldsymbol{u}) = \exp\left(-\frac{(\log(|\boldsymbol{u}|/2^k))^2}{2(\log(c))^2}\right) \tag{11}$$

where k is a constant indicating the center frequency and c is a constant indicating the bandwidth of the bandpass (e.g. $c = 0.55$ corresponds to two octaves). The transfer functions and impulse responses of some filters are illustrated in Fig. .

We applied the 2D SQF to some natural and synthetic images and the 3D filters to a synthetic image sequence[2]. The results of the 2D experiments (synthetic images) can be found in Fig. . The amplitude of the filter responses of the Siemens star and the modulated ring show the isotropy of the spherical quadrature filters. The local orientation and the local phase are represented as grey values which vary linearly in angular and radial direction, respectively.

The experiments with real images[3] are shown in Fig. . The local amplitude indicates where to find edges. The absolute value depends on both, the markedness of the structure and the local contrast. The images of the local orientation and the local phase are masked by a threshold of the local amplitude. In the

[2] The 3D results can be found as mpeg-movies at the homepage of the first author (URL: http://www.ks.informatik.uni-kiel.de/~mfe). The normal vector of the plane is estimated with an error of less than 0.1°.

[3] URL: http://www-syntim.inria.fr/syntim/analyse/images-eng.html

Fig. 2. Spherical quadrature filters (lognormal radial bandpass). Upper row: even filters, bottom row: odd filters. Left column: frequency domain, middle and right column: spatial domain with different c/k-ratio

Fig. 3. Experiments with synthetic 2D data. Siemens star (left column, top and bottom), signal and amplitude of filter response, modulated ring (middle column, top) and filter response (amplitude: bottom middle, local orientation: upper right, and local phase: bottom right)

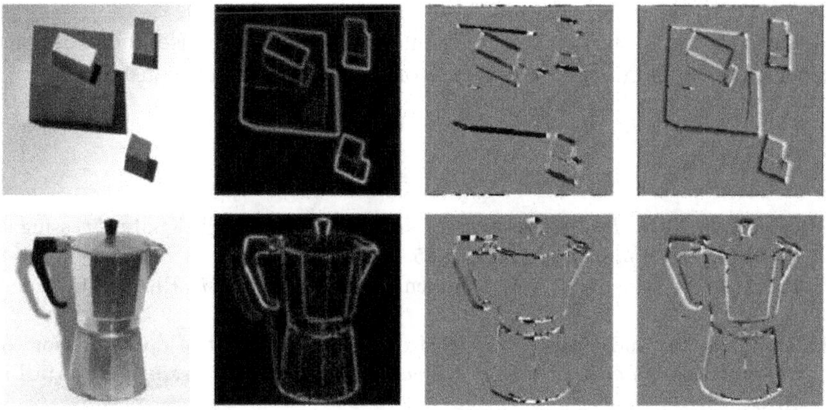

Fig. 4. Experiments with real 2D data. From left to right: original image, local amplitude, local orientation and local phase, images from INRIA-Syntim ©

areas of very low local amplitude the phase and orientation is irrelevant due to noise. Note that the phase and the orientation are cyclic. Therefore, white indicates nearly the same angle as black.

5 Conclusion

In our opinion, the monogenic signal is the consequent multidimensional generalization of the analytic signal. It is seamless embedded into the theory of Clifford analysis. The Riesz transform is an elegant way to overcome the problem of odd isotropic multi-dimensional filters and therefore, it is the best way to generalize the Hilbert transform, not only from a mathematician's point of view but also from the perspective of a signal-theorist.

The spherical quadrature filters are more capable than the classical quadrature filters for higher dimensions. Besides the complete representation of local structure, they are linear and of lower complexity than classical approaches (e.g. structure tensor). Due to linearity, they form a good basis for the design of linear and non-linear second level filters.

The algebraic representation of the filter response is geometrically insightful. The interpretation of the data is directly given by the involved geometry, all calculations can be vividly designed. This proves once more the power of geometric algebra.

Acknowledgments

The authors would like to thank Thomas Bülow for many useful discussions and especially for giving some references concerning the Riesz transform.

References

1. Bracewell, R. N. *Two-Dimensional Imaging.* Prentice Hall signal processing series. Prentice Hall, Englewood Cliffs, 1995.
2. Brackx, F., Delanghe, R., and Sommen, F. *Clifford Analysis.* Pitman, Boston, 1982.
3. Brady, J. M., and Horn, B. M. P. Rotationally symmetric operators for surface interpolation. *Computer Vision, Graphics, and Image Processing 22* 1 (April 1983). 70–94.
4. Felsberg, M., and Sommer, G. Structure multivector for local analysis of images. Tech. Rep. 2001, Institute of Computer Science and Applied Mathematics, Christian-Albrechts-University of Kiel, Germany, February 2000.
5. Freeman, W. T., and Adelson, E. H. The design and use of steerable filters. *IEEE Transactions on Pattern Analysis and Machine Intelligence 13,* 9 (September 1991), 891–906.
6. Granlund, G. H. Hierarchical computer vision. In *Proc. of EUSIPCO-90, Fifth European Signal Processing Conference, Barcelona* (1990), L. Torres, E. Masgrau, and M. A. Lagunas, Eds., pp. 73–84.
7. Granlund, G. H., and Knutsson, H. *Signal Processing for Computer Vision.* Kluwer Academic Publishers, Dordrecht, 1995.
8. Haglund, L. *Adaptive Multidimensional Filtering.* PhD thesis, Linkoping University, 1992.
9. Hahn, S. L. *Hilbert Transforms in Signal Processing.* Artech House, Boston, London,1996.
10. Hestenes, D., and Sobczyk, G. *Clifford algebra to geometric calculus, A Unified Language for Mathematics and Physics.* Reidel, Dordrecht, 1984.
11. Jahne, B. *Digitale Bildverarheitung.* Springer, Berlin, 1997.
12. Kovesi, P. *Invariant Measures of Image Features from Phase Information.* PhD thesis, University of Western Australia, 1996.
13. Kovesi, P. Image features from phase information. *Videre: Journal of Computer Vision Research 1,* 3 (1999).
14. Krieger, G., and Zetzsche, C. Nonlinear image operators for the evaluation of local intrinsic dimensionality. *IEEE Transactions on Image Processing* 5, 6 (June 1996). 1026–1041.
15. Merron, J., and Brady, M. Isotropic gradient estimation. In *IEEE Computer Vision and Pattern Recognition* (1996), pp. 652–659.
16. Michaelis, M. *Low Level Image Processing Using Steerable Filters.* PhD thesis, Christian-Albrechts-University of Kiel, 1995.
17. Nabighian, M. N. Toward a three-dimensional automatic interpretation of potential field data via generalized Hilbert transforms: Fundamental relations. *Geophysics 49,* 6 (June 1984), 780–786.
18. Oppenheim, A. and Lim, J. The importance of phase in signals. *Proc. of the IEEE 69,* 5 (May 1981), 529–541.

19. Porteous, I. R. *Clifford Algebras and the Classical Groups.* Cambridge University Press, 1995.
20. Sommer, G. The global algebraic frame of the perception-action cycle. In *Handbook of Computer Vision and Applications* (1999), B. Jahne, H. Haufiecker, and P. Geissler, Eds., vol. 3, Academic Press, San Diego, pp. 221–264.
21. Stein, E., and Weiss, G. *Introduction to Fourier Analysis on Euclidean Spaces.* Princeton University Press, New Jersey, 1971.

Sparse Feature Maps in a Scale Hierarchy

Per-Erik Forssén and Gösta Granlund

Computer Vision Laboratory, Department of Electrical Engineering
Linköping University, SE-581 83 Linköping, Sweden

Abstract. This article describes an essential step towards what is called a view centered representation of the low-level structure in an image. Instead of representing low-level structure (lines and edges) in one compact feature map, we will separate structural information into several feature maps, each signifying features at a *characteristic phase*, in a specific scale. By characteristic phase we mean the phases 0, π, and $\pm\pi/2$, corresponding to bright, and dark lines, and edges between different intensity levels, or colours. A lateral inhibition mechanism selects the strongest feature within each local region of scale represented. The scale representation is limited to maps one octave apart, but can be interpolated to provide a continous representation. The resultant image representation is *sparse*, and thus well suited for further processing, such as pattern detection.

Keywords: sparse coding, image representation, view centered representation, edge detection, scale hierarchy, characteristic phase

1 Introduction

From neuroscience we know that biological vision systems interpret visual stimuli by separation of image features into several retinotopic maps [ ]. These maps encode highly specific information such as colour, structure (lines and edges), motion, and several high-level features not yet fully understood. This feature separation is in sharp contrast to what many machine vision applications do when they synthesize image features into objects. We have earlier discussed these two approaches, which are called *view centered*, and *object centered* image representations [ ]. This report describes an attempt to move one step further towards a view centered representation of low level properties.

As we move upwards in the interpretation hierarchy in biological vision systems, the cells within each feature map tend to be increasingly selective, and consequently the high level maps tend to employ more sparse representations [ ]. There are several good reasons why biological systems employ sparse representations, many of which could also apply to machine vision systems.

Sparse coding tends to minimize the activity within an over-complete feature set, whilst maintaining the information conveyed by the features. This leads to representations in which pattern recognition, template storage, and matching are made easier [ ]. Compared to compact representations, sparse features convey

G. Sommer and Y. Y. Zeevi (Eds.): AFPAC 2000, LNCS 1888, pp. 186– , 2000.

more information when they are active, and contrary to how it might appear, the amount of computation will not be increased significantly, since only the *active* features need to be considered.

Most feature generation procedures employ filtering in some form. The outputs from these filters tell quantitatively more about the filters used than the structures they were meant to detect. We can get rid of this excessive load of data, by allowing only certain phases of output from the filters to propagate further. These characteristic phases have the property that they give invariant structural information rather than all the phase components of a filter response.

The feature maps we generate describe image structure in a specific scale, and at a specific phase. The distance between the different scales is one octave (i.e. each map has half the center frequency of the previous one.) The phases we detect are those near the *characteristic phases* [1] 0, π, and $\pm\pi/2$. Thus, for each scale, we will have three resultant feature maps (see figure).

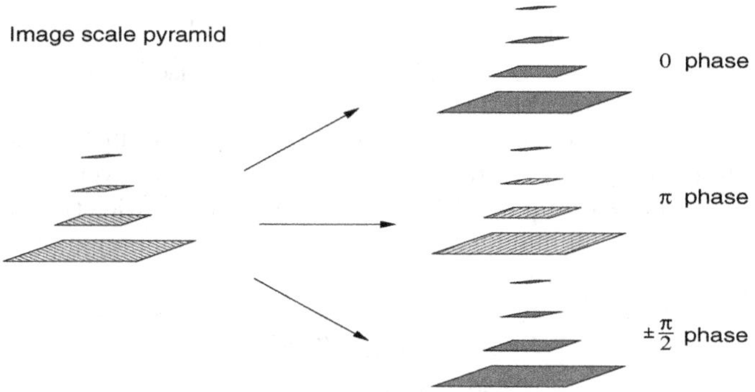

Fig. 1. Scale hierarchies

This approach touches the field of scale-space analysis pioneered by Witkin [ ]. See [ ] for a recent overview. Our approach to scale space analysis is somewhat similar to that of Reisfield [ ]. Reisfield has defined what he calls a *Constrained Phase Congruency Transform* (CPCT), that maps a pixel position and an orientation to an energy value, a scale, and a symmetry phase (0, π, $\pm\pi/2$, or none). We will instead map each image position, at a given scale, to three complex numbers, one for each of the characteristic phases. The argument of the complex numbers indicates the dominant orientation of the local image region at the given scale, and the magnitude indicates the local signal energy when the phase is near the desired one. As we move away from the characteristic phase, the magnitude will go to zero. This representation will result in a number of complex valued images that are quite sparse, and thus suitable for pattern detection.

[1] We will define the concept of characteristic phase in a following section.

2 Phase from Line and Edge Filters

For signals containing multiple frequencies, the phase is ambiguous, but we can always define the *local phase* of a signal, as the phase of the signal in a narrow frequency range.

The local phase can be computed from the ratio between a band-pass filter (even, denoted f_e) and it's quadrature complement (odd, denoted f_o). These two filters are usually combined into a complex valued *quadrature filter*, $\boldsymbol{f} = f_e + i f_o$ [ ]. The real and imaginary parts of a quadrature filter correspond to line, and edge detecting filters respectively. The local phase can now be computed as the argument of the filter response, $\boldsymbol{q}(x)$, or if we use the two real-valued filters separately, as the four quadrant inverse tangent; $\arctan(q_o(x), q_e(x))$.

To construct the quadrature pair, we start with a discretized lognormal filter function, defined in the frequency domain.

$$R_i(\rho) = \begin{cases} e^{-\dfrac{\ln^2(\rho/\rho_i)}{\ln 2}} & \text{if } \rho > 0 \\ 0 & \text{otherwise} \end{cases} \tag{1}$$

The parameter ρ_i determines the peak of the lognorm function, and is called the center frequency of the filter. We now construct the even and odd filters as the real and imaginary parts of an inverse discrete Fourier transform of this filter.

$$f_{e,i}(x) = \mathrm{Re}(\mathrm{IDFT}\{R_i(\rho)\}) \tag{2}$$

$$f_{o,i}(x) = \mathrm{Im}(\mathrm{IDFT}\{R_i(\rho)\}) \tag{3}$$

We write a filtering of a sampled signal, $s(x)$, with a discrete filter $f_k(x)$ as $q_k(x) = (s * f_k)(x)$, giving the response signal the same indices as the filter that produced it.

3 Characteristic Phase

By *characteristic phase* we mean phases that are consistent over a range of scales, and thus characterize the local image region. In practise this only happens at local magnitude peaks of the responses from the even, and odd filters. In other words, the characteristic phases are always one of 0, π, and $\pm\pi/2$.

Only some occurrences of these phases are consistent over scale though (see figure). First, we can note that band-pass filtering always causes ringings in

[2] Note that there are other ways to obtain spatial filters from frequency descriptions that, in many ways produce better filters [ ].

[3] A peak in the even response will always correspond to a zero crossing in the odd response, and vice versa, due to the quadrature constraint.

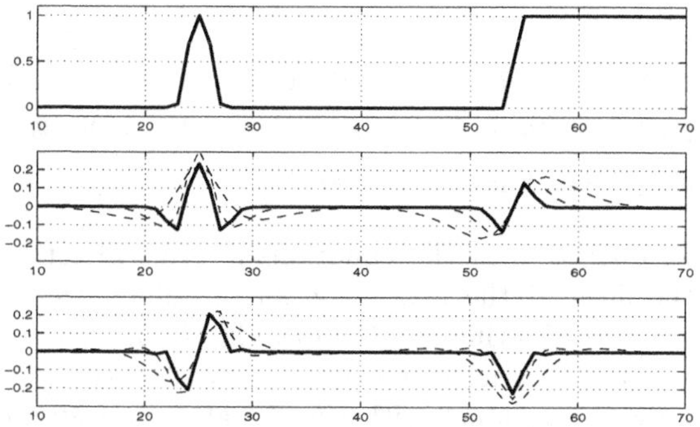

Fig. 2. Line and edge filter responses in 1D
Top: A one-dimensional signal.
Center: Line responses at $\rho_i = \pi/2$ (solid), and $\pi/4$ and $\pi/8$ (dashed)
Bottom: Edge responses at $\rho_i = \pi/2$ (solid), and $\pi/4$ and $\pi/8$ (dashed)

the response. For isolated line and edge events this will mean one extra magnitude peak (with the opposite sign) at each side of the peak corresponding to the event. These extra peaks will move when we change frequency bands, and consequently they do not correspond to characteristic phases. Second, we can note that each line event will produce one magnitude peak in the line response, and two peaks in the edge response. The peaks in the edge response, however, are not consistent over scale. Instead they will move as we change frequency bands. This phenomenon can be used to sort out the desired peaks.

4 Extracting Characteristic Phase in 1D

Starting from the line and edge filter responses at scale i, $q_{e,i}$, and $q_{o,i}$, we now define three *phase channels*:

$$p_{1,i} = \max(0, q_{e,i}) \tag{4}$$

$$p_{2,i} = \max(0, -q_{e,i}) \tag{5}$$

$$p_{3,i} = \text{abs}(q_{o,i}) \tag{6}$$

That is, we let $p_{1,i}$ constitute the positive part of the line filter response, corresponding to 0 phase, $p_{2,i}$, the negative part, corresponding to π phase, and $p_{3,i}$ the magnitude of the edge filter response, corresponding to $\pm\pi/2$ phase.

Phase invariance over scale can be expressed by requiring that the signal at the next lower octave has the same phase:

$$p_{1,i} = \max(0, q_{e,i} \cdot q_{e,i-1}/a_{i-1}) \cdot \max(0, \text{sign}(q_{e,i})) \quad (7)$$
$$p_{2,i} = \max(0, q_{e,i} \cdot q_{e,i-1}/a_{i-1}) \cdot \max(0, \text{sign}(-q_{e,i})) \quad (8)$$
$$p_{3,i} = \max(0, q_{o,i} \cdot q_{o,i-1}/a_{i-1}) \quad (9)$$

The first max operation in the equations above will set the magnitude to zero whenever the filter at the next scale has a different sign. This operation will reduce the effect of the ringings from the filters. In order to keep the magnitude near the characteristic phases proportional to the local signal energy, we have normalized the product with the signal energy at the next lower octave $a_{i-1} = \sqrt{q_{e,i-1}^2 + q_{o,i-1}^2}$. The result of this operation can be viewed as a phase description at a scale in between the two used. These channels are compared with the original ones in figure .

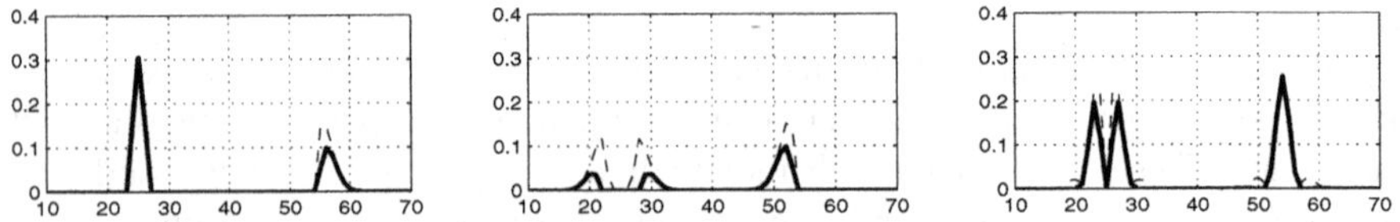

Fig. 3. Consistent phase in 1D. ($\rho_i = \pi/4$)
$p_{1,i}$, $p_{2,i}$, $p_{3,i}$ according to equations - (dashed), and equations - (solid)

We will now further constrain the phase channels in such a way that only responses consistent over scale are kept. We do this by inhibiting the phase channels with the complementary response in the third lower octave:

$$c_{1,i} = \max(0, p_{1,i} - \alpha \,\text{abs}(q_{o,i-2})) \quad (10)$$
$$c_{2,i} = \max(0, p_{2,i} - \alpha \,\text{abs}(q_{o,i-2})) \quad (11)$$
$$c_{3,i} = \max(0, p_{3,i} - \alpha \,\text{abs}(q_{e,i-2})) \quad (12)$$

We have chosen an amount of inhibition $\alpha = 2$, and the base scale, $\rho_i = \pi/4$. With these values we sucessfully remove the edge responses at the line event, and a the same time keep the rate of change in the resultant signal below the Nyquist frequency. The resultant characteristic phase channels will have a magnitude corresponding to the energy at scale i, near the corresponding phase. These channels are compared with the original ones in figure .

As we can see, this operation manages to produce channels that indicate lines and edges without any unwanted extra responses. An important aspect of this operation is that it results in a gradual transition between the description of a signal as a line or an edge. If we continuously increase the thickness of a line,

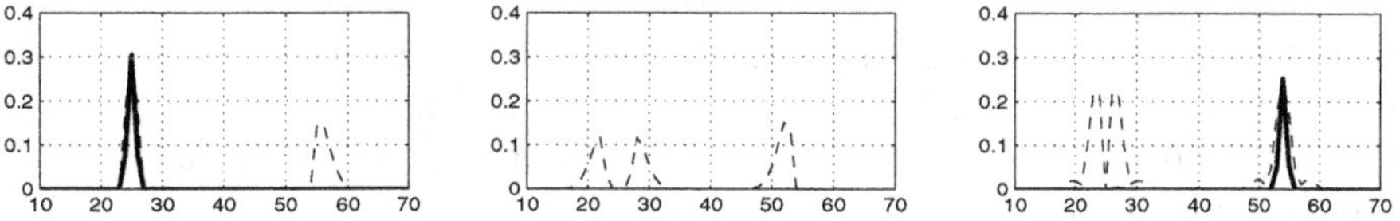

Fig. 4. Phase channels in 1D. ($\rho_i = \pi/4$, $\alpha = 2$)
$p_{1,i}$, $p_{2,i}$, $p_{3,i}$ according to equations - (dashed), and equations - (solid)

it will gradually turn into a bar that will be represented as two edges. This phenomenon is illustrated in figure .

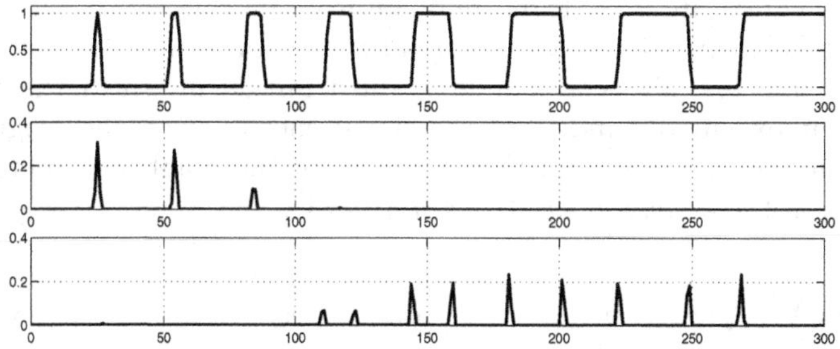

Fig. 5. Transition between line and edge description. ($\rho_i = \pi/4$)
Top: Signal Center: $c_{1,i}$ phase channel Bottom: $c_{3,i}$ phase channel

5 Local Orientation Information

The filters we employ in 2D will be the extension of the lognorm filter function (equation) to 2D [ ]:

$$\mathrm{F_{ki}}(\mathbf{u}) = \mathrm{R_i}(\rho)\,\mathrm{D_k}(\hat{\mathbf{u}}) \tag{13}$$

Where

$$\mathrm{D_k}(\hat{\mathbf{u}}) = \begin{cases} (\hat{\mathbf{u}} \cdot \hat{\mathbf{n}}_k)^2 & \text{if } \mathbf{u} \cdot \hat{\mathbf{n}}_k > 0 \\ 0 & \text{otherwise} \end{cases} \tag{14}$$

[4] Note that the fact that both the line, and the edge statements are low near the fourth event (positions 105 to 125) does not mean that this event will be lost. The final representation will also include other scales of filters, which will describe these events better.

We will use four filters, with directions $\hat{\mathbf{n}}_1 = (0\;1)^t$, $\hat{\mathbf{n}}_2 = (\sqrt{0.5}\;\sqrt{0.5})^t$, $\hat{\mathbf{n}}_3 = (1\;0)^t$, and $\hat{\mathbf{n}}_4 = (\sqrt{0.5}\;-\!\sqrt{0.5})^t$. These directions have angles that are uniformly distributed modulo π. Due to this, and the fact that the angular function decreases as $\cos^2\varphi$, the sum of the filter-response magnitudes will be orientation invariant [ ].

Just like in the 1D case, we will perform the filtering in the spatial domain:

$$(f_{e,ki} * p_{ki})(\mathbf{x}) \approx \mathrm{Re}(\mathrm{IDFT}\{\mathrm{F_{ki}}(\mathbf{u})\}) \tag{15}$$
$$(f_{o,ki} * p_{ki})(\mathbf{x}) \approx \mathrm{Im}(\mathrm{IDFT}\{\mathrm{F_{ki}}(\mathbf{u})\}) \tag{16}$$

Here we have used a filter optimization technique [ ] to separate the lognorm quadrature filters into two approximately one-dimensional components. The filter $p_{ki}(\mathbf{x})$, is a smoothing filter in a direction orthogonal to $\hat{\mathbf{n}}_k$, while $f_{e,ki}(\mathbf{x})$, and $f_{o,ki}(\mathbf{x})$ constitute a 1D lognorm quadrature pair in the $\hat{\mathbf{n}}_k$ direction.

Using the responses from the four quadrature filters, we can construct a *local orientation* image. This is a complex valued image, in which the magnitude of each complex number indicates the signal energy when the neighbourhood is locally one-dimensional, and the argument of the numbers denote the local orientation, in the *double angle representation* [ ].

$$\boldsymbol{z}(\mathbf{x}) = \sum_k a_{ki}(\hat{n}_{k1} + \boldsymbol{i}\hat{n}_{k2})^2 = a_{1i}(\mathbf{x}) - a_{3i}(\mathbf{x}) + \boldsymbol{i}(a_{2i}(\mathbf{x}) - a_{4i}(\mathbf{x})) \tag{17}$$

where $a_{ki}(\mathbf{x})$, the signal energy, is defined as $a_{ki} = \sqrt{q^2_{e,ki} + q^2_{o,ki}}$.

6 Extracting Characteristic Phase in 2D

To illustrate characteristic phase in 2D, we need a new test pattern. We will use the 1D signal from figure , rotated around the origin (see figure). The image has also been degraded with a small amount of Gaussian noise. The signal to noise ratio is 10 dB.

When extracting characteristic phases in 2D we will make use of the same observation as the local orientation representation does: Since visual stimuli can locally be approximated by a simple signal in the dominant orientation [ ], we can define the *local phase* as the phase of the dominant signal component.

To deal with characteristic phases in the dominant signal direction, we first synthesize responses from a filter in a direction, $\hat{\mathbf{n}}_z$, compatible with the local orientation.

$$\hat{\mathbf{n}}_z = (\,\mathrm{Re}(\sqrt{\boldsymbol{z}})\;\mathrm{Im}(\sqrt{\boldsymbol{z}})\,)^t \tag{18}$$

[5] Since the local orientation, $\boldsymbol{z}$, is represented with a double angle argument, we could just as well have chosen the opposite direction. Which one of these we choose does not really matter, as long as we are consistent.

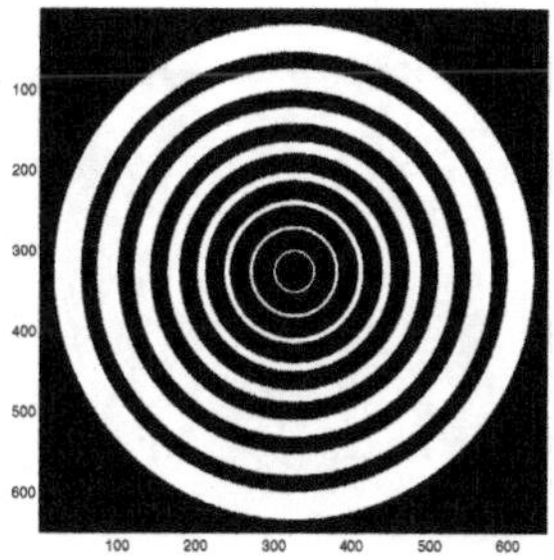

Fig. 6. A 2D test pattern. (10 dB SNR)

The filters will be weighted according to the value of the scalar product between the filter direction, and this orientation compatible direction.

$$w_k = \hat{\mathbf{n}}_k^t \hat{\mathbf{n}}_z \tag{19}$$

Thus, in each scale we synthesize one odd, and one even response projection as:

$$q_{e,i} = \sum_k q_{e,i,k} \operatorname{abs}(w_k) \tag{20}$$

$$q_{o,i} = \sum_k q_{o,i,k} w_k \tag{21}$$

This will change the sign of the odd responses when the directions differ more than π, but since the even filters are symmetric, they should always have a positive weight. In accordance with our findings in the 1D study (equations - , -), we now compute three phase channels, $c_{1,i}$, $c_{2,i}$, and $c_{3,i}$, in each scale.

The characteristic phase channels are shown in figure . As we can see, the channels exhibit a smooth transition from describing the white regions in the test pattern (see figure) as lines, and as two edges. Also note that the phase statements actually give the phase in the dominant orientation, and not in the filter directions, as was the case for CPCT [ ].

7 Local Orientation and Characteristic Phase

An orientation image can be be gated with a phase channel, $c_n(\mathbf{x})$, in the following way:

[6] The magnitude of lines this thin can be difficult to reproduce in print. However, the magnitudes in this plot *should* vary just like in figure .

Fig. 7. Characteristic phase channels in 2D. ($\rho_i = \pi/4$)
Left to right: Characteristic phase channels $c_{1,i}$, $c_{2,i}$, and $c_{3,i}$, according to equations - ($\alpha = 2$)

$$z_g(\mathbf{x}) = \begin{cases} 0 & \text{if } c_n(\mathbf{x}) = 0 \\ \frac{c_n(\mathbf{x}) \cdot z(\mathbf{x})}{\|z(\mathbf{x})\|} & \text{otherwise} \end{cases} \tag{22}$$

We now do this for each of the characteristic phase statements $c_{1,i}(\mathbf{x})$, $c_{2,i}(\mathbf{x})$, and $c_{3,i}(\mathbf{x})$, in each scale. The magnitude of the result is shown in figure . Notice for instance how the bridge near the center of the image changes from being described by two edges, to being described as a bright line, as we move through scale space.

8 Concluding Remarks

The strategy of this approach for low-level representation is to provide sparse, and reliable statements as much as possible, rather than to provide statements in all points.

Traditionally, the trend has been to combine or merge descriptive components as much as possible; mainly to reduce storage and computation. As the demands on performance are increasing it is no longer clear why components signifying different phenomena should be mixed. An edge is something separating two regions with different properties, and a line is something entirely different.

The use of sparse data representations in computation leads to a mild increase in data volume for separate representations, compared to combined representations.

Although the representation is given in discrete scales, this can be viewed as a conventional sampling, although in scale space, which allows interpolation between these discrete scales, with the usual restrictions imposed by the sampling theorem. The requirement of a good interpolation between scales determines the optimal relative bandwidth of filters to use.

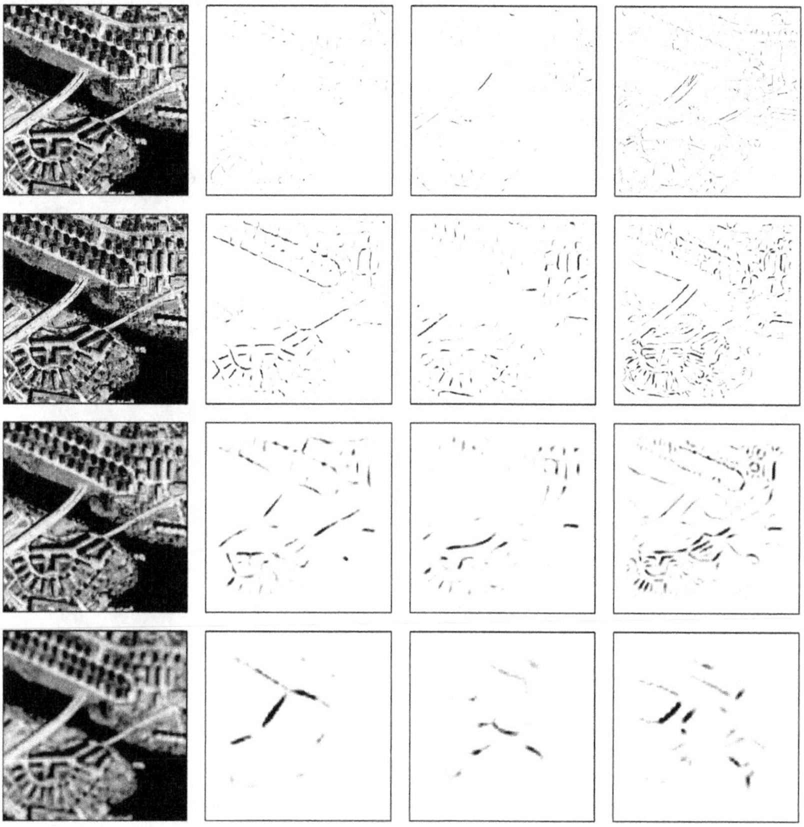

Fig. 8. Sparse feature hierarchy. ($\rho_i = \{\pi/2, \pi/4, \pi/8, \pi/16\}$)

Acknowledgements

The work presented in this paper was supported by WITAS, the Wallenberg laboratory on Information Technology and Autonomous Systems, which is greatfully acknowledged.

References

1. Andrew Witkin. (1983) *Scale-space filtering.* In Proc. International Joint Conference on Artificial Intelligence, Karlsruhe, 1983.
2. Tony Lindeberg. (1994) *Scale-space Theory in Computer Vision.* Kluwer Academic Publishers 1994. ISBN 0792394186.
3. Field, D. J. (1994) *What is the goal of sensory coding?* Neural Computation, 6:559-601, 1994.
4. Reisfeld, D. (1996) *The Constrained Phase Congruency feature detector: simultaneous localization, classification, and scale determination.* Pattern Recognition letters 17(11) 1996 pp. 1161-1169.

5. Bear, M. F. et al. (1996) *Neuroscience: Exploring the Brain.* Williams & Wilkins, ISBN 0-683-00488-3.
6. Granlund, G. H., Knutsson, H. (1995) *Signal Processing for Computer Vision.* Kluwer Academic Publishers, ISBN 0-7923-9530-1.
7. Knutsson, H., Andersson, M., Wiklund J. (1999) *Advanced Filter Design.* Proceedings of SCIA, 1999.
8. Granlund, G. (1999) *The complexity of vision.* Signal Processing 74, pp 101-126.

Estimation and Tracking of Articulated Motion Using Geometric Algebra

Hiniduam Udugama Gamage Sahan Sajeewa and Joan Lasenby

Department of Engineering, University of Cambridge
Trumpington Street, Cambridge CB2 1PZ, UK
ssh23,jl@eng.cam.ac.uk
http://www-sigproc.eng.cam.ac.uk/~ssh23,~jl/

Abstract. In this paper estimation and extraction of complex articulated rotational motion is addressed using the mathematical framework of geometric algebra (GA). The basis of the method outlined here is the ability to optimise expressions with respect to rotors (the quantities that perform rotations) and rotational bivectors – something which would be very difficult to do with conventional representations. As an application of these techniques we then look at the problem of tracking in optical motion capture and give some results on real and simulated data. It will also be illustrated how the same mathematics can be used for model extraction and inverse kinematics.

Keywords: Rotations, geometric algebra, articulated motion, motion estimation, motion modelling, tracking.

1 Introduction

Estimating rotational motion in 3D is important in many real-world applications; e.g. tracking, human motion analysis, biomechanics, robotics and animation. It is often the case in such applications that substantial effort is spent on finding efficient methods for representing and estimating rotations. In this paper we will show that the rotor formulation used in geometric algebra has distinct advantages over more conventional representations in terms of both efficiency and consistency. In particular, the multivector calculus, which enables us to differentiate wrt any element of the algebra (rotors, bivectors etc.) allows us to optimise over the correct manifolds.

2 An Introduction to Geometric Algebra

There are now many good introductions to GA; detailed treatments of GA can be found in [ ], [ ] and briefer introductions in [ ], [ ]. A tutorial introduction with a GA software package called GABLE , which is used extensively for simulations in this paper, is available at [ ]. Here, we therefore give only the briefest of introductions. In what follows we adopt the convention that vectors will be

G. Sommer and Y. Y. Zeevi (Eds.): AFPAC 2000, LNCS 1888, pp. 197– , 2000.

written as lower case roman letters (not bold, except where we discuss state and observation vectors in the tracking) while multivectors will be written as upper case roman letters.

Let $\mathcal{G}_n$ denote the geometric algebra of n-dimensions – this is a graded linear space. As well as vector addition and scalar multiplication we have a non-commutative product which is associative and distributive over addition – this is the **geometric** or **Clifford** product. A further distinguishing feature of the algebra is that any vector squares to give a scalar. The geometric product of two vectors a and b is written ab and can be expressed as a sum of its symmetric and antisymmetric parts

$$ab = a \cdot b + a \wedge b, \tag{1}$$

where the inner product $a \cdot b$ and the outer product $a \wedge b$ can therefore be defined in terms of the more fundamental geometric product. The crucial point here is that the geometric product is **invertible** unlike either the inner or outer products.

The inner product of two vectors is the standard *scalar* or *dot* product and produces a scalar. The outer or wedge product of two vectors is a new quantity we call a **bivector**. We think of a bivector as a directed area in the plane containing a and b, formed by sweeping a along b. The outer product of k vectors is a k-vector or k-blade, and such a quantity is said to have *grade* k. A general **multivector**, A, is made up of a linear combination of objects of different grade, i.e.

$$A = \langle A \rangle_0 + \langle A \rangle_1 + \langle A \rangle_2 + \ldots + \langle A \rangle_n \tag{2}$$

where $\langle A \rangle_r$ is a pure r-blade. GA provides a means of manipulating multivectors which allows us to keep track of different grade objects simultaneously – much as one does with complex number operations. For a general multivector A, the notation $\langle A \rangle$ will mean *take the scalar part of* A. $\tilde{A}$ is *reversion* and tells us to reverse the order of all vectors in A, i.e. $(abc)\tilde{} = cba$.

We can multiply together any two multivectors using the geometric product and we can also define the inner and outer products between arbitrary multivectors as the following grade-lowering and grade-raising operations: $A_r \cdot B_s = \langle A_r B_s \rangle_{|r-s|}$ and $A_r \wedge B_s = \langle A_r B_s \rangle_{|r+s|}$. Of particular interest to us in this paper are the GA quantities which perform rotations. It can be shown [ ] that in any dimension, the multivectors which represent rotations can be written as $R = \pm e^B$ where B is a pure bivector representing the plane in which the rotation occurs and R satisfies $R\tilde{R} = 1$. R is called a *rotor* and has a two-sided action, i.e. a vector a will be rotated into a vector a' where $a' = Ra\tilde{R}$. We can see here that the concept of rotation takes the same form in any dimension and that we can rotate any quantities, not simply vectors, e.g. $R(a \wedge b)\tilde{R} = (Ra\tilde{R}) \wedge (Rb\tilde{R})$.

We now have an algebra whose basic elements are these 'geometric' entities of different dimension. It is also possible to show that there is a well-defined linear algebra and calculus framework on the algebra.

3 Review of Geometric Calculus

We will see in later sections that when we formulate our models in terms of rotors we are led to optimisation processes over rotor and bivector manifolds. In order to carry out such optimisations we require some general results from geometric calculus which we discuss in this section.

3.1 Differentiation with Respect to Multivectors

If X is a mixed-grade multivector, $X = \sum_r X_r$, and $F(X)$ is a general multivector-valued function of X, then the derivative of F in the A 'direction' is written as $A * \partial_X F(X)$ (here we use $*$ to denote the scalar part of the product of two multivectors, i.e. $A * B \equiv \langle AB \rangle$), and is defined as (with τ a scalar)

$$A * \partial_X F(X) \equiv \lim_{\tau \to 0} \frac{F(X + \tau A) - F(X)}{\tau}. \tag{3}$$

For the limit on the right hand side to make sense, A must contain only grades which are contained in X and no others. If X contains no terms of grade-r and A_r is a homogeneous multivector, then we define $A_r * \partial_X = 0$. We can now use the above definition of the directional derivative to formulate a general expression for the multivector derivative ∂_X without reference to one particular direction. This is accomplished by introducing an arbitrary frame $\{e_j\}$ and extending this to a basis (vectors, bivectors, etc..) for the entire algebra, $\{e_J\}$. Then ∂_X is defined as

$$\partial_X \equiv \sum_J e^J (e_J * \partial_X) \tag{4}$$

where $\{e^J\}$ is an extended basis built out of the reciprocal frame (if $\{e_i\}$ is a basis for the space then $\{e^j\}$, where $e_i \cdot e^j = \delta_i^j$, is the reciprocal frame). The directional derivative, $e_J * \partial_X$, is only non-zero when e_J is one of the grades contained in X (as previously discussed) so that ∂_X inherits the multivector properties of its argument X.

Another useful notation is

$$\underline{f}(A) = (A * \partial_X) F(X) \tag{5}$$

where $\underline{f}$ is termed the *differential* of F and is a linear function of A acting at X – here we see the derivative as an approximation to the locally linearised form of the function.

3.2 The Taylor Expansion of Multivector Quantities

The Taylor expansion is useful in deriving identities in multivector calculus. Let X, A and $F(X)$ be as above and τ be a scalar, then the Taylor expansion of $F(X + \tau A)$ is defined as

$$F(X + \tau A) = \sum_{k=0}^{\infty} \frac{\tau^k}{k!} \left[(A * \partial_X)^k F(X) \right] \tag{6}$$

where $(A * \partial_X)^k F(X) \equiv [(A * \partial_X) ... (A * \partial_X) [(A * \partial_X) F(X)]]$. Here $(A * \partial_X)^k$ is a grade preserving operator therefore resulting in correct grade-matching in equation ( ). It can easily be verified that this form gives the correct expansion for some simple cases (e.g. use $F(X) = X^2$ with X a bivector).

3.3 The Chain Rule for Multivector Differentiation

In any practical analysis the use of the chain rule is essential. For multivector differentiation it is formulated as follows. Define

$$H(X) \equiv G[F(X)].$$

Using equation ( ) and equation ( ) the chain rule for multivector differentiation can be expressed [ ], [ ] as

$$(A * \partial_X) H(X) = \left[\{(A * \partial_X) F(X)\} * \partial_{F(X)}\right] G[F(X)] \tag{7}$$

Writing equation ( ) in terms of differentials

$$\underline{h}(A) = \underline{g}(\underline{f}(A)) \tag{8}$$

we have the remarkable result that when expressed in this way the chain-rule has a very simple intuitive form.

4 Rotor Estimation

In many tracking and computer vision problems it may be desirable to represent the set of rotations that describe the motion as a time evolving state. Suppose we have a vector which is undergoing some rotational motion in 3D and a series of measurements. If $R_n = e^{-B_n}$, where R_n and B_n are the rotor and bivector representing the rotation at time $t = n$, then it will generally be possible to write down Kalman-filter type equations governing the evolution of the state and the observations. If $\boldsymbol{v}_n^i$ is the ith ($i = 1, \ldots, L$) observation vector at time n and $\boldsymbol{u}_{n-1}^i$ is some estimate of $\boldsymbol{v}^i$ at the previous time-step, then one possible model is that our system evolves so that the changes in the rotational bivectors are small and the rotors approximately rotate $\boldsymbol{u}^i$ to $\boldsymbol{v}^i$, i.e.

$$B_n = B_{n-1} + \Delta B_n \tag{9}$$

$$\boldsymbol{v}_n^i = R_n \boldsymbol{u}_{n-1}^i \tilde{R}_n + \boldsymbol{q}_n^i \tag{10}$$

where ΔB_n is essentially the error in our smoothness model and $\boldsymbol{q}_n^i$ is the observation error. Since bivectors always square to give a scalar, we can write $B_n = \theta \hat{B}_n$ with $\hat{B}_n^2 = -1$, a *unit bivector.*

Given the state ( ) and observation ( ) equations we may now construct a weighted least squares approach in which we minimize the errors in these two equations. Under the assumption of Gaussian errors and subject to certain other conditions, [ ], such an approach would be optimal. Consider the minimisation of the following cost function:

$$C = \left[\sum_{i=1}^{L} \frac{(\boldsymbol{q}_n^i \cdot \boldsymbol{q}_n^i)}{w_i} \right] - \frac{(\Delta B_n)^2}{w_o} \tag{11}$$

Where w_i, w_o are weighting factors. Under certain conditions this is the same cost function as used in the Kalman Filter [ ] correction at a given time step. Essentially we have only a *correction* part and no prediction phase; however the simplicity of this formulation may be quite adequate for the particular tracking applications we will discuss and has been applied successfully in a number of similar cases [ ].

Differentiation of this expression with respect to B_n involves some straightforward but involved manipulation (details of the steps involved can be found in [ ]) eventually leading us to a bivector equation for the least squares solution to B_n;

$$\sum_{i=1}^{L} \frac{1}{w_i} \left\{ -\Gamma_i(B_n) + \frac{1}{\theta^2} \left\langle B_n \Gamma_i(B_n) \tilde{R}_n B_n R_n \right\rangle_2 \right\}$$
$$- \frac{\sin(\theta)}{w_i \theta^2} \left\langle \frac{B_n \Gamma_i(B_n) \tilde{R}_n B_n}{\theta} + \theta \Gamma_i(B_n) \tilde{R}_n \right\rangle_2 = \frac{1}{w_o}(B_n - B_{n-1}) \tag{12}$$

where $\Gamma_i(B_n) = \boldsymbol{u}_n^{i-1} \wedge \tilde{R}_n \boldsymbol{v}_n^i R_n$ and $\theta = |B_n| \equiv \sqrt{-B_n^2}$. Equation ( ) can be solved in practice either by simple iterative techniques or by some optimization method such as the Levenberg-Marquardt algorithm [ ]. Note here that we have found a least squares solution to the problem without introducing any coordinates or particular representation (Euler angles, direction cosines etc.) for the rotations – this guarantees that our solution will not depend on the coordinate frame that we choose to numerically solve our equations in. When we are talking about the evolution of a rotor field we are really talking about the evolution of the bivectors representing that field, it therefore seems natural that we should work with the bivectors as the basic quantities. Indeed it has been shown, [ ], [ ] that for rotational invariance we must interpolate rotations by interpolating the representative bivectors. It is also our view that fewer singularities will occur in this bivector approach.

5 Application to Articulated Motion

We now look at applying the mathematics of the previous section to the case where the rotations describe the movement of an articulated body. We begin by setting up the problem in the required form. Let us begin by considering the simple model of a number of linked rigid rods joined by revolute joints (with all rotational degrees of freedom). Let x_i^n be the position vector of the ith joint at time n, and define $x_{i,i+1}^n = x_{i+1}^n - x_i^n$. Let R_i^n be the rotor which represents

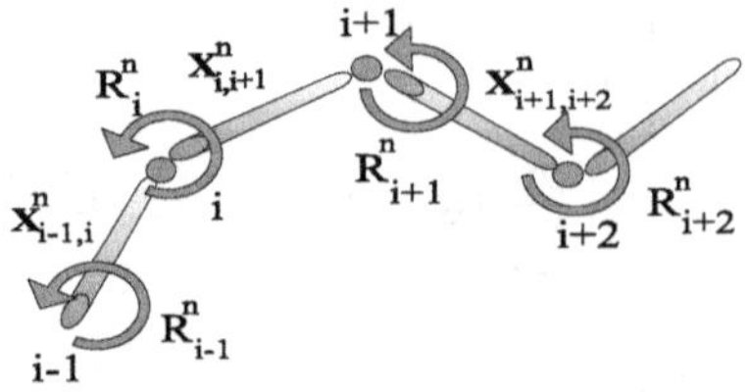

Fig. 1. A section of the articulated model

rotation about the ith joint at time n. If the motion at the ith joint is constrained so as only to allow rotations about a unit axis $\hat{n}_i$, then R_i^n would be a rotation of θ_i about the rotated $\hat{n}_i$ axis ($= R_{i-1}^n ... R_1^n \hat{n}_i \tilde{R}_1^n ... \tilde{R}_{i-1}^n$). In figure we would therefore be able to write, say $x_3^{n+1} = x_2^{n+1} + R_2^n R_1^n x_{2,3}^n \tilde{R}_1^n \tilde{R}_2^n$ – which can be generalised to give

$$x_{i,i+1}^{n+1} = R_i^n R_{i-1}^n \dots R_1^n (x_{i,i+1}^n) \tilde{R}_1^n \dots \tilde{R}_{i-1}^n \tilde{R}_i^n.$$

But it can be shown that [ ]

$$x_{i,i+1}^{n+1} = R_{1,0}^n R_{2,0}^n \dots R_{i-1,0}^n R_{i,0}^n (x_{i,i+1}^n) \tilde{R}_{i,0}^n \tilde{R}_{i-1,0}^n \dots \tilde{R}_{2,0}^n \tilde{R}_{1,0}^n, \tag{13}$$

where $R_{i,0}^n$ is about the $\hat{n}_i$ axis (i.e. previous rotations not being applied). We therefore see that the bivectors corresponding to the individual rotors are now independent.

5.1 Articulated Rotor Estimation

As in Section , we can now postulate an evolution model of the form

$$B_{i,0}^n = B_{i,0}^{n-1} + \Delta B_{i,0}^n \tag{14}$$
$$x_{i,i+1}^n = R_{1,0}^n R_{2,0}^n \dots R_{i-1,0}^n R_{i,0}^n x_{i,i+1}^{n-1} \tilde{R}_{i,0}^n \tilde{R}_{i-1,0}^n \dots \tilde{R}_{2,0}^n \tilde{R}_{1,0}^n$$

and an appropriate cost function (as in equation). Thus we are able to differentiate wrt each of the bivectors and iteratively solve the resulting equations as outlined in the previous section. We note here that if there are constraints at any of the joints (which, in practice is often the case), such constraints are easily allowed for in the model.

6 Results

In order to check the validity of the expressions above they were incorporated into a tracking problem. Here a real world articulated object was observed through a system of calibrated perspective cameras with arbitrary orientations and positions. Firstly the algorithms were applied to simulated data and secondly to real data taken with a 3-camera system. For the simulations we are able to compare the results with the known 'truth', whereas for the real data the performance of the algorithm was judged by its ability to track and reconstruct the object reliably.

6.1 Simulations

An articulated model which has four linked points with total rotational freedom at each point (except the last point) was simulated. Errors were added at the camera-plane level. The model was simulated with the dynamical behaviour described by ( ). Bivector values, lengths and the initial positions of the model were assigned randomly. The values ΔB^n and the projection of the observation noise onto the image plane were taken as pseudo-random sequences with Gaussian pdfs of variances a fraction of the maximum rotational bivector and a fraction of the maximum link length, respectively. Two experiments with the variance pairs given by $(0.001, 0.00001)$ and $(0.01, 0.01)$ were conducted and these two experiments will be referred to as *simul1* and *simul2*. Note that independence of the distributions wrt each other and wrt different time steps was assumed in these simulations. Although the assumptions about independence and Gaussianity do not strictly hold in practice, these formulations are generally a reasonable approximation [ ], [ ]. In the above experiments, correspondences in the image planes were achieved by a global point assignment scheme [ ]. Reconstruction of the 3D position was performed using the algorithm given in [ ]. Corrections to the bivectors were calculated using ( ) and these corrected bivectors were used to predict the world position at the next time instance. Only two iterations over the number of links were carried out, estimating each rotor/bivector value associated with each link. Several different weighting factors (w_i, w_0) were used and were assumed to be the same for every link. These weighting pairs were $(1, 1)$, $(1, 50)$ and $(50, 1)$ respectively. A set of experiments that only uses the previous observations as current predictions has results labelled as $(0, 0)$. The GA software package GABLE [ ] was used for the simulations. The sum of the squared difference between the actual world points and the predicted world points for each time instant (*sdiff*), the sum of link lengths at each time instant (*slink*) and the length of a single example link (*slength*) are plotted in Figure .

From Figure we can see that the weighting factors do not affect the results significantly for given variances. As expected the model preserves the lengths of the links. When the previous prediction is used as the current prediction, the lengths of links in the reconstruction change considerably, so we see that although it appears that this method gives smaller values of *sdiff*, it does so by violating the physical constraints of the system.

6.2 Real Data

The algorithms were tested on two sets of real optical motion capture data taken with a 50Hz 3-camera system; these data sets, referred to as *'golf1'* and *'jlwalk3'* represent the arm movements in a golf swing, and a person walking respectively. The captured data consisted of a number of bright points in each image, these coming from retroreflective markers on the subject.

Here the tracking algorithm used was as described for the simulated data but observed world points were used rather than real world points (now unknown).

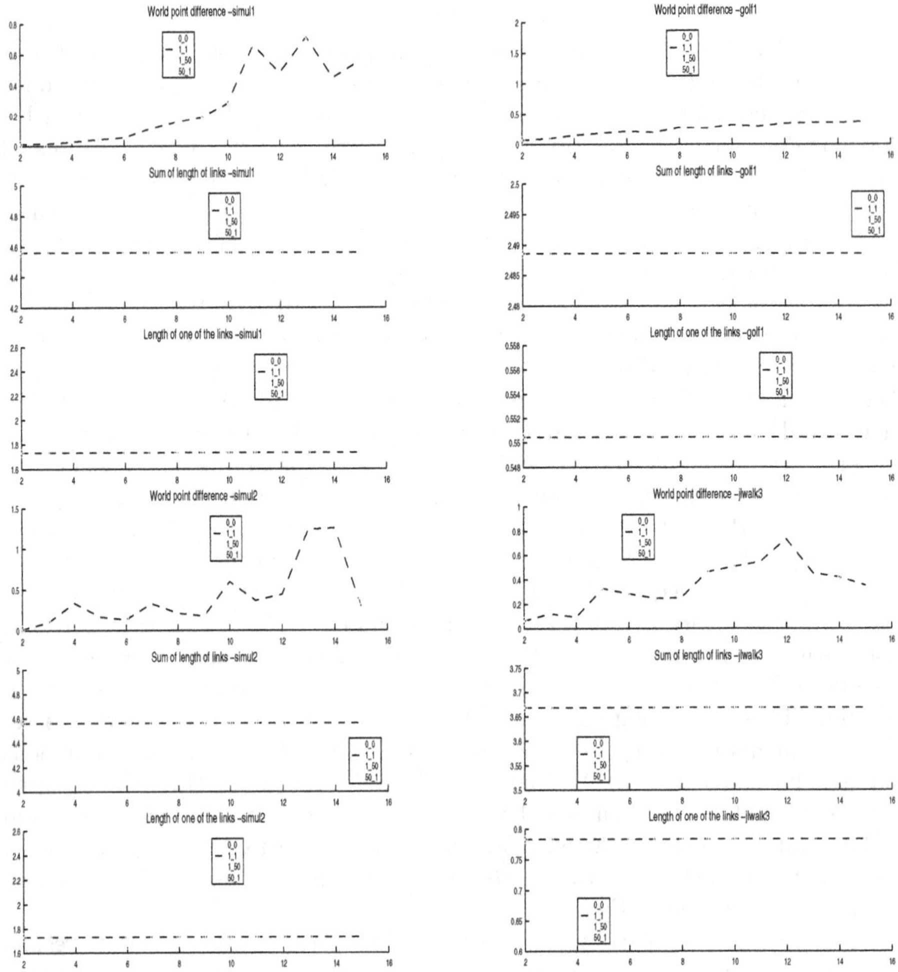

Fig. 2. First column shows the results of simulated data with variance pairs (0.001, 0.00001) and (0.01, 0.01) which has three plots for each variance pair. These correspond to *sdiff*, *slink* and *slength* plotted against frame number. Each graph contains plots for the different weighting schemes as shown in the legend. The 0_0 line indicates no estimation is performed and the previous observation is used as the current prediction. The second column has the results from real data sets *'golf1'* and *'jlwalk3'*. These also have two sets of three plots as in the simulated case

Initial rotor values were estimated using the method described in [ ], using 3 frames and manually identifying the correct association of links in images. Same quantities, *sdiff*, *slink* and *slength* are plotted in Figure except now for the calculation of *sdiff* the reconstructed world points (i.e. the world observation)

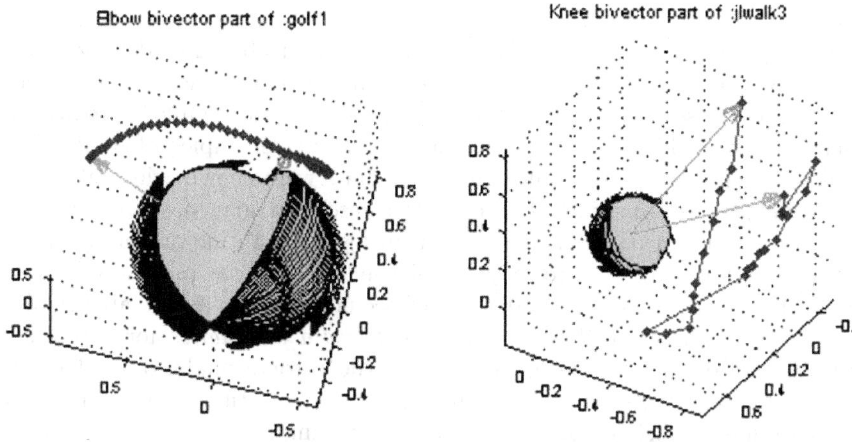

Fig. 3. Left-hand figure shows the magnitude and the direction of the bivector and the direction of the path taken by the normal unit vector to the bivector over time for the rotor at the elbow joint for *'golf1'* ((1, 1) weighting used). The right-hand figure is similar but shows the results for the knee joint rotor in *'jlwalk3'*

from the current image planes were used instead of the actual world point, which is of course, unknown.

It was found that in this real-data case, in order to achieve reliable tracking it was necessary to weight the *prediction* term more heavily. Again, it was found that using the previous observations as the current prediction led to poor reconstructions due to its inability to preserve the physical model. Overall the method performed well on the given data sets and was able to track and reconstruct all of the motions considered.

The advantage of estimating rotational bivectors is evident in Figure . It shows the rotational bivector estimated at the elbow and knee joints in *'golf1'* and *'jlwalk3'* respectively. It is possible to gain much more information from figures like these than one would get from scalar plots of Euler angles. For example, from figure it can be seen that the forearm has rotated roughly through the same angle between each frame wrt the elbow since the size of the bivectors shown (see the circular disc) have approximately the same diameter. We also see that the direction of the forearm in this estimated model changes smoothly over time and has a graceful transition when the arm starts to swing in the opposite direction. It is also interesting to look at the bivector evolution for the knee rotor; it is clear that in this case both the magnitude and orientation of the rotation plane change much more. This is indicative of real changes in the orientation of the knee and could present us with a useful visualisation tool for investigative studies on gait.

7 Conclusions and Future Work

In this paper we have discussed the use of geometric algebra in modelling evolving systems which can be described by rotors. In particular, the issue of tracking has been addressed where a least-squares technique which employed multivector calculus on the defining bivectors, was addressed. Examples of applying the resulting technique to real and simulated data were shown which illustrated the validity of the method. The concepts described may have many uses in motion analysis. In particular, as shown in figure , an important application of the techniques may be to extract relevant rotors from pre-tracked motion data – this can be used to understand specific motions, detect abnormalities etc. The method outlined has set up the *forward kinematics* for a particular articulated model – we also envisage that the rotor formulation will be very useful for *inverse kinematics* (inferring the state from the observations) which is of vital importance in control systems for walking robots etc.

References

1. C. Doran and A. Lasenby, "Physical applications of geometric algebra," 1999. Available at `http://www.mrao.cam.ac.uk/~clifford/ptIIIcourse/`.
2. D. Hestenes, *New Foundations for Classical Mechanics.* Kluwer Academic Publishers, 2 ed., 1999.
3. F. A. McRobie and J. Lasenby, "Simo-Vu Quoc rods using Clifford algebra," *International Journal for Numerical Methods in Engineering*, vol. 45, pp. 377–398, 1999.
4. J. Lasenby, W. Fitzgerald, A. Lasenby, and C. Doran, "New geometric methods for computer vision: An application to structure and motion estimation," *International Journal of Computer Vision*, vol. 26, no. 3, pp. 191–213, 1998.
5. J. Lasenby and A. Stevenson, "Using geometric algebra for optical motion capture," CUED Technical Report CUED/F-INENG/TR.374, Cambridge University Engineering Department, 2000, and to appear in *Geometric Algebra: A geometric approach to computer vision, neural and quantum computing, robotics and engineering*, Birkhauser Boston, 2000. ,
6. D. Hestenes, "New foundations for mathematical physics." Available at `http://modelingnts.la.asu.edu/html/NFMP.html`, 1998.
7. L. Dorst, S. Mann, and T. Bouma, "GABLE: A MATLAB tutorial for geometric algebra," 2000. Available at `http://www.carol.wins.uva.nl/~leo/clifford/gablebeta.html`. ,
8. D. Hestenes and G. Sobczyk, *Clifford Algebra to Geometric Calculus: A Unified Language for Mathematics and Physics.* D. Reidel: Dordrecht, 1984. ,
9. N. Ayache, *Artificial Vision for Mobile Robots: Stereo Vision and Multisensory Perception.* The MIT Press, Cambridge, Massachusetts, 1991.
10. B. D. Anderson and J. B. Moore, *Optimal Filtering.* Prentice-Hall, Englewood Cliffs, New Jersey, 1979. ,
11. J. M. Rehg and T. Kanade, "DigitEyes: Vision-Based human hand tracking," Tech. Rep. CMU-CS-93-220, School of Computer Science, Carnegie Mellon University, December 1993. ,

12. S. Gamage , "Applications of Multivector Calculus" CUED Technical Report CUED/F-INENG/TR.382, Cambridge University Engineering Department, 2000.
13. J. J. More, "The Levenberg-Marquardt algorithm: Implementation and theory," in *Numerical Analysis* (G. A. Watson, ed.), vol. 630 of *Lecture Notes in Mathematics*, pp. 105–116, Springer Verlag, 1977.
14. J. Clements. Beam buckling using geometric algebra. M.Eng. project report, Cambridge University Engineering Department, 1999.
15. M. A. Ringer and J. Lasenby. Tracking with Articulated Motion Models To appear in Proceedings of *Biosignal 2000, Czech Republic, July 2000.*

Geometric Properties of Central Catadioptric Projections

Christopher Geyer and Kostas Daniilidis
University of Pennsylvania, GRASP Laboratory*

Abstract. In this paper we consider all imaging systems that consist of reflective and refractive components –called catadioptric– and possessing a unique effective viewpoint. Conventional cameras are a special case of such systems if we imagine a planar mirror in front of them. We show that all unique viewpoint catadioptric systems can be modeled with a two-step projection: a central projection to the sphere followed by a projection from the sphere to an image plane. Special cases of this equivalence are parabolic projection, for which the second map is a stereographic projection, and perspective projection, for which the second map is central projection. Certain pairs of catadioptric projections are dual by the mapping which takes conics in the image plane to their foci. The foci of line images are points of another, dual, catadioptric projection; and vice versa, points in the image are foci of line images in the dual projection. The proved unifying model for all central catadioptric projections gives us further insight to practical advantages of catadioptric systems.

1 Introduction

In the past decade the vision community has seen a resurgence in the intelligent design of imaging sensors. It has been recognized that perspective cameras are not necessarily best suited to most tasks. Many tasks require constant and simultaneous omnidirectional vision. Vision for robotics, immersive telepresence, videoconferencing, and mosaicing, are all examples of tasks in which this is the case. A popular solution has been to use the already available technology (perspective or approximately orthographic lenses and rectangular CCD arrays) in combination with properly designed mirrors, thereby achieving the goal of omnidirectional vision. It is often desirable to choose systems whose locus of viewpoints is a single point. In doing so the complexity of interpreting the information obtained is reduced, and in addition it is possible to generate a more natural (to us) equivalent perspective image in an arbitrary direction (assuming that the image transformation is known and calibrated). Nayar et al. has studied the various configurations and shapes of mirrors which yield a single effective viewpoint. In general they are formed by a combination of mirrors which are surfaces of revolution and whose cross-section is a conic, and a perspective or

* The financial support by DOE-GAANN fellowship, ARO/MURI-DAAH04-96-1-0007, NSF-CISE-CDS-97-03220, DARPA-ITO-MARS-DABT63-99-1-001, and by Advanced Networks and Services is gratefully acknowledged.

G. Sommer and Y. Y. Zeevi (Eds.): AFPAC 2000, LNCS 1888, pp. 208-217, 2000.

orthographically projecting camera. They have shown that any combination of these mirrors and type of camera is equivalent to a single conical mirror and camera.

Let us briefly summarize recent activities on omnidirectional vision. A panoramic field of view camera was first proposed by Rees [9]. After 20 years the concept of omnidirectional sensing was reintroduced in robotics [12] for the purpose of autonomous vehicle navigation. In the last five years, several omnidirectional cameras have been designed for a variety of purposes. The rapid growth of multimedia applications has been a fruitful testbed for panoramic sensors [5, 6, 8] applied for visualization. Another application is telepresence [10, 1] where the panoramic sensor achieves the same performance as a remotely controlled rotating camera with the additional advantage of an omnidirectional alert awareness. Srinivasan [2] designed omnidirectional mirrors that preserve ratios of elevations of objects in the scene and Hicks [4] constructed a mirror-system that rectifies planes perpendicular to the optical axis. The application of mirror-lens systems in stereo and structure from motion has been prototypically described in [11, 3]. The fact that lines project to conics is mentioned in the context of epipolar lines by Svoboda [11] and Nayar [7].

In this paper we investigate the geometric properties induced in the image by a single viewpoint catadioptric projection. We show that in all cases the projection is equivalent to a parameterized projection of a sphere. Here we mean that a point in space is first projected to the sphere from the center, then projected from a point on an axis to an image plane perpendicular to this axis; the position along this axis is the parameter. This gives us a canonical representation of any catadioptric projection. In particular, in the case of a parabolic mirror, this is equivalent to projection from the pole of this axis, known as stereographic projection. Reflection by a planar mirror, i.e. perspective projection, is equivalent to projection from the center of the sphere. This enables us to more easily determine the set of conics which are images of lines, and also their invariants, by determining the images of great circles by these projections.

In addition we show that each of these catadioptric projections has a dual. The mapping between a projection and its dual is that which returns the foci of a conic. The foci of a line image are points in its dual projection and points in an image are foci of line images in the dual projection.

We will show that due to the nature of the geometry one may calibrate any catadioptric system in a single frame with as few as two lines except in singular cases where only three are necessary for the parabola, or impossible with a perspective camera.

2 Connection with the Projection of the Sphere

We will first provide formulas for point projection via parabolic and hyperbolic mirrors, then demonstrate their equivalence with projection of the sphere. Then we shall describe the induced projective geometries and their constituent points and lines.

2.1 P arabolicProjection

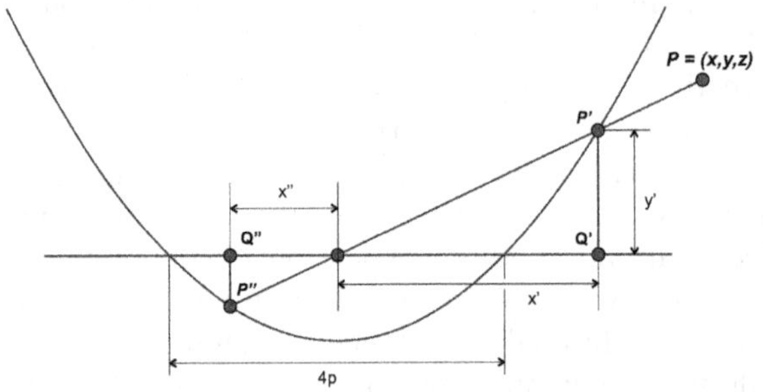

Fig. 1. Cross-section of a parabolic mirror. The image plane is through the focal point. The point in space P is projected to the antipodal points P' and P'', whic h are then *orthographic ally* projected to Q' and Q'' respectively.

Assume that a regular paraboloid is placed so that its focal point is the origin, and its axis is the z-axis, see figure 1. Let $4p$ be the diameter of the circle obtained by intersection of the paraboloid and the plane $z = 0$ ($4p$ is the latus rectum). The projection of a point is determined by the orthographic projection to $z = 0$ of the intersection of the line through the point P and focal point F. When the point lies on the z-axis, the image is a single point, otherwise the image is a set of two points. If $r = \sqrt{x^2 + y^2 + z^2}$, the point(s) of intersection are

$$\left(\pm\frac{2px}{r \mp z}, \pm\frac{2py}{r \mp z}, \pm\frac{2pz}{r \mp z}\right).$$

The orthographic projection yields,

$$\left(\pm\frac{2px}{r \mp z}, \pm\frac{2py}{r \mp z}\right).$$

Let X be the set of all such images of points, then $q_p : \mathbb{R}^3 \to X$ is defined by

$$q_p(x, y, z) = \left(\pm\frac{2px}{r \mp z}, \pm\frac{2py}{r \mp z}\right).$$

2.2 Hyperbolic and Elliptical Projections

Again assume the h yperboloid (of tw osheets) is placed so that its focal point F_1 is the origin and axis is the z-axis, see figure 2. Assume the other focal point F_2 is placed at $(0, 0, -d)$ and the latus rectum is $4p$. The projection of a point P is defined as the perspective projectionfrom F_2 to the plane $z = 0$, of the intersection of the line through P and F_1. If the Y is the set of image points,

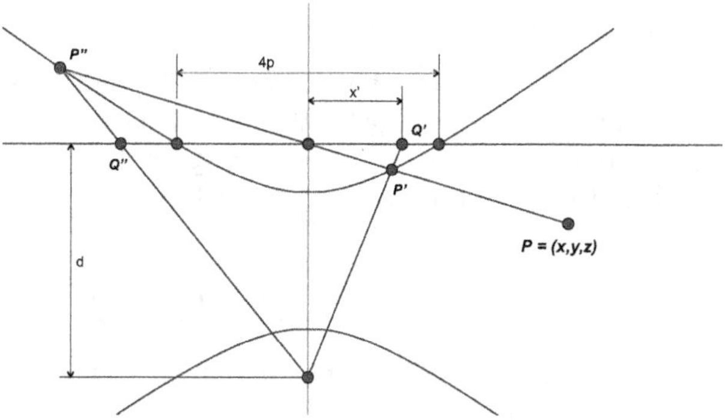

Fig. 2. Cross-section of a hyperbolic mirror, again the image plane is through the focal point. The point in space P is projected to the antipodal points P' and P'', which are then *perspectively* projected to Q' and Q'' from the second focal point.

and again $r = \sqrt{x^2 + y^2 + z^2}$, then the hyperbolic projection $r_{p,d} : \mathbb{R}^3 \to Y$ is defined by,

$$r_{p,d}(x, y, z) = \left(\pm \frac{2xdp}{dr \mp \alpha z}, \pm \frac{2ydp}{dr \mp \alpha z} \right),$$

where $\alpha = \sqrt{d^2 + 4p^2}$. Notice that the limit of this map as d approaches $+\infty$ is the map q_p, as expected.

Assuming the foci of an ellipse are $F_1 = (0, 0, 0)$ and $F_2 = (0, 0, -d)$, then the map $r'_{p,d}$ defined,

$$r'_{p,d}(x, y, z) = \left(\pm \frac{2xdp}{dr \pm \alpha z}, \pm \frac{2ydp}{dr \pm \alpha z} \right),$$

is the elliptical projection function. Note that $r_{p,d}(x, y, z) = r_{p,d}(x, y, -z)$. So hyperbolic and elliptical projections where p and d are equal differ only by a reflection by the $z = 0$ plane.

2.3 Perspective Projection

Perspective projection may be viewed as a special case of catadioptric projection; for perspective projection is unchanged, except by viewpoint, by the presence of a planar mirror. Assume that the focal point of a perspective projection is located at $(0, 0, -2f)$, where f is the focal length. Also assume that the image plane is at $z = -f$, but also within the same plane lies a planar mirror. The equivalent focal point is then $(0, 0, 0)$. The projection is a function $p_f : \mathbb{R}^3 \to \mathbb{R}^2$, and unlike the projections mentioned above maps a point in space to a single image point. It is given by

$$p_f(x, y, z) = \left(-\frac{xf}{z}, -\frac{yf}{z} \right).$$

2.4 Spherical Projection

Let a sphere of unit radius be centered at the origin. A point (x, y, z) is mapped by central projection to $(\frac{x}{r}, \frac{y}{r}, \frac{z}{r})$ (figure 3). We now wish to project the antipodal pair $(\pm\frac{x}{r}, \pm\frac{y}{r}, \pm\frac{z}{r})$ from the point $(0, 0, l)$, where $0 \le l \le 1$, to the plane $z = -m$, $m \neq -l$. We achiev e this with the maps$_{l,m}$ defined by

$$s_{l,m}(x, y, z) = \left(\pm\frac{x(l+m)}{lr \mp z}, \pm\frac{y(l+m)}{lr \mp z}\right),$$

If we change the plane, the projection differs only by a scale, as $s_{l,m} = \frac{l+m}{l+n} s_{l,n}$.

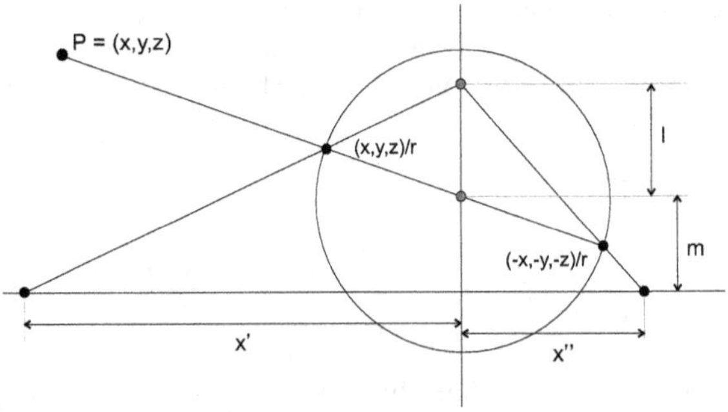

Fig. 3. A point $P = (x, y, z, w)$ is projected via s to tw o an tipodal points $(\pm x, \pm y, \pm z)/r$ on the sphere. The tw o an tipodal points are projected to the image plane $z = -m$ via projection from the point $(0, 0, l)$.

2.5 Unification

How are q_p, $r_{p,d}$, p_f, and $s_{l,m}$ related? It is a simple matter of substitution to prov e the following.

Theorem 1. *Projective Equivalence.* *Catadioptric projection with a single effective viewpoint is equivalent to projection to a sphere follow ed by projection to a plane from a point.*

Proof. We have the following relationships for the projection functions:

$$\begin{aligned} r_{p,d}(x, y, z) &= s_{\frac{d}{\sqrt{d^2+4p^2}}, \frac{d(1-2p)}{\sqrt{d^2+4p^2}}}(x, y, z) \\ r'_{p,d}(x, y, z) &= s_{\frac{d}{\sqrt{d^2+4p^2}}, \frac{d(1-2p)}{\sqrt{d^2+4p^2}}}(x, y, -z) \\ q_p(x, y, z) &= s_{1,2p-1}(x, y, z) \\ p_f(x, y, z) &= s_{0,f}(x, y, z) \end{aligned}$$

The first corresponds to hyperbolic projection, the second to elliptical projection, the third to parabolic projection, and the fourth to perspective projection. □

Now let l and m be chosen. We are now able to prove the following, where the fronto-parallel plane is the plane $z = 0$ and its horizon is the image of the points at infinity contained by this plane.

Proposition 1. *The image of a line in space is a conic which intersects the fronto-parallel horizon antipodally.*

Proof. A line in space is projected to a great circle on the sphere. To determine the projection of this great circle to the image plane we determine the cone whose vertex is the point of projection and which goes through the great circle. The intersection of this cone is the line image and is clearly a conic, being a section of a cone.

The fronto-parallel horizon is the projection of the equator and is a circle whose center is the image center and whose radius is $\frac{l+m}{l}$. Any great circle intersects any other great circle antipodally, and therefore also the equator. The projection of their intersection is also antipodal on the fronto-parallel horizon.

If the normal of the plane containing the great circle is $\hat{n} = (n_x, n_y, n_z)$ then the conic has the following characteristics,

$$\begin{aligned} f_{\pm} &= \left(\frac{(l+m)n_x}{n_z \mp \sqrt{1-l^2}}, \frac{(l+m)n_y}{n_z \mp \sqrt{1-l^2}} \right) \\ a &= \frac{l(l+m)n_z}{l^2 - n_x^2 - n_y^2} \\ b &= \frac{l+m}{\sqrt{l^2 - n_x^2 - n_y^2}}, \end{aligned} \tag{1}$$

where $f_{\pm}$ are the foci, a is the minor axis, and b is the major axis. Notice that the foci lie on a line through the origin (image center). □

Let

$$\Pi = \{(\pm x, \pm y, \pm z) \,|\, x^2 + y^2 + z^2 = 1\}$$

and let

$$\Lambda = \{[\pm n_x, \pm n_y, \pm n_z] \,|\, n_x^2 + n_y^2 + n_z^2 = 1\},$$

where

$$[n_x, n_y, n_z] = \{(x, y, z) \in \Pi \,|\, x n_x + y n_y + z n_z = 0\}.$$

Then Π is the set of antipodal point pairs on the sphere, and Λ is the set of great circles, where each $[n_x, n_y, n_z]$ is the set of points on the great circle. Now let $\pi_{l,m} = (s_{l,m}(\Pi), s_{l,m}(\Lambda))$. $\pi_{l,m}$ is the projective plane generated by $s_{l,m}$, and we call it a *catadioptric projective plane.* Let $\Pi(\pi_{l,m}) = s_{l,m}(\Pi)$ and $\Lambda(\pi_{l,m}) = s_{l,m}(\Lambda)$.

3 Duality

Let (n_x, n_y, n_z) be the normal of a plane; we found that the foci of the line image are given by equation 1. Notice that this is the projection of the same point (n_x, n_y, n_z) by $s_{l',m'}$, where $l' = \sqrt{1-l^2}$ and $m' = l + m - \sqrt{1-l^2}$.

Before giving the theorem on duality we define the operators *meet* ($\vee$) and *join* ($\wedge$), then we will give a proposition. Let $\hat{p} = (x_1, y_1, z_1)$ and $\hat{q} = (x_2, y_2, z_2)$ be points on the sphere. The great circle through them has normal $[\hat{p} \times \hat{q}]$. The image line through them is then the image of this great circle. Let $p \wedge q$ be this "line". Similarly, if $p = [\hat{m}] = [m_x, m_y, m_z]$ and $q = [\hat{n}] = [n_x, n_y, n_z]$ are the normals of tw o great circles, let$p \vee q$ be the the point $\hat{m} \times \hat{n}$, i.e. the intersection of the tw o lines. The composition of these operators with a projection from the sphere to the image plane yields binary operators in the catadioptric projective plane. For example, if $p, q \in \Pi(\pi_{l,m})$, then

$$p \vee q = s_{l,m}\left(s_{l,m}^{-1}(p) \vee s_{l,m}^{-1}(q)\right).$$

Proposition 2. *Let $\{[\hat{m}_i]\}$ be a set of line images all of which intersect a point $\hat{n}$, i.e. for all i, $\hat{m}_i \cdot \hat{n} = 0$. Then the locus of foci of the line images lie on a conic whose foci is the same as the point $\hat{n}$.*

Pr oof. Because of rotational symmetry, we may assume without loss of generality that $n_y = 0$. This implies that

$$\{[\hat{m}_i]\} = \{[-n_z \sin\theta_i, \cos\theta_i, n_x \sin\theta_i]\}.$$

Then the foci are,

$$f_{\pm}^{i} = \left(-\frac{(l+m)n_z \cos\theta}{n_x \cos\theta \mp \sqrt{1-l^2}}, \frac{(l+m)\sin\theta}{n_x \cos\theta \mp \sqrt{1-l^2}}\right).$$

But this is just one of the points $\hat{m}_i$ projected by

$$s_{\sqrt{1-l^2},\, l+m-\sqrt{1-l^2}},$$

and one finds that the second focus is the second projected point. Therefore this point is in the image of the line $\hat{n}$ by this same projection. Its foci are

$$f_{\pm} = \left(\frac{(l+m)n_x}{n_z \mp l}, 0\right),$$

which is the projection of $\hat{n}$ by $s_{l,m}$. □

We use this proposition to prove the following theorem on duality.

Theorem 2. *If $\pi_{l,m} = (\Pi_1, \Lambda_1)$ and $\pi_{l',m'} = (\Pi_2, \Lambda_2)$ are tw ocatadioptric planes such that*

$$l^2 + l'^2 = 1 \quad \text{and} \quad l + m = l' + m',$$

then $f_{l,m}$ which gives the foci of a line image in the context of some catadioptric plane $\pi_{l,m}$, maps as follows,

$$f_{l,m} : \Lambda_1 \to \Pi_2$$
$$f_{l',m'} : \Lambda_2 \to \Pi_1$$

and their inverses exist. In addition, incidence relationships are preserved by $f_{l,m}$*:*

$$f_{l,m}(p_1 \wedge p_2) = f^{-1}_{l',m'}(p_1) \vee f^{-1}_{l',m'}(p_2),$$
$$f^{-1}_{l,m}(l_1 \vee l_2) = f_{l',m'}(l_1) \wedge f_{l',m'}(l_2),$$

where $p_1, p_2 \in \Pi_1$ *and* $l_1, l_2 \in \Lambda_2$*. We therefore call the projective planes,* $\pi_{l,m}$ *and* $\pi_{l',m'}$*, dual catadioptric projective planes.*

Proof. We have already shown the first part of the theorem, it only remains to show that incidence relationships are preserved. But this follows from the proposition and the fact that incidence relationships are already known to be preserved on the sphere by the mapping taking antipodal points to great circles and vice versa. □

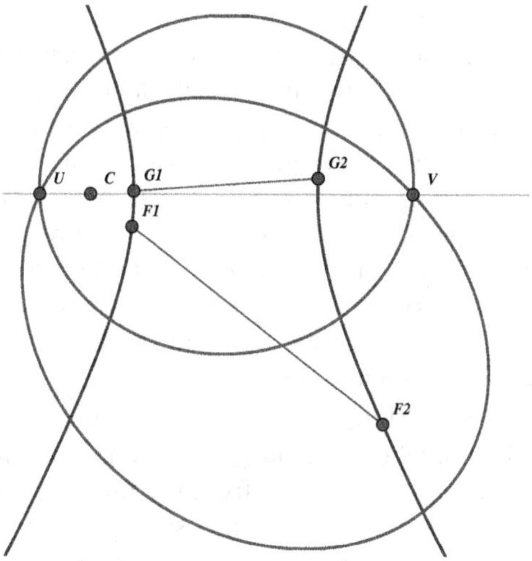

Fig. 4. The two ellipses are projections of two lines in space. Their foci $F1$, $F2$, and $G1$, $G2$ respectively lie on a hyperbola containing the foci of all *all* ellipses through U and V. The foci of this hyperbola are the points U and V.

4 Consequences

4.1 Stereographic Projection

Stereographic projection is a mapping from the sphere to the plane which is conformal; angles between great circles on the sphere are mapped to circles which have the same angle between them. This means that the horizons of planes which are perpendicular to each other will be orthogonal. It also implies the following.

Corollary 1. *In a stereo pair of parabolic catadioptric systems, angles between epipolar line images are constant and equal to the angles between the planes containing the points.*

4.2 Calibration

The geometric properties discovered here prove to be very useful in calibration. The first is the invariant property of each line image, namely that every line image intersects the fronto-parallel horizon antipodally. This is extremely useful because the line image can be used as the cross section of a surface of revolution about its major axis. Then every surface so created, either a hyperboloid, ellipsoid, or sphere, always contains the points $(0, 0, \pm 2p)$. Therefore intersection of at least three of these surfaces will yield the image center and focal length.

The second useful property is duality, by which we are able to determine the second intrinsic parameter, d, in the case of the hyperbolic mirror. When a line image is a conic, the dual points are its foci, and through these points there exists a line image in the dual projection whose foci are on the original conic. This second line image corresponds to the great circle perpendicular to that of the original line image, and their intersection are points on the equator. With a second line image in the dual projection for every original line image, we calibrate both the original projection *and* the dual projection simultaneously, under the assumption that the image centers are identical, using the intersection of surfaces described above. Together this pair of calibrated systems will encode both d and p, as well as the image center.

However, using this algorithm, it becomes very difficult to differentiate between hyperbolic mirrors, where d is large, and parabolic mirrors. For as d increases, the hyperboloid sheets obtained from the line images in the dual projection will become more and more planar and estimates of d will become less and less precise. Nevertheless, in general, it is possible to perform calibration with only two line images.

In the parabolic case we need a minimum of three line images. This is necessary, because, as we will see, the dual projection, which is perspective projection, cannot be calibrated with any number of lines in a single frame. In this case the surfaces of revolution are spheres whose equators are line images. Three spheres when intersected yield p and the image center, and the third, d we already assume to be infinite. In particular the intersection of those spheres will be the points $(0, 0, \pm 2p)$.

It is impossible to calibrate a perspective camera with lines in a single frame because each line introduces two unknowns (its orientation) and two equations (its position in the image), and thus the number of unknowns, which is at first three (focal length and image center), never decreases. There are zero constraints on the system, and no matter how many lines we obtain we will not be able to calibrate the perspective camera with lines in a single image.

5 Conclusion

Let us review the key points which we hae introduced here.

1. Every single viewpoint catadioptric system is equivalent to central projection to a sphere followed by projection from a point on the sphere's axis. In particular, parabolic projection is equivalent to stereographic projection, and is therefore conformal.
2. To every catadioptric projection there is a dual, the mapping between projections is by taking focal points of line images. This mapping preserves incidence relationships.
3. Calibration of a catadioptric projection is possible with only two lines, and in general three, except for perspective projection. In the parabolic case, calibration is performed by intersecting spheres whose equators are line images.

The natural next step is to extend this theory to multiple catadioptric views as well as a study of robustness of scene recovery using the principles described herein.

References

1. T.E. Boult. Remote reality demonstration. In *IEEE Conf. Computer Vision and Pattern Recognition*, pages 966–967, Santa Barbara, CA, June 23-25, 1998.
2. J.S. Chahl and M.V. Srinivasan. Range estimation with a panoramic sensor. *Journal Opt. Soc. Am. A*, 14:2144–2152, 1997.
3. J. Gluckman and S.K. Nayar. Ego-motion and omnidirectional cameras. In *Proc. Int. Conf. on Computer Vision*, pages 999–1005, Bombay, India, Jan. 3-5, 1998.
4. A. Hicks and R. Bajcsy. Reflective surfaces as computational sensors. In *CVPR-Workshop on Perception for Mobile Egents, Fort Collins, CO, June 26*, 1999.
5. V. Nalwa. Bell labs 360-degree panoramic webcam. News Release, http://www.lucent.com/press/0998/980901.bla.html, 1998.
6. S. Nayar. Catadioptric omnidirectional camera. In *IEEE Conf. Computer Vision and Pattern Recognition*, pages 482–488, Puerto Rico, June 17-19, 1997.
7. S.A. Nene and S.K. Nayar. Stereo with mirrors. In *Proc. Int. Conf. on Computer Vision*, pages 1087–1094, Bombay, India, Jan. 3-5, 1998.
8. Y. Onoe, K. Yamazawa, H. Takemura, and N. Yokoya. Telepresence by real-time view-dependent image generation from omnidirectional video streams. *Computer Vision and Image Understanding*, 71:588–592, 1998.
9. D. W. Rees. Panoramic television viewing system. United States Patent No. 3, 505, 465, Apr. 1970.
10. D. Southwell, A. Basu, and B. Vandergriend. A conical mirror pipeline inspection system. In *Proc. IEEE Int. Conf. on Robotics and Automation*, pages 3253–3258, 1996.
11. T. Svoboda, T. Padjla, and V. Hlavac. Epipolar geometry for panoramic cameras. In *Proc. 6th European Conference on Computer Vision*, pages 218–231, 1998.
12. Y. Yagi, S. Kawato, and S. Tsuji. Real-time omnidirectional image sensor (copis) for vision-guided navigation. *Trans. on Robotics and Automation*, 10:11–22, 1994.

Lie-Theory and Dynamical Illumination Changes

Reiner Lenz*

Dept. Science and Engineering, Campus Norrköping, Linköping University
Bredgatan, SE-60174 Norrköping, Sweden
reile@itn.liu.se
URL: www.itn.liu.se/~reile

Abstract. The description of the relation between the one-parameter groups of a group and the differential operators in the Lie-algebra of the group is one of the major topics in Lie-theory.
In this paper we use this framework to derive a partial differential equation which describes the relation between the time-change of the spectral characteristics of the illumination source and the change of the color pixels in an image.
In the first part of the paper we introduce and justify the usage of conical coordinate systems in color space. In the second part we derive the differential equation describing the illumination change and in the last part we illustrate the algorithm with some simulation examples.

Keywords: group theoretical frames in robotics, vision and neurocomputing, non-linear metrics and linearization, Lie-theory, color vision, invariance

1 Introduction and Notations

Color constancy is traditionally treated as a static problem which in a general form can be described as follows: Given the image $\boldsymbol{I}^{(0)}$ of a scene $\boldsymbol{R}$ under illumination $\boldsymbol{l}^{(0)}$: $\boldsymbol{I}^{(0)} = I(\boldsymbol{R}, \boldsymbol{l}^{(0)})$ and another illumination $\boldsymbol{l}^{(1)}$ estimate the image $\boldsymbol{I}^{(1)}$ of the same scene under the new illumination $\boldsymbol{I}^{(1)} = I(\boldsymbol{R}, \boldsymbol{l}^{(1)})$. In the most general case only the image data $\boldsymbol{I}^{(0)}$ and the illumination $\boldsymbol{l}^{(1)}$ is given. The estimation of the relation between the scene and the illumination is part of the problem to be solved. Many approaches to this problem have been suggested but in this general form the problem remains unsolved today. The method described in this paper differs in one essential point from these traditional approaches by assuming that we can observe the scene under a continuously changing illumination: $\boldsymbol{I}(t) = I(\boldsymbol{R}, \boldsymbol{l}(\boldsymbol{t}))$. From this image sequence we estimate the parameters which describe the evolution of the changing illumination condition. This can then be used to compensate the influence of the illumination change by computing a stable image of the scene which is independent of the illumination change under consideration.

* Reiner Lenz was supported by a grant from the CENIIT program at Linköping University and the grant "Signal Processing with Transformation Groups" financed by TFR.

G. Sommer and Y. Y. Zeevi (Eds.): AFPAC 2000, LNCS 1888, pp. 218– , 2000.

After the introduction of conical coordinate systems in color space we will show that the change of the color vector in a given image location is described by a partial differential equation with polynomial coefficients. The unknown curve parameters which control the illumination change enter these equations linearily and can thus be recovered by statistical estimation techniques.

The outline of the paper is as follows: in the first part we introduce conical coordinate systems as natural descriptors of color spaces. Then we use these coordinates to derive the partial differential equation which describes the influence of the illumination changes. In the last section we illustrate how to use this description to estimate the dynamic properties of the dynamically changing illumination properties. We will use the following notations: bold letters denote vectors $\boldsymbol{s}$ and matrices $\boldsymbol{M}$, λ is the wavelength variable and the identity matrix will be denoted by $\boldsymbol{E}$.

2 The Conical Structure of Spectral Color Spaces

Reflectance spectra measured from color chips of the Munsell and NCS color systems can be described by linear combinations of a few basis vectors. Usually the eigenvectors are taken as these basis vectors [, , , , , ,]. For a general spectral vector $\boldsymbol{s}(\lambda)$, basis vectors $\boldsymbol{b}^{(k)}(\lambda)$ and coefficients σ_k this gives:

$$\boldsymbol{s}(\lambda) \approx \sum_{k=1}^{K} \sigma_k \boldsymbol{b}^{(k)}(\lambda) \tag{1}$$

Collecting the coefficients σ_k in the vector $\boldsymbol{\sigma}$, the $\boldsymbol{b}^{(k)}(\lambda)$ in the matrix $\boldsymbol{B}$ Equation ( ) becomes: $\boldsymbol{s} \approx \boldsymbol{B}\boldsymbol{\sigma}$. Usually $\boldsymbol{\sigma}$ is treated as an element in $\mathbb{R}^K$. In [ ] we showed however that these vectors $\boldsymbol{\sigma}$ are located in a cone $\mathcal{C} = \left\{(\sigma_0, \sigma_1, \sigma_2) : \sigma_0^2 - \sigma_1^2 - \sigma_2^2 \geq 0\right\}$. There we used a database consisting of reflectance spectra of 2782 color chips, 1269 from the Munsell system and the rest from the NCS system. The eigenvectors computed from the database are collected in the matrix $\boldsymbol{B}_{KLT}$. The first three eigenvectors and the distribution of the coefficient vectors are shown in Figure (1a) and (1b). The conical structure of the eigenvector space is a special case of the following theorem:

Theorem 1. *For every system of $N+1$ basis vectors $\boldsymbol{B}$ of unit length such that $\min_k b_k^{(0)} > 0$ we can find positive constants $\tilde{b}_n, n = 1 \dots N$ such that the coefficients $\sigma_n = \langle \boldsymbol{b}^{(n)}, \boldsymbol{s} \rangle$ satisfy:*

$$\sigma_0^2 - \tilde{b_1}\sigma_1^2 - \tilde{b_2}\sigma_2^2 - \dots - \tilde{b_N}\sigma_N^2 \geq 0$$

for all spectral vectors $\boldsymbol{s}$. The coefficient vectors of all spectra are therefore located in a cone.

To see this observe that

$$|\langle \boldsymbol{b}, \boldsymbol{s} \rangle| = \left| \int b(\lambda) s(\lambda) d\lambda \right| = \left| \int \frac{b(\lambda) b^{(0)}(\lambda)}{b^{(0)}(\lambda)} s(\lambda) d\lambda \right| \leq \tilde{b} \int b^{(0)}(\lambda) s(\lambda) d\lambda \tag{2}$$

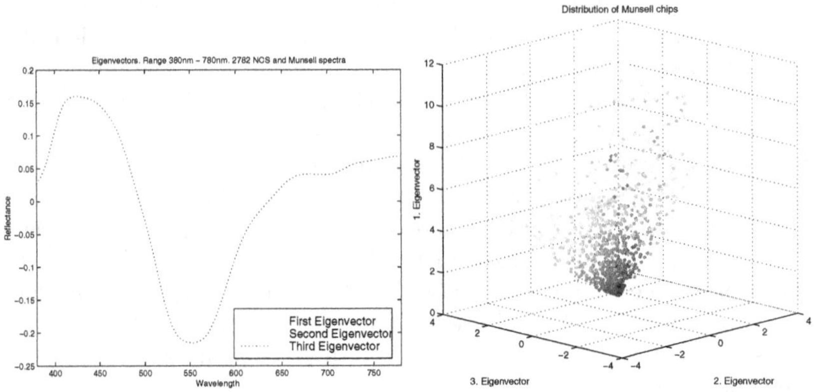

(a) (b)

Fig. 1. (a) The first three eigenvectors computed from the combined Munsell/NCS database. (b) The distribution of the expansion coefficients of the spectra in the Munsell/NCS database in the basis shown in (a)

where $\widetilde{b} = \max_\lambda |b(\lambda)|/b^{(0)}(\lambda)$ is finite since the first basis function is non-negative everywhere. In the following we will always assume a scaling of the basis functions which allows us to select $\widetilde{b}_n = 1$ for all n. Basis systems for which this theorem is valid include the eigenvector system $\boldsymbol{B}_{KLT}$ shown in Figure (1a) but also the CIE-XYZ system if we select $\boldsymbol{b}^{(0)} = Y$. From the positivity of the first basis vector follows that the coefficient σ_0 is related to the intensity of the corresponding color spectrum. The expression

$$\|\boldsymbol{\sigma}\| = \sigma_0^2 - \sigma_1^2 - \sigma_2^2 - \ldots - \sigma_N^2 \tag{3}$$

on the other hand is a measurement of the "grayness" of the spectrum. In [ ] it was shown that σ_0, $\sigma_1^2 + \sigma_2^2$ and $\arctan \frac{\sigma_1}{\sigma_2}$ have an intuitive explanation as intensity, saturation and hue when the basis system is computed from the eigenvectors of the spectral database mentioned above.

The grayness value $\|\boldsymbol{\sigma}\|$ in Equation ( ) can be used as a length measure since it is always non-negative. The linear transformations which preserve the grayness form the group $\mathrm{SO}(1,N)$. In the following we will only use $N = 2$ for which there are three basis functions. From general theory it is known that $\mathrm{SO}(1,2)$ is essentially the same group as $\mathrm{SU}(1,1)$ which contains all complex 2×2 matrices of the form

$$\boldsymbol{M} = \begin{pmatrix} a & b \\ \bar{b} & \bar{a} \end{pmatrix} \text{ with } |a|^2 - |b|^2 = 1 \tag{4}$$

Application of the elements in $\mathrm{SU}(1,1)$ transform only the chromaticity part of the colors. Therefore it is necessary to combine it with the positive real numbers and define the group $\mathrm{SU}(1,1)^+ = \mathbb{R}^+ \times \mathrm{SU}(1,1)$ as the direct product. It consists of all pairs $(\mathrm{e}^{\rho}, \boldsymbol{M})$ where $\rho \in \mathbb{R}$ and $\boldsymbol{M} \in \mathrm{SU}(1,1)$. The group operation is defined as:

$$(\rho_1, \boldsymbol{M}_1) \circ (\rho_2, \boldsymbol{M}_2) = (\rho_1 + \rho_2, \boldsymbol{M}_1 \boldsymbol{M}_2) \tag{5}$$

where $\boldsymbol{M}_1\boldsymbol{M}_2$ denotes ordinary matrix multiplication.

3 One Parameter Groups and Differential Operators

Given a matrix of basis vectors $\boldsymbol{B}$ and a spectrum $\boldsymbol{s}$ the expansion coefficients σ_0, σ_1 and σ_2 are obtained. Given their relation to intensity, saturation and hue we imitate traditional color science notations and define

$$L = \sigma_0, \qquad A = \frac{\sigma_1}{\sigma_0}, \qquad B = \frac{\sigma_2}{\sigma_0} \tag{6}$$

L is related to the intensity and A, B to the chromaticity of $\boldsymbol{s}$. For a coordinate vector (L, A, B) we also introduce τ and z as:

$$\tau = \log L \qquad \text{and} \qquad z = A + iB \tag{7}$$

In the following we consider functions of the coordinate vector $\boldsymbol{\sigma}$

$$f : (\sigma_0, \sigma_1, \sigma_2) \mapsto f(\sigma_0, \sigma_1, \sigma_2) \tag{8}$$

Partial differential operators in the $\boldsymbol{\sigma}$ coordinate system are denoted by D_k :

$$D_k f(\sigma_0, \sigma_1, \sigma_2) = \frac{\partial f(\sigma_0, \sigma_1, \sigma_2)}{\partial \sigma_k}, \qquad k = 0, 1, 2 \tag{9}$$

The group $\mathrm{SU}(1,1)^+$ acts on LAB-space by (see Equation ( )):

$$(\rho, \boldsymbol{M})(L, A, B) = (\rho, \boldsymbol{M})(L, z) = (\mathrm{e}^{\rho} L, \boldsymbol{M} z) = \left(\mathrm{e}^{\rho+\tau}, \frac{az + b}{\overline{b}z + \overline{a}}\right) \tag{10}$$

There is a close connection between certain subgroups and differential operators. In the simplest case it is obtained as follows: take the subgroup $\mathfrak{g}_0 = \{(\rho, \boldsymbol{E}) : \rho \in \mathbb{R}\}$ defining the mapping $(L, z) \mapsto (\rho, \boldsymbol{E})(L, z) = (\mathrm{e}^{\rho} L, z)$. The subgroup $\mathfrak{g}_0$ is an example of a one-parameter subgroup of $\mathrm{SU}(1,1)^+$ since it depends only on the parameter ρ. This is used to define the differential operator $D_{\mathfrak{g}_0}$ as:

$$\begin{aligned}
f \mapsto D_{\mathfrak{g}_0} f &= \frac{d}{d\rho} f\left((\rho, \boldsymbol{E})(L, z)\right)|_{\rho=0} \\
&= \frac{d}{d\rho} f\left(\mathrm{e}^{\rho} L, z\right)|_{\rho=0} = \frac{d}{d\rho} f\left(\mathrm{e}^{\rho} L, \mathrm{e}^{\rho} LA, \mathrm{e}^{\rho} LB\right)|_{\rho=0} \\
&= L D_0 f(L, LA, LB) + LA D_1 f(L, LA, LB) + LB D_2 f(L, LA, LB) \\
&= L\left(D_0 f + A D_1 f + B D_2 f\right)
\end{aligned} \tag{11}$$

which gives the operator identity:

$$D_{\mathfrak{g}_0} = L\left(D_0 + AD_1 + BD_2\right) = \sigma_0 D_1 + \sigma_1 D_1 + \sigma_2 D_2 \tag{12}$$

Using the same construction shows that each one-parameter subgroup $\mathfrak{g}$ defines a differential operator $D_{\mathfrak{g}}$. The main property of these differential operators is collected in the following theorem:

Theorem 2. *1. The differential operators $D_{\mathfrak{g}}$ defined by the one-parameter subgroups $\mathfrak{g}$ of $\mathrm{SU}(1,1)^{+}$ form a vector space of dimension four. This vector space is known as the Lie-algebra of the Lie-group $\mathrm{SU}(1,1)^{+}$. It will be denoted by $\mathfrak{su}(1,1)^{+}$.*

2. Four basis vectors are defined through the following one-parameter groups:

$$\mathfrak{g}_0 = \mathbb{R} \tag{13}$$

$$\mathfrak{g}_1 = \left\{\left(0, \begin{pmatrix} \cosh\frac{\alpha}{2} & \sinh\frac{\alpha}{2} \\ \sinh\frac{\alpha}{2} & \cosh\frac{\alpha}{2} \end{pmatrix}\right)\right\} = \{(0, \boldsymbol{M}_1(\alpha))\} \tag{14}$$

$$\mathfrak{g}_2 = \left\{\left(0, \begin{pmatrix} \cosh\frac{\alpha}{2} & i\sinh\frac{\alpha}{2} \\ -i\sinh\frac{\alpha}{2} & \cosh\frac{\alpha}{2} \end{pmatrix}\right)\right\} = \{(0, \boldsymbol{M}_2(\alpha))\} \tag{15}$$

$$\mathfrak{g}_3 = \left\{\left(0, \begin{pmatrix} e^{i\frac{\varphi}{2}} & 0 \\ 0 & e^{-i\frac{\varphi}{2}} \end{pmatrix}\right)\right\} = \{(0, \boldsymbol{M}_3(\alpha))\} \tag{16}$$

3. The corresponding differential operators are:

$$D_{\mathfrak{g}_0} = \sigma_0 D_0 + \sigma_1 D_1 + \sigma_2 D_2 \tag{17}$$

$$D_{\mathfrak{g}_1} = \frac{\sigma_0^2 - \sigma_1^2 + \sigma_2^2}{2\sigma_0} D_1 - \frac{\sigma_1\sigma_2}{\sigma_0} D_2 \tag{18}$$

$$D_{\mathfrak{g}_2} = -\frac{\sigma_1\sigma_2}{\sigma_0} D_1 + \frac{\sigma_0^2 + \sigma_1^2 - \sigma_2^2}{2\sigma_0} D_2 \tag{19}$$

$$D_{\mathfrak{g}_3} = -\sigma_2 D_1 + \sigma_1 D_2 \tag{20}$$

More information about this construction and other basic facts on Lie-groups and Lie-algebras can be found in books on Lie-theory (such as [, , ,]).

Each one-parameter subgroup $\mathfrak{g}$ of $\mathrm{SU}(1,1)^{+}$ defines a curve in $(\sigma_0, \sigma_1, \sigma_2)$ space and by differentiation an operator $D_{\mathfrak{g}}$. This is an element of the Lie-algebra $\mathfrak{su}(1,1)^{+}$ and thus there are constants $a_0, \ldots, a_3$ such that:

$$D_{\mathfrak{g}} = a_0 D_{\mathfrak{g}_0} + a_1 D_{\mathfrak{g}_1} + a_2 D_{\mathfrak{g}_2} + a_3 D_{\mathfrak{g}_3} \tag{21}$$

Now assume that the function $f\left(\sigma_0(t), \sigma_1(t), \sigma_2(t)\right)$ describes our measurements varying over some period of time. Assume furthermore that this variation (over a sufficiently small period of time) originates in a one-parameter subgroup $\mathfrak{g}$ of $\mathrm{SU}(1,1)^+$. From these measurements we can compute the derivative

$$D_{\mathfrak{g}} f = \frac{df\left(\sigma_0(t), \sigma_1(t), \sigma_2(t)\right)}{dt}\Big|_{t=0} \tag{22}$$

The operator $D_{\mathfrak{g}}$ is an element of the Lie-algebra and thus there are constants $a_0, \ldots, a_3$ for which

$$\frac{df\left(\sigma_0(t), \sigma_1(t), \sigma_2(t)\right)}{dt}\Big|_{t=0} = D_{\mathfrak{g}} f = a_0 D_{\mathfrak{g}_0} f + a_1 D_{\mathfrak{g}_1} f + a_2 D_{\mathfrak{g}_2} f + a_3 D_{\mathfrak{g}_3} f \tag{23}$$

All of the quantities $D_{\mathfrak{g}} f, D_{\mathfrak{g}_k} f, (k = 0, \ldots, 3)$ can be computed from the measured data and the unknown constants $a_0, \ldots, a_3$ can therefore be estimated.

Sometimes it is useful to do the same calculations in LAB-space as described in the next theorem:

Theorem 3. *A basis of the Lie-algebra $\mathfrak{su}(1,1)^+$ in LAB coordinates is given by the operators:*

$$D_{\mathfrak{g}_0} = L D_L \tag{24}$$

$$D_{\mathfrak{g}_1} = \frac{\left(1 - A^2 + B^2\right) D_A - (AB) D_B}{2} \tag{25}$$

$$D_{\mathfrak{g}_2} = \frac{-(AB) D_A + \left(1 + A^2 - B^2\right) D_B}{2} \tag{26}$$

$$D_{\mathfrak{g}_3} = -B D_A + A D_B \tag{27}$$

where D_L, D_A, D_B are the differential operators:

$$D_L g = \frac{\partial g(L,A,B)}{\partial L}, \qquad D_A g = \frac{\partial g(L,A,B)}{\partial A}, \qquad D_B g = \frac{\partial g(L,A,B)}{\partial B}$$

4 Application to Illumination Invariant Recognition

We will now show how the general theory can be used to recover characteristic illumination parameters from a sequence of images taken under changing illumination conditions.

Our simple model of the imaging process combines the illumination spectrum $\boldsymbol{l}(\lambda)$, the reflectance spectrum at position $\boldsymbol{x}$ given by $\boldsymbol{r}(\boldsymbol{x}, \lambda)$ and the spectral characteristic of an imaging sensor (for example a camera) $\boldsymbol{c}(\lambda)$ as:

$$m_\kappa(\boldsymbol{x}) = \int \boldsymbol{l}(\lambda)\, \boldsymbol{r}(\boldsymbol{x}, \lambda)\, \boldsymbol{c}^{(\kappa)}(\lambda)\; d\lambda \tag{28}$$

where $m_\kappa(\boldsymbol{x})$ is the value measured with sensor number κ at position $\boldsymbol{x}$. In most cases the $m_\kappa(\boldsymbol{x})$ are the measured R, G, B values and $\boldsymbol{c}^{(\kappa)}(\lambda)$ is the spectral sensitivity of the R, G or B channel.

We use different basis vectors $\boldsymbol{b}_\nu^{(R)}(\lambda), \boldsymbol{b}_\nu^{(I)}(\lambda), \nu = 0, 1, 2$ in the spaces of object reflectance and illumination spectra. For a distribution $\boldsymbol{r}(\boldsymbol{x}, \lambda)$ and an illumination spectrum $\boldsymbol{l}(\lambda)$ we get the coordinate vectors $\boldsymbol{\sigma}^{(r)}(\boldsymbol{x}), \boldsymbol{\sigma}^{(l)}$ leading to the approximation:

$$\boldsymbol{r}(\boldsymbol{x}, \lambda) \approx \sum_{\nu=0}^{2} \boldsymbol{b}_\nu^{(R)}(\lambda)\, \sigma_\nu^{(r)}(\boldsymbol{x}), \quad \boldsymbol{l}(\lambda) \approx \sum_{\mu=0}^{2} \boldsymbol{b}_\mu^{(I)}(\lambda)\, \sigma_\mu^{(l)} \tag{29}$$

Inserting these approximations into Equation ( ) gives:

$$\begin{aligned} m_\kappa(\boldsymbol{x}) &\approx \sum_{\nu=0}^{2}\sum_{\mu=0}^{2} \sigma_\nu^{(r)}(\boldsymbol{x})\, \sigma_\mu^{(l)} \int \boldsymbol{b}_\nu^{(r)}(\lambda)\, \boldsymbol{b}_\mu^{(l)}(\lambda)\, \boldsymbol{c}^{(\kappa)}(\lambda)\, d\lambda \\ &= \sum_{\nu=0}^{2}\sum_{\mu=0}^{2} \sigma_\nu^{(r)}(\boldsymbol{x})\, \sigma_\mu^{(l)} \gamma_{\nu\mu}^{(\kappa)} = \boldsymbol{\sigma}^{(r)}(\boldsymbol{x}')\boldsymbol{G}^{(\kappa)}\boldsymbol{\sigma}^{(l)} \end{aligned} \tag{30}$$

The known matrix $\boldsymbol{G}^{(\kappa)}$ characterizes channel κ of the sensor. The coordinate vectors $\sigma^{(r)}(\boldsymbol{x}), \sigma_\mu^{(l)}$ are in general unknown. The estimation of the coordinate vector(s) of the reflectance spectra involved is not the topic of this paper and therefore we assume that we can estimate the vectors $\boldsymbol{v}_{\boldsymbol{\kappa}}(\boldsymbol{x})' = \sigma^{(r)}(\boldsymbol{x})'\, \boldsymbol{G}^{(\kappa)}$. A general discussion of this type of bilinear calibration-estimation problems can be found in [ ].

Simplifying notations we avoid the superscript ${}^{(l)}$ and get for the measurements:

$$m_{\kappa,\boldsymbol{x}}(\boldsymbol{\sigma}) = \boldsymbol{v}_\kappa(\boldsymbol{x})'\boldsymbol{\sigma} = \langle \boldsymbol{v}_\kappa(\boldsymbol{x}), \boldsymbol{\sigma} \rangle \tag{31}$$

The measurements considered as functions of the illumination define special cases of the functions defined in Equation ( ). We write

$$f(\sigma_0, \sigma_1, \sigma_2) = f_{\kappa,\boldsymbol{x}}(\sigma_0, \sigma_1, \sigma_2) = \langle \boldsymbol{v}_\kappa(\boldsymbol{x}), \boldsymbol{\sigma} \rangle = m_{\kappa\boldsymbol{x}}(\boldsymbol{\sigma}) \tag{32}$$

Observing the same scene point with the same camera under changing illumination conditions produces a measurement series $m_{\kappa\boldsymbol{x}}(\boldsymbol{\sigma}, t)$ which is the raw input data. ¿From this we compute the time derivative

$$m'_{\kappa\boldsymbol{x}}(\boldsymbol{\sigma}, t) = \frac{\partial m_{\kappa,\boldsymbol{x}(\boldsymbol{\sigma},t)}}{\partial t} \tag{33}$$

Theorem shows that the mapping $m \mapsto m'$ defines a differential operator $D_\mathfrak{g}$ which is a linear combination of the known differential operators $D_{\mathfrak{g}_k}$ with unknown coefficients a_k. The algorithm to recover the unknown illumination parameters is thus as follows:

1. Select a number of points in the image
2. For each point/sensor combination $\pi = (\boldsymbol{x}, \kappa)$ measure the sequence $f_{\kappa,\boldsymbol{x}}(\sigma_0(t), \sigma_1(t), \sigma_2(t)) = f_\pi$
3. For each point/sensor combination π compute the derivative

$$m'_\pi = D_{\mathfrak{g}} f_\pi = \frac{df_{\kappa,\boldsymbol{x}}(\sigma_0(t), \sigma_1(t), \sigma_2(t))}{dt}|_{t=0}$$

4. Collect all values in the vector $\boldsymbol{M} = (m_\pi)$.
5. For each point/sensor combination π compute the derivatives

$$D_{\mathfrak{g}_k} f_\pi = \boldsymbol{u}_{k\pi}$$

   for $k = 0, \ldots, 3$ and collect them in the matrix $\boldsymbol{U}$.
6. Between each row $\boldsymbol{u}_\pi$ of $\boldsymbol{U}$ and the corresponding element m_π in $\boldsymbol{M}$ there is a relation

$$m_\pi = \boldsymbol{u}_\pi \boldsymbol{a}$$

   where the vector $\boldsymbol{a}$ collects the unknown coefficients $a_0, \ldots, a_3$.
7. The unknown coefficient vector can be estimated by solving in a statistical-sense the equation

$$\boldsymbol{M} = \boldsymbol{U}\boldsymbol{a} \tag{34}$$

5 Experiments

In the experiments we tested the algorithm by simulating a sequence of images which show the same scene under changing illumination conditions. These experiments show that two factors are of importance in the application of the framework developed so far.

- The first factor is the selection of basis vectors in the space of illumination and reflectance spectra. It is of advantage to use different basis systems in both spaces since prior knowledge about the general nature of the spectra involved can be used in the selection of these basis vectors.
- Furthermore it has to be decided how the estimates from different points and different channels should be combined. Treating all estimates equally ignores the fact that some estimates are less reliable than others.

In the simulations we use multispectral images described in [ ]. For a given multispectral image we use the spectral characteristics of a CCD-camera to simulate the color image captured by the camera. The basis vectors in the space of reflection functions are the eigenvectors computed from the Munsell and NCS systems described above. The basis vectors in the space of illumination spectra are the eigenvectors derived by generating a random mixture of 1000 spectra consisting of the CIE-light sources A, B, C, D65, a flat spectrum, 5 measured daylight spectra and 3 artificial daylight spectra.

In a typical experiment we generate a time varying series of illumination spectra as follows: at time $t = 0$ the illumination light is characterized by the flat illumination spectrum corresponding to white light. At time $t = 1$ the illumination is given by a pre-defined spectrum. We then simulate the changing illumination condition by connecting these two spectra by a one-parameter subgroup of $\mathrm{SU}(1,1)^+$. In a common simulation we use the CIE-A source as the light source at time $t = 1$. An image sequence consists of 3 to 10 frames corresponding to time parameters in the range $t = 0.2, \ldots 1.1$.

In the experiment illustrated in Figure (2a) we use 3 frames with time parameters $t = 0.2, 0.3, 0.4$. The light source at $t = 1$ is the A-source and in the estimation the 10 random points were tracked. The scene used is named "inlab-2" in the database. In the figure four spectra are shown. The original spectrum is shown as dotted line. The basis vectors for the space of illumination spectra was computed by a principal component analysis of 1000 randomly generated spectra as described above. The approximation of the original spectrum of the A-source in this basis system is shown as solid line in the figure. This is the best we can achive within the chosen system. The difference between the dotted and the solid line originates in the difference between the three-dimensional PCA-based description and the original.

The dashed and the dotted-dashed lines show two estimations from the image sequence. The dashed estimation was obtained using the mean over all estimations. In that case all observations are treated equally. The dashed-dotted estimation was obtained using a weighted sum. For each point $\boldsymbol{x}$ we compute first the determinant of the matrix formed by the vectors $\boldsymbol{v}_\kappa(\boldsymbol{x})$ introduced in Equation ( ). Low values of this determinant indicate that small variations in the measurement vector can lead to large deviations in the estimation of the parameter vector. Estimations based on observations of such a point are thus considered unreliable and weighted down. The figure shows that in this case the weighting leads to a considerable improvement of the estimation result. This is confirmed by the results shown in Figure (2b). In this series of simulations we used three input scenes (Inlab2, Ashton2 and Rwood) and tracked 5 points over three frames. This was done 100 times for each scene. The estimated spectrum was then compared with the approximation of original spectrum of the A-source in the coordinate system used in the simulation. The errors for the mean-based estimation were sorted and are shown as the solid line in Figure (2b). The corresponding errors for the estimation based on the weighted estimation is shown as an 'x' for each simulation. The diagram shows a clear improvement of the estimation results for the weighted mean based experiments.

6 Conclusions

We demonstrated the use of Lie-techniques in the estimation of illumination spectra from dynamical image sequences. This illustrates one application in which methods from the theory of Lie-groups and Lie-algebras can be used in the anal-

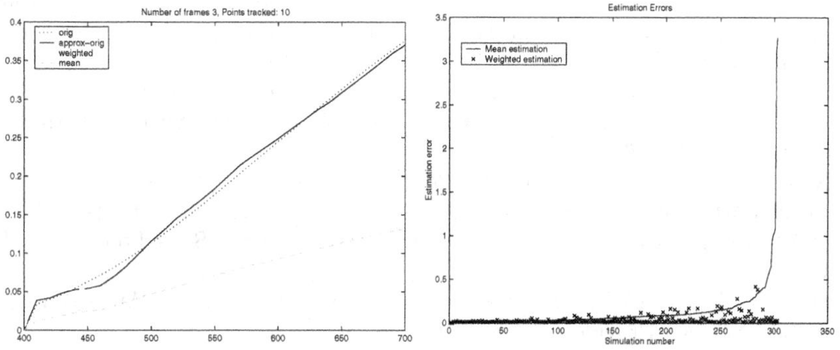

(a) 10 points tracked (b) Mean/Weighted estimation

Fig. 2. (a) Estimation results using Inlab2 with 10 points tracked over 3 frames (b) Comparison of estimation errors obtained by mean and weighted mean estimation

ysis of time-varying image sequences. Other problems involving different groups like the euclidean motion group can be analysed along the same lines.

References

1. J. Cohen, *Dependency of the spectral reflectance curves of the Munsell color chips*, Psychon Science **1** (1964), 369–370.
2. Michael D'Zmura and Geoffrey Iverson, *Color constancy. I. Basic theory of two-stage linear recovery of spectral descriptions for lights and surfaces*, Journal of the Optical Society of America A **10** (1993), 2148–2165.
3. Jan J. Koenderink and Andrea J. van Doorn, *The generic bilinear calibration-estimation problem*, International Journal of Computer Vision **23** (1997), no. 3, 217–234.
4. R. Lenz and P. Meer, *Non-euclidean structure of spectral color space*, Polarization and Color Techniques in Industrial inspection (E. A. Marszalec and E. Trucco, eds.), Proceedings Europto Series, vol. 3826, SPIE, 1999, pp. 101–112.
5. R. Lenz, M. Österberg, J. Hiltunen, T. Jaaskelainen, and J. Parkkinen, *Unsupervised filtering of color spectra*, Journal of the Optical Society of America A **13** (1996), no. 7, 1315–1324.
6. Laurence T. Maloney, *Evaluation of linear models of surface spectral reflectance with small numbers of parameters*, Journal of the Optical Society of America A **3** (1986), no. 10, 1673–1683.
7. Peter J. Olver, *Applications of Lie groups to differential equations*, Springer, New York, 1986.
8. J. P. S. Parkkinen, J. Hallikainen, and T. Jaaskelainen, *Characteristic spectra of Munsell colors*, Journal of the Optical Society of America A **6** (1989), no. 2, 318–322.
9. C. A. Párraga, G. Brelstaff, T. Troscianko, and errata I. R. Moorehead, *Color and luminance information in natural scenes*, J. Opt. Soc. Am. A **15** (1998), no. 12, 1708.

10. D. H. Sattinger and O. L. Weaver, *Lie groups and algebras with applications to physics, geometry and mechanics*, Applied Mathematical Sciences, vol. 61, Springer, 1986.
11. Shiro Usui, Shigeki Nakauchi, and Masae Nakano, *Reconstruction of Munsell color space by a five-layer neural network*, Journal of the Optical Society of America A **9** (1992), no. 4, 516–520, Color.
12. N.Ja. Vilenkin and A. U. Klimyk, *Representation of Lie groups and special functions*, Mathematics and its applications : 72, Kluwer Academic, 1991-1993.
13. G. Wyszecki and W. S. Stiles, *Color science*, 2 ed., Wiley & Sons, London, England, 1982.
14. D. P. Zelobenko, *Compact Lie groups and their representations*, American Mathematical Society, Providence, Rhode Island, 1973.

A Group Theoretical Formalization of Con tact Motion

Yanxi Liu

The Robotics Institute, Carnegie Mellon University, Pittsburgh 15213, USA,
yanxi@cs.cmu.edu,
http://www.cs.cmu.edu/~yanxi/www/home.html

Abstract. The work reported in this paper applies group theory to characterize **local** surface contact among solids, and makes such a formalization computationally tractable. This paper serves three purposes: (1) to report a new treatment of contacting surfaces as oriented sets and to verify the consistency of their corresponding symmetry groups in a group theoretical formalization framework; (2) to give a concise summary of this group theoretical approach, from theory, to algorithms, to applications; (3) to pinpoint some unsolved problems and possible future directions.

1 Introduction

Contact motion analysis of rigid bodies is one of the most useful yet difficult topics in robotics, design, assem bly planning and man ufacturing. Increasing level of automation demands more effectiv e computational method for represen ting and reasoning about contacts, and transforming high-level specifications to low-level executable commands. Admittedly, contact analysis is "a computational bottleneck in mechanical design", "it is especially challenging for curved parts with multiple, changing contacts" [5, 6]. The current state of art in this area has left much to be desired. The input to almost all the reported automatic assem bly planning systems, such as [24, 25, 16, 4], is one-static-state of the final assembly configuration regardless the assem bly is meant to be rigid or articulated. Most work in contact analysis is dealing with planar surfaces [22, 17, 7]. While the most impressive work on *higher pairs*[1] analysis and simulations [5, 21, 2] still need human in tervention, there exist no formal theory and algorithms for contact analysis of rigid bodies in general.

Hervè [3] and Popplestone [19] are among the few pioneers who con tributed to the group theoretical formalization of mechanical engineering practice. Hervé has introduced a rational classification of mechanisms by applying the theory of continuous groups. Since each lower-pair allows a set of relative motions of two coupled bodies, these motions can be regarded as subgroups of $\mathcal{E}^+$. The

[1] **Lower pairs** are mechanical joints that have surface to surface contact (area contact). otherwise they are called **Higher pairs** (line/curve/point contact) in mechanical engineering.

G. Sommer and Y. Y. Zeevi (Eds.): AFPAC 2000, LNCS 1888, pp. 229-240, 2000.

number of independent variables required to define the relative position of two coupled links is referred to as their *degree of freedom*, which can be extended to a subgroup of $\mathcal{E}^+$ in 3-space, the corresponding concept being that of *dimension*. In [3], Hervè gave a table (Table III) listing the intersection and composition of continuous subgroups of the proper Euclidean group. Only pairs of continuous subgroups are considered. In [19] Popplestone relates robotics and group theory, in particular he pointed out that 1) the symmetry group of a feature (of a solid) is perhaps more important than that of the whole solid, and 2) not only continuous groups (as treated in [3]) but also finite and discrete groups should be handled.

In Liu's Ph.D. thesis [9,11], the group theoretical formalization of surface contact among solids was further extended and solidified. The novelties in her work include, (1) both finite and continuous groups are treated under a uniform general formalization; (2) a simple geometric algorithm for combining different symmetry groups of local contacting surfaces, to determine relative motions of the contacting solids, is developed and implemented with proven tractability. (3) the approach carries through: starting from CAD models of solids, followed by automatic determination of relative locations/motions of multi-contact-assembly-parts, to the generation of assembly procedures for robots [12–14]. Liu's notations and algorithms can deal with symmetry groups which is itself a semidirect product of a continuous group and a discrete group, thus providing the power of reasoning about both degrees of freedom and combinatoric configurations of contacting solids simultaneously. A list of some important canonical subgroups of $\mathcal{E}^+$ treated in [9] is given in Table 1. Further development after [9] includes,

Table 1. Some important subgroups of $\mathcal{E}^+$

Canonical Groups	Definition
$\mathcal{G}_{id}$	$\{1\}$
$\mathcal{T}^1$	$\mathbf{gp}\{\mathbf{trans}(0,0,z) \mid z \in \Re\}$
$\mathcal{T}^2$	$\mathbf{gp}\{\mathbf{trans}(x,y,0) \mid x,y \in \Re\}$
$\mathcal{T}^3$	$\mathbf{gp}\{\mathbf{trans}(x,y,z) \mid x,y,z \in \Re\}$
$SO(3)$	$\mathbf{gp}\{\mathbf{rot}(\mathbf{i},\theta)\mathbf{rot}(\mathbf{j},\sigma)\mathbf{rot}(\mathbf{k},\phi) \mid \theta,\sigma,\phi \in \Re\}$
$SO(2)$	$\mathbf{gp}\{\mathbf{rot}(\mathbf{k},\theta) \mid \theta \in \Re\}$
$O(2)$	$\mathbf{gp}\{\mathbf{rot}(\mathbf{k},\theta)\mathbf{rot}(\mathbf{i},n\pi) \mid \theta \in \Re, n \in \mathcal{N}\}$
$\mathcal{G}_{cyl}$	$\mathbf{gp}\{\mathbf{trans}(0,0,z)\mathbf{rot}(\mathbf{k},\theta)\mathbf{rot}(\mathbf{i},n\pi) \mid n \in \mathcal{N}, \theta, z \in \Re\}$
$\mathcal{G}_{dir\_cyl}$	$\mathbf{gp}\{\mathbf{trans}(0,0,z)\mathbf{rot}(\mathbf{k},\theta) \mid z,\theta \in \Re\}$
$\mathcal{G}_{plane}$	$\mathbf{gp}\{\mathbf{trans}(x,y,0)\mathbf{rot}(\mathbf{k},\theta) \mid x,y,\theta \in \Re\}$
$\mathcal{G}_{screw}(p)$	$\mathbf{gp}\{\mathbf{trans}(0,0,z)\mathbf{rot}(\mathbf{k},2z\pi/p) \mid z \in \Re\}$
$\mathcal{G}_{T_1C_2}$	$\mathbf{gp}\{\mathbf{trans}(0,0,z)\mathbf{rot}(\mathbf{i},n\pi) \mid n \in \mathcal{N}, z \in \Re\}$
D_{2n}	$\mathbf{gp}\{\mathbf{rot}(\mathbf{k},2\pi/n)\mathbf{rot}(\mathbf{i},m\pi) \mid m,n \in \mathcal{N}\}$
C_n	$\mathbf{gp}\{\mathbf{rot}(\mathbf{k},2\pi/n) \mid n \in \mathcal{N}\}$
$\mathcal{E}^+$	$\mathbf{gp}\{\mathbf{trans}(x,y,z)\mathbf{rot}(\mathbf{i},\theta)\mathbf{rot}(\mathbf{j},\sigma)\mathbf{rot}(\mathbf{k},\phi) \mid x,y,z,\theta,\sigma,\phi \in \Re\}$

transforming high-level spatial relation descriptions to low-level compliant motions [15]; treating each *oriented* surface on a solid as a primitive feature; and obtaining an explicit expression for general contact relationships in terms of the symmetry groups of the contacting surfaces. The last two parts are described in this paper. We are now ready to go beyond surface contact, to venture into more general algebraic surfaces, higher pairs and group products. Different from the study of solids in local contact, e.g., [8, 18], our aim is to have a precise and complete description of the intended, possibly articulated, final assembly configuration of solids where each part usually has multiple contacts with the rest of the assembly; and our approach is algebraic in nature. Also different from [20, 23] in that a group theoretical formalism is embedded in a concise representation of contact motions not involving extensive algebraic equation manipulation.

2 Basic Group Theoretical Formalism of Contact Analysis

Since contacts among solids happen via the contacts of the surfaces of the solids, the representation and characterization of each surface constitutes the foundation of any formalization for solid contacts.

2.1 Oriented Surface and Its symmetry Group

The surfaces which we have treated mathematically as subsets of $\Re^3$ [9, 11] have no intrinsic *inside and outside*. To remedy this we introduce the concept of *oriented features* by defining a set of outward-pointing normal vectors for each surface point of a solid. The polynomial used to express an algebraic surface implicitly defines such normal vectors. Let $\mathcal{S}^2$ be the unit sphere at the origin embedded in $\Re^3$, each point of $\mathcal{S}^2$ corresponds to a unit vector in $\Re^3$.

Definition 2.1.1 *An* **oriented primitive feature** $F = (S, \rho)$ *of a solid* M *is an oriented surface where*

1) $S \subset \Re^3$ *is a connected, irreducible*[2] *and continuous algebraic surface which partially or completely coincides with one or more finite oriented faces of* M*;*
2) $\rho \subset S \times \mathcal{S}^2$ *is a continuous relation. For each* $s \in S$ *if* s *is a non-singular point of surface* S *(p.78 [1]) then* $v \in \mathcal{S}^2$ *is one of two opposing normals of the tangent plane at point* s *such that* $(s, v) \in \rho$*; if* s *is a singular point of* S *(e.g. at the apex of a cone) then, for all* v*, where* $v \in \mathcal{S}^2$ *is the limit of the orientations of its neighborhood,* $(s, v) \in \rho$*.*
3) *For all* $s \in M, (s, v) \in \rho, v$ *points away from* M*.*

Intuitively speaking, a feature is composed of both "skin", S, and "hair", the set of normal vectors which correspond to the points on $\mathcal{S}^2$. Each element of

[2] Here *irreducible* implies that a primitive feature cannot be composed of any other *more basic* surfaces.

relation ρ is a correspondence between a point on S and a vector on $\mathcal{S}^2$. Note, there may be more than one 'normal vector' at one point of a surface, e.g. at the apex of a conic shaped surface.

Let $\mathcal{E}^+$ be the proper Euclidean group which contains all the rotations and translations in $\Re^3$, and $\mathbf{T}^3$ be the maximum translation subgroup of $\mathcal{E}^+$.

Definition 2.1.2 *Any isometry* $g = tr$ *of* $\mathcal{E}^+, t \in \mathbf{T}^3, r \in SO(3)$ *acts on* ρ *in such a way that* $(s, v) \in \rho \Leftrightarrow (gs, rv) \in g * \rho$.

Definition 2.1.3 *An isometry* $g \in \mathcal{E}^+$ *is a* **proper symmetry of a feature** $F = (S, \rho)$ *if and only if* $g(S) = S$ **and** $g * \rho = \rho$.

Note, there is an extra demand on a symmetry for an oriented feature — it has to preserve the orientations ρ of the feature as well as the point set S. Since orientations are points on $\mathcal{S}^2$, symmetries of an oriented feature have to keep **two** sets of points in $\Re^3$ setwise invariant. One can prove[3] that the symmetries for an oriented surface form a group.

2.2 Multiple Contacts: Compound Surface Features and Their Symmetry Groups

An assembly or mechanism is a manifestation of multiple surface interactions of its subparts, albeit the physical property of each individual part (rigid or deformable) or the nature of the contact (static or articulated). Thus the representation of an assembly or mechanism is reduced to how to specify a set of contact constraints which dictate the configuration of a set of solids. Given two solids B_1 and B_2 in contact via surfaces F_1 and F_2 respectively, the relative motions of B_2 respect to B_1 can be expressed as:

$$l_1^{-1} l_2 \in f_1 G_1 G_2 f_2^{-1}, \tag{1}$$

where $l_1^{-1} l_2$ is the relative position of solid 2 w.r.t. solid 1, G_1, G_2 are symmetry groups of F_1 and F_2 respectively, l_1, l_2 specify the locations of solids B_1, B_2 in the world coordinate system and f_1 and f_2 specify the locations of F_1, F_2 in their respective body coordinates. A more specific form for the relative positions of two solids under n *surface contacts* (two contacting surfaces coincide):

$$l_1^{-1} l_2 \in f_1 G f_2^{-1} \tag{2}$$

has shown clearly that the possible motions of a solid or a subassembly S in an assembly are described precisely by the symmetry group G of the multiple contacting oriented surfaces of S. Note, G here is usually **not** the symmetry group of the whole solid/subassembly S but the symmetry group of those surfaces of S that are in **contact** with other solids. If G is an identity group, i.e. $l_1^{-1} l_2 = f_1 f_2^{-1}$ gives a fixed position for S. If G is a finite rotation group, then $f_1 G f_2^{-1}$ contains

[3] Due to space limit, we only give results without proofs. Interested readers can find details in the references listed.

a finite number of positions reflecting the existence of discrete symmetries in the collection of contacting surfaces. If G is a continuous group then there exists relative continuous motions between S and the rest of the assembly. It is then the degrees of freedom of such a configuration can be spoken of.

Let us first give a denotation for such a collection of contacting surfaces from a single solid, and then determine what effect the symmetry group of this surface-set will impose on the relative motions of contacting solids.

Definition 2.2.1 *A* **compound feature** $F = (S, \rho)$ *of primitive features* $F_1 = (S_1, \rho_1), ..., F_n = (S_n, \rho_n)$*, is defined to be* $S = S_1 \cup ... \cup S_n$ *and* $\rho = \rho_1 \cup ... \cup \rho_n$*.*

Pairwise Relationship of Oriented Features In order to determine the symmetry group of a compound feature systematically, we start with the simplest compound feature — a compound feature composed of only one pair of primitiv e features. See Figure 1(a), 1(b) and Figure 1(c) for examples of these simple compound features (Note that only a finite face on each primitive feature is drawn).

Given a pair of primitive features, what kind of relationship holds between the two features and what is the effect of such a relationship in terms of determining their symmetry group? The following definition gives a characterization of four relationships between a pair of primitive features:

Definition 2.2.2 *Two oriented primitive features* $F_1 = (S_1, \rho_1), F_2 = (S_2, \rho_2)$ *are said to be*

- **Distinct:** if for any open subsets $S_1' \subset S_1, S_2' \subset S_2$, no $g = tr \in \mathcal{E}^+$ exists such that $g(S_1') \subset S_2$ or $g(S_2') \subset S_1$. See Figure 1(a) for an example of a pair of distinct features F_1, F_2.
- **1-congruent**: if there exists at least one $g \in \mathcal{E}^+$ such that $g(S_1) = S_2$ and $g * \rho_1 = \rho_2$, but for all such $g, g(S_2) \neq S_1$. For an example see Figure 1(b). Another example is two parallel planar surfaces with normal vectors pointing in the same direction.
- **2-congruent**: if there exists $g_c \in \mathcal{E}^+$ such that $g_c(S_1) = S_2, g_c(S_2) = S_1, g_c * \rho_1 = \rho_2$ and $g_c * \rho_2 = \rho_1$. For an example, consider two parallel cylindrical surfaces having the same radius and normal v ectors pointing away from their center lines, as in Figure 1(c). Also, two parallel planar surfaces with normal vectors pointing to the opposite directions serve as examples of a pair of 2-congruent features.
- **Complemen tary:**if there exists $g \in \mathcal{E}^+$ such that $g(S_1) = S_2$ and $g * \rho_1 = -\rho_2$ where $-\rho_2 = \{(s, -v) | (s, v) \in \rho_2\}$; in other words, $\forall (s, v) \in g * \rho_1, \exists (s, -v) \in \rho_2$, and $\forall (s, v) \in \rho_2, \exists (s, -v) \in g * \rho_1$. See Figure 1(d) for an example.

It is easy to verify that these relationships are symmetrical relations. Immediately we can prove that this characterization has exhaustively enumerated all the possible cases between a pair of oriented primitive features.

Proposition 2.2.3 Distinct, 1-congruent, 2-congruent *and* complementary *are the only possible relationships between a pair of primitive features.*

Corollary 2.2.4 *Except for a pair of planar surface primitive features,* distinct, 1-congruent, 2-congruent *and* complementary *relationships are mutually exclusive relations between a pair of primitive features.*

Proposition 2.2.5 *If features $F_1 = (S_1, \rho_1), F_2 = (S_2, \rho_2)$ are complementary of each other, where $a(S_1) = S_2, a \in \mathcal{E}^+$, and G_1, G_2 are the symmetry groups of F_1, F_2 respectively, then $aG_1a^{-1} = G_2$. In particular, if $S_1 = S_2$ then $G_1 = G_2$* **(the necessary condition for surface contact)**.

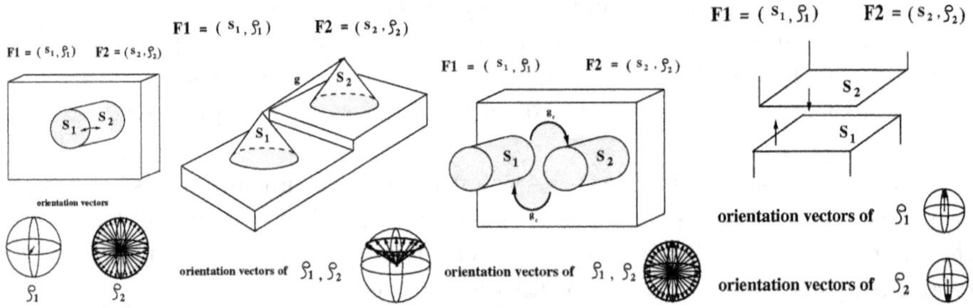

(a) distinct (b) 1-congruent (c) 2-congruent (d) complementary

Fig. 1. Four types of surface relations between a pair of solids.

Symmetry Group of Multiple Oriented Surfaces In the next few propositions we shall explore how the symmetry group of a compound feature is expressed by the symmetry groups of its component primitive features. The first case we consider is when a compound feature F is composed of n pairwise *distinct* features.

Proposition 2.2.6 *Given a compound feature $F = (S, \rho)$ of primitive features $F_1 = (S_1, \rho_1), ..., F_n = (S_n, \rho_n)$ where $F_1, ..., F_n$ are pairwise distinct primitive features with symmetry groups $G_1, ...G_n$ respectively. Then the symmetry group G of F is $G = G_1 \cap ... \cap G_n$.*

Proposition 2.2.7 *Let a compound feature $F = (S, \rho)$ be composed of a pair of primitive features $F_1 = (S_1, \rho_1)$ and $F_2 = (S_2, \rho_2)$ which are 1-congruent of each other. If G_1, G_2 are the symmetry groups of F_1, F_2 respectively, and G is the symmetry group of F then $G = G_1 \cap G_2$.*

Proposition 2.2.8 *Let a compound feature $F = (S, \rho)$ be composed of a pair of primitive features F_1 and F_2 which are 2-congruent of each other via g_c (Definition 2.2.2). If $F_1 = (S_1, \rho_1), F_2 = (S_2, \rho_2)$ have symmetry groups G_1, G_2*

respectively, and G is the symmetry group of F then $G = \langle g_c \rangle (G_1 \cap G_2)$ where $\langle g_c \rangle$ denotes the subgroup of $\mathcal{E}^+$ generated by g_c.

In general, the symmetry group G of a compound feature F can be found from the intersection of the symmetry groups G_i of its primitive features. The only exception is when a mapping which flips a pair of 2-congruent features in F is also a member of G. These kinds of mappings are new symmetries that do not exist in any G_i and the new group they generate is a discrete group.

2.3 Surface Contact: Computationally Tractable

As one can observe from the proven results for surface contact, the intersection of symmetry groups of the primitive features is one of the crucial operations in determining the relative motions of contacting solids. We face two computational problems: (1) How to denote symmetry groups, which can be finite, infinite, discrete or continuous, on computers? (2) How to do intersections of subgroups of $\mathcal{E}^+$ on computers efficiently? We have successfully implemented an efficient **TR**[4] subgroup intersection algorithm using geometric invariants denotation of the groups [9, 10] (Figures 2 and 3). The basic symmetry group of each surface of a solid is obtained by a straightforward mapping from the boundary (surface) file of the solid to their respective canonical symmetry groups. The

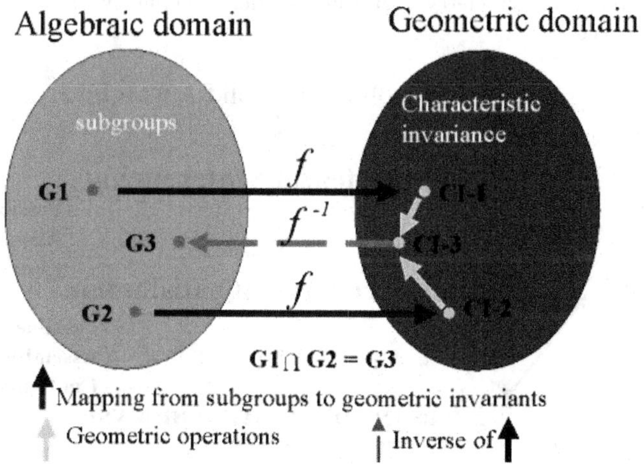

Fig. 2. A geometric representation, characteristic invariants, for the subgroups of the Euclidean group

group theoretical formalization of surface contact has also been embedded into an assembly planning system *KA3* (Figure 4).

[4] A symmetry group G (a set of motions) that can be divided into a translation subgroup T and a rotation subgroup R — a *semidirect product* $G = TR$. We call such a group a **TR**subgroup of $\mathcal{E}^+$.

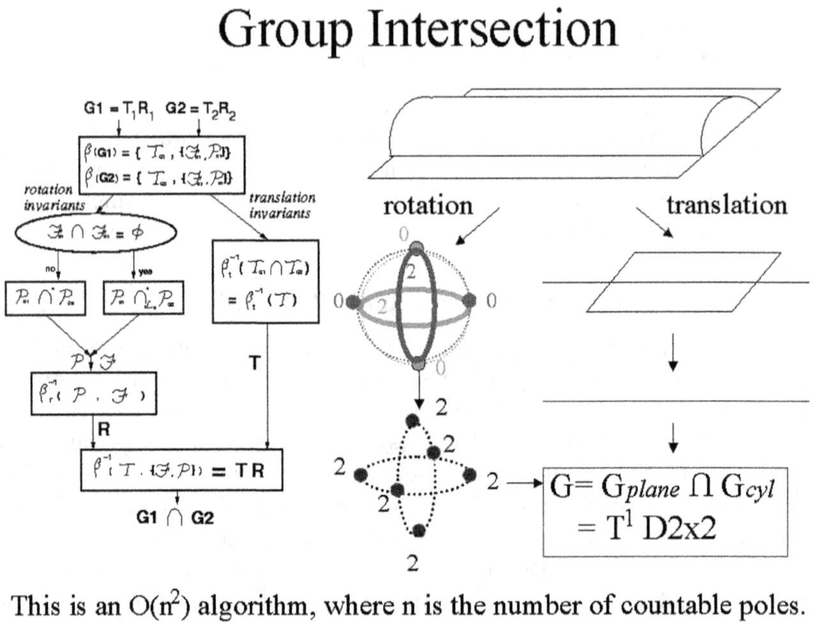

Fig. 3. Left: TR group intersection algorithm using translation and rotation invariants. Right: an example of symmetry group intersection (symmetry groups of planar and cylindrical surfaces)

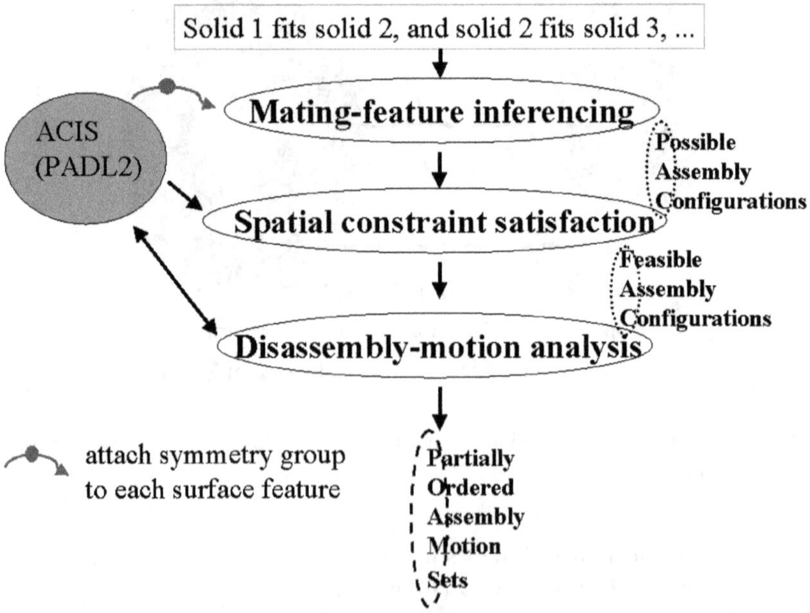

Fig. 4. An assembly planner *KA3* takes high-level task specifications and generates robot executable assembly instructions.

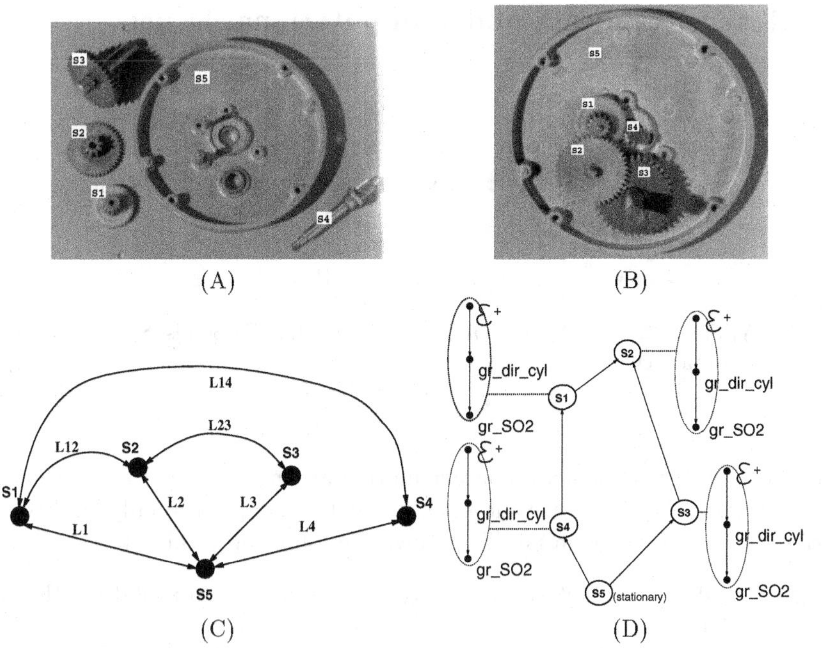

Fig. 5. (A) A five-part gearbox assembly. (B) The top view of the gearbox configuration automatically computed by *KA3*. (C) Representation of the gearbox assembly in terms of contacting compound feature symmetry groups, see text for details. (D) The output POAMS for the gearbox assembly. Each solid S_i is associated with a sequence of motions in terms of the symmetry group of its contacting surfaces with the rest of the assembly. Notice, during disassembly each solid is going through a 2-surface contact ($SO(2)$, assembled configuration), 1-surface contact ($\mathcal{G}_{dir\_cyl}$) and finally, no contact ($\mathcal{E}^+$, disassembled configuration).

As an example of assembly specification using symmetry groups, see Figure 5 for a five-part gearbox. The representation of the assembly is shown in Figure 5(C), where $L_i, i = 1..4$ is the symmetry group of the contacting compound feature between solids S_i and S_5. $L_i = a_i SO(2) a_i^{-1}, i = 1..4$, $SO(2)$ is a one degree rotation group resulted from the intersection of the symmetry group of a plane with that of a cylinder (the compound feature composed of two surfaces of the shaft of a gear). $L_{ij} = L_i L_j = a_{ij} SO(2) b_{ij} SO(2) c_{ij}, i, j = 1..4$ indicates that the relative positions between gears (non-surface contact) are simply determined by rotations in $SO(2)$ and some specific translations a_{ij}, b_{ij}, c_{ij}, where the relative gear pitch ratio is also embedded. This representation of the gearbox (Figure 5(C)) specifies precisely the articulated gearbox assembly. After a disassembly analysis by moving one solid away from this contacting network (Figure 5(C)) at a time, a *partially ordered assembly motion set* (POAMS) is found (Figure 5 (D)).

3 General Con tact and Computational Issues

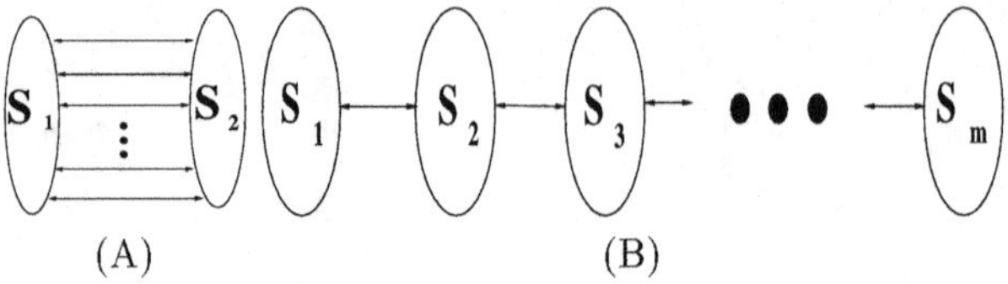

Fig. 6. (A) Solids S_1 and S_2 have n parallel contacts. (B) Solids $S_1, S_2, ...S_m$ form a sequential contact chain

We have divided the general contact motion among solids in to surface contact (lower pairs), which we have successfully treated computationally, and the non-surface-contact (higher pairs). They have the following specific forms:

1. Two solids have n surface contact, the relative position of solid 2 with respect to solid 1:
$$l_1^{-1}l_2 \in f_1 G f_2^{-1} \tag{3}$$
where G is the symmetry group of the compound feature composed of *all* the contact primitive features of S_1 or S_2. If F is composed of n pairwise distinct features F_i, then $G = \cap G_i$, where G_i is the symmetry group of F_i.
2. Two solids have n general contact (Figure 6 (A)), the relative position of solid 2 with respect to solid 1:
$$l_1^{-1}l_2 \in f_{11}G_{11}\sigma_1 G_{21} f_{21}^{-1} \cap f_{12}G_{12}\sigma_2 G_{22} f_{22}^{-1} \cap ... \cap f_{1n}G_{1n}\sigma_n G_{2n} f_{2n}^{-1} \tag{4}$$
where G_{ij} is the symmetry group of primitiv e feature j of S_i and f_{ij} is its feature coordinates.
3. m solids have a chaining general contact (Figure 6 (B)), the relative location of solid m with respect to solid 1:
$$l_1^{-1}l_m \in f_1 G_{12}\sigma_1 G_{21} f_{21}^{-1} f_2 G_{23}\sigma_2 G_{32} f_{32}^{-1} ... f_{m-1} G_{(m-1)m}\sigma_{m-1} G_{m(m-1)} f_{m(m-1)}^{-1} \tag{5}$$
where G_{ij} is the symmetry group of the surface on solid i in contact with solid j.

Encouraged by our existing work on making the group theoretical formalizatio n of surface contact (case 1. above) tractable, our goal is to seek the computational means to deal with the general con tact using group theoretical formalization (cases 2. and 3.).

Man y open problems remain: (1) *Can the geometric representation used for* **TR** *groups be used for group product?* It is known that in general, a product of groups is not a group. For **TR** groups, the **TR**-restriction presented in [9,

10] provides us with a constructive way to judge from its invariants whether a product remains a **TR** group. However we do not yet know how to compute a product from the invariants of its subgroups. (2) *Given an algebraic surface, how can one find its symmetry group computationally?* We intend to investigate the possibility of using the Gröbner basis of a given polynomial to find its symmetry group. An alternative is to use semi-algebraic set to represent the contacting surfaces and find the continuous symmetries using Lie algebra[5], but under this formalism it is not clear how to deal with discrete symmetries effectively (3) *Given a set of n algebraic surfaces and their respective symmetry groups, what is the exact algorithm to find the symmetry group for the whole set?* [9–11] have only proved results for subcases, i.e. n distinct surfaces, a pair of 1-cong or 2-cong surfaces. No proven result for the most general case yet exists. We are going to further study those compound features with more complicated inner structures. For example, one may define a concept of n-congruence on n features $F_1 \dots F_n$ as requiring that there exists $g \in \mathcal{E}^+$ such that $g(F_i) = F_{(i \bmod n)+1}$; this is a natural extension of 2-congruence. Such congruences will give rise to new symmetries of the compound feature.

4 Conclusion

This work provides a good example of applying algebra to solve a fundamental problem in robotics: solids in contact. It reflects both the power of group theory, and the effort one has to expend to make a mathematical theory computationally feasible. We establish a group theoretical formalization of general contact motion. The generality of this approach allows the treatment of solids in contact as subgroup manipulations, and provides a uniform computational platform for both continuous and discrete groups. The next challenge is to construct tools enabling computations of higher pairs, which will make this work computationally complete. Our previous work has shed some light on the feasibility of achieving this goal.

References

1. M.P. Do Carmo. *Differential Geometry of Curves and Surfaces.* Prentice Hall, New Jersey, 1976.
2. I. Drori, L. Joskowicz, and E. Sacks. Contact analysis of spatial fixed-axes pairs using configuration spaces. In *IEEE International Conference on Robotics and Automation*, pages 578,584. IEEE, 1999.
3. J.M. Hervé. Analyse structurelle des mécanismes par groupe des déplacements. *Mechanism and Machine Theory*, 13(4):437 – 450, 1977.
4. L.S. Homem de Mello. *Task Sequence Planning for Robotic Assembly.* PhD thesis, Carnegie Mellon University, 1989.
5. L. Joskowicz and E. Sacks. Computer-aided mechanical design using configuration spaces. *IEEE Computers in Science and Engineering*, page 14/21, Nov/Dec 1999.

[5] Thanks to an anonymous reviewer for pointing this out

6. L. Joskowicz, E. Sacks, and V. Kumar. Selecting an effective task-specific contact analysis algorithm. *IEEE Workshop on New Directions in Contact Analysis and Simulation*, 1998.
7. J.C. Latombe. *Robot Motion Planning.* Kluwer, 1993.
8. C. Laugier. Planning fine motion strategies by reasoning in the contact space. In *IEEE International Conference on Robotics and Automation*, pages 653–661, Washington, DC, 1989. IEEE Computer Society Press.
9. Y. Liu. *Symmetry Groups in Robotic Assembly Planning.* PhD thesis, University of Massachusetts, Amherst, MA., September 1990.
10. Y. Liu. A Geometric Approach for Denoting and Intersecting TR Subgroups of the Euclidean Group. *DIMACS Technical Report, Rutgers University*, 93-82:1–52, 1993.
11. Y. Liu and R. Popplestone. A Group Theoretical Formalization of Surface Contact. *International Journal of Robotics Research*, 13(2):148 – 161, April 1994.
12. Y. Liu and R.J. Popplestone. Assembly planning from solid models. In *IEEE International Conference on Robotics and Automation*, Washington, DC, 1989. IEEE Computer Society Press.
13. Y. Liu and R.J. Popplestone. Symmetry constraint inference in assembly planning. In *Eighth National Conference on Artificial Intelligence*, Boston, Mass., July/August 1990.
14. Y. Liu and R.J. Popplestone. Symmetry groups in analysis of assembly kinematics. In *IEEE International Conference on Robotics and Automation*, Washington, DC, May 1991. IEEE Computer Society Press.
15. Y. Liu and R.J. Popplestone. Symmetry groups in solid model-based assembly planning. In *Artificial Intelligence Applications in Manufacturing*, pages 103,131, Cambridge,Massachusetts, 1992. AAAI Press and The MIT Press.
16. T. Lozano-Pérez, J.L. Jones, E. Mazer, P.A. O'Donnell, W.E.L. Grimson, P. Tournassoud, and A. Lanusse. Handey: A robot system that recognizes, plans, and manipulates. In *IEEE International Conference on Robotics and Automation*, pages 843–849, Washington, DC, March 1987. IEEE Computer Society Press.
17. A.T. Miller and P.K. Allen. Examples of 3d grasp quality computations. In *IEEE International Conference on Robotics and Automation*, pages 1240,1246. IEEE, 1999.
18. David Montana. The kinematics of contact and grasp. *The International Journal of Robotics Research*, 7(3):17–32, June 1988.
19. R.J. Popplestone. Group theory and robotics. In M. Brady and R. Paul, editors, *Robotics Research, The First Int. Symp.* MIT Press, Cambridge,Massachusetts, 1984.
20. R.J. Popplestone, A.P. Ambler, and I. Bellos. An interpreter for a language for describing assemblies. *Artificial Intelligence*, 14(1):79–107, 1980.
21. E. Sacks. Practical sliced configuration spaces for curved planar pairs. *Internaltional Journal of Robotics Research*, 18(1):59–63, January 1999.
22. C.A. Tanaglia, D.E. Orin, R.A. LaFarge, and C. Lewis. Toward development of a generalized contact algorithm for polyhedral objects. In *IEEE International Conference on Robotics and Automation*, pages 2887,2892. IEEE, 1999.
23. F. Thomas and C. Torras. Inferring feasible assemblies from spatial constraints. *IEEE Transactions on Robotics and Automation*, 8(2):228,239, April 1992.
24. Randall Wilson and Jean-Claude Latombe. Geometric reasoning about mechanical assembly. *Artificial Intelligence*, 71(2), December 1994.
25. J.D. Wolter. *On the Automatic Generation of Plans for Mechanical Assembly.* PhD thesis, University of Michigan, 1988.

Periodic Pattern Analysis under Affine Distortions Using Wallpaper Groups

Yanxi Liu and Robert T. Collins

The Robotics Institute, Carnegie Mellon University, Pittsburgh 15213, USA,
yanxi,rcollins@cs.cmu.edu,
http://www.cs.cmu.edu/~yanxi/www/home.html

Abstract. In this paper, the mathematical theory of wallpaper groups is used to construct a computational tool for symmetry analysis of periodic patterns. Starting with a novel peak detection algorithm based on "regions of dominance", an input periodic pattern can be automatically classified into one of the 17 wallpaper groups. The orbits of stabilizer subgroups within the group lead to a small set of candidate motifs that exhibit local symmetry consistent with the global symmetry of the entire pattern. We further consider affine distorted periodic patterns and show that each such pattern can be classified into a small set of symmetry groups that describe the patterns' potential symmetries under affine transformation.

1 Introduction

Symmetry is pervasive in both natural and man-made environments. Humans have an innate ability to perceive symmetry, but it is not obvious how to automate this powerful insight. It is a continuous effort of the authors to find proper computational tools for dealing with symmetry. Symmetries of periodic patterns in a plane are of particular interest in computer vision. This is because the symmetry group of a pattern is independent of scale, absolute color, lighting, density and orientation/position of the pattern. Periodic patterns can be found in regular textures, indoor and outdoor scenes (e.g. brick walls, tiled floors, wallpapers, ceilings, clothes, windows on buildings, cars in a parking lot), or in intermediate data representations (e.g. periodicity analysis of human and animal gaits in the spatio-temporal domain).

A mature mathematical theory for periodic patterns has been known for over a century [1,2]. For monochrome planar periodic patterns, there are seven *frieze groups* for 2D patterns repeated along one dimension, and seventeen *wallpaper groups* describing patterns extended by two linearly independent translational generators. Despite an infinite variety of instantiations, this finite set of symmetry groups completely characterizes the possible structural symmetry of any periodic pattern.

We have developed a computational model of periodic pattern perception composed of: generating the underlying translational lattice from the image of a periodic pattern, classifying the symmetry group of the periodic pattern, and

G. Sommer and Y. Y. Zeevi (Eds.): AFPAC 2000, LNCS 1888, pp. 241-250, 2000.

identifying the preferred "motif" of the pattern. Our work, initially inspired by [7], appears to be the first to use the theory of frieze and wallpaper groups for automated analysis of periodic patterns, although there exist flowcharts and computer programs that allow humans to interactively generate and identify periodic patterns for educational purposes [8,3]. Due to space limitations, this paper concentrates only on wallpaper groups. Furthermore, we assume that the translational lattice of the 2D periodic pattern has already been extracted. The reader is referred to [5] to find our algorithm for performing robust lattice extraction. Figure 1 shows one sample result produced by this algorithm.

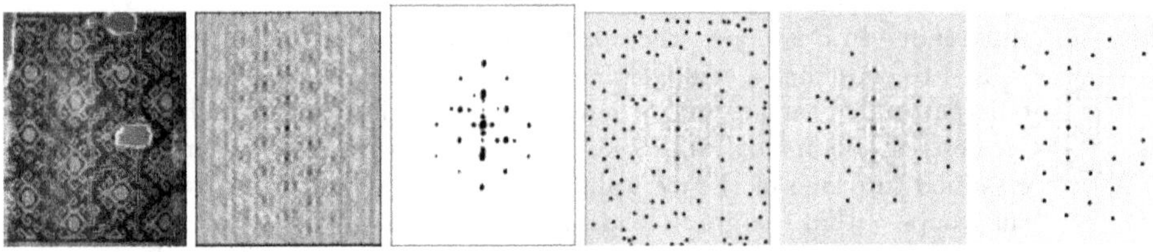

Fig. 1. An oriental rug image and A) its autocorrelation surface, B) peaks found using a global threshold, C) peaks extracted using the threshold-free method of Lin, et al. [4], D) the highest 32 peaks from those returned by Lin, et al., E) the 32 most-dominant peaks found using our approach described in [5].

2 Symme try Group Classification under Euclidean Transformations

A 2D *repeated* or *periodic* pattern has the following property: there exists a finite region bounded by two linearly independent translations which, when acted upon by the group generated by the translations, produces simultaneously a covering (no gaps) and a packing (no overlaps) of the original image [7,2]. The smallest such bounded region is called a *unit of the pattern* or *lattice unit*, since the translational orbit of any single point on the plane is a lattice. A *symmetry* of a subset S of Euclidean space is an isometry that keeps S setwise invariant. All symmetries of S form the *symmetry group* of S under composition. It has been proven that there are seventeen *wallpaper groups* (Figure 2) describing patterns extended by two linearly independent translational generators [7,2]. Mathematically, wallpaper groups are defined only for infinite patterns that cover the whole plane. In practice, we analyze a periodic pattern P of a finite area, and use the phrase "symmetry group G of P" to mean that G is the symmetry group of the infinite periodic pattern that has P as a finite patch.

Figure 2 depicts unit lattices for the 17 distinct wallpaper groups (from [7]). Each unit is characterized in terms of its translation generators, rotation, reflec-

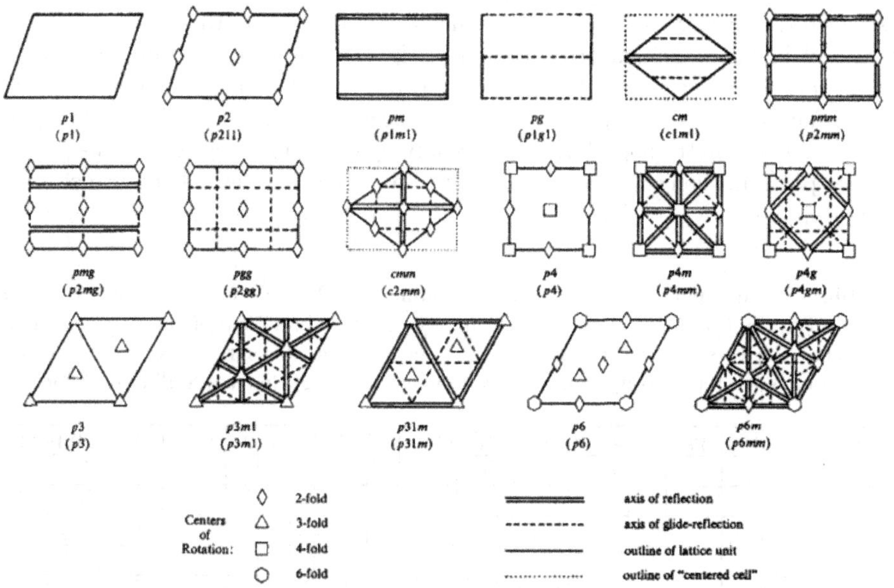

Fig. 2. The generating regions for the 17 Wallpaper groups (from [7])

tion and glide-reflection symmetries. The two linearly independent translations of minimum length are the two basic generators of each group, and they construct a lattice for the group. Even though the variety of pattern instantiations is endless, the underlying relationship between translation, rotation, reflection and glide-reflection in any 2D periodic pattern **must** conform to one of these seventeen cases.

Since a symmetry of a 2D periodic pattern has to map the lattice associated with the pattern onto itself, i.e., map centers of rotation to new centers of rotation having the same order, the only possible rotation symmetries are $2, 3, 4, 6$-fold rotations. This restriction is often referred to as the *crystallographic restriction*. Furthermore, reflection axes can only be oriented parallel, diagonal, or perpendicular to the lattice translation vectors. Under these constraints, there are only five possible lattice unit shapes: (1) parallelogram (two groups: $p1, p2$), (2) rectangular (five groups: pm, pg, pmm, pmg, pgg), (3) rhombic (two groups:cm, cmm), (4) square (three groups:$p4, p4m, p4g$) and (5) hexagonal (five groups:$p3, p3ml, p3lm, p6, p6m$). All lattice units are parallelograms. Rectangular units have angles of 90°. Rhombic units have equal-length edges. Square units are a special case of both (2) and (3), and hexagonal units are a special case of (3).

We have constructed an algorithm that can automatically classify which symmetry group a 2D periodic pattern under Euclidean transformations belongs to. The practical value of understanding the 17 wallpaper groups is that correct pattern classification can be performed after verifying the existence of only a small set of rotation and/or reflection symmetries. Table 1 lists the eight symmetries checked in the classification algorithm. It is clear that each group corresponds to a unique sequence of values listed in Table 1, and all are mutually exclusive from each other. The determination of a specific rotation or reflection or glide-reflection symmetry is performed b y applying the symmetry to be tested to the entire pattern, then checking the similarity between the original and transformed images.

Table 1. Wallpaper group classification: numbers 2,3,4 or 6 denote n-fold rotational symmetry, Tx (or Dx) denotes reflectional symmetry about one of the translation (or diagonal) vectors of the unit lattice. "Y" means that the symmetry exists for that particular symmetry group; empty space means no. Y(g) denotes a glide reflection.

	p1	p2	pm	pg	cm	pmm	pmg	pgg	cmm	p4	p4m	p4g	p3	p3m1	p31m	p6	p6m
2		Y				Y	Y	Y	Y	Y	Y	Y				Y	Y
3													Y	Y	Y	Y	Y
4										Y	Y	Y					
6																Y	Y
T1			Y	Y(g)		Y	Y(g)	Y(g)			Y	Y(g)			Y		Y
T2						Y	Y	Y(g)			Y	Y(g)			Y		Y
D1					Y				Y		Y	Y		Y	Y		Y
D2									Y		Y	Y					Y

3 Extracting Representative Motifs

Although other work has addressed detection of the translational lattice of a periodic pattern, ours is the first to seek a principled method for determining a representative motif. The issue here is that consideration of translational symmetry alone fixes the size, shape and orientation of the lattice, but leaves open the question of where the lattice is located in the image. Any offset of the lattice carves the pattern into a set of identical tiles, but these tiles typically provide no computational insight, and appear nonintuitive to a human observer (Figure 3). Choosing a good motif should help one see, from a single tile, what the whole pattern looks like. From work in perceptual grouping, it is known that the human perceptual system often has a preference for symmetric figures. Our contribution in this section is to show how a small set of tiles can be chosen, in a principled way, such that the symmetry of the pattern fragmen ts on them is maxim ized.

If we entertain the idea that the most represen tative motif is the one that is most symmetrical, one plausible strategy for generating motifs is to align the motif center with the center of the highest-order of point symmetry in the pattern. This is the point fixed by the largest stabilizer subgroup of the symmetry group of the pattern. If we choose the centers of the highest order of rotational

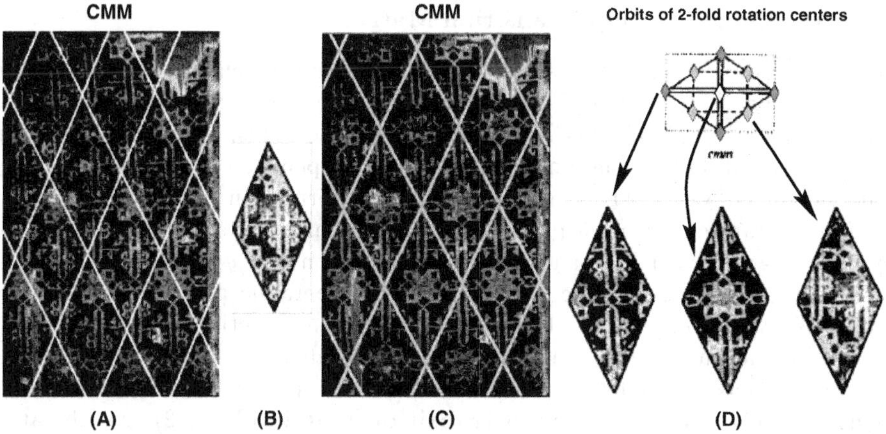

Fig. 3. (A) and (B) show an automatically extracted lattice and the tile that it implies. The tile is not a good representation of the pattern motif. (C) and (D) show the lattice position in terms of one of the three most-symmetric motifs found for the oriental rug image. The latter was generated automatically by an algorithm that analyzes pattern symmetry based on knowledge of the 17 wallpaper groups.

symmetries, candidate motifs can then be determined systematically by enumerating each distinct center point of the highest-order rotation. Two rotation centers are distinct if they lie in different **orbits** of the symmetry group, that is, if one cannot be mapped into the other by applying any translation, rotation, reflection or glide-reflection symmetries in its own symmetry group.

Figure 3 shows an example of an automatically extracted lattice, and an arbitrary tile that it carves out, followed by three symmetrical tiles centered on 2-fold rotation centers More examples and explanations can be found in [5].

4 Symmetry Group Classification Under Affine Transformations

When a 2D pattern undergoes a rigid transformation, its symmetry group remains. Strictly speaking, its symmetry group is conjugated by the transformation that acts on the pattern. Since there exists a bijection between the original symmetry group and the conjugated symmetry group, the two groups are considered equivalent (isomorphic). If one imagines a coordinate system fixed on the pattern, the translation, rotation, reflection and glide-reflection symmetries are unchanged under this coordinate system when the pattern is undergoing rigid transformations. This situation will no longer be true when the pattern undergoes a non-rigid transformation. However, certain symmetries of a periodic pattern may survive.

4.1 Wallpaper Group Transition Matrix

If g is a symmetry of a 2D periodic pattern P, by the definition of symmetry $g(P) = P$. Let $T(g(P)) = T(P)$, here T is a transformation. Then $T(gT^{-1}T(P)) = T(P) \Rightarrow TgT^{-1}(T(P)) = T(P)$. A useful question to ask is: Does TgT^{-1} remain a symmetry of $T(P)$? The answers, of course, depend on what g and T are. The answer is "yes" if (1) T is a similarity transformation (a proof that under similarity transformation, a periodic pattern remains the same in terms of its symmetry group) (2) T is an affine transformation *and* g is either a translation (a proof that a periodic pattern remains a periodic pattern under affine transformation) or g is a 2-fold rotation; (3) g is a reflection (glide-reflection) and T is a non-uniform scaling parallel or perpendicular to g's reflection axis. Relevant proofs can be found in [6]. Based on these proven results, we can construct a 17x17 **wallpaper group transition matrix** (Table 2) that dictates how the symmetry group of a periodic pattern can be transformed in to other groups under non-rigid transformations. It turns out that only certain groups can be associated with a pattern under affine distortions. This matrix leads to a new way of evaluating a periodic pattern affine deformation: we should not only consider the symmetry group of the pattern as given, but also all the possible symmetry groups that can be associated with that pattern when it transformed affinely. Table 2 tells us that these transitions form well-defined small, finite orbits. For example, there are two large orbits of the 17 groups: the $p1$-orbit and the $p2$-orbit. This comes from the fact that 2-fold rotation always survives any nonsingular affine distortion. Figure 4 shows one example of symmetry group transition as a pattern undergoes a series of affine deformations.

4.2 Symmetry Group Classification Algorithm

When the 2D pattern undergoes an affine transformation that preserves the shortest vector property, the same Euclidean algorithm (Table 1) can be applied for determining the lattice unit and classifying its symmetry group [1]. From Table 2, only those entries with P need to be further checked for possible "higher symmetries".

The implementation of this idea is carried out as follows: Once the lattice unit is decided, the input unit lattice is simultaneously deformed into a hexagonal lattice and a square lattice, with the pattern deformed accordingly. Hexagonal and square lattices are the most symmetrical lattices, therefore these deformations allow the most symmetrical potential patterns to form. Meanwhile, the original symmetries of the pattern are guaranteed to be preserved under at least one of these two deformations, because hexagonal and square lattices are special cases of the more general lattices (rhombus, rectangular and parallelogram, see Section 2). The group classification procedure can then proceed in the same way

[1] When the affine distortion is so large that the nearest neighboring lattice points no longer form the boundary of a proper generating region, additional information is needed to locate the lattice unit. These include finding an axis of skewed symmetry, which is beyond scope of this paper.

Table 2. W allpaper Group T ransition Matrix
Empty entries mean that there exist no transformations between the two groups (to the left and to the top). **S**: similarity transformation, **N**: non-uniform scaling ⊥ or ∥ to all reflection axes in the group to the left, **A**: general affine transformation other than **S** or **N**, and **P**: possible affine transformation (pattern dependent).

	p1	p2	pm	pg	cm	pmm	pmg	pgg	cmm	p4	p4m	p4g	p3	p3m1	p31m	p6	p6m
p1	A		P	P	P								P	P	P		
p2		A				P	P	P	P	P	P	P				P	P
pm	A		N														
pg	A			N													
cm	A				N									P	P		
pmm		A				N					P						
pmg		A					N										
pgg		A						N				P					
cmm		A							N		P	P					P
p4		A								S							
p4m		A				N			N		S						
p4g		A						N	N			S					
p3	A												S				
p3ml	A				N									S			
p3lm	A				N										S		
p6		A														S	
p6m		A							N								S

as stated in Section 2. A diagram version of the algorithm is shown in Figure 5.

4.3 Symmetry Group Classification Experimen tal Results

We have successfully processed all seventeen wallpaper group patterns[2]. Here we provide one example to illustrate how our algorithm works.

The first step is to determine the underlying translational lattice structure of the original image, in the form of t wo independent generating vectors t_1 and t_2. Since we are assuming that the wallpaper pattern has been previously isolated, the lattice points are determined by finding significant peaks in the pattern's autocorrelation surface (Figure 6a-c). The lattice of dots is decomposed into two generating vectors by finding the two shortest difference vectors t_1 and t_2 such that the angle between them is between 60 and 90 degrees. The second step involves transforming the lattice to a square grid, aligned with the horizontal and vertical axes (Figure 6d-f). This is performed by applying an affine transformation to the image and its autocorrelation surface. The transformation used is the unique affine transform lea ving the origin (0,0) fixed and taking t_1 to $(L, 0)$

[2] For a more complete set of results on all 17 wallpaper groups see [6].

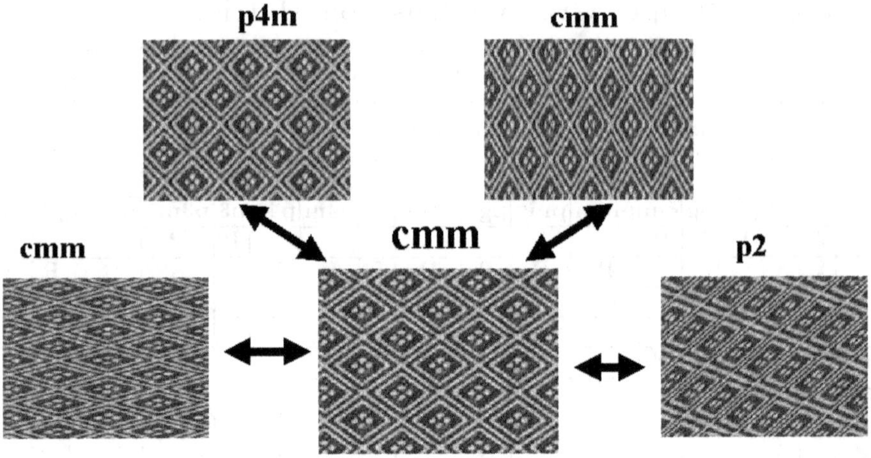

Fig. 4. The original periodic pattern has symmetry group *cmm* (middle). Its symmetry group migrates to different groups within its orbit in the *wallpaper group transition matrix* (Table 2) while the pattern is being affinely transformed.

and t_2 to $(0, L)$, where L is the larger of the two generating vectors lengths $||t_1||$ and $||t_2||$.

After transforming to a square lattice, a square generating region (with dimensions $L \times L$) is cropped from the transformed image. This is used as a template, the rotated and reflected versions of which are correlated with the transformed image to determine what, if any, type of rotation and reflection symmetry it has. In the location determined by the highest correlation peak, a match score between the rotated/reflected template and the image is computed as the mean of the absolute difference between corresponding intensity values. The lower the value of this match score, the more likely it is that the image has that particular rotational/reflectional symmetry. This yields a set of "typical" match scores for that pattern – the mean and standard deviation of these scores are used as an adaptive threshold tailored for this pattern. Match scores associated with rotated/reflected templates are compared to this threshold to determine whether that particular symmetry holds.

An example is shown in Figure 6. The processed values for both square and hexagon lattices are shown below:

	rot180	rot120	rot90	rot60	T1 refl	T2 refl	D1 refl	D2 refl
square	**0.040**	0.279	0.296	0.269	0.272	0.275	0.269	0.268
hexag	**0.040**	**0.038**	0.310	**0.043**	0.269	0.271	0.271	0.271

We find that the pattern only has two-fold rotation symmetry when represented using a square lattice grid, which signifies group $p2$ (Table 1). To transform the image to a hexagonal lattice structure, the affine transformation is used that

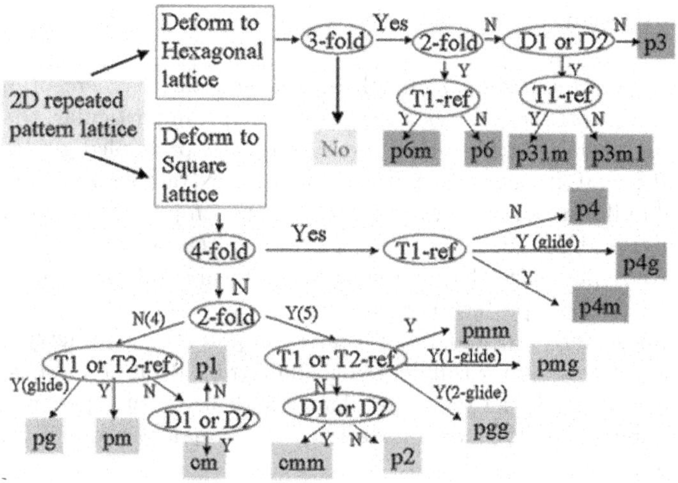

Fig. 5. An algorithm for symmetry group classification of 2D periodic pattern under affine transformation: Y(glide) means the reflection symmetry must be a non-trivial glide reflection. Y(n) / N(n) means the test result is positive/negative and n is the possible number of symmetry groups need to be further distinguished.

leaves the origin (0,0) fixed while mapping t_1 to $(L, 0)$ and t_2 to $(L/2, L*(\sqrt{3}/2))$, where L is a length chosen as before. The row labeled "hexag" in the table shows rotation and reflection results for the hexagonally transformed pattern. We see that now, in addition to two-fold symmetry, the pattern also has 60 and 120 degree rotational symmetry. The pattern is uniquely classified as being from the *p6* wallpaper symmetry group (Table 1). One can also verify this transition between *p*6 and *p*2 in the wallpaper group transition matrix (Table 2).

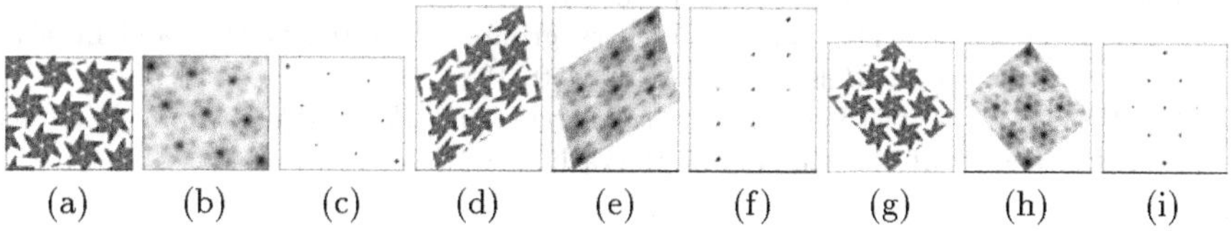

Fig. 6. (a) original image. (b) autocorrelation of image. (c) detected lattice points. (d) transformed image. (e) transformed autocorrelation. (f) transformed lattice points, now a square grid. (g) hexagonal transformed image. (h) transformed autocorrelation. (i) transformed lattice points, now a hexagonal grid.

5 Conclusion

We propose a computational model for periodic pattern perception based on the mathematical theory of crystallographic groups, in particular, the wallpaper groups. This mature mathematical theory provides principled guidelines for analyzing and classifying periodic patterns, and for extracting a patterns' visually meaningful building blocks, namely motifs. This computational model has been implemented and tested on both synthetic and real-world images of periodic patterns. We hypothesize that symmetric tiles form good candidates for human and machine periodic pattern perception.

More importantly, an understanding of the potential symmetry group transitions of a periodic pattern undergoing affine transformation opens a door for us to apply this method to new problems, such as texture perception and replacement, localization, robot navigation, and human perceptual organization, among others.

References

1. E.S. Fedorov. The elements of the study of figures. [Russian] (2) 21. In *Zapiski Imperatorskogo S. Peterburgskogo Mineralogichesgo Obshchestva [Proc. S. Peterb. Mineral. Soc.]*, pages 1–289, 1885.
2. B. Grünbaum and G.C. Shephard. *Tilings and Patterns.* W.H. Freeman and Company, New York, 1987.
3. Kali. Programs that can automatically generate 2d planar crystallographic patterns. *http://www.geom.umn.edu/apps/kali/.*
4. Hsin-Chih Lin, Ling-Ling Wang, and Shi-Nine Yang. Extracting periodicity of a regular texture based on autocorrelation functions. *Pattern Recognition Letters*, 18:433–443, 1997.
5. Y. Liu and R. T. Collins. A Computational Model for Repeated Pattern Perception using Frieze and Wallpaper Groups. In *Computer Vision and Pattern Recognition Conference*, June 2000.
6. Y. Liu and R.T. Collins. Frieze and wallpaper symmetry groups classification under affine and perspective distortion. Technical Report CMU-RI-TR-98-37, The Robotics Institute, Carnegie Mellon University, Pittsburgh, PA, 1998.
7. D. Schattschneider. The plane symmetry groups: their recognition and notation. *American Mathematical Monthly*, 85:439–450, 1978.
8. D.K. Washburn and D.W. Crowe. *Symmetries of Culture: Theory and Practice of Plane Pattern Analysis.* University of Washington Press, 1991.

Wavelet Filter Design via Linear Independent Basic Filters

Kai Neckels*

Institut für Informatik und Praktische Mathematik
Christian-Albrecht-Universität Kiel
Preußerstraße 1–9
24105 Kiel, Germany
kn@ks.informatik-uni-kiel.de

Abstract. A new point of view for wavelet filters is presented. This leads to a description of wavelet filters in terms of certain linear independent *basic filters* which can be designed to construct wavelets with special properties. Furthermore, it is shown, that this approach makes explicit closed form descriptions for higher order DAUBECHIES wavelet filters (at least for D_8 and D_{10}) possible, which were unaccessible before. Additionally, some biorthogonal examples are discussed and finally, a conceptual generalization to the twodimensional case is given.

1 Introduction

Since its introduction in the early 1980s, the evolution of wavelet analysis caused a deep impact in nearly all tasks of signal processing as well as computer vision applications and related questions (e.g. image compression, feature detection, optic flow estimation, treatment of PDEs). Though, the onedimensional theory has grown rapidly in the last two decades, there are several open questions concerning the general, multidimensional wavelet theory, for example the lack of factorization theorems like the FEJER-RIESZ-Lemma, which makes the design of scaling (and wavelet) filters with *desirable* properties in more than one dimension quite tricky. The aim of this paper is the presentation of a framework for onedimensional scaling filter design, which can be easily generalized to higher dimensions and may therefore help to overcome some of the existing problems. The reason for this is the fact, that a direct design method is used, which is independent of factorization questions. Finally, we shall mention that similar results were presented in the article [AHC93], which the author was unaware of during the first writing of this text. However, in [AHC93] the concept of linear independence was not used and no multidimensional generalization was intended; additionally, the closed form descriptions for higher order maximally flat orthogonal wavelet filters are a new contribution (although, they are mostly of theoretical interest).

* The author is supported by the DEUTSCHE FORSCHUNGSGEMEINSCHAFT (DFG) within the Graduiertenkolleg 357.

G. Sommer and Y. Y. Zeevi (Eds.): AFPAC 2000, LNCS 1888, pp. 251-258, 2000.

2 Basic Material

All of the following investigations are restricted to wavelets that come from a dyadic multiresolution analysis (however, the generalization to arbitrary integer dilations is straightforward). Consider a scaling function $\varphi \in L_2(\mathbb{R})$ and suppose, that the related scaling filter symbol $m_0(\omega)$ is given by

$$m_0(\omega) = \sum_{k=0}^{n} \gamma_k \cdot e^{i\omega k} \qquad , a_k \in \mathbb{R}.$$

To yield an orthonormal basis for $L_2(\mathbb{R})$, the symbol has to satisfy the *orthogonality criterion*

$$\begin{aligned} 1 &\equiv |m_0(\omega)|^2 + |m_0(\omega+\pi)|^2 \\ &= 2 \cdot \sum_{k=0}^{n} \gamma_k^2 + 4 \cdot \sum_{k=1}^{\frac{n-1}{2}} \sum_{j=0}^{n-2k} \gamma_j \cdot \gamma_{j+2k} \cdot \cos(2k\omega). \end{aligned}$$

From this, we can directly derive the following $(n+1)/2$ constraint equations of order two:

$$\sum_{j=0}^{n} \gamma_j^2 = 1 \qquad \text{and} \qquad \sum_{j=0}^{n-2k} \gamma_j \cdot \gamma_{j+2k} = 0, \quad k = 1, \dots, (n-1)/2. \qquad (1)$$

For most applications, wavelets with a sufficient high regularity and a number of vanishing moments (this gives polynomial reproducibility) are desirable. Moreover, it is a well known fact, that both of the mentioned properties are in some sense connected to the zero order, say m, of the scaling filter symbol $m_0(\omega)$ at the aliasing frequency $\omega = \pi$. These zeros can be characterized by the STRANG-FIX *conditions* or *sum rules* of order $m-1$, that is

$$\sum_{k=0}^{n} (-1)^k \, k^l \, h_k = 0 \qquad \text{for } l = 0, 1, \dots, m-1.$$

From the constraints in (1) one easily shows that an orthogonal scaling filter symbol of length $n+1$ can have at most a zero of order $(n+1)/2$ at π.

3 The Framework

The main idea of our framework is the following: consider a linear combination of linear independent (in vectorial sense) *basic filters* and solve the equation system (1) for the coefficients of the linear combination; from this point of view the linear independence means that no redundancy can occur and all solutions (if they exist) are accessible. In this section we will now successively build such families of linear independent basic filters. These will additionally be chosen such that they satisfy the STRANG-FIX conditions up to a certain order.

Lemma 1. *Suppose, the filter* $[\ \gamma_0\ \gamma_1\ \dots\ \gamma_n\ ]$ *satisfies the sum rules exactly up to order* $n-1$. *Then the filter*

$$[\ \tilde{\gamma}_0\ \ \tilde{\gamma}_1 \dots \tilde{\gamma}_{n+1}\ ] := [\ 1\ \ 1\ ] * [\ \gamma_0\ \ \gamma_0\ \ \gamma_1\ \ \dots\ \ \gamma_n\ ]$$

satisfies the sum rules exactly up to order n.

Pr oof. The proof is straightforward. Just evaluate

$$\sum_{k=0}^{n+1} (-1)^k \cdot k^p \cdot \tilde{\gamma}_k = -\sum_{j=0}^{p-1} \binom{p}{j} \cdot \sum_{k=0}^{n} (-1)^k \cdot k^j \cdot \gamma_k .$$

The inner sum on the right side vanishes for $p = 0, 1 \dots n$ and is different from zero for $p = n+1$ by the assumptions that were made. □

Corollary 1. *F or everyn* $\in \mathbb{N}_*$ *the filter*

$$h^n := [\ 1\ \ 1\ ]^{*n}$$

satisfies the sum rules up to order $n-1$, *where* γ^{*n} *denotes the* n *times subsequently repeated discrete convolution of* γ.

Lemma 2. *Define*

$$g^{l,m} := [\ 1\ \ -1\ ]^{*l} * [\ 1\ \ 1\ ]^{*m} .$$

Then, for all $m, l \in \mathbb{N}_*$ *the filter* $g^{m,l}$ *satisfies exactly the sum rules of order* $m-1$.

The filters h^n and the con volutionfilters $g^{l,m}$ with $l+m=n$ form a linear independent family of basic filters and will be very useful in the design of several scaling filters, as we shall see in the following.

Proposition 1. *Let* $\mathbf{A}_n$ *be the* $(n+1) \times (n+1)$ *matrix*

$$\mathbf{A}_n = \begin{bmatrix} h_0^n\ g_0^{1,n-1} & g_0^{2,n-2} & \cdots & g_0^{n,0} \\ h_1^n\ g_1^{1,n-1} & g_1^{2,n-2} & \cdots & g_1^{n,0} \\ \vdots \quad \vdots & \vdots & \ddots & \vdots \\ h_n^n\ g_n^{1,n-1} & g_n^{2,n-2} & \cdots & g_n^{n,0} \end{bmatrix},$$

then

$$\det \mathbf{A}_n = (-2)^{\frac{n(n+1)}{2}} .$$

Especially, $\det \mathbf{A}_n \neq 0$ for all $n \in \mathbb{N}_*$ and from this, we directly deduce that the $n+1$ filters

$$\{h^n, g^{1,n-1}, g^{2,n-2}, \dots g^{n,0}\}$$

form a linear independent family of basic filters and moreover, every subfamily

$$\{h^n, g^{1,n-1}, g^{2,n-2}, \dots g^{n-k,k}\}$$

additionally satisfies all sum rules up to order $k-1$ by Corollary 1 and Lemma 2. Furthermore, this family of basic filters yields a natural decomposition of every scaling filter that satisfies the sum rules up to order $k-1$ into an even (symmetric) and an odd (antisymmetric) part since h^n is always even and $g^{j,n-j}$ is even for j even and odd for j odd. Note further that the sum rule order decreases by one with each symmetry switch.

Proof of Proposition 1. We will make use of the induction principle. For $n = 1$ we obtain $\mathbf{A}_1 = \left[\begin{smallmatrix} 1 & 1 \\ 1 & -1 \end{smallmatrix}\right]$ and $\det \mathbf{A}_1 = -2$. In the second step, we will evaluate $\det \mathbf{A}_{n+1}$ from $\det \mathbf{A}_n$ by elementary matrix operations. In particular, we acquire

$$\det \mathbf{A}_{n+1} = (-2)^{n+1} \cdot \det \mathbf{A}_n.$$

By induction, we obtain the desired relation. □

4 Examples

Maximally flat filters. We will start with an example, that leads to the classical D_8 filter, i.e. an orthogonal filter with a zero of fourth order at the aliasing frequency $\omega = \pi$. Therefore, consider a combination of linear independent basic filters of length eight, that satisfy the STRANG-FIX conditions up to order three. By Proposition 1 such a filter is given by

$$\gamma = \lambda_0 \cdot h^7 + \lambda_1 \cdot g^{1,6} + \lambda_2 \cdot g^{2,5} + \lambda_3 \cdot g^{3,4}.$$

Solving (1) for this filter yields the solution

$$\lambda_0 = \frac{1}{128}$$

$$\lambda_1 = \frac{1}{384} \cdot \left(\sqrt{21 + 3\mu - 42\mu^{-1}} + \sqrt{42 - 3\mu + 42\mu^{-1} + 18\sqrt{105} \cdot (7 + \mu - 14\mu^{-1})^{-1/2}}\right)$$

$$\lambda_2 = 64 \cdot \lambda_1^2 - \frac{7}{256}$$

$$\lambda_3 = \frac{\sqrt{35}}{128},$$

where we used the abbreviation $\mu = \sqrt[3]{154 + 42\sqrt{15}}$. Thus we found a closed form description for a DAUBECHIES filter of length eight, which was impossible using other filter design methods. In the same manner we can also find an explicit analytical form for D_{10}. For bigger filters, the complexity increases too much and permits explicit forms. However, if one considers a filter

$$\gamma = \lambda_0 \cdot h^n + \lambda_1 \cdot g^{1,n-1} + \lambda_2 \cdot g^{2,n-2} \dots + \lambda_{(n-1)/2} \cdot g^{(n-1)/2,(n+1)/2}$$

of arbitrary even length and solves the system (1), one can at least verify that

$$\lambda_0 = 2^{-n}$$
$$\lambda_2 = 2^{n-1} \cdot \lambda_1^2 - n \cdot 2^{-n-1}$$
$$\lambda_{(n-1)/2} = 2^{-n} \cdot \sqrt{\binom{n}{(n+1)/2}}.$$

We additionally remark, that the solutions for $\lambda_3, \ldots, \lambda_{(n-1)/2-1}$ can all be written as rational functions in λ_1, while λ_1 itself is a root of a polynomial of degree $2^{(n-3)/2}$ for $n \geq 7$.

Biorthogonal wavelets. Our framework can also be used to design biorthogonal filters, which have some advantages over orthogonal filters in special applications (e.g. symmetry for image compression). We differ between two cases of biorthogonal filters: if a primal filter is given, dual filters can always be found by solving a system of linear equations; this is the *easy* case and not considered here (since this solutions can be obtained by several other design methods). On the other hand, one can take two linear combinations of even basic filters and solve their coefficients for the *biorthogonality constraints*

$$\tilde{m}_0(\omega) \cdot \overline{m_0(\omega)} + \tilde{m}_0(\omega + \pi) \cdot \overline{m_0(\omega + \pi)} \equiv 1, \qquad \tilde{m}_0(0) = m_0(0) = 1,$$

which again leads to a quadratic equation system. For example, considering a symmetric pair of length 9 and 7 and imposing the maximal number of sum rules on these filters, in particular, we take

$$\gamma = \lambda_0 \cdot h^9 + \lambda_1 \cdot g^{2,7} + \lambda_2 \cdot g^{4,5} \qquad \text{and} \qquad \tilde{\gamma} = \mu_0 \cdot h^7 + \mu_1 \cdot g^{2,5},$$

we obtain the *classical* and till today widely used 9/7 image compression filter, that was first presented in [ABMD92]. Estimating the joint spectral radius of the associated linear operators $(\mathbf{T}_0)_{jk} = \gamma_{2j-k-1}$ and $(\mathbf{T}_1)_{jk} = \gamma_{2j-k}$ reduced to a certain invariant subspace E, we obtain the smoothness values $\alpha \approx 1.068$ and $\tilde{\alpha} \approx 1.701$ in terms of the HÖLDER exponent (these techniques are discussed in detail in [DL92] and [Gri96]). To obtain better smoothness results, one could give up one zero order of $\tilde{\gamma}$ (by adding $\mu_2 \cdot g^{4,3}$), and use this degree of freedom to find *better* filters. Another wish could be the property, that the coefficients are rationals (as in the easy case), because this can reduce the computational amount of the wavelet transform. In order to achieve these requirements, we use a numerical heuristic that approximates

$$\min_{\mu_2} \left\{ \max_{|\lambda|} \left\{ \lambda \in \text{Spectrum}\left(\mathbf{T}_{0|E}(\mu_2), \mathbf{T}_{1|E}(\mu_2), \tilde{\mathbf{T}}_{0|\tilde{E}}(\mu_2), \tilde{\mathbf{T}}_{1|\tilde{E}}(\mu_2) \right) \right\} \right\}$$

at dyadic rational values μ_2. Thereby one finds

$$\gamma = \left[\begin{array}{ccccccccc} \frac{9}{320} & \frac{-3}{160} & \frac{-3}{40} & \frac{43}{160} & \frac{19}{32} & \frac{43}{160} & \frac{-3}{40} & \frac{-3}{160} & \frac{9}{320} \end{array} \right]$$

and the dual filter

$$\tilde{\gamma} = \left[\begin{array}{ccccccc} \frac{-3}{64} & \frac{-1}{32} & \frac{19}{64} & \frac{9}{16} & \frac{19}{64} & \frac{-1}{32} & \frac{-3}{64} \end{array} \right].$$

This new pair of filters is indeed very promising in applications such as image compression. Its smoothness values are $\alpha \approx 1.48409$ and $\tilde{\alpha} \approx 1.67807$, respectively. Note that $\tilde{\alpha}$ is only minimally worse than in the classical 9/7 case, while the value for α is significantly better and additionally, all the filter coefficients are rational. Finally, we shall mention that this heuristic method does not guarantee, that there exist no *better* solutions than the given one.

5 The 2D Case

In the same manner one can build tw odimensional filters from linear combinations of basic filters. The main ideas of this conceptual generalization will be described in this section. First, w ewill state a similar result to Lemma 1. In that case, repeated con volutionswith the sequences $[\,1\ \ 1\,]$ and $[\,1\ \ -1\,]$ were used to successively build longer filters with a higher sum rule order and it turns out that a similar thing can be done in higher dimensions.

Lemma 3. *Suppose, the twodimensional filter* $\left[\,\{\gamma_{jk}\}_{j\in 1\ldots n_x, k\in 1\ldots n_y}\,\right]$ *satisfies the (twodimensional) sum rules up to order* $m-1$. *Then the filter*

$$[\,\gamma_{jk}\,] * \left[\begin{array}{ccc} & \alpha & \\ \beta & 2\alpha+2\beta & \beta \\ & \alpha & \end{array} \right]$$

satisfies the sum rules at least up to order m, *if* $\alpha, \beta \neq 0$.

The *proof* is similar to the onedimensional case and omitted. However, these filters will not be sufficient; w eadditionally need some antisymmetric filters, which we will get from the following Lemma.

Lemma 4. *Suppose, the* symmetric *filter* $\left[\,\{\gamma_{jk}\}_{j\in 1\ldots n_x, k\in 1\ldots n_y}\,\right]$ *satisfies the sum rules up to order* $m-1$. *Then for* $\alpha \neq 0$ *both of the* an tisymmetric *filters*

$$[\,\gamma_{jk}\,] * \left[\begin{array}{ccc} & -\alpha & \\ -\alpha & 0 & \alpha \\ & \alpha & \end{array} \right] \qquad \textit{and} \qquad [\,\gamma_{jk}\,] * \left[\begin{array}{ccc} & -\alpha & \\ \alpha & 0 & -\alpha \\ & \alpha & \end{array} \right]$$

fulfill the sum rules at least up to order m.

T aking a little care about some possible redundancies (since specific choices of α and β may lead to linear dependent filters) while using these Lemmata, the linear independence of the basic filters directly carries ov er to the t w odimensional case. The thing that makes everything more difficult is the orthogonality constraint, which now becomes

$$|m_0(\omega_x, \omega_y)|^2 + |m_0(\omega_x + \pi, \omega_y + \pi)|^2 \equiv 1.$$

Assuming that the filter is rhombic shaped (this is in some sense the most convenient and applicable case) and consists of ν double diagonals each of length μ, this leads us to $2\nu\mu - \nu - \mu + 1$ equations of the form

$$\sum_{k=1}^{\nu}\sum_{j=1}^{\mu} \gamma^2_{j-k,j+k-1} + \gamma^2_{j-k+1,j+k-1} = 1,$$

$$\sum_{k=1}^{\nu}\sum_{j=1}^{\mu-1} \gamma_{j-k,j+k-1} \cdot \gamma_{j-k+\tau,j+k+\sigma-1} + \gamma_{j-k+1,j+k-1} \cdot \gamma_{j-k+\tau+1,j+k+\sigma-1} = 0$$

with $(\tau, \sigma) \in \{[-\nu+1, -\nu+2, \ldots, \nu-1] \times [0, 1, \ldots, \mu-1]\} \setminus \{(0,0)\}$.

Example. We will now give an example of a tw odimensional orthogonal scaling filter, that satisfies the STRANG-FIX conditions up to order one. Therefore, we take the simple HAAR scaling filters

$$[1\ \ 1] \qquad \text{and} \qquad [1\ -1]$$

and apply the Lemmata 3 and 4 to them. This gives us a linear combination

$$\lambda_0 \cdot \begin{bmatrix} & 1 & 1 & \\ 1 & 5 & 5 & 1 \\ & 1 & 1 & \end{bmatrix} + \lambda_1 \cdot \begin{bmatrix} & -1 & -1 & \\ 1 & 1 & 1 & 1 \\ & -1 & -1 & \end{bmatrix} + \lambda_2 \cdot \begin{bmatrix} & 1 & -1 & \\ 1 & 3 & -3 & -1 \\ & 1 & -1 & \end{bmatrix} + \lambda_3 \cdot \begin{bmatrix} & -1 & -1 & \\ -1 & -1 & 1 & 1 \\ & 1 & 1 & \end{bmatrix}$$

of four basic filters. Solving the orthogonality criterion for the coefficients λ_i, we obtain

$$\lambda_0 = \frac{1}{16}, \qquad \lambda_1 = \frac{-1}{8}, \qquad \lambda_2 = \frac{\pm\sqrt{3}}{16} \qquad \text{and} \qquad \lambda_3 = \frac{\pm\sqrt{3}}{8},$$

which reproduces the KO V ČEVIĆ-VETTERLI scaling filter (see [KV92]), the first known orthogonal 2D-filter that leads to a continouus wa velet for the important quincunx sampling grid. Note, that we again found a decomposition of the filter into an even and an odd part, where the ev en part satisfies the sum rules up to order tw o and the odd part up to order one — everything is very similar to the 1D case. We only need more basic filters because more constraints are to be considered. We should remark, that since $\mu = \nu = 2$ in the previous example, we would have to satisfy five constraint equations and thus we should use five basic filters instead of four, but it turns out that one of the coefficients always gets zero. For filters that satisfy the sum rules up to a higher order (e.g. for order tw o, one has to choose at least $\nu \geq 3$ and $\mu \geq 4$ or vice versa), the orthogonality constraints seem to be solvable only numerically, because of the rapidly increasing complexity of the related nonlinear equation system.

Finally, w eshall remark, that the presented framework could also be used to design tw odimensional biorthogonal filters. But due to the symmetry properties of these, the MCCLELLAN transform can be used to derive 2D-filters directly from their 1D-*prototypes*, which is much faster to implement. Thus, the direct usage of basic filters seems to make less sense if one is interested in tw odimensional biorthogonal filters.

6 Discussion and Conclusion

A framework for the design of wavelet filters was presented, which can be generalized to higher dimensions. There are very few different approaches to direct multidimensional orthogonal filter design. The most important among these is the paraunitary polyphase decomposition due to VAIDYANATHAN ([VH88]). But since his building matrices do not commute in general, the a priori ordering of these matrices is not clear and thus there is no unique representation of all possible orthogonal filters of a given shape, which can be obtained by the proposed method. However, numerical experiments lead to the conjecture, that both methods yield the same filter families. It is intended to apply the multidimensional wavelets, that stem from these approaches to optic flow estimations and to image feature detection within the scope of the authors further research. The presented variations for 1D biorthogonal wavelets and their 2D counterparts (built via MCCLELLAN transform) seem to have nice properties for image compression and some cooperation with researchers from this area is planned.

Acknowledgements. The author would like to thank G. SOMMER, B. ENGELKE and S. VUKOVAC

References

[ABMD92] M. Antonini, M. Barlaud, P. Mathieu, and I. Daubechies. Image Coding Using Wavelet Transforms. *IEEE Trans. on Image Process.*, 1:205–220, 1992.

[AHC93] Ali N. Akansu, Richard A. Haddad, and Hakan Caglar. The Binomial QMF-Wavelet Transform for Multiresolution Signal Decomposition. *IEEE Trans. on Signal Processing*, 41(1):13–19, 1993.

[Dau92] Ingrid Daubechies. *Ten Lectures on Wavelets.* No. 61 in CBMS-NSF Regional Conference Series in Applied Mathematics, SIAM Publishing, Philadelphia, 1992.

[DL92] I. Daubechies and J. Lagarias. Two-scale Difference Equations II. Local Regularity, Infinite Products of Matrices and Fractals. *SIAM J. Math. Anal.*, 23(4):1031–1079, 1992.

[Gri96] Gustaf Gripenberg. Computing the Joint Spectral Radius. *Linear Algebra and its Applications*, 234:43–60, 1996.

[KV92] J. Kovačević and M. Vetterli. Nonseparable Multidimensional Perfect Reconstruction Filter Banks and Wavelet Bases for $\mathbb{R}^n$. *IEEE Trans. Inform. Theory, Special Issue on Wavelet Transforms and Multiresolution Signal Analysis*, 38(2):533–555, 1992.

[VH88] P.P. Vaidyanathan and P.-Q. Hoang. Lattice Structures for Optimal Design and Robust Implementation of Two-Channel Perfect Reconstruction Filter Banks. *IEEE Trans. Acoust., Speech and Signal Proc.*, 36(1):81–94, 1988.

Lie Group Modeling of Nonlinear Point Set Shape Variability

Niels Holm Olsen[1] and Mads Nielsen[2]

[1] 3D - Lab, Dept. of Pediatric Dentistry, University of Copenhagen
Nørre Alle 20, DK-2200 Copenhagen N., Denmark
nielsho@lab3d.odont.ku.dk
http://www.lab3d.odont.ku.dk/
[2] The IT University of Copenhagen
Glentevej 67, DK-2400 København NV, Denmark
malte@itu.dk http://www.itu.dk/

Abstract. Linear statistical models of shape variability of identifiable point sets have previously been described and applied successfully to the empirical modeling of appearance variability in natural images. One of the limitations of these linear models has been demonstrated in the nonlinear "bending" shape variability of point sets where a length ratio is constant.
We point out that modeling point set variability with groups of transformations generated by linear vector fields constitute an algebraic frame for modeling simple nonlinear point set variability suitable for the modeling of shape variability. As an example, the very simple "bending" shape variability of three points in the complex plane is in this way generated by a linear vector field described by a complex 3×3 matrix.

Keywords: Point Sets, Nonlinear Variability, Shape Space, Lie Groups, Linear Vector Fields.

1 Introduction

Shape is often defined as whatever is left when position, size and orientation are ignored [,]. Abstractly the set of shapes can be defined as a set of equivalence classes, where the equivalence is "equal except for translation, scaling and rotation". Often these equivalence classes are modeled by a choice of a canonical element from each class.

When one considers a finite set of uniquely identifiable points in the plane, space or generally $\mathbf{R}^m$, an explicit assumption of invariance of variability under centroid translation, scaling and rotation leads to the statistical theory of shape as introduced by David G. Kendall in 1977 [ ]. In this theory the space of shapes is modeled as a Riemannian manifold. For an in-depth coverage please consult one of the two books "The statistical Theory of shape" [ ] by Christopher G. Small and "Statistical Shape Analysis" [ ] by Ian L. Dryden and Kanti V. Mardia.

G. Sommer and Y. Y. Zeevi (Eds.): AFPAC 2000, LNCS 1888, pp. 259– , 2000.

To keep things simple only point sets in the plane will be considered. Though this restriction avoids the non-trivial generalizations of rotations and the complications of singular point set configurations invariant under non-trivial subgroups of rotations [, p. 84], it is not essential for the modeling by 1-parameter groups of transformations.

The set of n-point sets in the plane $(\mathbf{R}^2)^n$ is conveniently modeled by the complex vector space $\mathbf{C}^n$, since the rotation of an n-point set is then simply given by multiplication with a unit complex number. This identification is only introduced to get a short and convenient notation, and we will still consider $\mathbf{C}^n$ as a $2n$-dimensional real vector space.

1.1 Kendall's Manifold of Shapes of Point Sets in the Plane

To fix terminology and notation we shortly review Kendall's shape space of n-point sets in the plane. First position normalization is done by orthogonal projection onto the subspace of "centered point sets" $C_0^{2n-2} = \{(z_1, \ldots, z_n) \in \mathbf{C}^n | \Sigma_{i=1}^n z_i = 0\}$ with real dimension $2n-2$. Size normalization is done by scaling onto the manifold of "pre-shapes" $S_0^{2n-3} = \{p_0 \in C_0^{2n-2} | \|p_0\|_2 = 1\}$. S_0^{2n-3} is thus a sphere of real dimension $2n-3$ in the linear subspace C_0^{2n-2} of $\mathbf{C}^n$.

Now the manifold Σ_2^n of shapes of n-point sets in the plane can be identified with the complex projective space $\mathbf{CP}^{n-2} \equiv \{\{zp_0 | z \in \mathbf{C}\} | p_0 \in C_0^{2n-2} \setminus \{0\}\}$. Kendall's "Shape Space" Σ_2^n is a Riemannian manifold with the Procrustes metric:

$$d(\Sigma(p), \Sigma(q)) = cos^{-1}(|\sum_{k=1}^{n} p_k q_k^*|) \quad , \quad p, q \in S_0^{2n-3}, \tag{1}$$

where $\Sigma(p) = \{\{zp | z \in \mathbf{C}\}\} \in \Sigma_2^n$ is the equivalence class representing the shape of p [, p. 13]. In the following we shall simply refer to C_0^{2n-2} as C_0 without explicitly noting the dimension.

1.2 Linear Point Set Models of Local Shape Variation

In "Active Shape Models" [ ] Cootes *et al.* use a linear model of local point set variation. The example point sets of which they model the variation have all been translated, scaled and rotated to match as well as possible a mean point set, which has been given a standard position, size and orientation and is found iteratively. They use a local linear model of the variation of the matched point sets:

$$p = \overline{p} + \underline{P}b, \tag{2}$$

1 We have not yet analyzed our compatibility requirement of commutativity with the group of centroid centered 3D-rotations, but expect that it is possible to find non-trivial 1-parameter groups generated by linear vector fields commuting with it.

2 When all points are coincident this is not possible. Thus these point sets are excluded.

3 They use a non-trivial weighting of the distances of corresponding points. This is not essential for the model here.

where $p \in (\mathbf{R}^2)^n$ is a point set and $\underline{P}$ is a matrix, which consists of the eigenvectors of the covariance matrix describing the example point sets deviations from the mean point set $\overline{p}$. The variable b is a column vector of parameters describing the point set deviation from the mean $\overline{p}$. It is a linear model of local shape variation in the sense that the point set deviation from the mean depends linearly on the parameters b.

They give an example of shape variability where this local linear model is inappropriate. The example is point sets along the outline of worms, and the problem is that they can bend at the middle. This bending is intuitively a 1-dimensional shape variability, so even though translation, scaling and rotation have been explicitly removed, there is one significant variation left.

The problem is that it is not possible to model the bending exactly by a 1-dimensional local linear model. When the bending is large, 2 eigenvectors are needed to span the example point set configurations. This problem of bending can be studied in the very simple setting of only three points in the plane allowed to vary freely under the constraint that a length ratio is preserved.

This is an inherent problem of modeling the variation of point sets on the unit sphere of pre-shapes S_0^{2n-3}. This sphere is curved and thus any large variation will be of a nonlinear nature. To overcome this difficulty Kent [ ] uses a tangent space approximation to the pre-shape manifold.

However it is still possible to think of a 1-dimensional shape variability which after projection on a tangent space has a nonlinear image. In the section on the bending of three points in the plane we shall see an example where the preserved length ratio is different from 1 and where the projected shape variation is nonlinear.

1.3 This Article

We have thus been inspired to study differential geometric modeling of nonlinear point set and shape variability. The theory of Lie groups and their Lie algebras provide a framework for modeling nonlinear continuous 1-dimensional modes of variation (variability) by 1-parameter groups of transformations. These 1-parameter groups are generated by vector fields describing the modeled variability. The linear point set models are in this framework generated by the constant vector fields. However, in the context of modeling shape variability the requirement of commutativity with scaling naturally leads to the study of point set variability generated by linear vector fields.

This article not only deals with the subject of modeling nonlinear point set variability, but also provides an analysis of the modeling of point set variability in the context of the inherently nonlinear shape space of point sets.

2 Variability Modeled by Group of Transformations

A 1-dimensional point set variability may be modeled by a 1-parameter group $(T_t)_{t \in R}$ of transformations of $(\mathbf{R}^2)^n$. That is, it is assumed that all variations originates from the same variability and are additively parameterized:

$$T_t \circ T_s = T_{t+s}. \tag{3}$$

Thus, a variation corresponding to a finite change is modeled by a transformation, while a variability is given by a continuous family of transformations corresponding to the continuum of degrees of variation describing a continuous course of change [ ].

In this frame the linear model is written as a 1-parameter group $(T_t)_{t\in R}$ of translation transformations of $(\mathbf{R}^2)^n$

$$p = T_t(\overline{p}) = \overline{p} + t\Delta p, \tag{4}$$

where $\Delta p \in (\mathbf{R}^2)^n$ specifies the direction of the linear variability.

Higher dimensional variabilities may be modeled by independent 1-dimensional variabilities. Two 1-parameter groups $(X_t)_{t\in R}$ and $(Y_s)_{s\in R}$ are capable of describing independent variabilities if they commute:

$$\forall t \in R, \forall s \in R : X_t \circ Y_s = Y_s \circ X_t. \tag{5}$$

2.1 Continuous Variability Generated by a Vector Field

Just as the linear variability above was described by $\Delta p \in (\mathbf{R}^2)^n$, a nonlinear variability may intuitively be described by a vector field on $(\mathbf{R}^2)^n$, which everywhere points in the direction of change under the studied variability. Consider the integral curves $(\gamma_p)_{p\in(R^2)^n}$ defined by the action of a 1-parameter group $(T_t)_{t\in R}$:

$$\gamma_p(t) = T_t(p) \quad , \quad t \in R, p \in (\mathbf{R}^2)^n. \tag{6}$$

The derivative of these curves define a vector field $X : (\mathbf{R}^2)^n \to (\mathbf{R}^2)^n$ satisfying

$$\gamma_p'(t) = X(\gamma_p(t)), \ \forall t \in R. \tag{7}$$

On the other hand, such a vector field uniquely identifies a 1-parameter group [, p. 37]. In this way the vector field X is said to generate a 1-parameter group of transformations $(X_t)_{t\in R}$, where the parameter indicates how far along the variability the point set should be transformed.

We now observe that when logarithmically parameterized the group of point set scalings $\overline{\sigma_s}(p) = e^s p$ is generated by the linear vector field given by the identity transformation I of $(\mathbf{R}^2)^n$:

$$\frac{\partial}{\partial s}|_{s=0}\overline{\sigma_s}(p) = p = I(\overline{\sigma_0}(p)). \tag{8}$$

[4] For simplicity we have here excluded variabilities with other topologies and inherent non-commutativity. Such a variability can be modeled by a non-commutative Lie group of dimension higher than 1.

Similarly using complex notation it is seen that rotation $\overline{\rho_\theta}(p) = e^{i\theta}p$ is generated by the linear vector field given by iI:

$$\frac{\partial}{\partial\theta}|_{\theta=0}\overline{\rho_\theta}(p) = ip = iI(\overline{\rho_0}(p)). \tag{9}$$

It is well known that point set centroid translations $\overline{\tau_a}(p) = (p_1 + a, \ldots, p_n + a)$ do not commute with scaling and rotation. We observe that this holds in general for the (non-trivial) linear variabilities generated by the constant vector fields.

2.2 Well Defined Shape Variability Generated by Linear Vector Fields

In the context of point set shape variability it is natural to consider only point set variabilities commuting with the shape similarity group of point set translations, scalings, and rotations. Such point set variabilities induce well defined shape variabilities. The induced shape transformations are defined by applying the corresponding point set transformations on all the point sets in a shape equivalence class. Because of commutativity with the shape similarity group the resulting point sets will all belong to one and the same shape equivalence class, thus providing a well defined shape transformation.

The problem of non-commutativity between the linear variabilities and scaling and the wish to model nonlinear bending naturally lead to higher order modeling of the generating vector fields. As a first simple step towards the general case it is natural to consider 1. order vector fields.

Since commutativity between 1-parameter groups corresponds to commutativity of the generating vector fields [, lemma 13 p. 5-35] and the Lie bracket between linear vector fields X, Y is given by $[X, Y] = X \circ Y - Y \circ X$ [, p. 87], it is seen that the requirement of commutation with scaling (generated by the identity vector field I) is automatically fulfilled when considering real linear vector fields on $(\mathbf{R}^2)^n$. In order to secure commutation with centroid translations $(\overline{\tau_a})_{a\in R^2}$ and because we only want variabilities which do not change the centroid position we only consider real linear vector fields on the subspace of centered point sets C_0 and extended to zero on the orthogonal subspace $C_0^\perp$ modeling the centroid position.

It remains to analyze the requirement of commutativity with point set rotation. Not all real linear vector fields on $C_0 \subset (\mathbf{R}^2)^n$ commute with the vector field generating point set rotation which is most easily expressed as iI using the

[5] One may without changing the defined shape space choose to consider centroid centered scaling and rotation which do commute with centroid translation. These centered scalings and rotations are generated by the linear vector fields given by the identity I_{C_0} on the subspace of centered point sets and iI_{C_0} [ ].

[6] Two vector fields commute when their Lie bracket is zero.

[7] This is an orthogonality constraint stating that the vector field should be everywhere orthogonal to the two constant vector fields $(1, \ldots, 1)$ and $(i, \ldots, i)$ generating centroid translations.

complex representation $(\mathbf{R}^2)^n = \mathbf{C}^n$. From this it is seen that the complex linear vector fields do commute with rotation.

To summarize [ ]:

- Commutativity with point set translations - Is obtained by working in the subspace of centered point sets.
- Commutativity with centered scaling - Is obtained by real linearity of the vector field on the subspace $C_0 \subset (\mathbf{R}^2)^n$ of centered point sets.
- Commutativity with centered rotation - Is obtained by complex linearity of the vector field on the subspace $C_0 \subset \mathbf{C}^n$ of centered point sets.

The above arguments have only resulted in sufficient conditions for a vector field to describe a well defined shape variability. They are not necessary.

2.3 1-Parameter Groups Generated by Complex Linear Vector Fields

Consider the Lie group $GL(n, C)$ of nonsingular $n \times n$ complex matrices. This has Lie algebra $gl(n, C)$ which consists of all $n \times n$ complex matrices. A complex linear vector field $X : \mathbf{C}^n \mapsto \mathbf{C}^n$ is represented by a complex $n \times n$ matrix $\underline{X} \in gl(n, C)$. The corresponding 1-parameter group of transformations $(X_t)_{t \in R}$ is obtained by considering the usual matrix exponential map $\exp : gl(n, C) \to GL(n, C)$ [, p. 283]:

$$\underline{X_t} = \exp(t\underline{X}) = I + t\underline{X} + (1/2!)(t\underline{X})^2 + (1/3!)(t\underline{X})^3 + \cdots. \tag{10}$$

3 Preserved Length Ratio - Nonlinear "Bending"

Complex linear vector fields commute with both scaling and rotation, and they can describe the nonlinear bending of three points in the plane as a one dimensional variability.

The simplest non-trivial example of a shape space is the shape space of three points in the plane, Σ_2^3 and is a 2-dimensional Riemannian manifold isometric to the sphere in $\mathbf{R}^3$ with radius $\frac{1}{2}$, $S^2(1/2)$ [, p. 70, 73]. It can thus be visualized by orthogonal projections on three orthogonal planes. In the following we will describe how the nonlinear bending shape variability of three points is generated by a complex linear vector field.

3.1 Centered Bending of 3 Points in the Plane

Centered bending of 3 points in the plane $p_1, p_2, p_3 \in \mathbf{R}^2$ with centroid position $p_c \in \mathbf{R}^2$ and the constant distances $|p_1 - p_2| = l_1$ and $|p_2 - p_3| = l_2$ can be parameterized by the angle θ:

$$p_1 = \frac{-2l_1 e^{i\theta} - l_2 e^{-i\theta}}{3} + p_c, \tag{11}$$

$$p_2 = \frac{l_1 e^{i\theta} - l_2 e^{-i\theta}}{3} + p_c, \tag{12}$$

$$p_3 = \frac{l_1 e^{i\theta} + 2l_2 e^{-i\theta}}{3} + p_c. \tag{13}$$

The above family of point sets can be considered as the result of "centered bending" of $(p_1(0), p_2(0), p_3(0))$ by the angle θ. But it can also be considered as the result of centered bending of $(p_1(\theta_0), p_2(\theta_0), p_3(\theta_0))$ by the angle $\theta - \theta_0$. In this way a 1-parameter group $(B_\theta)_{\theta \in R}$ of "centered bending" transformations is defined. The vector field $B : (\mathbf{R}^2)^3 \to (\mathbf{R}^2)^3$, generating the 1-parameter group of centered bending of a 3-point set is seen to be given by $(\frac{\partial}{\partial\theta}|_{\theta=0})p(\theta)$:

$$\frac{\partial}{\partial\theta}p_1(\theta) = \frac{-2l_1ie^{i\theta} + l_2ie^{-i\theta}}{3} = \frac{i}{3}(2p_1(\theta) - 3p_2(\theta) + p_3(\theta)), \tag{14}$$

$$\frac{\partial}{\partial\theta}p_2(\theta) = \frac{l_1ie^{i\theta} + l_2ie^{-i\theta}}{3} = \frac{i}{3}(-p_1(\theta) + p_3(\theta)), \tag{15}$$

$$\frac{\partial}{\partial\theta}p_3(\theta) = \frac{l_1ie^{i\theta} - 2l_2ie^{-i\theta}}{3} = \frac{i}{3}(-p_1(\theta) + 3p_2(\theta) - 2p_3(\theta)). \tag{16}$$

This is seen to be a linear vector field, given by the matrix

$$\underline{B} = \frac{i}{3}\begin{bmatrix} 2 & -3 & 1 \\ -1 & 0 & 1 \\ -1 & 3 & -2 \end{bmatrix} = \frac{1}{3}\begin{bmatrix} 0 & -2 & 0 & 3 & 0 & -1 \\ 2 & 0 & -3 & 0 & 1 & 0 \\ 0 & 1 & 0 & 0 & 0 & -1 \\ -1 & 0 & 0 & 0 & 1 & 0 \\ 0 & 1 & 0 & -3 & 0 & 2 \\ -1 & 0 & 3 & 0 & -2 & 0 \end{bmatrix}. \tag{17}$$

The above equations use first complex and then real notation corresponding to the discrimination between $\mathbf{C}^3$ and $(\mathbf{R}^2)^3 = \mathbf{R}^6$.

Both the odd and the even columns of $\underline{B}$ add to $0 \in \mathbf{R}^6$ in agreement with the fact that $\underline{B}$ defines a vector field which is zero on the subspace modeling the position of the point set centroid. Similarly both the even and the odd rows add to $0 \in \mathbf{R}^6$, proving that this vector field is orthogonal to the directions for translating the centroid. The vector field B thus commutes with and is orthogonal to the vector fields for translation in the x and y directions. Since B is linear and represented by a complex matrix, it also commutes with the vector fields generating scaling and rotation.

As an indirect illustration of the vector field represented by $\underline{B}$, the generated 1-parameter group $(B_\theta)_{\theta \in R}$ has been evaluated for a few different angles using the matrix exponential. These have been applied to two different 3-point configurations (see Figure). Since the linear vector field B commutes with point set translation, scaling and rotation, it generates a 1-parameter group of point set transformations inducing a well defined shape variability by acting on the point sets in the equivalence class representing the shape. This shape variability is illustrated in Figure . The figure shows orthogonal projections of $S^2(1/2)$ on three orthogonal planes. The two coordinate axes in the projection plane and the axis coming up from the paper have been illustrated with a small corresponding point set configuration. In the left column a close sampling of the shape of $B_\theta(-1, 1, \sqrt{3}) = \exp(\theta\underline{B})(-1, 1, \sqrt{3}) \in \mathbf{C}^3 = (\mathbf{R}^2)^3$, for $\theta = 0, \pi/60, \ldots, \pi$ has been marked by a "+". In the right column 7 samples ($\theta = 0, \pi/6, \ldots, \pi$) along the bending have been marked by small point-set configurations. The first

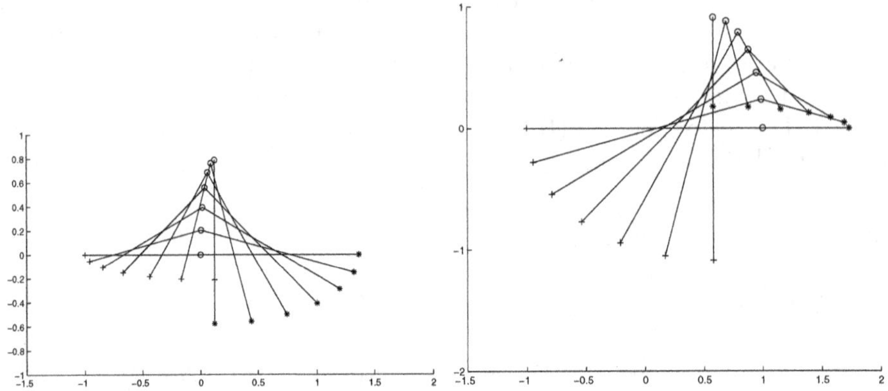

Fig. 1. Centered bending of 3-point set in the plane. The figure shows the point sets $B_\theta(-1, 0, (1+\sqrt{3})/2) = \exp(\theta \underline{B})(-1, 0, (1+\sqrt{3})/2) \in \mathbf{C}^3 = (\mathbf{R}^2)^3$, and $B_\theta(-1, 1, \sqrt{3}))$ for $\theta = 0, \pi/12, \ldots, \pi/2$

and the last of these 7 shapes are the same, but the point set figures have been rotated 180 degrees relatively to each other.

The starting point set configuration has been chosen so that the bending passes through the point $(x, y, z) = (0, 1/2, 0)$. This is most easily seen in the bottom left projection, which can be considered a projection on Kent's tangent subspace approximation to Σ_2^3 at the shape of $(0, 1/2, 0)$. It is thus seen that the bending shape variability of a shape with a preserved ratio of $2/(1-\sqrt{3})$ does not have a linear image in this subspace. You may also note that the non-equidistant spacing of the bended shapes is not in harmony with the Procrustes distance, but corresponds to the choice of additive parameterization.

4 Conclusions

We have described how Lie groups provide a framework for modeling general variability. The classical linear models of point set variability are in this framework modeled by the Lie groups generated by the constant (0-order) vector fields.

We have found that in the context of shape variability, the natural constraint of commutativity with the similarity group provides an inherent need for non-linear modeling of point set variability. By considering Lie groups generated by 1-order vector fields, we are able to model:

- Nonlinear point set variabilities commuting with scaling and rotation, thus inducing well defined shape variabilities.
- A larger class of variabilities which include the nonlinear bending of 3 points in the plane.

Future research will focus on methods of inferring these variabilities.

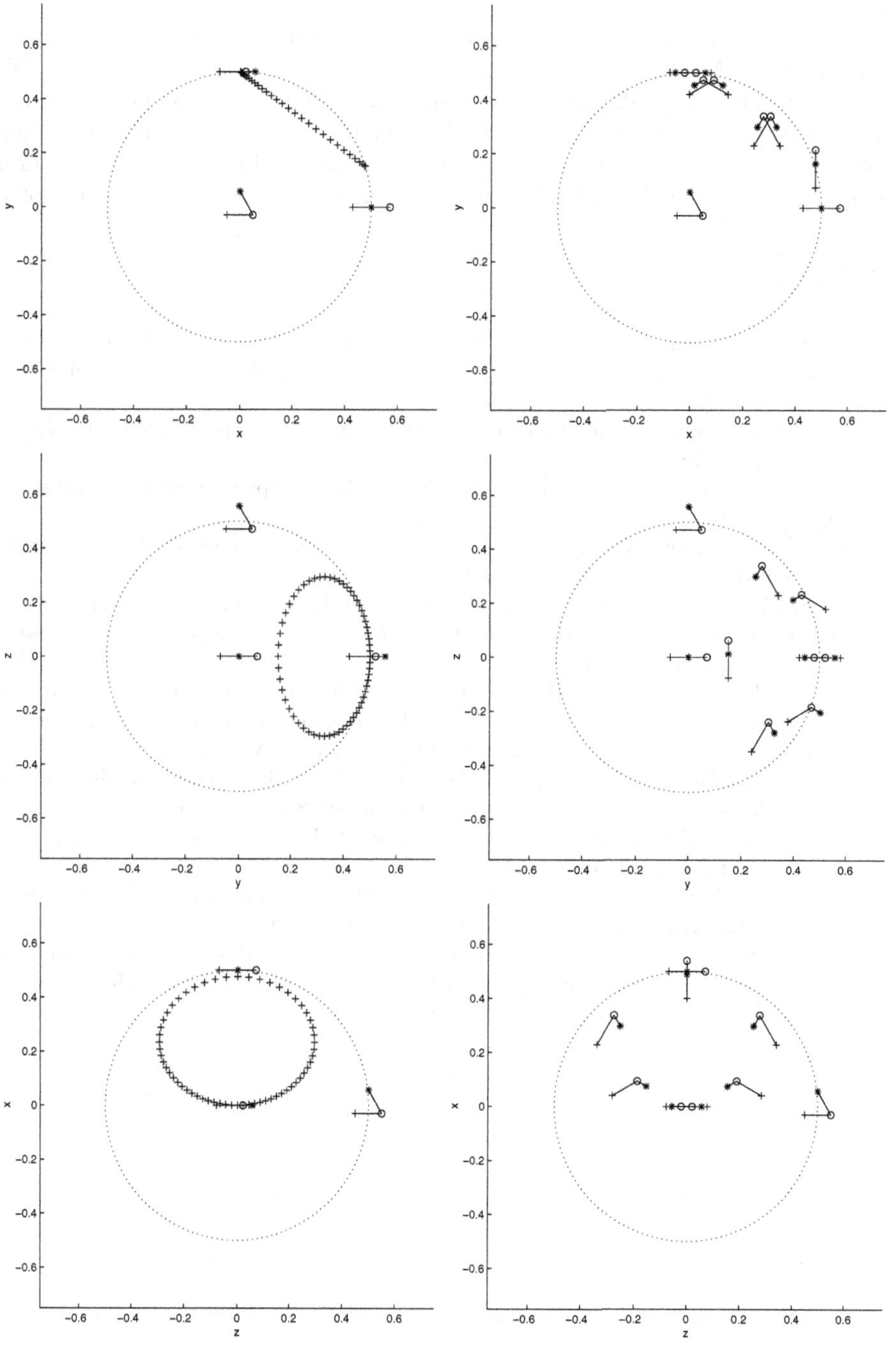

Fig. 2. "Bending" of the shape of a 3-point set in the plane

Acknowledgments

We gratefully acknowledge the very detailed and constructive comments from Jon Sporring, Marianne Christiansen, and the two anonymous reviewers. We would also like to thank Peter Johansen, Henrik Pedersen, Jens Gravesen and Andrew Swann in the Natural Shape Project for fruitful discussions and help with the mathematics. This work has been sponsored in part by the Natural Shape Project, which is funded by the Danish Research Agency, project 9900095.

References

1. Fred L. Bookstein. Shape and the information in medical images: A decade of the morphometric synthesis. *Computer Vision and Image Understanding*, 66(2):97–118, May 1997.
2. David G. Kendall. A survey of the statistical theory of shape. *Statistical Science*, 4(2):87–120, 1989.
3. Christopher G. Small. *The Statistical Theory of Shape*. Springer Series in Statistics. Springer, 1996.
4. Ian L. Dryden and Kanti V. Mardia. *Statistical Shape Analysis*. Wiley Series in Probability and Statistics. Wiley, 1998.
5. T. F. Cootes, C. J. Taylor, D. H. Cooper, and J. Graham. Active shape models - their training and application. *Computer Vision and Image Understanding*, 61(1):38–59, January 1995.
6. John T. Kent. Concluding address. In K. V. Mardia and C. A. Gill, editors, *Current Issues for Statistical Inference in Shape Analysis*, pages 167–175. Department of Statistics, University of Leeds, Leeds University Press, April 1995.
7. Niels Holm Olsen. Statistisk formbeskrivelse vha. differentialgeometrisk modellering. Master's thesis, Department of Computer Science, University of Copenhagen, Datalogisk Institut, Universitetsparken 1, DK-2100 Copenhagen Ø, Denmark, November 1999. In Danish.
8. Frank W. Warner. *Foundations of Differentiable Manifolds and Lie Groups*. Number 94 in Graduate Texts in Mathematics. Springer, 1983.
9. Michael Spivak. *A Comprehensive Introduction to Differential Geometry*, volume 1. Publish or Perish, 6 Beacon street, Boston, Mass. 02108 (U. S. A.), 1970.
10. Michael Artin. *Algebra*. Prentice-Hall, Inc., 1991.

Symmetries in World Geometry and Adaptive System Behaviour*

Robin Popplestone and Roderic A. Grupen

Laboratory for Perceptual Robotics, Department of Computer Science
University of Massachusetts, Amherst, MA 01003
{coelho,piater,grupen}@cs.umass.edu
http://www-robotics.cs.umass.edu/

Abstract. We characterise aspects of our worlds (great and small) in formalisms that exhibit symmetry; indeed symmetry is seen as a fundamental aspect of any physical theory. These symmetries necessarily have an impact on the way systems exhibit reactive behaviour in a given world for a symmetry determines an *equivalence* between states making it appropriate for an reactive system to respond identically to equivalent states. We develop the concept of a General Transfer Function (GTF) considered as a building block for reactive systems, define the concept of *full symmetry operator* acting on a GTF, and show how such symmetries induce a quotient structure which simplifies the process of building an invertible domain model for control.

1 Introduction

This paper explores the relationship between symmetries of the world and symmetries of the *(generalised) transfer functions* which are used to characterise the response of a reactive system. It is written from the perspective of Artificial Intelligence to the extent that we consider how some principles of automating problem reductions which might in many cases be "obvious" to humans.

A symmetry of a physical theory is an invertible mapping of the space in which the theory is expressed to itself under which the theory is invariant. For example, the symmetries of Newtonian mechanics are drawn from the *Galilean group* of symmetries of space. This consists of translations and rotations and uniform translatory motion (but not of rotatory motion).

In characterising a reactive system it is adequate to take a special case of a physical theory — for example we characterise the gravitational field as uniform rather than using the full Newtonian formulation of gravitation. However, a reactive (or adaptive) system does not sense the world directly, but only through its sensors so that taking such a limited view has advantage for the explanation of the behaviour of *adaptive* biological systems: the world is modelled at a level closer to what may be perceived. In general, sensor space is not isomorphic

* This work was supported in part by the NSF (CDA-9703217), by AFRL/IFTD (F30602-97-2-0032), and by DARPA/ITO/SDR DABT63-99-1-0022.

G. Sommer and Y. Y. Zeevi (Eds.): AFPAC 2000, LNCS 1888, pp. 269– , 2000.

to world-space. The question then naturally arises, how do world-symmetries relate to symmetries of transfer functions which are used to characterise reactive systems? In particular, is there a useful concept of the symmetry of a controller and of a plant?

The disposition of matter in space has the effect of reducing the symmetry of the world from the point of view of reactive systems existing therein. Nevertheless, residual symmetry frequently remains and is important for determining the possible behaviour of an reactive system. When we say that an object or world feature has a symmetry we are in essence identifying an invertible mapping from the object (or feature) to itself under which it is invariant. In the case of rigid bodies, all symmetries are members of the Special Euclidean group of translations and rotations in 3-space.

Why should we be interested in symmetric controllers? The advantage lies in the possibility of being able to handle discrete event dynamic systems by combining a repetoire of controllers. We can regard a controller as establishing a property (such as maintaining stability in a gravitational field). If a controller establishes just that property and no other, then it may be combined with another controller which establishes another property (forward locomotion, say). Thus a space of behaviours can be spanned by combining elementary controllers.

But with a given property may be associated world-transformations that preserve the property. A controller that maintains stability in a gravitational field should be invariant with respect to the symmetries of that field. In the case of a linear controller, our concept of *input symmetry* can be related to that of the null-space of the controller. Our group-theoretic formulation has the advantage of being a generalisation to systems that may be non-linear, non-differentiable or even non-continuous.

2 Previous Work

The idea of classifying physical theories in terms of symmetry groups is due to Noether[ ]. Noether's Theorem is a very general result which shows that any physical theory couched in variational terms necessarily has conservation laws related to the symmetries of the space in which the theory is expressed. While this has applications to the more exotic groups associated with modern physical theories, the historical development of physics can also be seen as a progression from theories of more restricted symmetry to those of less restricted symmetry. Thus Newtonian mechanics, characterised by the Galilean group, provides a more symmetric world-view than the Aristotelian.

In the 1980's one of us, cognisant of the importance of group theory in physics, sought to apply it to robotics — specifically to the characterisation of spatial relationships between body features established during assembly [ ]. Subsequently Liu[ ] demonstrated the practicality of this approach by developing a computationally tractable representation of subgroups of the Euclidean group, providing a software implementation thereof together with theoretical justification. Earlier Zahnd and Nair [ ] had provided a more limited approach. It should be noted that what needs to be represented is not a *member* of the Euclidean group but a subgroup embedded in it.

In psychology, Michael Leyton [ ] is a pioneer in the use of group theory as a basis for understanding perception. His view is that the mind perceives a shape in terms of a causal history of how it was formed, so that a deformed can is perceived as resulting from the act of denting it. Such deformations are not members of the Euclidean Group; indeed they are drawn from a group of diffeomorphisms which is much larger than the Euclidean group and therefore more challenging to represent computationally.

By relating the study of shape to the study of symmetry, Leyton is able to argue that symmetry is crucial to cognitive processing. Thus perception is seen by Leyton as the creation of a causal history which explains the sense-data in terms of a process that extends over time — what we would call a general transfer function. Thus, in our terms, Leyton sees a primary skill of the human mind as being the synthesis of general transfer functions. Crucial to this skill is the understanding of the characteristic symmetries of such functions. We perceive a pot in terms of the rotational symmetry induced by the potter's wheel.

In this paper we shall draw, upon the concept of "naive physics" expounded by Hayes[ ] in the general sense that it can be desirable to make use of a simplified physics for understanding the functioning of biological adaptive systems. Such simplified physical systems may in general be characterised by having more restricted symmetry groups than do the standard models of physics.

While our formalism does not in general require that mappings be differentiable, important examples are differentiable and characterisable by differential equations. In that case invariance under groups of symmetries is recognised to be an important characteristic of a set of equations, see [ ]. Our treatment of *output symmetries* is related to the topological concept of *homotopy*.

A discussion of the Missionaries and Cannibals problem is found in Amarel[ ]; our treatment of the quotient GTF of this problem is closely related to his.

Over discrete domains, our work has a strong relationship with *model checking*. Chapter 14 of [ ] entitled "Symmetry" contains a definition of an automorphism group of a Kripke structure, and develops the concept of a quotient structure. This is closely related to our discussion of the symmetries of a GTF. One view of our work is that it is an approach building a synthesis of classical Control Theory (over a continuous domain) with model checking (over a discrete domain).

3 Notation: Operators and Groups

By an *operator* σ we mean an entity drawn from an arbitrary set Σ (which may be infinite). A *multiplication* is defined on operators. If σ_1 and σ_2 are operators, then their product is written $\sigma_1\sigma_2$.

This product is associative, that is

$$(\sigma_1\sigma_2)\sigma_3 = \sigma_1(\sigma_2\sigma_3)$$

The *identity operator* denoted by 1 has the property that $1\sigma = \sigma = \sigma 1$ for all operators σ.

If σ is an operator it may have an inverse, written writen σ^{-1} with the property that $\sigma\sigma^{-1} = \sigma^{-1}\sigma = 1$.

A set Σ of operators for which every operator has an inverse, which includes the identity operator, and which is closed under multiplication and inverse is called a *group* of operators.

Given a set U and a set Σ of operators, we say that the operators of Σ *act* on U if each operator of Σ is associated with a mapping from U to U. If σ is an operator, we write the effect of applying σ to $u \in U$ by $u.\sigma$. We require that

- the identity operator acts as the identity mapping, that is $u.1 = u$.
- the product of operators acts as the product of the corresponding mappings, so that $u.(\sigma_1\sigma_2) = (u.\sigma_1).\sigma_2$.
- if an operator σ has an inverse, then mapping corresponding to σ is invertable, and the mapping $u \mapsto u.(\sigma^{-1})$ is the mapping inverse to $u \mapsto u.\sigma$, that is $(u.\sigma^{-1}).\sigma = u$.

If U is any finite set, then we denote the *symmetric group* of all permutations of the members of U by S_U . We shall use S_U as an operator set.

4 General Transfer Functions and Their Symmetries

Transfer functions have long been used by control theorists to characterise the behaviour of systems. Essentially, a transfer function characterises the input-output relationship of a system that may have internal state (for example the charge on the capacitor of an integrator). As such, they are necessarily *functionals* or *higher order functions*, mapping from a specification of how the input to a (sub)system evolves over time to a specification of how its output evolves over time. A specification of initial state is also required.

It should be noted that we regard a Finite State Automaton (FSA) with outputs as a generalised transfer function, so that we are not restricting ourselves to mappings that are differentiable or even continuous. We may regard the evolution of a system over time as discrete or continuous (but not, in our current formulation, hybrid).

[1] Operators are related to the concept of a universal algebra. They also resemble the *methods* of object-oriented programming languages. We chose to speak of operators rather than work with sets of mappings over our domains for much the same reason that object-orientation is used — we can discuss the operators and their properties before introducing all the sets they operate on. For example, bilateral symmetry, in which a reflection about a central plane is a symmetry operator, is very common. Since a reflection of a reflection is the identity operation, the bilateral symmetry operator σ_{LR} should obey the law $\sigma_{LR}^2 = 1$. But how it operates on a given domain of values is application specific.

[2] Group theorists identify symmetric groups by their isomorphism class, and speak, for example, of S_3 as the group of all permutations on 3 elements. This identification is inappropriate for our purposes.

Definition 4.1 *We use $\mathcal{I}$ to denote an index-set used to characterise the passage of time. $\mathcal{I}$ will be the real numbers ≥ 0 (for a continuous system) or the integers ≥ 0 (for a discrete system).*

In general, the index set will be totally ordered, with a least element written as 0.

4.1 Domains of Values

Control systems have traditionally been defined to act on real valued variables or vectors thereof. Here we use the concept of a *domain* of values as a general set on which components of a control system may act. In particular, elements of domains of values are not necessarily real numbers. We can think of discrete domains as supporting the concept of *logical sensors* and *logical behaviours* abstracted from actual sensors and behaviours by software layers.

We shall generally use U or U_{in} to denote a domain of inputs for a given transfer function. We may use X or U_{out} to denote a domain which is an outputs of a given transfer function.

We extend operators over a domain to act on functions over that domain. Thus if U is a domain, and σ acts on U, and $\mathbf{u} : \mathcal{I} \to U$ then $\mathbf{u}.\sigma$ is defined by $(\mathbf{u}.\sigma)(t) = \mathbf{u}(t).\sigma$

4.2 Initialiser Domains

A general transfer function may have an *initial state*, which must be specified. For example, in a classical control system, integrators may be given an initial value determined by the system designer. Likewise a robot may be activated in one of many possible initial states. We use P to denote a set of initialiser values, referring to P as the *initialiser domain*.

In order to support the *cascading* of generalised transfer functions, initialiser values need to be finite sequences, possibly empty, possibly of length one. Cascading general transfer functions will involve concatenation of their initialiser values, which we'll write as a product $p_1 p_2$. For a given initialiser domain P the sequences must be all of the same length.

Definition 4.2 *Let U and X be domains which we will call the* input domain *and the* output domain *respectively. Let P be an initialiser domain. Then a general transfer function T is a mapping in $T : ((\mathcal{I} \to U) \times P) \to (\mathcal{I} \to X)$.*

We will use the abbreviation "GTF" for "general transfer function". The idea is that a GTF maps an input function (of time) whose values range over an input-space U to an output function (of time) whose values range over an output space X. However this mapping may also depend on initialisation determined by a member of P.

Definition 4.3 *A GTF T with preset domain P and output domain X is said to be* presettable *if*

$$T(\mathbf{u}, p)(0) = p$$

Necessarily $P \subset X$; thus the initial output of a presettable GTF can be directly specified to be $p \in X$.

Definition 4.4 *An invertible operator σ acting on input-space U is said to be an input-symmetry of a transfer-function T if $T(\mathbf{u}.\sigma, x_0) = T(\mathbf{u}, x_0)$ for all $\mathbf{u} \in \mathbf{U}$.*

The idea being captured here is that there may be transformations of the input space to which a given transfer function is insensitive. This can amount to saying that there is state which is hidden, at least from the given function. Whether this is a problem depends on whether what is hidden is relevant. A more positive view of input symmetries is that they provide a way of defining properties of the input space of a transfer function which may be kept invariant.

For example, suppose we have a robot with a sideways pointing range sensor. Suppose the range sensor is sensing a planar wall. Then the transfer function for this sensor (mapping from robot coordinates to a real-valued distance) has an input symmetry with respect to translation parallel to the wall, thereby enabling wall-following behaviour.

Definition 4.5 *An output-symmetry of a presettable transfer-function T with output domain X is an invertable operator σ which acts on X and for which*

$$T(\mathbf{u}, x_0.\sigma) = T(\mathbf{u}, x_0).\sigma$$

5 Examples of GTF's

Typically, the GTF that represents the electro-mechanics of a robot will be a represented as a presettable transfer function if we regard it as being possible to initialise its position. On the other hand, incorporating an output-transducer as complex as a digitised TV camera leads to a system that not presettable: we can define images, (that is luminance functions defined on an image plane) which are not images of any scene realisable in a given world; to define $P \subset X$ would characterise the system as one which could be set up with an *exactly determined* image on the image plane, something we can't do in practice for we don't have an exact model of the imaging process.

A Mobile Robot Consider, for example, a mobile robot that is placed, initially at rest, on an infinite plane. Its input space is the cartesian product $U_{MR} = \mathcal{R} \times \mathcal{R}$. We shall write (u_a, u_c) for a typical member of U_{MR}. u_a is acceleration, and u_c is path-curvature, determined by a conventional steering mechanism. Its output space is $X_{MR} = \mathcal{R} \times \mathcal{R} \times \mathcal{R}$. We shall write (x, y, θ) for a typical member of X_{MR}, where (x, y) is the position of the robot in the plane, and θ is its orientation. All of these are functions of time.

Its GTF, T_{MR}, can be characterised by the differential equations

$$\frac{dx}{dt} = v(t)\cos\theta, \frac{dy}{dt} = v(t)\sin\theta$$

$$\frac{d\theta}{dt} = v(t)u_c(t), \frac{dv}{dt} = u_a(t)$$

For example, if we apply T_{MR} to the simple input function defined by $u_{simple}(t) = (0.1, 0)$, then

$$T_{MR}(u_{simple}, (10, 5, 0))(t) = (0.05t^2 + 10, 5, 0)$$

that is uniform acceleration along a straight line through the starting point (10,5,0) and parallel to the X-axis.

The Guards and Prisoners Problem Three guards and three prisoners are on the left bank of a river, and need to cross over to the right. A boat is available on the left bank. It holds two people. Prisoners must not be allowed to outnumber guards. How can the party cross the river?

To express the problem formally we will need the following notation:

- if s is a finite sequence (or tuple), then we use $s_{i \leftarrow v}$ to mean that finite sequence which differs from s only at index i, where it has the value v. This is extended to multiple successive modifications. For example $s_{i \leftarrow v, j \leftarrow w}$ is $s'_{j \leftarrow w}$ where $s' = s_{i \leftarrow v}$. We write () for the empty sequence.
- There is a set of 2 boolean values $\{\mathbf{t}, \mathbf{f}\}$.
- There is a set of 3 *guards* $\{\mathbf{g1}, \mathbf{g2}, \mathbf{g3}\}$.
- There is a set of 3 *prisoners* $\{\mathbf{p1}, \mathbf{p2}, \mathbf{p3}\}$.
- An *occupant* can be EITHER a guard OR a prisoner OR **n** (indicating that a place is unoccupied). The first 3 places on each bank will, if occupied, be occupied by a guard. The remaining 3 places will, if occupied, be occupied by a prisoner.
- A *bank* is a sextuple of occupants.
- A *state* can be EITHER
  - A triple $(c, bank_1, bank_2)$ where the condition c is a boolean indicating whether the boat is on the left bank $c = \mathbf{t}$ or the right bank $c = \mathbf{f}$ and $bank_1$ and $bank_2$ each specify the occupants of the left and right banks respectively.

[3] Following Amarel, we are deliberately not using a concise representation of the state-space, for we wish to discuss how state-space can be contracted by the recognition of its symmetries. Our representation is arguably a natural one for a graphical presentation of the problem which is to be solved by a human interacting with a computer.

[4] This formalisation was guided by a definition of aspects of the problem written in the SML language [ ].

- OR **b** indicating a *bad state* resulting from a physically impossible transition such as trying to move the occupant of an empty location. Other states are referred to as "good".

We denote the set of states by U_{GP}

To specify a move, we select one or two "places", that is indices into of the state vector, whose occupants may cross the river to the opposite bank. If any of the selected places is unoccupied, the move will be deemed physically impossible.

$$U_{move} = \{(i|i \in 1\ldots6)\} \cup \{(i,j|i,j \in 1\ldots6, i \neq j)\}$$

We may now define the presettable GTF which characterises the physically possible moves of the problem. We'll call it T_{moveGP}. To define it, we need a *Move* function, which moves a single occupant from one bank to the other. Here $Move(i,x) = x'$ means that the output state x' is obtained from the output-state x by moving the occupant $o = b_i$ to the other side of the river, where b is the appropriate bank, indexed by i.

There are three main cases

- Case 1: $x = \mathbf{b}$ In this case $Move(i,x) = \mathbf{b}$. In other words, once a state is bad, it remains so.
- Case 2: $x = (\mathbf{t}, b, b')$ where b, b' are banks. So the boat is on the left bank since the first component of the state is $\mathbf{t}$.
  Let $o = b_i$. Let $o' = b'_i$ There are three sub-cases
  - Case 2.1: $o = \mathbf{n}$. To move a non-existent occupant is physically impossible, so $Move(i,x) = \mathbf{b}$.
  - Case 2.2: $o' \neq \mathbf{n}$. To move an occupant into an occupied location is impossible , so $Move(i,x) = \mathbf{b}$
  - Case 2.3 $o \neq \mathbf{n}$ $Move(i,x) = (\mathbf{f}, b_{i \leftarrow \mathbf{n}}, b_{i \leftarrow o})$
- Case 3: $x = (\mathbf{f}, b, b')$ where b, b' are banks.
  Let $o = b_i$. Let $o' = b'_i$ There are three sub-cases
  - Case 3.1: $o' = \mathbf{n}$. To move a non-existent occupant is physically impossible, so $Move(i,x) = \mathbf{b}$.
  - Case 3.2: $o \neq \mathbf{n}$. To move an occupant into an occupied location is impossible, so $Move(i,x) = \mathbf{b}$
  - Case 3.3 $o' \neq \mathbf{n}$ $Move(i,x) = (\mathbf{f}, b_{i \leftarrow o}, b_{i \leftarrow \mathbf{n}})$

Now let's define *MoveAll* which operates on the members of U_{move}.

$$\begin{aligned} MoveAll((),x) = x, \quad & MoveAll((i, u_1 \ldots u_n, x) \\ & = Move(i, MoveAll((u_1 \ldots u_n), x)) \end{aligned}$$

[5] Good states and bad states are specified by a discriminated union in the SML formulation. The two cases given correspond to the two cases in the **datatype** declaration in the program

[6] With a standard initial condition in which everybody is on one bank it's not possible to move an occupant into an occupied location.

We can now define the GTF T_{moveGP} with index-set the non-negative integers as follows:

$$T_{moveGP}(\mathbf{u},(p_1,p_2,p_3,p_4,p_5,p_6))(0) = (\mathbf{t},(p_1,p_2,p_3,p_4,p_5,p_6)(\mathbf{n},\mathbf{n},\mathbf{n},\mathbf{n},\mathbf{n},\mathbf{n}))$$

Provided that $p_1 \ldots p_3 \in \mathit{Guards}$ and $p_4 \ldots p_6 \in \mathit{Prisoners}$ — otherwise $T_{moveGP}(\mathbf{u},(p_1,p_2,p_3,p_4,p_5,p_6))(0) = \mathbf{b}$ That is, initially everybody is on the left bank.

For $t > 0$ let's suppose that $x = T_{moveGP}(\mathbf{u},p)(t-1)$.

$$T_G P(\mathbf{u},p)(t) = \mathit{MoveAll}(\mathbf{u}(t-1),x)$$

To meet the conditions of the problem we can classify a state as either legal $\mathbf{l}$ (so that neither on the left bank nor on the right bank are the guards outnumbered) or illegal $\mathbf{i}$ (in which the guards are outnumbered on the left bank or on the right bank), or terminal $\mathbf{t}$ which is a legal state with everybody on the right bank.

The domain

$$U_{eval} = \{\mathbf{l},\mathbf{i},\mathbf{t}\}$$

will be the output domain of a GTF which classifies a given situation as either legal, illegal or terminal.

We can also define the GTF T_{evalGP} which evaluates a state arising from a move in the guards-and-prisoners problem.

To evaluate $T_{evalGP}(\mathbf{x},p)(t)$, let $(c,b,b') = \mathbf{x}(t)$. Then

- $T_{evalGP}(\mathbf{x},p)(t) = \mathbf{t}$ if $b = \emptyset$
- $T_{evalGP}(\mathbf{x},p)(t) = \mathbf{i}$ if $\mathit{OutNumbered}(b) = \mathbf{t}$
- $T_{evalGP}(\mathbf{x},p)(t) = \mathbf{l}$ otherwise

The *OutNumbered* function applied to a bank b evaluates to $\mathbf{t}$ if there are guards on b, and they are fewer in number than the prisoners on b.

Input Symmetries of T_{evalGP} Suppose our set of operators Σ contains the symmetric group S_{Guards}, which acts on the space U_{GP} by permuting the guards on each bank. Then S_{Guards} is a group of input symmetries of T_{evalGP}. Likewise if Σ contains the symmetric group $S_{Prisoners}$ then $S_{Prisoners}$ is a group of input symmetries of T_{evalGP}.

That is to say, if we take any member of $U_{stateGP}$ and permute the prisoners and/or the guards, that state will receive the same evaluation under T_{evalGP}.

5.1 Full Symmetries

We have seen so far symmetries of the inputs and of the outputs of GTF's. However it is frequently the case that a GTF has a symmetry that affects both input and output. For example, any device with a bilateral symmetry will have a symmetry operator that, in some sense, interchanges left and right. Applying such an operator on the input space (so that left and right are interchanged on input) will give rise to an output that has left and right interchanged. A left-right interchange on the preset specification will also be needed.

For example, in the problem of balancing an inverted pendulum, a left-right interchange will involve changing the sign of the input command to the motor driving the system. Provided the preset (the initial position and velocity of the pendulum) is also mapped by a left-right interchange, the output behaviour will also exhibit a left-right interchange.

In our mobile robot example the transformation $u_c \mapsto -u_c$ is a left-right symmetry operator applied to the input. The output-space, which is also the preset-space, may be mapped by $y \mapsto -y$, $\theta \mapsto -\theta$. If we apply these mappings to the input-space and the ouput-space the behaviour of the system is described by the same transfer function. This is not the only such symmetry operator — any reflection operator on the output-space gives rise to a left-right symmetry, provided we correctly map θ.

Definition 5.1 *Let T be a GTF. Let σ be an invertible operator which acts on the input domain U_{in} of T, the output domain U_{out} of T, and the preset domain P of T. We say that σ is a* full symmetry *of T if*

$$T(\mathbf{u}.\sigma, p.\sigma) = T(\mathbf{u}, p).\sigma$$

Proposition 5.2 *Let T be a GTF with input domain U_{in} and output domain U_{out} and preset domain P. Let Σ be a set of operators acting on these domains. Then the set of full symmetry operators on T form a group.*

Proof: Let $\sigma_1, \sigma_2 \in \Sigma$ be full symmetry operators on T. Then

$$\begin{aligned} T(\mathbf{u}.(\sigma_1\sigma_2), p.(\sigma_1\sigma_2)) &= T((\mathbf{u}.\sigma_1).\sigma_2, (p.\sigma_1).\sigma_2) \\ &= T(\mathbf{u}.\sigma_1, p.\sigma_1).\sigma_2 \\ &= T(\mathbf{u}, p).\sigma_1).\sigma_2 \\ &= T(\mathbf{u}, p).(\sigma_1\sigma_2) \end{aligned}$$

Hence $\sigma_1\sigma_2$ is a full symmetry operator on T.

Let σ be a full symmetry operator on T

$$T(\mathbf{u}.\sigma^{-1}.\sigma, p.\sigma^{-1}.\sigma) = T(\mathbf{u}.\sigma^{-1}, p.\sigma^{-1}).\sigma$$

Now consider

$$\begin{aligned} T(\mathbf{u}.\sigma^{-1}.\sigma, p.\sigma^{-1}.\sigma).\sigma^{-1} &= T(\mathbf{u}.\sigma^{-1}, p.\sigma^{-1}).\sigma.\sigma^{-1} \\ &= T(\mathbf{u}.\sigma^{-1}, p.\sigma^{-1}).(\sigma\sigma^{-1}) \\ &= T(\mathbf{u}.\sigma^{-1}, p.\sigma^{-1}) \end{aligned}$$

However

$$\begin{aligned} T(\mathbf{u}.\sigma^{-1}.\sigma, p.\sigma^{-1}.\sigma).\sigma^{-1} &= T(\mathbf{u}.(\sigma^{-1}\sigma), p.(\sigma^{-1}\sigma)).\sigma^{-1} \\ &= T(\mathbf{u}.1, p.1).\sigma^{-1} \\ &= T(\mathbf{u}, p).\sigma^{-1} \end{aligned}$$

So we've shown that

$$T(\mathbf{u}.\sigma^{-1}, p.\sigma^{-1}) = T(\mathbf{u}, p).\sigma^{-1}$$

Hence σ^{-1} is a full symmetry operator on T.

We will write $\mathcal{G}_\Sigma T$ for the group of full symmetries of T.

We can regard input symmetries of a GTF as a special case of full symmetries in which the operator acts as the identity operation on the output and preset spaces. Likewise we can regard output symmetries of a presettable GTF as a special case of full symmetries in which the operator acts as the identity operation on the input space.

In the Prisoners and Guards world, $\Sigma = S_{Guards} \cup S_{Prisoners} \cup S_{\{1\ldots3\}} \cup S_{\{4\ldots6\}}$ so that Σ includes operators permutating the guards, the prisoners, the guards' places and the prisoners' places. All three of the operator subgroups above naturally map the domains U_{GP}, U_{move} and (trivially) U_{eval}. Each is a group of full symmetries of the GTF's T_{moveGP} and T_{evalGP}.

6 Cascaded GTF's

Definition 6.1 *A transfer function T is said to be a* transducer *if there is a function f for which $T(\mathbf{u}, x_0)(t) = f(\mathbf{u})$.*

Thus a transducer (in our sense) is a transfer function whose output depends only on the instantaneous value of its input.

Definition 6.2 *The identity transducer is the map defined by*

$$I(\mathbf{u}, ()) = \mathbf{u}$$

Definition 6.3 *Let T_1, T_2 be general transfer functions. Then the product T_1T_2 is defined by*

$$(T_1T_2)(\mathbf{u}, p_1p_2) = T_1(T_2(\mathbf{u}, p_2), p_1)$$

Note that the factorisation of a sequence p into p_1p_2 is unique because the sequence-length in a given initialiser domain is fixed. Clearly, the product of GTFs is a GTF.

Proposition 6.4 *The product of transfer functions is associative, with the identity transducer as its identity.*

Proof:

$$(T_1(T_2T_3))(\mathbf{u}, p_1p_2p_3)) = T_1((T_2T_3)(\mathbf{u}, p_2p_3), p_1) = T_1(T_2(T_3(\mathbf{u}, p_3), p_2), p_1)$$

while

$$((T_1T_2)T_3)(\mathbf{u}, p_1p_2p_3)) = (T_1T_2)(T_3(\mathbf{u}, p_3), p_1p_2) = T_1(T_2(T_3(\mathbf{u}, p_3), p_2), p_1)$$

Moreover $(IT)(\mathbf{u}, p) = I(T(\mathbf{u}, p), ()) = T(\mathbf{u}, p)$, $(TI)(\mathbf{u}, p) = T(I(\mathbf{u}, ()), p) = T(u, p)$.

Proposition 6.5 *Let T_1, T_2 be GTF's for which U_1 is the input space of T_1, U_2 is the output space of T_1 and the input space of T_2, while U_3 is the output space of T_2. Let Σ_1 be a group of full symmetries of T_1, while Σ_2 is a group of full symmetries of T_2. Then $\Sigma_1 \cap \Sigma_2$ is a group of full symmetries of T_2T_1.*

Proof: Let $\sigma \in \Sigma_1 \cap \Sigma_2$. Consider

$$\begin{aligned}(T_2T_1)(\mathbf{u}.\sigma, (p_1p_2).\sigma) &= T_2(T_1(\mathbf{u}.\sigma, p_1.\sigma), p_2.\sigma)\\ &= T_2(T_1(\mathbf{u}, p_1).\sigma, p_2.\sigma)\\ &= T_2(T_1(\mathbf{u}, p_1), p_2).\sigma\\ &= (T_2T_1)(\mathbf{u}, p_1p_2).\sigma\end{aligned}$$

Thus σ is a full symmetry of T_1T_2.

For example consider a mobile robot, equipped with a camera, on an infinite plane on which a straight line is painted. A symmetry group of the whole system is the intersection of the output-symmetries of the mechanics with the input symmetries of the camera as it views the line.

7 Taking the Quotient Simplifies Transfer Functions

Proposition 7.1 *Any group of operators Σ' defines an equivalence relationship on a domain on which it operates.*

$$u \equiv u' \Leftrightarrow \exists \sigma \in \Sigma', \quad u' = u.\sigma$$

Definition 7.2 *Let U be a domain, P be a preset domain, $\Sigma' \subset \Sigma$ a group of operators on U. Then we write U/Σ' for the set of classes of members of U equivalent under Σ'. Also we write $(U \times P)/\Sigma'$ for the set of classes of members of $U \times P$ equivalent under Σ' .*

[7] We are extending the operator set to act on the cartesian product in the obvious way

We write $\bar{u}$ for the equivalence class corresponding to u The mapping $u \mapsto \bar{u}$ is treated as the operator $1/\Sigma'$.

For finite domains, the advantage of going to quotient domains is that the size of the search space required to invert a transfer function for the purpose of creating a regulator or an open-loop controller is reduced. In [ ] being able to take the quotient of the Special Euclidean Group by the group of translations proved a useful way of simplifying the robotic assembly problems studied in that paper.

Proposition 7.3 *Let $\Sigma' \subset \Sigma$ be a group of full symmetries of a transfer function T whose input space is U, whose output space is X and whose preset space is P. Let $\theta = 1/\Sigma'$. Then the function T/Σ' defined by $(T/\Sigma')(\mathbf{u}.\theta, p.\theta) = T(\mathbf{u}, p).\theta$ is a GTF.*

Proof: The only thing we have to show is that the mapping is well defined. Suppose $(\mathbf{u}, p) \equiv (\mathbf{u}', p')$. Then $\mathbf{u}' = \mathbf{u}.\sigma, p' = p.\sigma$

$$(T/\Sigma')(\mathbf{u}'.\theta, p'.\theta) = T(\mathbf{u}', p').\theta = T(\mathbf{u}.\sigma, p.\sigma).\theta = T(\mathbf{u}, p).\sigma.\theta = T(\mathbf{u}, p).\theta$$

7.1 The Quotient of the Guards and Prisoners Problem

In taking the quotient domain U_{move}/Σ' the guards' places are equivalent and so are the prisoners' places. Thus the possible moves are $(\bar{1})$ meaning "1 guard crosses", $(\bar{1}, \bar{1})$ meaning "2 guards cross", $(\bar{1}, \bar{4})$ and $(\bar{4}, \bar{1})$ meaning "1 guard and one prisoner cross", $(\bar{4}$ meaning "one prisoner crosses" and $(\bar{4}, \bar{4})$ meaning "two prisoners cross". Thus the size of the input space is reduced from 36 for T_{moveGP} to 6 for T_{moveGP}/Σ'.

Reducing the size of the input domain makes a significant reduction in the size of the search-space for a sequence of inputs that will produce the desired terminal output $\mathbf{t}$ — the fan-out is divided by 6 at each stage.

Moreover the state-space is also shrunk by taking the quotient. In the original formulation there were $2 \times (6 \times 2^3)^2 = 4608$ states that could be reached from possible initial states. These are shrunk down to $3 * 3 * 2 = 18$ possible states in the quotient domain, for any two states that have the same number of guards on the left bank and have the same number of prisoners on the left bank and have the boat in the same place will be equivalent under Σ'.

8 Summary – Future Work

In this paper we have developed the concept of a generalised transfer function, illustrating how the concept encompasses both discrete and continuous systems. We have developed the concept of symmetry of a GTF which, in its most general

[8] Further condensation of the input space is possible if we note that permutation of the entries in an input-tuple is an input symmetry of T_{moveGP}

form, tells us how modifications of the input (and preset) of a GTF will affect its output. Symmetry thus has the potential to play the role of differentiation in classical control theory.

A strong motivation for generalisation is that biological adaptive systems appear to operate both in continuous and discrete domains. While there is a continuum of configurations of the human body, human communication takes place in a discretised vocabulary of words which are used *inter alia* to discuss actions and which arguably may characterise aspects of the mental processes underlying actions.

A potential bridge between continuous and discrete domains is the *quotient* operation, since it supports the collapse of a continuous domain into a discrete one, for example on the basis of topological invariants .

The value of the *generalisation* depends on whether it can be used as the basis of synthesis and analysis of reactive systems. We might wish to analyse a GTF from and *extensive* or *intensive* point of view. The extensive (or "white box") approach supposes we have a definition of a GTF that is open for inspection, so that its symmetries can be inferred from its definition. The intensive (or "black box") approach requires the formulation of a characterisation of a GTF by observing its behaviour.

One question that has been left unexplored is "Of what class of inputs is a given GTF a symmetry operator?" This is crucial to the understanding of linear systems whose analysis depends on the observation that a linear GTF is a scale symmetry of the exponential function over the complex domain.

References

1. Amarel, S. "On Representations of Problems of Reasoning About Actions", Machine Intelligence 3, Michie (ed), Edinburgh University Press, 1968; reprinted in "Readings in Artificial Intelligence", by B. L.Webber and N. J.Nilsson, Tioga, 1981.
2. Arnold,V. I. [1988] "Geometrical Methods in the Theory of Ordinary Differential Equations" (in English translation) Springer Verlag.
3. Clarke E. M., Grumberg O, Peled D. A.[1999] "Model Checking", MIT Press.
4. Hayes P,[1979] "The Naive Physics Manifesto" in D. Michie (Ed.) Expert Systems. Edinburgh U. Press.
5. Leyton M [1992] "Symmetry, Causality, Mind", MIT Press.
6. Milner R., Tofte M., Harper R., MacQueen D [1997] The Definition of Standard ML (Revised)
7. Liu Y., Popplestone R.[1994] "A Group Theoretic Formalization of Surface Contact", I. J. R. R. Vol. 13 No. 2.
8. E. Noether [1918], "Invariante Variationsprobleme", Nachr. d. König. Gesellsch. d. Wiss. zu Gottingen, Math-phys. Klasse, 235-257.
9. Popplestone R. J., Liu Y., and Weiss R.[1990] "A Group Theoretic Approach to Assembly Planning." *Artificial Intelligence Magazine*, Spring 1990, Vol. 11, No. 1, pp. 82-97.

[9] Our definition of symmetry includes homeomorphisms

10. Y.Liu, "Symmetry Groups in Robotic Assembly Planning", *COINS Technical Report 90-83* Computer and Information Sciece Dept, University of Massachusetts.
11. Popplestone,R. J., Ambler A. P., and Bellos, I. 1980, "An Interpreter for a Language for Describing Assemblies", *Artificial Intelligence* Vol. 14 No. 1, pp 79-107
12. Popplestone,R.J, 1984, Group Theory and Robotics, Robotics Research: The First International Symposium, (eds Brady,M. and Paul,R.), MIT Press, Cambridge MA and London.
13. Zahnd,A., Nair,S. and Popplestone, R. J. [1989] "Symmetry Inference in Planning Assembly", Proc. 5th ISRR, Tokyo Japan.

Pose Estimation in the Language of Kinematics

Bodo Rosenhahn, Yiwen Zhang, Gerald Sommer

Institut für Informatik und Praktische Mathematik
Christian-Albrechts-Universität zu Kiel
Preußerstrasse 1-9, 24105 Kiel, Germany
bro,yz,gs@ks.informatik.uni-kiel.de

Abstract. The paper concerns 2D-3D pose estimation in the algebraic language of kinematics. The pose estimation problem is modelled on the base of sev eral geometric constraint equations. In that way the projective geometric aspect of the topic is only implicitly represented and thus, pose estimation is a pure kinematic problem. The dynamic measurements of these constrain tsare either points or lines. The authors propose the use of motor algebra to introduce constrain tequations, which keep a natural distance measurement, the Hesse distance. The motor algebra is a degenerate geometric algebra in which line transformations are linear ones. The experiments aim to compare the use of different constraints and different methods of optimal estimating the pose parameters.

1 Introduction

The paper describes the estimation of pose parameters of known rigid objects in the framework of kinematics. Pose estimation is a basic visual task. In spite of its importance it has been identified for a long time (see e.g. Grimson [5]), and although there is published an ov erwhelming number of papers with respect to that topic [9], up to now there is no unique and general solution of the problem. P ose estimation means to relate several coordinate frames of measurement data and model data by finding out the transformations between, which can subsume rotation and translation. Since w e assume our measurement data as 2D and model data as 3D, w e are concerned with a 2D-3D pose estimation problem. Camera self-localization and navigation are typical examples of such types of problems. The coupling of projective and Euclidean transformations, both with nonlinear representations in Euclidean space, is the main reason for the difficulties to solve the pose problem. In this paper we attend to a pose estimation related to estimations of line motion as a problem of kinematics. The problem can be linearly represented in motor algebra [8] or dual quaternion algebra [7]. Instead of using invariances as an explicit formulation of geometry as often has been done in projective geometry, we are using implicit formulations of geometry as geometric constraints. We will demonstrate that geometric constraints are well conditioned, in contrast to in variances.

The paper is organized as follows. In section two we will introduce the motor algebra as representation frame for either geometric entities, geometric constraints, and Euclidean transformations. In section three we introduce the geometric constraints and their changes in an observation scenario. Section four is dedicated to the geometric analysis of these constraints. In section five we show some results for constraint based pose estimation with real images.

2 The motor algebra in the frame of kinematics

A geometric algebra $\mathcal{G}_{p,q,r}$ is a linear space of dimension 2^n, $n = p + q + r$, with a rich subspace structure, called blades, to represent so-called multiv ectors

G. Sommer and Y. Y. Zeevi (Eds.): AFPAC 2000, LNCS 1888, pp. 284-293, 2000.

as higher order algebraic entities in comparison to vectors of a vector space as first order entities. A geometric algebra $\mathcal{G}_{p,q,r}$ results in a constructive way from a vector space $\mathbb{R}$, endowed with the signature (p,q,r), $n = p+q+r$ by application of a geometric product. The geometric product consists of an outer ($\wedge$) and an inner ($\cdot$) product, whose role is to increase or to decrease the order of the algebraic entities, respectively.

To make it concretly, a motor algebra is the 8D even algebra $\mathcal{G}^+_{3,0,1}$, derived from $\mathbb{R}^4$, i.e. $n = 4$, $p = 3$, $q = 0$, $r = 1$, with basis vectors γ_k, $k = 1, ..., 4$, and the property $\gamma_1^2 = \gamma_2^2 = \gamma_3^2 = +1$ and $\gamma_4^2 = 0$. Because $\gamma_4^2 = 0$, $\mathcal{G}^+_{3,0,1}$ is called a degenerate algebra. The motor algebra $\mathcal{G}^+_{3,0,1}$ is of dimension eight and spanned by qualitative different subspaces with the following basis multivectors:

one scalar : 1
six bivectors : $\gamma_2\gamma_3, \gamma_3\gamma_1, \gamma_1\gamma_2, \gamma_4\gamma_1, \gamma_4\gamma_2, \gamma_4\gamma_3$
one pseudoscalar : $\boldsymbol{I} \equiv \gamma_1\gamma_2\gamma_3\gamma_4$.

Because $\gamma_4^2 = 0$, also the unit pseudoscalar squares to zero, i.e. $\boldsymbol{I}^2 = 0$. Remembering that the hypercomplex algebra of quaternions $\mathbb{H}$ represents a 4D linear space with one scalar and three vector components, it can simply be verified that $\mathcal{G}^+_{3,0,1}$ is isomorphic to the algebra of dual quaternions $\hat{\mathbb{H}}$ [11]. The geometric product of bivectors $\boldsymbol{A}$, $\boldsymbol{B} \in \langle\mathcal{G}^+_{3,0,1}\rangle_2$, $\boldsymbol{AB}$, splits into $\boldsymbol{AB} = \boldsymbol{A}\cdot\boldsymbol{B} + \boldsymbol{A}\times\boldsymbol{B} + \boldsymbol{A}\wedge\boldsymbol{B}$, where $\mathbf{A}\cdot\mathbf{B}$ is the inner product, which results in a scalar $\boldsymbol{A}\cdot\boldsymbol{B} = \alpha$, $\boldsymbol{A}\wedge\boldsymbol{B}$ is the outer product, which in this case results in a pseudoscalar $\boldsymbol{A}\wedge\boldsymbol{B} = \boldsymbol{I}\beta$, and $\boldsymbol{A}\times\boldsymbol{B}$ is the commutator product, which results in a bivector $\boldsymbol{C}$, $\boldsymbol{A}\times\boldsymbol{B} = \frac{1}{2}(\boldsymbol{AB} - \boldsymbol{BA}) = \boldsymbol{C}$. In a general sense, motors are called all the entities existing in motor algebra. They are constituted by bivectors and scalars. Thus, any geometric entity as points, lines, and planes have a motor representation. Changing the sign of the scalar and bivector in the real and the dual parts of the motor leads to the following variants of a motor

$$\boldsymbol{M} = (a_0 + \boldsymbol{a}) + \boldsymbol{I}(b_0 + \boldsymbol{b}) \qquad \widetilde{\boldsymbol{M}} = (a_0 - \boldsymbol{a}) + \boldsymbol{I}(b_0 - \boldsymbol{b})$$
$$\overline{\boldsymbol{M}} = (a_0 + \boldsymbol{a}) - \boldsymbol{I}(b_0 + \boldsymbol{b}) \qquad \widetilde{\overline{\boldsymbol{M}}} = (a_0 - \boldsymbol{a}) - \boldsymbol{I}(b_0 - \boldsymbol{b}) \ .$$

We will use the term motor in a more restricted sense to call with it a screw transformation, that is an Euclidean transformation embedded in motor algebra. Its constituents are rotation and translation (and dilation in case of non-unit motors). In line geometry we represent rotation by a rotation line axis and a rotation angle. The corresponding entity is called a unit rotor, $\boldsymbol{R}$, and reads as follows

$$\boldsymbol{R} = r_0 + r_1\gamma_2\gamma_3 + r_2\gamma_3\gamma_1 + r_3\gamma_1\gamma_2 = \cos\left(\tfrac{\theta}{2}\right) + \sin\left(\tfrac{\theta}{2}\right)\boldsymbol{n} = \exp\left(\tfrac{\theta}{2}\boldsymbol{n}\right).$$

Here θ is the rotation angle and $\boldsymbol{n}$ is the unit orientation vector of the rotation axis, spanned by the bivector basis.

If on the other hand, $\boldsymbol{t} = t_1\gamma_2\gamma_3 + t_2\gamma_3\gamma_1 + t_3\gamma_1\gamma_2$ is a translation vector in bivector representation, it will be represented in motor algebra as the dual part of a motor, called translator $\boldsymbol{T}$ with

$$\boldsymbol{T} = 1 + \boldsymbol{I}\tfrac{\boldsymbol{t}}{2} = \exp\left(\tfrac{\boldsymbol{t}}{2}\boldsymbol{I}\right).$$

Thus, a translator is also a special kind of rotor.

Because rotation and translation concatenate multiplicatively in motor algebra, a motor $\mathbf{M}$ reads

$$\boldsymbol{M} = \boldsymbol{TR} = \boldsymbol{R} + \boldsymbol{I}\tfrac{\boldsymbol{t}}{2}\boldsymbol{R} = \boldsymbol{R} + \boldsymbol{IR}'.$$

A motor represents a line transformation as a screw transformation. The line $\boldsymbol{L}$ will be transformed to the line $\boldsymbol{L}'$ by means of a rotation $\boldsymbol{R}_s$ around a line $\boldsymbol{L}_s$

by angle θ, followed by a translation t_s parallel to $\boldsymbol{L}_s$. Then the screw motion equation as motor transformation reads

$$\boldsymbol{L}' = \boldsymbol{T}_s \boldsymbol{R}_s \boldsymbol{L} \widetilde{\boldsymbol{R}}_s \widetilde{\boldsymbol{T}}_s = \boldsymbol{M} \boldsymbol{L} \widetilde{\boldsymbol{M}}.$$

For more detailed introductions see [8] and [10]. Now we will introduce the description of the most important geometric entities [8].

A point $\boldsymbol{x} \in \mathbb{R}^3$, represented in the bivector basis of $\mathcal{G}^+_{3,0,1}$, i.e. $\boldsymbol{X} \in \mathcal{G}^+_{3,0,1}$, reads $\boldsymbol{X} = 1 + x_1\gamma_4\gamma_1 + x_2\gamma_4\gamma_2 + x_3\gamma_4\gamma_3 = 1 + \boldsymbol{I}\boldsymbol{x}$.

A line $\boldsymbol{L} \in \mathcal{G}^+_{3,0,1}$ is represented by $\boldsymbol{L} = \boldsymbol{n} + \boldsymbol{I}\boldsymbol{m}$ with the line direction $\boldsymbol{n} = n_1\gamma_2\gamma_3 + n_2\gamma_3\gamma_1 + n_3\gamma_1\gamma_2$ and the moment $\boldsymbol{m} = m_1\gamma_2\gamma_3 + m_2\gamma_3\gamma_1 + m_3\gamma_1\gamma_2$.

A plane $\boldsymbol{P} \in \mathcal{G}^+_{3,0,1}$ will be defined by its normal $\boldsymbol{p}$ as bivector and by its Hesse distance to the origin, expressed as the scalar $d = (\boldsymbol{x} \cdot \boldsymbol{p})$, in the following way $\boldsymbol{P} = \boldsymbol{p} + \boldsymbol{I}d$.

In case of screw motions $\boldsymbol{M} = \boldsymbol{T}_s \boldsymbol{R}_s$ not only line transformations can be modelled, but also point and plane transformations. These are expressed as follows.

$$\boldsymbol{X}' = \boldsymbol{M}\boldsymbol{X}\widetilde{\overline{\boldsymbol{M}}} \qquad \boldsymbol{L}' = \boldsymbol{M}\,\boldsymbol{L}\,\widetilde{\boldsymbol{M}} \qquad \boldsymbol{P}' = \boldsymbol{M}\,\boldsymbol{P}\,\widetilde{\overline{\boldsymbol{M}}}$$

We will use in this study only point and line transformations because points and lines are the entities of our object models.

3 Geometric constraints and pose estimation

First, we make the following assumptions. The model of an object is given by points and lines in the 3D space. Furthermore we extract line subspaces or points in an image of a calibrated camera and match them with the model of the object. The aim is to find the pose of the object from observations of points and lines in the images at different poses. Figure 1 shows the scenario with respect to observed line subspaces. The method of obtaining the line subspaces is out of scope of this paper. Contemporary we simply got line segments by marking certain image points by hand. To estimate the pose, it is necessary to relate the observed lines in the image to the unknown pose of the object using geometric constraints.

The key idea is that the observed 2D entities together with their corresponding 3D entities are constraint to lie on other, higher order entities which result from the perspective projection. In our considered scenario there are three constraints which are attributed to two classes of constraints:

1. Collinearity: A 3D point has to lie on a line (projection ray) in the space
2. Coplanarity: A 3D point or line has to lie on a plane (projection plane).

With the terms projection ray or projection plane, respectively, we mean the image-forming ray which relates a 3D point with the projection center or the infinite set of image-forming rays which relates all 3D points belonging to a 3D line with the projection center, respectively. Thus, by introducing these two entities, we implicitly represent a perspective projection without necessarily formulating it explicitly. The most important consequence of implicitly representing projective geometry is that the pose problem is in that framework a pure kinematic problem. A similar approach of avoiding perspective projection equations by using constraint observations of lines has been proposed in [2].

To be more detailed, in the scenario of figure 1 we describe the following situation: We assume 3D points $\boldsymbol{A}'_i$ and lines $\boldsymbol{L}'_{Ai}$ of an object model. Further we extract line subspaces $\boldsymbol{l}_{ai}$ in an image of a calibrated camera and match them with the model.

Three constraints can be depicted:

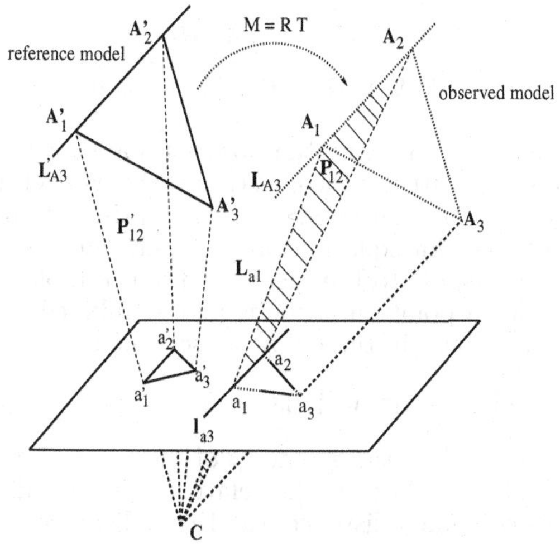

Fig. 1. The scenario. The solid lines at the left hand describe the assumptions: the camera model, the model of the object and the initially extracted lines on the image plane. The dashed lines at the right hand describe the actual pose of the model, which leads to the best fit of the object with the actual extracted lines.

1. A transformed point, e.g. $\boldsymbol{A_1}$, of the model point $\boldsymbol{A'_1}$ must lie on the projection ray $\boldsymbol{L_{a1}}$, given by $\boldsymbol{C}$ and the corresponding image point $\boldsymbol{a_1}$.
2. A transformed point, e.g. $\boldsymbol{A_1}$, of the model point $\boldsymbol{A'_1}$ must lie on the projection plane $\boldsymbol{P_{12}}$, given by $\boldsymbol{C}$ and the corresponding image line $\boldsymbol{l}_{a3}$.
3. A transformed line, e.g. $\boldsymbol{L}_{A3}$, of the model line $\boldsymbol{L'}_{A3}$ must lie on the projection plane $\boldsymbol{P_{12}}$, given by $\boldsymbol{C}$ and the the corresponding image line $\boldsymbol{l}_{a3}$.

constrain t	en tities	dual quaternion algebra	motor algebra
point-line	point $\boldsymbol{X} = 1 + I\boldsymbol{x}$ line $\boldsymbol{L} = \boldsymbol{n} + I\boldsymbol{m}$	$\boldsymbol{LX} - \boldsymbol{X}\overline{\boldsymbol{L}} = 0$	$\boldsymbol{XL} - \overline{\boldsymbol{L}}\boldsymbol{X} = 0$
point-plane	point $\boldsymbol{X} = 1 + I\boldsymbol{x}$ plane $\boldsymbol{P} = \boldsymbol{p} + I\boldsymbol{d}$	$\boldsymbol{P}\overline{\boldsymbol{X}} - \boldsymbol{X}\overline{\boldsymbol{P}} = 0$	$\boldsymbol{PX} - \overline{\boldsymbol{XP}} = 0$
line-plane	line $\boldsymbol{L} = \boldsymbol{n} + I\boldsymbol{m}$ plane $\boldsymbol{P} = \boldsymbol{p} + I\boldsymbol{d}$	$\boldsymbol{LP} - \boldsymbol{P}\overline{\boldsymbol{L}} = 0$	$\boldsymbol{LP} + \boldsymbol{P}\overline{\boldsymbol{L}} = 0$

Table 1. The geometric constraints expressed in motor algebra and dual quatenion algebra, respectively.

Table 1 gives an overview on the formulations of these constraints in motor algebra, taken from Blaschke [4], who used expressions in dual quaternion algebra. Here we adopt the terms from section 2.

The meaning of the constraint equations is immediately clear. In section 4 w ewill proceed to analyse them in detail. They represent the ideal situation, e.g. ac hiev ed as the result of the pose estimation procedure with respect to the observation frame. With respect to the previous reference frame, indicated by primes, these constraints read

$$(\boldsymbol{MX'}\widetilde{\overline{\boldsymbol{M}}})\boldsymbol{L} - \overline{\boldsymbol{L}}(\boldsymbol{MX'}\widetilde{\overline{\boldsymbol{M}}}) = 0$$

$$P(MX'\widetilde{M}) - \overline{(MX'\widetilde{M})\overline{P}} = 0$$
$$(ML'\widetilde{M})P + \overline{P(ML'\widetilde{M})} = 0.$$

These compact equations subsume the pose estimation problem at hand: find the best motor M which satisfies the constraint. We will get a convex optimization problem. Any error measure $|\epsilon| > 0$ of the optimization process as actual deviation from the constraint equation can be interpreted as a distance measure of misalignment with respect to the ideal situation of table 1. That means e.g. that the constraint for a point on a line is almost fulfilled for a point near the line. This will be made clear in the following section 4.

4 Analysis of the constraints

In this section we will analyse the geometry of the constraints introduced in the last section. We want to show that the relations between different entities are controlled by their orthogonal distance, the Hesse distance.

4.1 Point-line constraint

Evaluating the constraint of a point $X = 1 + Ix$ collinear to a line $L = n + Im$ leads to

$$\begin{aligned} 0 &= XL - \overline{L}X = (1 + Ix)(n + Im) - (n - Im)(1 + Ix) \\ &= n + Im + Ixn - n + Im - Inx = I(2m + xn - nx) \\ &= 2I(m - n \times x) \\ \Leftrightarrow 0 &= I(m - n \times x). \end{aligned}$$

Since $I \neq 0$, although $I^2 = 0$, the aim is to analyze the bivector $m - n \times x$. Suppose $X \notin L$. Then, nonetheless, there exists a decomposition $x = x_1 + x_2$ with $X_1 = (1 + Ix_1) \in L$ and $X_2 = (1 + Ix_2) \perp L$. Figure 2 shows the scenario.

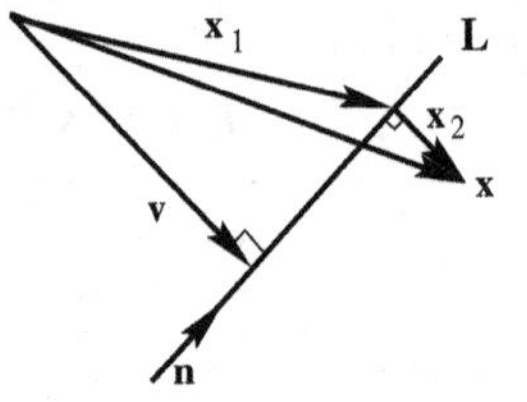

Fig. 2. The line L consists of the direction n and the moment $m = n \times v$. Further, there exists a decomposition $x = x_1 + x_2$ with $X_1 = (1 + Ix_1) \in L$ and $X_2 = (1 + Ix_2) \perp L$, so that $m = n \times v = n \times x_1$.

Then we can calculate

$$\begin{aligned} \|m - n \times x\| &= \|m - n \times (x_1 + x_2)\| = \|m - n \times x_1 - n \times x_2\| \\ &= \| - n \times x_2\| = \|x_2\|. \end{aligned}$$

Thus, satisfying the point-line constraint means to equate the bivectors m and $n \times x$, respectively making the Hesse distance $\|x_2\|$ of the point X to the line L to zero.

4.2 Point-plane constraint

Evaluating the constraint of a point $\boldsymbol{X} = 1 + \boldsymbol{I}\boldsymbol{x}$ coplanar to a plane $\boldsymbol{P} = \boldsymbol{p} + \boldsymbol{I}d$ leads to

$$\begin{aligned} 0 &= \boldsymbol{P}\boldsymbol{X} - \overline{\boldsymbol{X}\boldsymbol{P}} = (\boldsymbol{p} + \boldsymbol{I}d)(1 + \boldsymbol{I}\boldsymbol{x}) - (1 - \boldsymbol{I}\boldsymbol{x})(\boldsymbol{p} - \boldsymbol{I}d) \\ &= \boldsymbol{p} + \boldsymbol{I}\boldsymbol{p}\boldsymbol{x} + \boldsymbol{I}d - \boldsymbol{p} + \boldsymbol{I}d + \boldsymbol{I}\boldsymbol{x}\boldsymbol{p} = \boldsymbol{I}(2d + \boldsymbol{p}\boldsymbol{x} + \boldsymbol{x}\boldsymbol{p}) \\ \Leftrightarrow 0 &= \boldsymbol{I}(d + \boldsymbol{p} \cdot \boldsymbol{x}). \end{aligned}$$

Since $\boldsymbol{I} \neq 0$, although $\boldsymbol{I}^2 = 0$, the aim is to analyze the scalar $d + \boldsymbol{p} \cdot \boldsymbol{x}$. Suppose $\boldsymbol{X} \notin \boldsymbol{P}$. The value d can be interpreted as a sum so that $d = d_{01} + d_{02}$ and $d_{01}\boldsymbol{p}$ is the orthogonal projection of $\boldsymbol{x}$ onto $\boldsymbol{p}$. Figure 3 shows the scenario. Then we

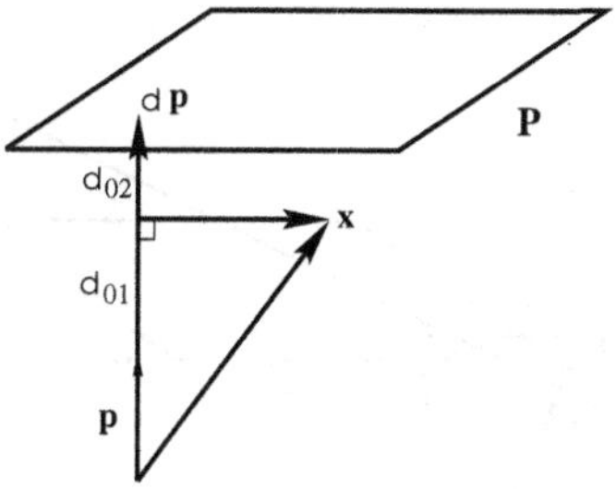

Fig. 3. The value d can be interpreted as a sum $d = d_{01} + d_{02}$ so that $d_{01}\boldsymbol{p}$ corresponds to the orthogonal projection of $\boldsymbol{x}$ onto $\boldsymbol{p}$.

can calculate

$$d + \boldsymbol{p} \cdot \boldsymbol{x} = d_{01} + d_{02} + \boldsymbol{p} \cdot \boldsymbol{x} = d_{01} + \boldsymbol{p} \cdot \boldsymbol{x} + d_{02} = d_{02}.$$

The value of the expression $d + \boldsymbol{p} \cdot \boldsymbol{x}$ corresponds to the Hesse distance of the point $\boldsymbol{X}$ to the plane $\boldsymbol{P}$.

4.3 Line-plane constraint

Evaluating the constraint of a line $\boldsymbol{L} = \boldsymbol{n} + \boldsymbol{I}\boldsymbol{m}$ coplanar to a plane $\boldsymbol{P} = \boldsymbol{p} + \boldsymbol{I}d$ leads to

$$\begin{aligned} 0 &= \boldsymbol{L}\boldsymbol{P} + \boldsymbol{P}\overline{\boldsymbol{L}} = (\boldsymbol{n} + \boldsymbol{I}\boldsymbol{m})(\boldsymbol{p} + \boldsymbol{I}d) + (\boldsymbol{p} + \boldsymbol{I}d)(\boldsymbol{n} - \boldsymbol{I}\boldsymbol{m}) \\ &= \boldsymbol{n}\boldsymbol{p} + \boldsymbol{I}\boldsymbol{m}\boldsymbol{p} + \boldsymbol{I}\boldsymbol{n}d + \boldsymbol{p}\boldsymbol{n} + \boldsymbol{I}\boldsymbol{n}d - \boldsymbol{I}\boldsymbol{p}\boldsymbol{m} \\ &= \boldsymbol{n}\boldsymbol{p} + \boldsymbol{p}\boldsymbol{n} + \boldsymbol{I}(2d\boldsymbol{n} - \boldsymbol{p}\boldsymbol{m} + \boldsymbol{m}\boldsymbol{p}) \\ \Leftrightarrow 0 &= \boldsymbol{n} \cdot \boldsymbol{p} + \boldsymbol{I}(d\boldsymbol{n} - \boldsymbol{p} \times \boldsymbol{m}) \end{aligned}$$

Thus, the constraint can be partitioned in one constraint on the real part of the motor and one constraint on the dual part of the motor. The aim is to analyze the scalar $\boldsymbol{n} \cdot \boldsymbol{p}$ and the bivector $d\boldsymbol{n} - (\boldsymbol{p} \times \boldsymbol{m})$ independently. Suppose $\boldsymbol{L} \notin \boldsymbol{P}$. If $\boldsymbol{n} \not\perp \boldsymbol{p}$ the real part leads to

$$\boldsymbol{n} \cdot \boldsymbol{p} = -\|\boldsymbol{n}\|\|\boldsymbol{p}\| \cos(\alpha) = -\cos(\alpha),$$

where α is the angle between $\boldsymbol{L}$ and $\boldsymbol{P}$, see figure 4. If $\boldsymbol{n} \perp \boldsymbol{p}$, we have $\boldsymbol{n} \cdot \boldsymbol{p} = 0$. Since the direction of the line is independent of the translation of the rigid body motion, the constraint on the real part can be used to generate equations with the parameters of the rotation as the only unknowns. The constraint on the dual part can then be used to determine the unknown translation. In other words, since the motor to be estimated, $\boldsymbol{M} = \boldsymbol{R} + \boldsymbol{I}\boldsymbol{R}\boldsymbol{T} = \boldsymbol{R} + \boldsymbol{I}\boldsymbol{R}'$, is determined in its real part only by rotation, the real part of the constraint allows to estimate the rotor $\boldsymbol{R}$, while the dual part of the constraint allows to estimate the rotor $\boldsymbol{R}'$. So it is possible to sequentially separate equations on the unknown rotation from equations on the unknown translation without the limitations, known from the embedding of the problem in Euclidean space [7]. This is very useful, since the tw osmaller equation systems are easier to solv ethan one larger equation system. To analyse the dual part of the constraint, we interpret the moment $\boldsymbol{m}$ of the line representation $\boldsymbol{L} = \boldsymbol{n} + \boldsymbol{I}\boldsymbol{m}$ as $\boldsymbol{m} = \boldsymbol{n} \times \boldsymbol{s}$ and choose a vector $\boldsymbol{s}$ with $\boldsymbol{S} = (1 + \boldsymbol{I}\boldsymbol{s}) \in \boldsymbol{L}$ and $\boldsymbol{s} \perp \boldsymbol{n}$. By expressing the inner product as the anti-commutator product, it can be shown ([1]) that $-(\boldsymbol{p} \times \boldsymbol{m}) = (\boldsymbol{s} \cdot \boldsymbol{p})\boldsymbol{n} - (\boldsymbol{n} \cdot \boldsymbol{p})\boldsymbol{s}$. Now we can evaluate

$$d\boldsymbol{n} - (\boldsymbol{p} \times \boldsymbol{m}) = d\boldsymbol{n} - (\boldsymbol{n} \cdot \boldsymbol{p})\boldsymbol{s} + (\boldsymbol{s} \cdot \boldsymbol{p})\boldsymbol{n}.$$

Figure 4 shows the scenario. Further, we can find a vector $\boldsymbol{s}_1 \parallel \boldsymbol{s}$ with

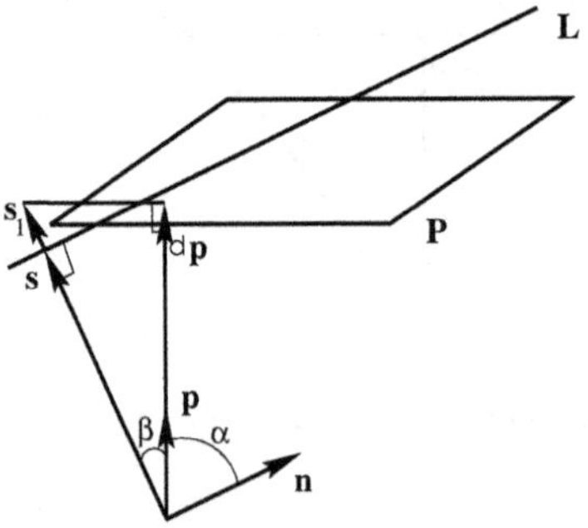

Fig. 4. The plane $\boldsymbol{P}$ consists of its normal $\boldsymbol{p}$ and the Hesse distance d. Furthermore we choose $\boldsymbol{S} = (1 + \boldsymbol{I}\boldsymbol{s}) \in \boldsymbol{L}$ with $\boldsymbol{s} \perp \boldsymbol{n}$. The angle of $\boldsymbol{n}$ and $\boldsymbol{p}$ is α and the angle of $\boldsymbol{s}$ and $\boldsymbol{p}$ is β. We choose the vector $\boldsymbol{s}_1$ with $\boldsymbol{s} \parallel \boldsymbol{s}_1$ so that $d\boldsymbol{p}$ is the orthogonal projection of $(\boldsymbol{s} + \boldsymbol{s}_1)$ onto $\boldsymbol{p}$.

$0 = d - (||\boldsymbol{s}|| + ||\boldsymbol{s}_1||)\cos(\beta)$. The vector $\boldsymbol{s}_1$ might also be antiparallel to $\boldsymbol{s}$. This leads to a change of the sign, but does not affect the constraint itself. Now we can evaluate

$$d\boldsymbol{n} - (\boldsymbol{n} \cdot \boldsymbol{p})\boldsymbol{s} + (\boldsymbol{s} \cdot \boldsymbol{p})\boldsymbol{n} = d\boldsymbol{n} - ||\boldsymbol{s}||\cos(\beta)\boldsymbol{n} + \cos(\alpha)\boldsymbol{s} = ||\boldsymbol{s}_1||\cos(\beta)\boldsymbol{n} + \cos(\alpha)\boldsymbol{s}.$$

The error of the dual part consists of the vector $\boldsymbol{s}$ scaled by the angle α and the direction $\boldsymbol{n}$ scaled by the norm of $\boldsymbol{s}_1$ and the angle β.
If $\boldsymbol{n} \perp \boldsymbol{p}$, then $\boldsymbol{p} \parallel \boldsymbol{s}$ and thus, we will find

$$||d\boldsymbol{n} - (\boldsymbol{p} \times \boldsymbol{m})|| = ||d\boldsymbol{n} + (\boldsymbol{s} \cdot \boldsymbol{p})\boldsymbol{n} - (\boldsymbol{n} \cdot \boldsymbol{p})\boldsymbol{s}|| = ||(d + \boldsymbol{s} \cdot \boldsymbol{p})\boldsymbol{n}|| = |(d + \boldsymbol{s} \cdot \boldsymbol{p})|.$$

This means, in agreement to the point-plane constraint, that $(d + \boldsymbol{s} \cdot \boldsymbol{p})$ describes the Hesse distance of the line to the plane. This analysis shows that the considered constraints are not only qualitative constraints, but also quantitative ones. This is very important, since w ewant to measure the extend of fulfillment of these constraints in the case of noisy data.

5 Experiments

In this section we present some experiments with real images. We expect that both the special constraint and the algorithmic approach of using it may influence the results. In our experimental scenario w etook a B21 mobile robot equipped with a stereo camera head and positioned it tw o meters in front of a

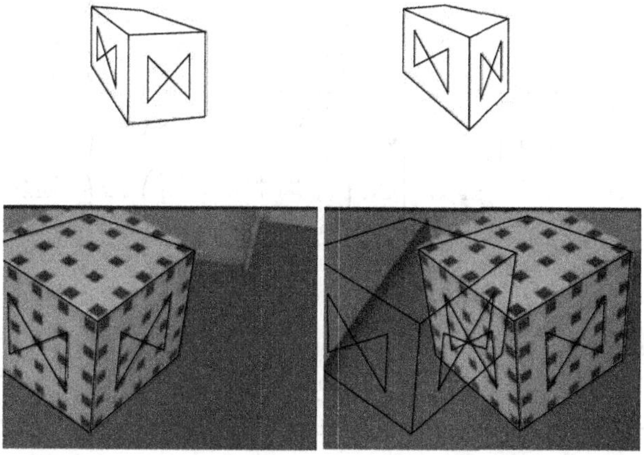

Fig. 5. The scenario of the experiment: In the top row two perspectives of the 3D object model are shown. In the second row (left) the calibration is performed and the 3D object model is projected on the image. Then the camera moved and corresponding line segments are extracted.

calibration cube. We focused one camera on the calibration cube and took an image. Then we moved the robot, focused the camera again on the cube and took another image. The edge size of the calibration cube is 46 cm and the image size is 384 × 288 pixel. Furthermore, we defined on the calibration cube a 3D object model. Figure 5 shows the scenario. In the first row two perspective views of the 3D object model are shown. In the left image of the second row the calibration is performed and the 3D object model is projected onto the image. Then the camera is moved and corresponding line segments are extracted. To visualize the movement, we also projected the 3D object model on its original position. The aim is to find the pose of the model and so the motion of the camera. In this experiment we actually selected certain points by hand and from these the depicted line segments are derived and, by knowing the camera calibration by the cube of the first image, the actual projection ray and projection plane parameters are computed. In table 2 we show the results of different algorithms for pose estimation. In the second column of table 2 EKF denotes the use of an extended Kalman filter. The design of the extended Kalman filters is described in [6]. MAT denotes matrix algebra, SVD denotes the singular value decomposition of a matrix to ensure a rotation matrix as a result. In the third column the used constraints, point-line (XL), point-plane (XP) and line-plane (LP) are indicated. The fourth column shows the results of the estimated rotation matrix $\mathcal{R}$ and the translation vector $\boldsymbol{t}$, respectively. Since the translation vectors are in mm, the results differ at around 2-3 cm. The fifth column shows the error of the equation system. Since the error of the equation system describes the Hesse distance of the entities, the value of the error is an approximation of the squared average distance of the entities. It is easy to see, that the results obtained with the different approaches are close to each other, though the implementation leads to different algorithms. Furthermore the EKF's perform more stable than the matrix solution approaches.

no.	$\mathcal{R}$ — t	Constraint	Experiment 1		Error
1	RtEKF — RtEKF	XL-XL	$\mathcal{R} = \begin{pmatrix} 0.987 & 0.089 & -0.138 \\ -0.117 & 0.969 & -0.218 \\ 0.115 & 0.231 & 0.966 \end{pmatrix}$	$\mathbf{t} = \begin{pmatrix} -58.21 \\ -217.26 \\ 160.60 \end{pmatrix}$	5.2
2	SVD — MAT	XL-XL	$\mathcal{R} = \begin{pmatrix} 0.976 & 0.107 & -0.191 \\ -0.156 & 0.952 & -0.264 \\ 0.154 & 0.287 & 0.945 \end{pmatrix}$	$\mathbf{t} = \begin{pmatrix} -60.12 \\ -212.16 \\ 106.60 \end{pmatrix}$	6.7
3	RtEKF — RtEKF	XP-XP	$\mathcal{R} = \begin{pmatrix} 0.987 & 0.092 & -0.133 \\ -0.118 & 0.973 & -0.200 \\ 0.111 & 0.213 & 0.970 \end{pmatrix}$	$\mathbf{t} = \begin{pmatrix} -52.67 \\ -217.00 \\ 139.00 \end{pmatrix}$	5.5
4	RtEKF — MAT	XP-XP	$\mathcal{R} = \begin{pmatrix} 0.986 & 0.115 & -0.118 \\ -0.141 & 0.958 & -0.247 \\ 0.085 & 0.260 & 0.962 \end{pmatrix}$	$\mathbf{t} = \begin{pmatrix} -71.44 \\ -219.34 \\ 124.71 \end{pmatrix}$	3.7
5	SVD — MAT	XP-XP	$\mathcal{R} = \begin{pmatrix} 0.979 & 0.101 & -0.177 \\ -0.144 & 0.957 & -0.251 \\ 0.143 & 0.271 & 0.952 \end{pmatrix}$	$\mathbf{t} = \begin{pmatrix} -65.55 \\ -221.18 \\ 105.87 \end{pmatrix}$	5.3
6	SVD — MAT	LP-XP	$\mathcal{R} = \begin{pmatrix} 0.976 & 0.109 & -0.187 \\ -0.158 & 0.950 & -0.266 \\ 0.149 & 0.289 & 0.945 \end{pmatrix}$	$\mathbf{t} = \begin{pmatrix} -66.57 \\ -216.18 \\ 100.53 \end{pmatrix}$	7.1
7	MEKF — MEKF	LP-LP	$\mathcal{R} = \begin{pmatrix} 0.985 & 0.106 & -0.134 \\ -0.133 & 0.969 & -0.208 \\ 0.107 & 0.229 & 0.969 \end{pmatrix}$	$\mathbf{t} = \begin{pmatrix} -50.10 \\ -212.60 \\ 142.20 \end{pmatrix}$	2.9
8	MEKF — MAT	LP-LP	$\mathcal{R} = \begin{pmatrix} 0.985 & 0.106 & -0.134 \\ -0.133 & 0.968 & -0.213 \\ 0.108 & 0.228 & 0.968 \end{pmatrix}$	$\mathbf{t} = \begin{pmatrix} -67.78 \\ -227.73 \\ 123.90 \end{pmatrix}$	2.7
9	SVD — MAT	LP-LP	$\mathcal{R} = \begin{pmatrix} 0.976 & 0.109 & -0.187 \\ -0.158 & 0.950 & -0.266 \\ 0.149 & 0.289 & 0.945 \end{pmatrix}$	$\mathbf{t} = \begin{pmatrix} -80.58 \\ -225.59 \\ 93.93 \end{pmatrix}$	6.9

Table 2. The experiment 1 results in different qualities of derived motion parameters, depending on the used constraints and algorithms to evaluate their validity.

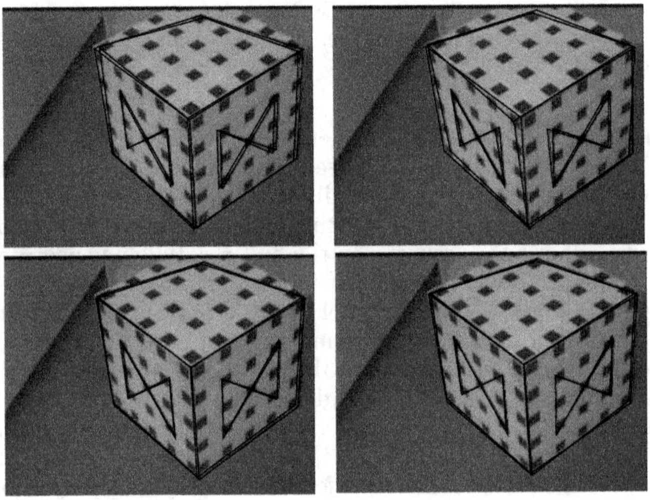

Fig. 6. Visualization of some errors. We calculate the motion of the object and project the transformed object in the image planes. The extracted line segments are also shown. In the first and second row, the results of nos. 5, 3 and nos. 7, 8 of table 2 are visualised respectiv ely

The visualization of some errors is done in figure 6. We calculated the motion of the object and projected the transformed object in the image plane. The extracted line segments are overlayed in addition. Figure 6 shows in the first row, left the results of nos. 5, 3 and nos. 7, 8 of table 2 respectively. The results of no. 7 and 8 are very good, compared with the results of the other algorithms.

These results are in agrement with the well known behavior of error propagation in case of matrix based rotation estimation. The EKF performs more stable. This is a consequence of the estimator themselves and of the fact that in our approach rotation is represented as rotors. The concatenation of rotors is more robust than that of rotation matrices.

6 Conclusions

The main contribution of the paper is to formulate 2D-3D pose determination in the language of kinematics as a problem of estimating rotation and translation from geometric constraint equations. There are three such constraints which relate the model frame to an observation frame. The model data are either points or lines. The observation frame is constituted by lines or planes. Any deviations from the constraint correspond the Hesse distance of the involv edgeometric entities. F rom this starting point as a useful algebraic frame for handling line motion, the motor algebra has been introduced. This is an eight-dimensional linear space with the property of representing rigid movements in a linear manner. The use of the motor algebra allows to subsume the pose estimation problem by compact equations, since the entities, the transformation of the entities and the constraints for collinearity or coplanarity of entities can be described very economically. F urthermore the introduced constraints con tain a natural distance measurement, the Hesse distance. This is the reason why the geometric constrain ts are well conditioned (in contrast to invariances) and, thus behave more robust in case of noisy data.

References

1. C. Perw ass and J. Lasenby. A no vel axiomatic derivation of geometric algebra. Technical Report CUED/F - INFENG/TR.347, Cambridge University Engineering Department, 1999.
2. Shevlin F. Analysis of orientation problems using Plüc ker lines. *International Conference on Pattern Recognition, Brisbane*, 1: 685–689, 1998.
3. Horaud R., Phong T.Q. and P.D. Tao. Object pose from 2-d to 3-d point and line correspondences. *International Journal of Computer Vision*, 15: 225–243, 1995.
4. Blaschke W. Mathematische Monographien 4, Kinematik und Quaternionen. *Deutscher Verlag der Wissenschaften*, 1960.
5. Grimson W. E. L. Object Recognition by Computer. *The MIT Press, Cambridge, MA*, 1990.
6. Sommer G., Rosenhahn B. and Zhang Y. P ose Estimation Using Geometric Constraints *T echnical R eprt 2003, Institut für Informatik und Praktische Mathematik, Christian-Albrechts-Universität zu Kiel*
7. Daniilidis K. Hand-eye calibration using dual quaternions. *Int. Journ. R obotics Res*, 18: 286–298, 1999.
8. Bayro-Corrochano E. The geometry and algebra of kinematics. *In Sommer G., editor, Ge ometric Computing with Clifford Algebra. Springer Verlag*, to be published, 2000.
9. Carceroni R. L. and C. M. Brown. Numerical Methods for Model-Based Pose Reco very. *T echn. Rept. 659, Comp. Sci. Dept., The Univ. of Rochester, Rochester, N. Y.*, August 1998.
10. Zhang Y., Sommer G., and E. Bayro-Corrochano. The motor extended Kalman filter for dynamic rigid motion estimation from line observations. *In G. Sommer, editor, Ge ometric Computing with Clifford Algebra. Springer Verlag*, to be published, 2000.
11. Hestenes D., Li H. and A. Rockwood. New algebraic tools for classical geometry. *In Sommer G., editor, Geometric Computing with Clifford Algebra. Springer Verlag*, to be published, 2000.

Algebraic Frames for Commutative Hyperharmonic Analysis of Signals and Images

Ekaterina Rundblad, Valeri Labunets, Jaakko Astola, and Karen Egiazarian

Tampere University of Technology, Signal Processing Laboratory,
Tampere, Finland

Abstract. Integral transforms and the signal representations associated with them are important tools in applied mathematics and signal theory. The Fourier transform and the Laplace transform are certainly the best known and most commonly used integral transforms. However, the Fourier transform is just one of many ways of signal representation and there are many other transforms of interest. In the past 20 years, other analytical methods have been proposed and applied, for example, wavelet, Walsh, Legendre, Hermite, Gabor, fractional Fourier analysis, etc. Regardless of their particular merits they are not as useful as the classical Fourier representation that is closely connected to such powerful concepts of signal theory as linear and nonlinear convolutions, classical and high–order correlations, invariance with respect to shift, ambiguity and Wigner distributions, etc. To obtain the general properties and important tools of the classical Fourier transform for an arbitrary orthogonal transform we associate to it generalized shift operators and develop the theory of abstract harmonic analysis of signals and linear and nonlinear systems that are invariant with respect to these generalized shift operators.

1 Generalized Convolutions and Correlations

The integral transforms and the signal representation associated with them are important concepts in applied mathematics and in signal theory. The Fourier transform is certainly the best known of the integral transforms and with the Laplace transform also the most useful. Since its introduction by Fourier in early 1800s, it has found use in innumerable applications and has, itself, led to the development of other transforms. We recall the most important properties of the classical Fourier transform: theorems of translation, modulation, scaling, convolution, correlation, differentiation, integration, etc. However, the Fourier transform is just one of many ways of signal representation, there are many other transforms of interest.

In the past 20 years, other analytical tools have been proposed and applied. An important aspect of many of these representations is the possibility to extract relevant information from a signal; information that is actually present but hidden in its complex representation. But, they are not efficient analysis tools compared to the group–theoretical classical Fourier representation, since the latter one is based on such useful and powerful tools of signal theory as linear and

G. Sommer and Y. Y. Zeevi (Eds.): AFPAC 2000, LNCS 1888, pp. 294– , 2000.

nonlinear convolutions, classical and higher–order correlations, invariance with respect to shift, ambiguity and Wigner distributions, etc. The other integral representations have no such tools.

The ordinary group shift operators $(T_t^\tau x)(t) := x(t \oplus \tau)$, $(T_t^{-\tau} x)(t) := x(t \ominus \tau)$ play the leading role in all the properties and tools of the Fourier transform mentioned above. In order to develop for each orthogonal transform a similar wide set of tools and properties as the Fourier transform has, we associate with each orthogonal transform a family of commutative generalized shift operators. Such families form *commutative hypergroups.* Only in particular cases hypergroups are well–known abelian groups. We will show that many well-known harmonic analysis theorems extend to the commutative hypergroups associated with arbitrary Fourier transforms.

1.1 Generalized Shift Operators

Let $y = f(t) : \Omega \longrightarrow \mathbf{C}$ be a complex–valued signal. Usually $\Omega = \mathbf{R}^n \times \mathbf{T}$, where $\mathbf{R}^n$ is n–D vector spaces, $\mathbf{T}$ is compact (temporal) subset from $\mathbf{R}$. Let Ω^* be the space dual to Ω. The first one will be called *spectral domain*, the second one – *signal domain*, keeping the original notion of $t \in \Omega$ as "time" and $\omega \in \Omega^*$ as "frequency". Let

$$\mathbf{L}(\Omega, \mathbf{C}) := \{f(t)|\ f(t) : \Omega \longrightarrow \mathbf{C}\} \quad \text{and} \quad \mathbf{L}(\Omega^*, \mathbf{C}) := \{F(\omega)|F(\omega) : \Omega^* \longrightarrow \mathbf{C}\}$$

be two vector spaces of square–integrable functions. In the following we assume that the functions satisfy certain general properties so that pathological cases where formulas would not hold are avoided. Let $\{\varphi_\omega(t)\}$ be an orthonormal system of functions in $\mathbf{L}(\Omega, \mathbf{C})$. Then for any function $f(t) \in \mathbf{L}(\Omega, \mathbf{C})$ there exists such a function $F(\omega) \in \mathbf{L}(\Omega^*, \mathbf{C})$, for which the following equations hold:

$$F(\omega) = \mathcal{F}\{f\}(\omega) = \int_{t \in \Omega} f(t) \bar{\varphi}_\omega(t) d\mu(t), \tag{1}$$

$$f(t) = \mathcal{F}^{-1}\{F\}(t) = \int_{\omega \in \Omega^*} F(\omega) \varphi_\omega(t) d\mu(\omega), \tag{2}$$

where $\mu(t), \mu(\omega)$ are certain suitable measures on the signal and spectral domains, respectively.

The function $F(\omega)$ is called $\mathcal{F}$–spectrum of a signal $f(t)$ and the expressions ( )–( ) are called the pair of *abstract Fourier transforms* (or $\mathcal{F}$–transforms). In the following we will use the notation $f(t) \underset{\mathcal{F}}{\longleftrightarrow} F(\omega)$ in order to indicate $\mathcal{F}$–transforms pair.

Along with the "time" and "frequency" domains we will work with "time–time" $\Omega \times \Omega$, "time–frequency" $\Omega \times \Omega^*$, "frequency–time" $\Omega^* \times \Omega$ and "frequency–frequency" $\Omega^* \times \Omega^*$, domains and with four joint distributions, which are denoted by double letters $\mathbf{ff}(t, \tau) \in \mathbf{L}_2(\Omega \times \Omega, \mathbf{C})$, $\mathbf{Ff}(\omega, \tau) \in \mathbf{L}_2(\Omega^* \times \Omega, \mathbf{C})$, $\mathbf{fF}(t, \nu) \in \mathbf{L}_2(\Omega \times \Omega^*, \mathbf{C})$, $\mathbf{FF}(\omega, \nu) \in \mathbf{L}_2(\Omega^* \times \Omega^*, \mathbf{C})$.

A fundamental and important tool of the signal theory are time–shift and frequency–shift operators. They are defined as $(T_t^\tau f)(t) := f(t+\tau), (D_\omega^\nu F)(\omega) := F(\omega+\nu)$. For $f(t) = e^{j\omega t}$ and $F(\omega) = e^{-j\omega t}$ we have $T_t^\tau e^{j\omega t} = e^{j\omega(t+\tau)} = e^{j\omega\tau}e^{j\omega t}$, $D_\omega^\nu e^{-j\omega t} = e^{-j(\omega+\nu)t} = e^{-j\nu t}e^{-j\omega t}$, i.e. functions $e^{j\omega t}$, $e^{-j\omega t}$ are eigenfunctions of time–shift and frequency–shift operators T^τ and D^ν corresponding to eigenvalues $\lambda_\tau = e^{j\omega\tau}$, $\omega \in \Omega^*$ and $\lambda_\nu = e^{-j\nu t}$, $t \in \Omega$, respectively. We now generalize this result.

Definition 1. *The following operators (with respect to which all basis functions $\varphi_\omega(t)$ are invariant eigenfunctions)*

$$(T_t^\tau \varphi_\omega)(t) = \int_{\sigma\in\Omega} T_{t\sigma}^\tau \varphi_\omega(\sigma) d\mu(\sigma) = \varphi_\omega(t)\varphi_\omega(\tau), \tag{3}$$

$$(\tilde{T}_t^\tau \varphi_\omega)(t) = \int_{\sigma\in\Omega} \tilde{T}_{t\sigma}^\tau \varphi_\omega(\sigma) d\mu(\sigma) = \varphi_\omega(t)\bar{\varphi}_\omega(\tau), \tag{4}$$

$$(D_\omega^\nu \bar{\varphi}_\omega)(t) = \int_{\alpha\in\Omega^*} D_{\omega\alpha}^\nu \varphi_\alpha(t) d\mu(\alpha) = \bar{\varphi}_\omega(t)\bar{\varphi}_\nu(t), \tag{5}$$

$$(\tilde{D}_\omega^\nu \bar{\varphi}_\omega)(t) = \int_{\alpha\in\Omega^*} \tilde{D}_{\omega\alpha}^\nu \varphi_\alpha(t) d\mu(\alpha) = \bar{\varphi}_\omega(t)\varphi_\nu(t). \tag{6}$$

are called commutative $\mathcal{F}$–generalized "time"–shift and "frequency"–shift operators (GSO's), respectively, where $\varphi_\omega(\tau)$, $\bar{\varphi}_\omega(\tau)$ are eigenvalues of GSO's T^τ and $\tilde{T}^\tau$, respectively.

For these operators we introduce the following designations:

$$(T_t^\tau \varphi_\omega)(t) = \varphi_\omega(t \boxplus \tau), \quad (\tilde{T}_t^\tau \varphi_\omega)(t) = \varphi_\omega(t \boxminus \tau),$$

$$(D_\omega^\nu \bar{\varphi}_\omega)(t) = \bar{\varphi}_{\omega\oplus\nu}(t), \quad D_\omega^\nu \bar{\varphi}_\omega)(t) = \bar{\varphi}_{\omega\ominus\nu}(t).$$

Here the symbols $\boxplus, \boxminus$ $\oplus, \ominus$ denote the quasi–sum and quasi–difference, respectively. The expressions ( )–( ) are called *multiplication formulae* for basis functions $\varphi_\omega(t)$. They show that the set of basis functions form two hypergroups with respect to multiplication rules ( ) and ( ), respectively. We see also that two families of time and frequency GSOs form two hypergroups.

For $f(t) \in \mathbf{L}(\Omega, \mathbf{C})$, $F(\omega) \in \mathbf{L}(\Omega^*, \mathbf{C})$ we define

$$f(t \boxplus \tau) := \int_{\omega\in\Omega^*} [F(\omega)\varphi_\omega(\tau)]\varphi_\omega(t)d\mu(\omega), \quad f(t \boxminus \tau) := \int_{\omega\in\Omega^*} [F(\omega)\bar{\varphi}_\omega(\tau)]\varphi_\omega(t)d\mu(\omega),$$

$$F(\omega \oplus \nu) := \int_{t\in\Omega} [f(t)\bar{\varphi}_\nu(t)]\bar{\varphi}_\omega(t)d\mu(t), \quad F(\omega \ominus \nu) := \int_{t\in\Omega} [f(t)\varphi_\nu(t)]\bar{\varphi}_\omega(t)d\mu(t).$$

In particular, it follows the first and second main theorems of generalized harmonic analysis: "theorems of generalized shifts and generalized modulations", respectively

$$f(t \boxplus \tau) \underset{\mathcal{F}}{\longleftrightarrow} F(\omega)\varphi_\omega(\tau), \quad f(t \boxminus \tau) \underset{\mathcal{F}}{\longleftrightarrow} F(\omega)\bar{\varphi}_\omega(\tau) \tag{7}$$

$$f(t)\bar{\varphi}_\nu(t) \underset{\mathcal{F}}{\longleftrightarrow} F(\omega \oplus \nu), \quad f(t)\varphi_\nu(t) \underset{\mathcal{F}}{\longleftrightarrow} F(\omega \ominus \nu). \tag{8}$$

1.2 Generalized Convolutions and Correlations

It is well known that any stationary linear dynamic systems (LDS) is described by the well–known convolution integral. Using the notion GSO, we can formally generalize the notions of convolution and correlation.

Definition 2. *The following functions*

$$y(t) := (h\diamondsuit x)(t) = \int_{\tau \in \Omega} h(\tau)x(t \boxminus \tau)d\mu(\tau), \tag{9}$$

$$Y(\omega) := (H\heartsuit X)(\omega) = \int_{\nu \in \Omega^*} H(\nu)X(\omega \ominus \nu)d\mu(\nu) \tag{10}$$

are called the T*–invariant time and* D*–invariant spectral convolutions (or* $\diamondsuit$*–convolution and* $\heartsuit$*–convolution), respectively.*

The spaces $\mathbf{L}(\Omega, \mathbf{C})$ and $\mathbf{L}(\Omega^*, \mathbf{C})$ equipped multiplications $\diamondsuit$ and $\heartsuit$ form commutative Banach signal and spectral convolution algebras $\langle\langle \mathbf{L}(\Omega, \mathbf{C}), \diamondsuit \rangle\rangle$ and $\langle\langle \mathbf{L}(\Omega^*, \mathbf{C}), \heartsuit \rangle\rangle$, respectively. The classical convolution algebras $\langle\langle \mathbf{L}(\Omega, \mathbf{C}), * \rangle\rangle$ and $\langle\langle \mathbf{L}(\Omega^*, \mathbf{C}), * \rangle\rangle$ are obtained if Ω is abelian group and $\varphi_\omega(t)$ are its characters.

Definition 3. *The following expressions*

$$(f\clubsuit g)(\tau) := \int_{t \in \Omega} f(t)\bar{g}(t \boxminus \tau)d\mu(t), \quad (F\spadesuit G)(\nu) := \int_{\omega \in \Omega^*} F(\omega)\overline{G}(\omega \ominus \nu)d\mu(\omega) \tag{11}$$

are referred to as T*–invariant and* D*–invariant cross–correlation functions of the signals and of the spectra, respectively.*

The measures indicating the similarity between a **tF**– distributions and **Ft**–distributions and its time– and frequency–shifted versions are their crosscorrelation functions.

Definition 4. *The following expressions*

$$(\mathbf{fF}\clubsuit\spadesuit\mathbf{gG})(\tau, \nu) := \int_{t \in \Omega}\int_{\omega \in \Omega^*} \mathbf{fF}(t, \omega)\overline{\mathbf{gG}}(t \boxminus \tau, \omega \ominus \nu)d\mu(t)d\mu(\omega), \tag{12}$$

$$(\mathbf{Ff}\spadesuit\clubsuit\mathbf{Gg})(\nu,\tau) := \int\limits_{\nu\in\Omega^*}\int\limits_{\tau\in\Omega} \mathbf{Ff}(\omega,t)\overline{\mathbf{Gg}}(\omega\ominus\nu,t\boxminus\tau)d\mu(t)d\mu(\omega). \qquad (13)$$

are referred to as TD– and DT–invariant cross-correlation functions of the distributions, respectively. If $\mathbf{fF}(t,\omega)=\mathbf{gG}(t,\omega)$ *and* $\mathbf{Ff}(\omega,t)=\mathbf{Gg}(\omega,t)$ *then the cross correlation functions are simply the autocorrelation functions.*

Theorem 1. Third main theorem of generalized harmonic analysis. *Generalized Fourier transforms ( ) and ( ) map linear* $\diamondsuit$*–and* $\heartsuit$*–convolutions and linear* $\clubsuit$*– and* $\spadesuit$*–correlations into the products of spectra and signals*

$$\mathcal{F}\{(h\diamondsuit x)(t)\}=\mathcal{F}\{h(t)\}\mathcal{F}\{x(t)\},\quad \mathcal{F}\{(f\clubsuit g)(t)\}=\mathcal{F}\{f(t)\}\overline{\mathcal{F}\{g(t)\}},$$

$$\mathcal{F}^{-1}\{(H\heartsuit X)(\omega)\}=\mathcal{F}^{-1}\{H(\omega)\}\mathcal{F}^{-1}\{X(\omega)\},$$

$$\mathcal{F}^{-1}\{(F\spadesuit G)(\omega)\}=\mathcal{F}^{-1}\{F(\omega)\}\overline{\mathcal{F}^{-1}\{G(\omega)\}}.$$

Taking special forms of the GSO's one can obtain known types of convolutions and crosscorrelations: arithmetic, cyclic, dyadic, m–adic, etc. Signal and spectral algebras have many of the properties, associated with classical group convolution algebras, many of them are catalogued in [ ]–[ ].

2 Generalized the Weyl Convolutions and Correlations

2.1 Generalized the Weyl Convolutions and Correlations

Linear time–invariant filtering may be viewed as an evaluation of weighted superpositions of time–shifted versions of the input signal

$$y(t)=(h*x)(t)=\int\limits_{-\infty}^{+\infty}h(\tau)x(t-\tau)d\tau=\int\limits_{-\infty}^{+\infty}h(\tau)(T_\tau^t x)(\tau)d\tau. \qquad (14)$$

Every linear time–invariant filter is a weighted superposition of time- -shifts, and conversely, every weighted superposition of time–shifts is a linear time–invariant filter.

The frequency convolution

$$Y(\omega)=(H*X)(\omega)=\int\limits_{-\infty}^{+\infty}H(\nu)X(\omega-\nu)d\nu=\int\limits_{-\infty}^{+\infty}H(\nu)(T_\nu^\omega X)(\nu)d\nu, \qquad (15)$$

is expressed as a weighted superposition of frequency–shifts versions of the spectrum.

We can combine expressions ( )–( ) into three time–frequency weighted superpositions of time–frequency shifts of the signal $e^{j\omega\tau}x(t-\tau)$, spectrum $e^{-j\nu t}\times$ $\times X(\omega-\nu)$ and frequency–time distribution $\mathbf{Xx}(\omega-\nu,t-\tau)e^{-j(\nu t+\omega\tau)}$, respectively, which are called *Weyl convolutions*:

1. The Weyl convolution of a FTD $\mathbf{Ww}(\nu,\tau)$ with a time–frequency shifted signal $x(t-\tau)e^{j\nu t}$

$$y(t)=\mathbf{Weyl}[\mathbf{Ww};x](t):=\int\limits_{-\infty}^{+\infty}\int\limits_{-\infty}^{+\infty}\mathbf{Ww}(\nu,\tau)x(t-\tau)e^{j\nu t}d\tau d\nu. \qquad (16)$$

2. The Weyl convolution of a TFD $\mathbf{wW}(\tau,\nu)$ with a frequency–time shifted spectrum $X(\omega-\nu)e^{-j\omega\tau}$

$$Y(\omega)=\mathbf{Weyl}[\mathbf{wW};X](\omega):=\int\limits_{-\infty}^{+\infty}\int\limits_{-\infty}^{+\infty}\mathbf{wW}(\tau,\nu)X(\omega-\nu)e^{-j\omega\tau}d\tau d\nu. \qquad (17)$$

3. The Weyl convolution of a FTD $\mathbf{Wt}(\omega,t)$ and $\mathbf{Xx}(\omega-\nu,t-\tau)e^{-j(\omega\tau-\nu t)}$

$$\mathbf{Weyl}[\mathbf{Ww};\mathbf{Xx}](\omega,t):=\int\limits_{-\infty}^{+\infty}\int\limits_{-\infty}^{+\infty}\mathbf{Ww}(\omega,t)\mathbf{Xx}(\omega-\nu,t-\tau)e^{-j(\omega\tau-\nu t)}d\nu d\tau. \qquad (18)$$

Convolutions ( )–( ) were proposed by H. WEYL [ ]. Now they are called *Weyl's convolutions.* The operators, $\mathbf{Weyl}[\mathbf{Ww};\circ]$, defined in ( ) and ( ) are time– and frequency– varying operators. The rules ( ) and ( ) which relate time–frequency symbols $\mathbf{wW}(t,\omega)$ and $\mathbf{Ww}(\omega,t)$ two a unique operators $y(t)=\mathbf{Weyl}[\mathbf{wW};\circ]$ and $Y(\omega)=\mathbf{Weyl}[\mathbf{wW};\circ]$, are called *Weyl correspondences.* It is easy to see that Weyl convolutions are invariant with respect to classical 2–D Heisenberg group consisting of all time–frequency shifts. The Weyl correspondence may be used to form linear time–varying filters in the two different ways. In each case, the filter is defined by choosing a mask, or symbol, in the time–frequency plane.

Analogously, we can introduce the following Weyl correlations:

1. The Weyl correlation between a 2–D TFD $\mathbf{fF}(t,\omega)$ and a time–frequency shifted signal $g(t-\tau)e^{j\nu t}$:

$$\mathbf{weylCor}[\mathbf{fF},g](\tau,\nu):=\int\limits_{-\infty}^{+\infty}\int\limits_{-\infty}^{+\infty}\mathbf{fF}(t,\omega)\bar{g}(t-\tau)e^{-j\nu t}e^{j\omega\tau}d\omega dt. \qquad (19)$$

2. The Weyl correlation between a 2–D FTD $\mathbf{Ff}(\omega,t)$ and a time–frequency shifted spectrum $G(\omega-\nu)e^{-j\omega\tau}$:

$$\mathbf{Weylcor}[\mathbf{Ff};G](\nu,\tau):=\int\limits_{-\infty}^{+\infty}\int\limits_{-\infty}^{+\infty}\mathbf{Ff}(\omega,t)\overline{G}(\omega-\nu)e^{j\omega\tau}e^{-j\nu t}dtd\omega. \qquad (20)$$

3. The Weyl correlation between 2–D TFDs $\mathbf{gG}(t-\tau,\omega-\nu)e^{j\nu t}e^{-j\omega\tau}$ and $\mathbf{fF}(t,\omega)$:

$$\mathbf{WEYLcor}[\mathbf{fF};\mathbf{gG}](\nu,\tau):=\int_{-\infty}^{+\infty}\int_{-\infty}^{+\infty}\mathbf{fF}(t,\omega)\overline{\mathbf{gG}}(t-\tau,\omega-\nu)e^{-j\nu t}e^{j\omega\tau}d\omega dt. \tag{21}$$

We generalize relations ( )–( ) for arbitrary pair of families of generalized shift operators.

Definition 5. *The following expressions are called $\mathcal{F}$–generalized Weyl convolutions and are expressed as weighted superpositions of time–frequency shifts of the signal $\bar{\varphi}_\omega(\tau)x(t\boxminus\tau)$, spectrum $\varphi_\nu(t)X(\omega\ominus\nu)$ and frequency–time distribution $\mathbf{Xx}(\omega\ominus\nu,t\boxminus\tau)\bar{\varphi}_\omega(\tau)\varphi_\nu(t)$, respectively:*

1. *The Weyl convolution of a FTD $\mathbf{Ww}(\nu,\tau)$ with a time–frequency shifted signal $x(t\boxminus\tau)\varphi_\nu(t)$*

$$\mathbf{Weyl}^{\diamondsuit}[\mathbf{Ww};x](t):=\int_{\tau\in\Omega}\int_{\nu\in\Omega^*}\mathbf{Ww}(\nu,\tau)x(t\boxminus\tau)\varphi_\nu(t)d\mu(\tau)d\mu(\nu). \tag{22}$$

2. *The Weyl convolution of a TFD $\mathbf{wW}(\tau,\nu)$ with a frequency–time shifted spectrum $X(\omega\ominus\nu)\bar{\varphi}_\omega(\tau)$*

$$\mathbf{Weyl}^{\heartsuit}[\mathbf{wW};X](\omega)=\int_{\nu\in\Omega^*}\int_{\tau\in\Omega}\mathbf{wW}(\tau,\nu)X(\omega\ominus\nu)\bar{\varphi}_\omega(\tau)d\mu(\tau)d\mu(\nu). \tag{23}$$

3. *The Weyl convolution of two FTDs $\mathbf{Xx}(\omega\ominus\nu,t\boxminus\tau)\overline{\varphi}_\omega(\tau)\varphi_\nu(t)$ and $\mathbf{Wt}(\omega,t)$:* $\mathbf{Weyl}[\mathbf{Ww};\mathbf{Xx}](\omega,t)=$

$$=\int_{\nu\in\Omega^*}\int_{\tau\in\Omega}\mathbf{Ww}(\omega,t)\mathbf{Xx}(\omega\ominus\nu,t\boxminus\tau)\bar{\varphi}_\omega(\tau)\varphi_\nu(t)d\mu(\nu)d\mu(\tau) \tag{24}$$

(in particular, here can be $\mathbf{Xx}(\omega,t):=X(\omega)x(t)$).

By analogy with ( )–( ) we can design generalized Weyl correlation functions.

Definition 6. *The following expressions are called $\mathcal{F}$–generalized Weyl correlations:*

1. *The Weyl correlation between a 2–D TFD $\mathbf{fF}(t,\omega)$ and a time–frequency shifted signal $g(t\boxminus\tau)\varphi_\nu(t)\bar{\varphi}_\omega(\tau)$:*

$$\mathbf{weylCor}^{\clubsuit}[\mathbf{fF};g](\tau,\nu):=\int_{t\in\Omega}\int_{\omega\in\Omega^*}\mathbf{fF}(t,\omega)\bar{g}(t\boxminus\tau)\bar{\varphi}_\nu(t)\varphi_\omega(\tau)d\mu(\omega)d\mu(t). \tag{25}$$

2. *The Weyl correlation between a 2–D FTD* $\mathbf{Ff}(\omega, t)$ *and a time–frequency shifted spectrum* $G(\omega \ominus \nu)\bar{\varphi}_\omega(\tau)\varphi_\nu(t)$:

$$\mathbf{Weylcor}^{\spadesuit}[\mathbf{Ff}; G](\nu, \tau) := \int\limits_{t\in\Omega}\int\limits_{\omega\in\Omega^*} \mathbf{Ff}(\omega, t)\bar{G}(\omega \ominus \nu)\varphi_\omega(\tau)\bar{\varphi}_\nu(t)d\mu(t)d\mu(\omega). \tag{26}$$

3. *The Weyl correlation between two 2–D TFDs* $\mathbf{gG}(t \boxminus \tau, \omega\ominus)\varphi_\nu(t)\bar{\varphi}_\omega(\tau)$ *and* $\mathbf{fF}(t, \omega)$: $\mathbf{Weylcor}^{\clubsuit\spadesuit}[\mathbf{fF}; \mathbf{gG}](\nu, \tau) :=$

$$= \int\limits_{t\in\Omega}\int\limits_{\omega\in\Omega^*} \mathbf{fF}(t, \omega)\overline{\mathbf{gG}}(t \boxminus \tau, \omega \ominus \nu)\bar{\varphi}_\nu(t)\varphi_\omega(\tau)d\mu(\omega)d\mu(t). \tag{27}$$

2.2 Generalized Ambiguity Functions and Wigner Distributions

The Wigner distribution was introduced in 1932 be E. WIGNER [ ] in the context of quantum mechanics, where he defined the probability function of the simultaneous values of the spatial coordinates and impulses. Wigner's idea was introduced in signal analysis in 1948 by J. VILLE [ ], but it did not receive much attention there until 1953 when P. WOODWARD [ ] reformulated it in the context of radar theory. Woodward proposed treating the question of radar signal ambiguity as a part of the question of target resolution. For that, he introduced a function that described the correlation between a radar signal and its Doppler–shifted and time–translated version. Physically, the ambiguity function represents the energy in received signal as a function of time delay and Doppler frequency. This function describes the local ambiguity in locating targets in range (time delay τ) and in velocity (Doppler frequency ν). Its absolute value is called *uncertainty function* as it is related to the *uncertainty principle* of radar signals. We can generalize this notion the following way.

Definition 7. *The* $\mathcal{F}$*–generalized symmetric and asymmetric cross–ambiguity functions of two pairs of functions* f, g *and of* F, G *are defined by*

$$\mathbf{aF}^s[f, g](\tau, \nu) := \mathop{\mathcal{F}}_{t\to\nu}\left\{\mathbf{fg}^s(\tau, t)\right\} = \int\limits_{t\in\Omega}\left[f\left(t \boxplus \frac{\tau}{2}\right)\bar{g}\left(t \boxminus \frac{\tau}{2}\right)\right]\bar{\varphi}_\nu(t)d\mu(t),$$

$$\mathbf{Af}^s[F, G](\nu, \tau) := \mathop{\mathcal{F}^{-1}}_{\omega\to\tau}\left\{\mathbf{FG}^s(\nu, \omega)\right\} = \int\limits_{\omega\in\Omega^*}\left[F\left(\omega \oplus \frac{\nu}{2}\right)\bar{G}\left(\omega \ominus \frac{\nu}{2}\right)\right]\varphi_\omega(\tau)d\mu(\omega),$$

$$\mathbf{aF}^a[f, g](\tau, \nu) := \mathop{\mathcal{F}}_{t\to\nu}\left\{\mathbf{fg}^a(\tau, t)\right\} = \int\limits_{t\in\Omega}\left[f(t)\bar{g}(t \boxminus \tau)\right]\bar{\varphi}_\nu(t)d\mu(t),$$

$$\mathbf{Af}^a[F, G](\nu, \tau) := \mathop{\mathcal{F}^{-1}}_{\omega\to\tau}\left\{\mathbf{FG}^a(\nu, \omega)\right\} = \int\limits_{\omega\in\Omega^*}\left[F(\omega)\bar{G}(\omega \ominus \nu)\right]\varphi_\omega(\tau)d\mu(\omega).$$

Definition 8. *The $\mathcal{F}$-generalized symmetric and asymmetric cross–Wigner FTDs* $\mathbf{Wd}^s[f,g](\omega,t)$, $\mathbf{Wd}^a[f,g](\omega,t)$ *and cross–Wigner TFDs* $\mathbf{wD}^s[F,G](t,\omega)$, $\mathbf{wD}^a[F,G](t,\omega)$ *of two pairs of functions* f,g *and of* F,G *are defined by*

$$\mathbf{Wd}^s[f,g](\omega,t) := \underset{\tau\to\omega}{\mathcal{F}}\{\mathbf{fg}^s(\tau,t)\} = \int\limits_{\tau\in\Omega}\left[f\left(t\boxplus\frac{\tau}{2}\right)\bar{g}\left(t\boxminus\frac{\tau}{2}\right)\right]\bar{\varphi}_\omega(\tau)d\mu(\tau),$$

$$\mathbf{wD}^s[F,G](t,\omega) := \underset{\nu\to t}{\mathcal{F}^{-1}}\{\mathbf{FG}^s(\nu,\omega)\} = \int\limits_{\nu\in\Omega^*}\left[F\left(\omega\oplus\frac{\nu}{2}\right)\bar{G}\left(\omega\oplus\frac{\nu}{2}\right)\right]\varphi_\nu(t)d\mu(\nu),$$

$$\mathbf{Wd}^a[f,g](\omega,t) := \underset{\tau\to\omega}{\mathcal{F}}\{\mathbf{fg}^a(\tau,t)\} = f(t)\overline{F}(\omega)\bar{\varphi}_\omega(t),$$

$$\mathbf{wD}^a[F,G](t,\omega) := \underset{\nu\to t}{\mathcal{F}^{-1}}\{\mathbf{FG}^a(\nu,\omega)\} = F(\omega)\bar{f}(t)\varphi_\omega(t),$$

$\mathcal{F}$–generalized asymmetrical Wigner distributions are called also *$\mathcal{F}$–generalized Richaczek's distributions.*

3 High–Order Volterra Convolutions and Correlations

The study of nonlinear systems $y(t) = \mathbf{SYST}[x(t)]$ was started by Volterra [ ] who investigated analytic operators and introduced the representation

$$y(t) = \mathbf{SYST}[x(t)] = \mathbf{Volt}[h_1,h_2,\ldots,h_q,\ldots;x](t) = \mathbf{Volt}[\mathbf{h};x](t) =$$

$$= \int\limits_{-\infty}^{\infty} h_1(\sigma_1)x(t-\sigma_1)d\sigma_1 + \int\limits_{-\infty}^{\infty}\int\limits_{-\infty}^{\infty} h_2(\sigma_1,\sigma_2)x(t-\sigma_1)x(t-\sigma_2)d\sigma_1 d\sigma_2 + \ldots$$

$$\ldots + \int\limits_{-\infty}^{\infty}\cdots\int\limits_{-\infty}^{\infty} h_q(\sigma_1,\ldots,\sigma_q)x(t-\sigma_1)\cdots x(t-\sigma_q)d\sigma_1\cdots d\sigma_q, \qquad (28)$$

where $q = 1,2,\ldots$; signals $x(t)$ and $y(t)$ are the input and output, respectively, of the system **SYST** at time t, $h_q(\sigma_1,\ldots,\sigma_q)$ is the q–th order Volterra kernel, and the set of kernels $\mathbf{h} := (h_1,h_2,\ldots,h_q,\ldots)$ is full characteristic of nonlinear system **SYST**. Equation ( ) is also known as a Volterra series.

3.1 Generalized Volterra Convolutions and Correlations

By analogy with the classical high–order convolutions and correlations we introduce generalized T–stationary and D–stationary high–order convolutions and correlations.

Definition 9. *The following expressions*

$$y^{(q)}(t) = (h_q \Diamond^q x)(t) := \int\limits_{\sigma_1 \in \Omega} \cdots \int\limits_{\sigma_q \in \Omega} h_q(\sigma_1, \ldots, \sigma_q) \left[\prod_{i=1}^{q} x(t \boxminus \sigma_i) d\mu(\sigma_i) \right], \quad (29)$$

$$Y^{(p)}(\omega) = (H_p \heartsuit^p X)(\omega) := \int\limits_{\alpha_1 \in \Omega^*} \cdots \int\limits_{\alpha_p \in \Omega^*} H_p(\alpha_1, \ldots, \alpha_p) \left[\prod_{j=1}^{p} X(\omega \ominus \alpha_j) d\mu(\alpha_j) \right], \quad (30)$$

$$(f \clubsuit^q g)(\tau_1, \ldots, \tau_q) := \int\limits_{t \in \Omega} f(t) \left[\prod_{i=1}^{q} \bar{g}(t \boxminus \tau_i) \right] d\mu(t), \quad (31)$$

$$(F \spadesuit^p G)(\nu_1, \ldots, \nu_s) := \int\limits_{\omega \in \Omega^*} F(\omega) \left[\prod_{i=1}^{q} \bar{G}(\omega \ominus \nu_j) \right] d\mu(\omega) \quad (32)$$

are called the T–invariant q–th order time convolution (correlation) and D–invariant p–th order frequency convolution (correlation), respectively.

Obviously, nonlinear operators

$$y(t) = \mathbf{Volt}^{\Diamond}[\mathbf{h}; x](t) = \sum_{q=1}^{\infty} (h_q \Diamond^q x)(t),$$

$$Y(\omega) = \mathbf{Volt}^{\heartsuit}[\mathbf{H}; X](\omega) = \sum_{p=1}^{\infty} (H_p \heartsuit^p X)(\omega),$$

$$\mathbf{Voltcor}^{\clubsuit}[f; g](\tau_1, \ldots, \tau_q, \ldots) := \sum_{q=1}^{\infty} (f \clubsuit^q g)(\tau_1, \ldots, \tau_q),$$

$$\mathbf{VoltCor}^{\spadesuit}[F, G](\nu_1, \nu_2, \ldots, \nu_p, \ldots) := \sum_{p=1}^{\infty} (F \spadesuit^p G)(\nu_1, \nu_2, \ldots, \nu_s)$$

one can call *full T–invariant and D–invariant Volterra convolution and Volterra correlation operators*, where $\mathbf{h} := (h_1, ..., h_q, ...)$, $\mathbf{H} := (H_1, ..., H_q, ...)$. The T–invariant Volterra operators describe nonlinear T–stationary dynamic systems. The T–stationarity means the following. If $y(t)$ is the output of such system for the input signal $f(t)$, then signal $y(t \boxplus s)$ will be the output for $f(t \boxplus s)$: $y(t \boxplus s) = \mathbf{Volt}^{\Diamond}[\mathbf{h}; x(t \boxplus s)]$. Analogous statement is true and for D–invariant Volterra convolutions.

Signal and spectrum can are processed not only separately but also jointly giving nonlinear Volterra TF and FT distributions as a result.

Definition 10. *The following TF and FT distributions*

$$\mathbf{yY}^{(qp)}(t, \omega) = (\mathbf{hH}_{qp} \Diamond^q x \heartsuit^p X)(t, \omega) :=$$

$$= \int\limits_{\sigma_1\in\Omega} \cdots \int\limits_{\sigma_q\in\Omega} \int\limits_{\alpha_1\in\Omega^*} \cdots \int\limits_{\alpha_p\in\Omega^*} \mathbf{hH}_{qp}(\sigma_1,\ldots,\sigma_q;\alpha_1,\ldots,\alpha_p) \left[\prod_{i=1}^{q} x(t \boxminus \sigma_i)\ d\mu(\sigma_i)\right] \times$$

$$\times \left[\prod_{j=1}^{p} X(\omega \ominus \alpha_j) d\mu(\alpha_j)\right], \tag{33}$$

$$\mathbf{Yy}^{(pq)}(\nu,\tau) = \mathbf{Hh}_{pq}\heartsuit^p x \diamondsuit^q X)(\nu,\tau) :=$$

$$= \int\limits_{\alpha_1\in\Omega} \cdots \int\limits_{\alpha_p\in\Omega^*} \int\limits_{\sigma_1\in\Omega^*} \cdots \int\limits_{\sigma_q\in\Omega} \mathbf{Hh}_{pq}(\alpha_1,\ldots,\alpha_p;\sigma_1,\ldots,\sigma_q) \left[\prod_{j=1}^{p} X(\omega \ominus \alpha_j) d\mu(\alpha_j)\right] \times$$

$$\times \left[\prod_{i=1}^{q} x(t \boxminus \sigma_i)\ d\mu(\sigma_i)\right], \tag{34}$$

$$\mathbf{yY}_{qp}^{\clubsuit\spadesuit}(\tau_1,...,\tau_q;\nu_1,...,\nu_p) := (\mathbf{wW}\clubsuit^q f \spadesuit^p G)(\tau_1,...,\tau_q;\nu_1,...,\nu_p) =$$

$$= \int\limits_{t\in\Omega} \int\limits_{\omega\in\Omega^*} \mathbf{wW}(t,\omega) \left[\prod_{i=1}^{q} f(t \boxminus \tau_i)\right] \left[\prod_{j=1}^{p} G(\omega \ominus \nu_j)\right] d\mu(t) d\mu(\omega), \tag{35}$$

$$\mathbf{Yy}(\nu_1,...,\nu_p;\tau_1,...,\tau_q) = (\mathbf{Ww}\spadesuit^p F \clubsuit^q g)(\nu_1,...,\nu_s;\tau_1,...,\tau_q) :=$$

$$= \int\limits_{\omega\in\Omega^*} \int\limits_{t\in\Omega} \mathbf{Ww}(\omega,t) \left[\prod_{j=1}^{q} F(\omega \ominus \nu_j)\right] \left[\prod_{i=1}^{q} g(t \boxminus \tau_i)\right] d\mu(\omega) d\mu(t). \tag{36}$$

are called the $p+q$–th order TF and FT Volterra convolutions and correlations of signals and spectra, respectively.

Adding TF and FT Volterra convolutions and correlations of all orders we obtain full TF and FT Volterra convolutions

$$\mathbf{yY}(t,\omega) = \mathbf{Volt}^{\diamondsuit\heartsuit}\left[\widehat{\mathbf{hH}};\ x, X\right](t,\omega) = \sum_{q=1}^{\infty}\sum_{p=1}^{\infty}(\mathrm{hH}_{pq}\diamondsuit^q x \heartsuit^p X)(t,\omega),$$

$$\mathbf{Yy}(\nu,\tau) = \mathbf{Volt}^{\heartsuit\diamondsuit}\left[\widehat{\mathbf{Hh}};\ X, x\right](\nu,\tau) = \sum_{p=1}^{\infty}\sum_{q=1}^{\infty}(\mathbf{Hh}_{qp}\heartsuit^q X \diamondsuit^p x)(\nu,\tau),$$

$$\mathbf{Voltcor}^{\clubsuit\spadesuit}\left[\widehat{\mathbf{hH}}; f, G\right](\tau_1,\ldots,\tau_q,\ldots;\nu_1,\ldots,\nu_p,\ldots) :=$$

$$= \sum_{q=1}^{\infty}\sum_{p=1}^{\infty}(\mathbf{hH}\clubsuit^q f \spadesuit^p G)(\tau_1,\ldots,\tau_q;\nu_1,\ldots,\nu_p),$$

$$\mathbf{VoltCor}^{\spadesuit\clubsuit}[\widehat{\mathbf{Hh}}; F, g](\nu_1,\nu_2,\ldots,\nu_p,\ldots;\tau_1,\ldots,\tau_q,\ldots) :=$$

$$= \sum_{p=1}^{\infty}\sum_{q=1}^{\infty}(\mathbf{Hh}\spadesuit^p F\clubsuit^q g)(\nu_1,\nu_2,\ldots,\nu_s;\tau_1,\ldots,\tau_q)$$

where $\widehat{\mathbf{hH}} := [\mathbf{hH}_{pq}]$, $\widehat{\mathbf{Hh}} := [\mathbf{Hh}_{pq}]$ are infinity matrices.

Similarly, we can define TD–invariant Volterra convolutions of TF and FT distributions.

Definition 11. *The following functions are called the TD–invariant and DT–invariant p–th and q–th order Volterra convolutions of TF and FT distributions:*

$$\mathbf{yY}^{(p)}(t,\omega) := (\mathbf{hH}_{qq}(\diamondsuit\heartsuit)^q\mathbf{xX})(t,\omega) =$$

$$= \int\limits_{\sigma_1\in\Omega}\cdots\int\limits_{\sigma_q\in\Omega}\int\limits_{\alpha_1\in\Omega^*}\cdots\int\limits_{\alpha_q\in\Omega^*}\mathbf{hH}_r(\sigma_1,\ldots,\sigma_q;\alpha_1,\ldots,\alpha_q)\times$$

$$\times\left[\prod_{k=1}^{q}\mathbf{xX}(t\boxminus\sigma_k,\omega\ominus\alpha_k)\ d\mu(\sigma_k)d\mu(\alpha_k)\right], \tag{37}$$

$$\mathbf{Yy}^{(p)}(\nu,\tau) := (\mathbf{Hh}_p(\heartsuit\diamondsuit)^p\mathbf{Xx})(\nu,\tau) =$$

$$= \int\limits_{\alpha_1\in\Omega^*}\cdots\int\limits_{\alpha_p\in\Omega^*}\int\limits_{\sigma_1\in\Omega}\cdots\int\limits_{\sigma_p\in\Omega}\mathbf{Hh}_q(\alpha_1,\ldots,\alpha_p;\sigma_1,\ldots,\sigma_p)\times$$

$$\times\left[\prod_{k=1}^{p}\mathbf{Xx}(\nu\ominus\alpha_k,t\boxminus\sigma_k)d\mu(\alpha_k)\ d\mu(\sigma_k)\right], \tag{38}$$

$$\mathbf{yY}(\tau_1,\ldots,\tau_q;\nu_1,\ldots,\nu_p) = (\mathbf{fF}_{qp}\clubsuit^q\spadesuit^p\mathbf{gG})(\tau_1,\ldots,\tau_q;\nu_1,\ldots,u_p) :=$$

$$:= \int\limits_{t\in\Omega}\int\limits_{\omega\in\Omega^*}\mathbf{gG}(t,\omega)\mathbf{fF}(t\boxminus\tau_1,\ldots,t\boxminus\tau_q;\omega\ominus\nu_1,\ldots,\omega\ominus\nu_p)d\mu(t)d\mu(\omega),$$

$$\mathbf{Yy}(\nu_1,\ldots,\nu_p;\tau_1,\ldots,\tau_q) = (\mathbf{Ff}_{pq}\spadesuit^p\clubsuit^q\mathbf{Gg})(\nu_1,\ldots,\nu_p;\tau_1,\ldots,\tau_q) :=$$

$$:= \int\limits_{\omega\in\Omega^*}\int\limits_{t\in\Omega}\mathbf{Ff}(\omega,t)\mathbf{Gg}(\omega\ominus\nu_1,\ldots,\omega\ominus\nu_p;t\boxminus\tau_1,\ldots,q\boxminus\tau_s)d\mu(\omega)d\mu(t).$$

respectively, and

$$\mathbf{yY}(t,\omega) = \mathbf{Volt}^{\diamondsuit\heartsuit}[\mathbf{hH};\mathbf{xX}](t,\omega) = \sum_{p=1}^{\infty}[\mathbf{hH}_p(\diamondsuit\heartsuit)^p\mathbf{xX}](t,\omega),$$

$$\mathbf{Yy}(\nu,\tau) = \mathbf{Volt}^{\heartsuit\diamondsuit}[\mathbf{Hh};\mathbf{Xx}](\nu,\tau) = \sum_{k=1}^{\infty}[\mathbf{Hh}_q(\diamondsuit\heartsuit)^q\mathbf{Xx})(\nu,\tau),$$

$$\mathbf{Voltcor}^{\clubsuit\spadesuit}[\mathbf{fF},\mathbf{gG}](\tau_1,\ldots,\tau_q,\ldots;\nu_1,\ldots,\nu_p,\ldots) =$$

$$= \sum_{q=1}^{\infty}\sum_{p=1}^{\infty}(\mathbf{fF}_{qp}\clubsuit^q\spadesuit^p\mathbf{gG})(\tau_1,\ldots,\tau_r;\nu_1,\ldots,u_s)$$

$$\mathbf{VoltCOR}^{\spadesuit,\clubsuit}[\mathbf{Ff},\mathbf{Gg}](\nu_1,\ldots,\nu_p,\ldots;\tau_1,\ldots,\tau_q,\ldots) =$$

$$= \sum_{p=1}^{\infty}\sum_{q=1}^{\infty}(\mathbf{Ff}_{pq}\spadesuit^p\clubsuit^q\mathbf{Gg})(\nu_1,\ldots,\nu_p;\tau_1,\ldots,\tau_q)$$

will be called full $\mathcal{F}$-generalized Volterra convolutions and correlations TF and FT distributions,respectively.

3.2 Generalized Time–Frequency Higher–Order Distributions

In this section we propose a new higher order time–frequency distribution associated with arbitrary orthogonal Fourier transform: Higher–Order Generalized Ambiguity Functions (HOG–AF) and Higher–Order Generalized Wigner Distributions (HOG–WD). The HOG–WD is an q–th order uni–time/multi–frequency (UT/MF) distribution that is based on the q–th order time–varying moments of deterministic signals.

Definition 12. *The higher–order $\mathcal{F}$-generalized multi–frequency/uni–time and multi–time/uni–frequency symmetric and asymmetric cross– Wigner distribution are defined as the q–D Fourier $\mathcal{F}$– transforms of a q–th order symmetrical and asymmetrical local cross– correlation functions, respectively,*

$$\mathbf{Wd}_q^s[f,g](\omega_1,\ldots,\omega_q,t) := \underset{\tau_1\to\omega_1}{\mathcal{F}}\cdots\underset{\tau_q\to\omega_q}{\mathcal{F}}\{\mathbf{fg}^s(\tau_1,\ldots,\tau_q,t)\} =$$

$$= \int\limits_{\tau_1\in\Omega}\cdots\int\limits_{\tau_q\in\Omega}\left[f(t\boxminus\langle\tau\rangle)\prod_{i=1}^{q}(\mathcal{C}_i g)(t\boxminus\langle\tau\rangle\boxplus\tau_i)\right]\left[\prod_{i=1}^{q}\varphi_{\omega_i}(\tau_i)d\mu(\tau_i)\right],$$

$$\mathbf{wD}_p^s[F,G](t_1,\ldots,t_p,\omega) := \underset{\nu_1\to t_1}{\mathcal{F}^{-1}}\cdots\underset{\nu_q\to t_q}{\mathcal{F}^{-1}}\{\mathbf{FG}(\nu_1,\ldots,\nu_p,\omega)\} =$$

$$= \int\limits_{\nu_1\in\Omega^*}\cdots\int\limits_{\nu_p\in\Omega^*}\left[F(\omega\ominus\langle\nu\rangle)\prod_{j=1}^{p}(\mathcal{C}_j G)(\omega\ominus\langle\nu\rangle\ominus\nu_j)\right]\left[\prod_{j=1}^{p}\varphi_{\nu_j}(\tau_j)d\nu_j\right],$$

where $\langle\tau\rangle := \frac{1}{q+1}\sum_{i=1}^{q}\tau_i$, $\langle\nu\rangle := \frac{1}{p+1}\sum_{j=1}^{p}\nu_j$ are centered time and frequency, respectively, $\mathcal{C}_i$ is the i– th conjugation operators (it conjugates the signal or spectrum if the index i is even) and

$$\mathbf{Wd}_q^a[f,g](\omega_1,...,\omega_q,t) := \underset{\tau_1\to\omega_1}{\mathcal{F}}\cdots\underset{\tau_q\to\omega_q}{\mathcal{F}}\left\{f(t)\prod_{i=1}^{q}g(t\boxminus\tau_i)\right\} =$$

$$= f(t)\left[\prod_{i=1}^{q}\overline{G}(\omega_i)\right]\varphi_{\omega_1\oplus\ldots\oplus\omega_p}(t),$$

$$\mathbf{wD}_p^a[F,G](t_1,...,t_p,\omega) := \mathop{\mathcal{F}^{-1}}_{\nu_1\to t_1}\cdots\mathop{\mathcal{F}^{-1}}_{\nu_q\to t_q}\left\{F(\omega)\prod_{j=1}^{p}G(\omega\ominus\nu_j)\right\} =$$

$$= F(\omega)\left[\prod_{j=1}^{p}\overline{g}(t_j)\right]\varphi_\omega(\tau_1\boxplus...\boxplus\tau_p).$$

Definition 13. *The higher–order $\mathcal{F}$-generalized multi–time/uni–fre quency and multi–frequency/uni–time symmetric and asymmetric ambiguity function are defined as the 1–D Fourier $\mathcal{F}$– transform of q–th order symmetrical and asymmetrical local cross–correlation functions, respectively,*

$$\mathbf{aF}_q^s[f,g](\tau_1,\ldots,\tau_q,\nu) := \mathop{\mathcal{F}}_{t\to\nu}\{\mathbf{fg}^s(\tau_1,\ldots,\tau_q,t)\} =$$

$$= \int\limits_{t\in\Omega} f(t\boxminus\langle\tau\rangle)\left[\prod_{i=1}^{q}(\mathcal{C}_i g)(t\boxminus\langle\tau\rangle\boxminus\tau_i)\right]\overline{\varphi}_\nu(t)d\mu(t),$$

$$\mathbf{Af}_p^s[F,G](\nu_1,\ldots,\nu_p,\tau) := \mathop{\mathcal{F}^{-1}}_{\omega\to\tau}\{\mathbf{FG}^s(\nu_1,\ldots,\nu_p,\omega)\} =$$

$$= \int\limits_{\omega_1\in\Omega^*} F(\omega\ominus\langle\nu\rangle)\left[\prod_{j=1}^{p}(\mathcal{C}G)_j(\omega\ominus\langle\nu\rangle\ominus\nu_j)\right]\varphi_\omega(\tau)d\mu(\omega),$$

where $\langle\tau\rangle := \frac{1}{q+1}\sum_{i=1}^{q}\tau_i$, $\langle\nu\rangle := \frac{1}{p+1}\sum_{j=1}^{p}\nu_j$ *are centered time and frequency, respectively, and*

$$\mathbf{aF}_q^a[f,g](\tau_1,...,\tau_q,\nu) := \mathop{\mathcal{F}}_{t\to\nu}\{\mathbf{fg}^a(\tau_1,\ldots,\tau_q,t)\} =$$

$$= \int\limits_{t\in\Omega} f(t)\left[\prod_{i=1}^{q}g(t\boxminus\tau_i)\right]\varphi_\nu(t)d\mu(t),$$

$$\mathbf{Af}_p^a[F,G](\nu_1,...,\nu_p,\tau) := \mathop{\mathcal{F}^{-1}}_{\omega\to\tau}\{\mathbf{FG}^a(\nu_1,\ldots,\nu_p,\omega)\} =$$

$$= \int\limits_{\omega\in\Omega^*} F(\omega)\left[\prod_{j=1}^{p}\overline{G}(\omega\ominus\nu_j)\right]\varphi_\omega(\tau)d\mu(\omega).$$

Fig. 2 contains block diagrams relating different generalized higher–order $(q+1)$D distributions. We can obtain generalized Cohen's class distributions as generalized convolution of the Wigner distribution $\mathbf{Wd}(\omega,t)$ with $\mathbf{Co}(\omega,t)$

$$\mathbf{Ft}(\omega,t) := \int\limits_{\nu\in\Omega^*}\int\limits_{\tau\in\Omega}\mathbf{Co}(\nu,\tau)\mathbf{Wd}(\omega\ominus\nu,t\boxminus\tau)d\mu(\nu)d\mu(\tau).$$

The purpose of the Cohen's kernel $\mathbf{Co}(\omega,t)$ as in the classical case is to filter out cross terms and maintain the resolution of the auto terms.

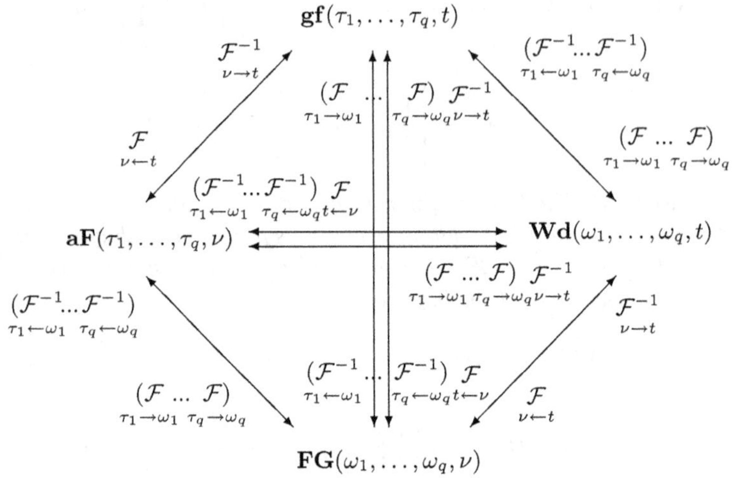

Fig. 1. Diagram of relations between the different generalized higher–order $(q+1)$D distributions

4 Conclusion

In this paper we have examined the idea of a generalized shift operator, associated with an arbitrary orthogonal transform and generalized linear and nonlinear convolutions based on these generalized shift operators. Such operators permit unify and generalize the majority of known methods and tools of signal processing based on classical Fourier transform.

References

1. Labunets, V. G. (1984): *Algebraic Theory of Signals and Systems,* (Russian), Krasnojarsk State University, Krasnojarsk
2. Labunets, V. G. (1986): Double orthogonal functions in generalized harmonic analysis. (Digital methods in control, radar and telecommunication systems). (Russian), Urals State Technical University, Sverdlovsk, 4–15
3. Weyl, H. (1931): The Theory Group and Quantum Mechanics, London: Methuen
4. Wigner, E. R. (1932): On the quantum correction for thermo–dynamic equilibrium. Physics Review, **40**, 749–759
5. Ville, J., (1948): Theorie et Applications de la Notion de Signal Analytique. Gables et Transmission. **2A**, 61– 74
6. Woodward, P. M. (1951): Information theory and design of radar receivers. Proceedings of the Institute of Radio Engineers, **39**, pp. 1521–1524
7. Volterra, V. (1959): Theory of Functionals and of Integral and Integro–Differential Equations. Dover Publications. New York

Gabor-Space Geodesic Active Contours

Chen Sagiv[1], Nir A. Sochen[1,2], and Yehoshua Y. Zeevi[1]

[1] Department of Electrical Engineering, Technion - Israel Institute of Technology
Technion City, Haifa 32000, Israel
chen@tiger.technion.ac.il, zeevi@ee.technion.ac.il
[2] Department of Applied Mathematics, University of Tel Aviv
Ramat-Aviv, Tel-Aviv 69978, Israel
sochen@math.tau.ac.il

Abstract. A novel scheme for texture segmentation is presented. Our algorithm is based on generalizing the intensity-based geodesic active contours model to the Gabor spatial-feature space of images. First, we apply the Gabor-Morlet transform to the image using self similar Gabor functions, and then implement the geodesic active snakes mechanism in this space. The spatial-feature space is represented, via the Beltrami framework, as a Riemannian manifold. The stopping term, in the geodesic snake mechanism, is generalized and is derived from the metric of the Gabor spatial-feature manifold. Experimental results obtained by applying the scheme to test images are presented.

Keywords: Texture segmentation, Gabor analysis, Geodesic active contours, Beltrami framework, Anisotropic diffusion, image manifolds.

1 Introduction

Image segmentation is an important issue in image analysis. Usually it is based on intensity features, e.g. gradients. However, real life images usually contain additional features such as textures and colors that determine image structure. In order to achieve texture segmentation (detecting the boundary between textural homogeneous regions), it is necessary to generalize the definition of segmentation to features other than intensity.

Since real world textures are difficult to model mathematically, no exact definition for texture exists. Therefore, ad-hoc approaches to the analysis of texture have been used, including local geometric primitives [8], local statistical features [3] and random field models [7,4]. A more general theory, based on the human visual system has emerged, in which texture features are extracted using Gabor filters [20].

The motivation for the use of Gabor filters in texture analysis is double folded. First, it is believed that simple cells in the visual cortex can be modeled by Gabor functions [16,5], and that the Gabor scheme provides a suitable representation for visual information in the combined frequency-position space [19]. Second, the Gabor representation has been shown to be optimal in the sense of minimizing the joint two-dimensional uncertainty in the combined spatial-frequency

G. Sommer and Y. Y. Zeevi (Eds.): AFPAC 2000, LNCS 1888, pp. 309–318, 2000.

space [6]. The analysis of Gabor filters was generalized to multi-window Gabor filters [23] and to Gabor-Morlet wavelets [19,23,17,12], and studied both analytically and experimentally on various classes of images [23]. A first attempt to use the Gabor feature space for segmentation was done by Lee et al [13] who use a variant of the Mumford-Shah functional adapted to some features in the Gabor space. Our method differs from theirs in using the entire information obtained by the Gabor analysis and in using a different segmentation technique.

In the last ten years, a great deal of attention was given to the "snakes", or active contours models which were proposed by Kaas et al [9] for intensity based image segmentation. In this framework an initial contour is deformed towards the boundary of an object to be detected. The evolution equation is derived from minimization of an energy functional, which obtains a minimum for a curve located at the boundary of the object.

The geodesic active contours model [2] offers a different perspective for solving the boundary detection problem; It is based on the observation that the energy minimization problem is equivalent to finding a geodesic curve in a Riemannian space whose metric is derived from image contents. The geodesic curve can be found via a geometric flow. Utilization of the Osher and Sethian level set numerical algorithm [21] allowed automatic handling of changes of topology.

It was shown recently that the Gaborian spatial-feature space can be described, via the Beltrami framework [22], as a 4D Riemannian manifold [11] embedded in $\mathbb{R}^6$. Based on this Riemannian structure we generalize the intensity based geodesic active contours method and apply it to the Gabor-feature space of images. Similar approaches, where the geodesic snakes scheme is applied to some feature space of the image, were studied by Lorigo et al [14] who used both intensity and its variance for MRI images' segmentation, and by Paragios et al [18] who generates the image's texture feature space by filtering the image using Gabor filters. Texture information is then expressed using statistical measurements. Texture segmentation is achieved by application of geodesic snakes to obtain the boundaries in the statistical feature space.

The aim of our study is to generalize the intensity-based geodesic active snakes method and apply it to the actual Gabor-feature space of images.

2 Geodesic Active Contours

In this section we review the geodesic active contours method for non-textured images [2]. The generalization of the technique for texture segmentation is described in section 4.

Let $\mathbf{C}(\mathbf{q}) : [0,1] \rightarrow \mathbb{R}^2$ be a parametrized curve, and let $I : [0,a] \times [0,b] \rightarrow \mathbb{R}^+$ be the given image. Let $E(r) : [0,\infty[\rightarrow \mathbb{R}^+$ be an inverse edge detector, so that E approaches zero when r approaches infinity. Visually, E should represent the edges in the image, so that we can judge the "quality" of the stopping term E by the way it represents the edges and boundaries in an image. Thus, the stopping term E has a fundamental role in the geodesic active snakes mechanism; if it does not well represents the edges, application of the snakes mechanism is likely to fail.

Minimizing the energy functional proposed in the classical snakes is generalized to finding a geodesic curve in a Riemannian space by minimizing:

$$L_R = \int E(|\nabla I(\mathbf{C}(q))|)\,|\mathbf{C}'(q)|dq. \tag{1}$$

We may see this term as a weighted length of a curve, where the Euclidean length element is weighted by $E(|\nabla I(C(q))|)$. The latter contains information regarding the boundaries in the image. The resultant evolution equation is the gradient descent flow:

$$\frac{\partial \mathbf{C}(t)}{\partial t} = E(|\nabla I|)k\mathbf{N} - (\nabla E \cdot \mathbf{N})\,\mathbf{N} \tag{2}$$

where k denotes curvature.

If we now define a function U, so that $\mathbf{C} = ((x,y)|U(x,y) = 0)$, we may use the Osher-Sethian Level-Sets approach [21] and replace the evolution equation for the curve $\mathbf{C}$, with an evolution equation for the embedding function U:

$$\frac{\partial U(t)}{\partial t} = |\nabla U|\mathrm{Div}\left(E(|\nabla I|)\frac{\nabla U}{|\nabla U|}\right). \tag{3}$$

A popular choice for the stopping function $E(|\nabla I|)$ is given by:

$$E(I) = \frac{1}{1+|\nabla I|^2}.$$

3 Feature Space and Gabor Transform

The Gabor scheme and Gabor filters have been studied by numerous researchers in the context of image representation, texture segmentation and image retrieval. A Gabor filter centered at the 2D frequency coordinates (U, V) has the general form of:

$$h(x,y) = g(x',y')\exp(2\pi i(Ux + Vy)) \tag{4}$$

where

$$(x',y') = (x\cos(\phi) + y\sin(\phi), -x\sin(\phi) + y\cos(\phi)), \tag{5}$$

and

$$g(x,y) = \frac{1}{2\pi\sigma^2}\exp\left(-\frac{x^2}{2\lambda^2\sigma^2} - \frac{y^2}{2\sigma^2}\right) \tag{6}$$

where, λ is the aspect ratio between x and y scales, σ is the scale parameter, and the major axis of the Gaussian is oriented at angle ϕ relative to the x-axis and to the modulating sinewave gratings.

Accordingly, the Fourier transform of the Gabor function is:

$$H(u,v) = \exp\left(-2\pi^2\sigma^2((u'-U')^2\lambda^2 + (v'-V')^2)\right) \tag{7}$$

where, (u', v') and (U', V') are rotated frequency coordinates. Thus, $H(u, v)$ is a bandpass Gaussian with minor axis oriented at angle ϕ from the u-axis, and the radial center frequency F is defined by : $F = U^2 + V^2$, with orientation $\theta = \arctan(V/U)$. Since maximal resolution in orientation is wanted, the filters whose sine gratings are cooriented with the major axis of the modulating Gaussian are usually considered ($\phi = \theta$ and $\lambda > 1$), and the Gabor filter is reduced to: $h(x, y) = g(x', y')exp(2\pi iFx')$.

It is possible to generate Gabor-Morlet wavelets from a single mother-Gabor-wavelet by transformations such as: translations, rotations and dilations. We can generate, in this way, a set of filters for a known number of scales, S, and orientations K. We obtain the following filters for a discrete subset of transformations: $h_{mn}(x, y) = a^{-m}g(x', y')$, where (x', y') are the spatial coordinates rotated by $\frac{\pi n}{K}$ and $m = 0...S - 1$. Alternatively, one can obtain Gabor wavelets by logarithmicaly distorting the frequency axis [19] or by incorporating multiwindows [23]. In the latter case one obtains a more general scheme wherein subsets of the functions constitute either wavelet sets or Gaborian sets.

The feature space of an image is obtained by the inner product of this set of Gabor filters with the image:

$$W_{mn}(x, y) = R_{mn}(x, y) + iJ_{mn}(x, y) = I(x, y) * h_{mn}(x, y). \tag{8}$$

4 Application of Geodesic Snakes to the Gaborian Feature Space of Images

The proposed approach enables us to use the geodesic snakes mechanism in the Gabor spatial feature space of images by generalizing the inverse edge indicator function E, which attracts in turn the evolving curve towards the boundary in the classical and geodesic snakes schemes. A special feature of our approach is the metric introduced in the Gabor space, and used as the building block for the stopping function E in the geodesic active contours scheme.

Sochen et al [22] proposed to view images and image feature space as Riemannian manifolds embedded in a higher dimensional space. For example, a gray scale image is a 2-dimensional Riemannian surface (manifold), with (x, y) as local coordinates, embedded in $\mathbb{R}^3$ with (X, Y, Z) as local coordinates. The embedding map is $(X = x, Y = y, Z = I(x, y))$, and we write it, by abuse of notations, as (x, y, I). When we consider feature spaces of images, e.g. color space, statistical moments space, and the Gaborian space, we may view the image-feature information as a N-dimensional manifold embedded in a $N + M$ dimensional space, where N stands for the number of local parameters needed to index the space of interest and M is the number of feature coordinates. For example, we may view the Gabor transformed image as a 2D manifold with local coordinates (x,y) embedded in a 6D feature space. The embedding map is $(x, y, \theta(x, y), \sigma(x, y), R(x, y), J(x, y))$, where R and J are the real and imaginary parts of the Gabor transform value, and θ and σ as the direction and scale for which a maximal response has been achieved. Alternatively, we can represent the transform space as a 4D manifold with coordinates (x, y, θ, σ) embedded in the same 6D feature space. The embedding map, in this case, is $(x, y, \theta, \sigma, R(x, y, \theta, \sigma), J(x, y, \theta, \sigma))$. The main difference between the two approaches is whether θ and σ are considered to be local coordinates or feature coordinates. In any case, these manifolds can evolve in their embedding spaces via some geometric flow.

A basic concept in the context of Riemannian manifolds is distance. For example, we take a two-dimensional manifold Σ with local coordinates (σ_1, σ_2). Since the local coordinates are curvilinear, the distance is calculated using a positive definite symmetric bilinear form called the metric whose components are denoted by $g_{\mu\nu}(\sigma_1, \sigma_2)$:

$$ds^2 = g_{\mu\nu} d\sigma^\mu d\sigma^\nu, \tag{9}$$

where we used the Einstein summation convention : elements with identical superscripts and subscripts are summed over.

The metric on the image manifold is derived using a procedure known as pullback. The manifold's metric is then used for various geometrical flows. We shortly review the pullback mechanism. More detailed information can be found in [22].

Let $X : \Sigma \to M$ be an embedding of Σ in M, where M is a Riemannian manifold with a metric h_{ij} and Σ is another Riemannian manifold. We can use the knowledge of the metric on M and the map X to construct the metric on Σ. This pullback procedure is as follows:

$$(g_{\mu\nu})_\Sigma(\sigma^1, \sigma^2) = h_{ij}(X(\sigma^1, \sigma^2)) \frac{\partial X^i}{\partial \sigma^\mu} \frac{\partial X^j}{\partial \sigma^\nu}, \tag{10}$$

where we used the Einstein summation convention, $i, j = 1, \ldots, dim(M)$, and σ^1, σ^2 are the local coordinates on the manifold Σ.

If we pull back the metric of a 2D image manifold from the Euclidean embedding space (x,y,I) we get:

$$(g_{\mu\nu}(x, y)) = \begin{pmatrix} 1 + I_x^2 & I_x I_y \\ I_x I_y & 1 + I_y^2 \end{pmatrix}. \tag{11}$$

The determinant of $g_{\mu\nu}$ yields the expression : $1 + {I_x}^2 + {I_y}^2$. Thus, we can rewrite the expression for the stopping term E in the geodesic snakes mechanism as follows:

$$E(|\nabla I|) = \frac{1}{1 + |\nabla I|^2} = \frac{1}{\det(g_{\mu\nu})}.$$

We may interpret the Gabor transform of an image as a function assigning for each pixel's coordinates, scale and orientation, a value (W). Thus, we may view the Gabor transform of an image as a 4D manifold with local coordinates (x, y, θ, σ) embedded in $\mathbb{R}^6$ of coordinates $(x, y, \theta, \sigma, R, J)$. We may pull back the metric for the 4D manifold from the 6D space, and use it to generate the stopping function E for the geodesic snakes mechanism. The metric derived for the 4D manifold is:

$$(g_{\mu\nu}) = \begin{pmatrix} 1+R_x^2+J_x^2 & R_xR_y+J_xJ_y & R_xR_\theta+J_xJ_\theta & R_xR_\sigma+J_xJ_\sigma \\ R_xR_y+J_xJ_y & 1+R_y^2+J_y^2 & R_yR_\theta+J_yJ_\theta & R_yR_\sigma+J_yJ_\sigma \\ R_xR_\theta+J_xJ_\theta & R_yR_\theta+J_yJ_\theta & 1+R_\theta^2+J_\theta^2 & R_\theta R_\sigma+J_\theta J_\sigma \\ R_xR_\sigma+J_xJ_\sigma & R_yR_\sigma+J_yJ_\sigma & R_\theta R_\sigma+J_\theta J_\sigma & 1+R_\sigma^2+J_\sigma^2 \end{pmatrix} \quad (12)$$

The resulting stopping function E is the inverse of the determinant of $g_{\mu\nu}$. Here $g_{\mu\nu}$ is a function of four variables (x, y, θ and σ), therefore, we obtain an evolution of a 4D manifold in a 6D embedding space.

Alternative approach is to derive a stopping term E which is a function of x and y only. One way to achieve this is to get the scale and orientation for which we have received the maximum amplitude of the transform for each pixel. Thus, for each pixel, we obtain: W_{max}, the maximum value of the transform, θ_{max} and σ_{max} – the orientation and scale that yielded this maximum value. This approach results in a 2D manifold (with local coordinates (x, y)) embedded in a 6D space (with local coordinates $(x, y, R(x, y), J(x, y), \theta(x, y), \sigma(x, y))$. If we use the pullback mechanism described above we get the following metric:

$$(g_{\mu\nu}) = \begin{pmatrix} 1+R_x^2+J_x^2+\sigma_x^2+\theta_x^2 & R_xR_y+J_xJ_y+\sigma_x\sigma_y+\theta_x\theta_y \\ R_xR_y+J_xJ_y+\sigma_x\sigma_y+\theta_x\theta_y & 1+R_y^2+J_y^2\sigma_x^2+\theta_x^2 \end{pmatrix} \quad (13)$$

Again, we use the fact that the determinant of the metric is a positive definite edge indicator to determine E as the inverse of the determinant of $g_{\mu\nu}$. Here $g_{\mu\nu}$ is a function of the two spatial variables only x and y, therefore, we obtain an evolution of a 2D manifold in a 6D embedding space.

5 Results and Discussion

Geodesic snakes provide an efficient geometric flow scheme for boundary detection, where the initial conditions include an arbitrary function U which implicitly represents the curve, and a stopping term E which contains the information regarding the boundaries in the image. Gabor filters are optimally tuned to localized scale and orientation, and can therefore represent textural information. We actually generalize the definition of gradients which usually refers to intensity gradients over (x, y) to other possible gradients in scale and orientation. This gradient information is the input function E to the newly generalized geodesic snakes flow.

In our application of geodesic snakes to textural images, we have used the mechanism offered by [15] to generate the Gabor wavelets for five scales and four orientations in a frequency range of $0.1 - 0.4$ cycles per pixel. We note that this choice is different from the usual scheme in vision, where there are four scales and at least six orientations in use. In the geodesic snakes mechanism U was initiated to be a signed distance function [2].

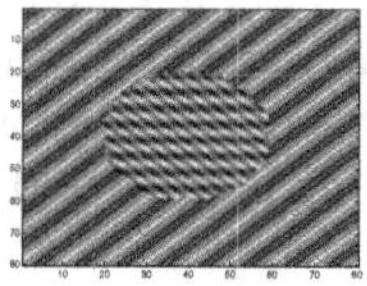

Fig. 1. A synthetic image made up of 2D sinewave gratings of different frequencies and orientations

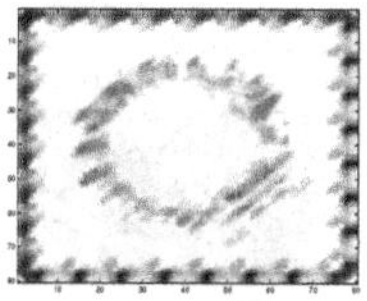

Fig. 2. The stopping function E calculated by means of the 2D manifold metric

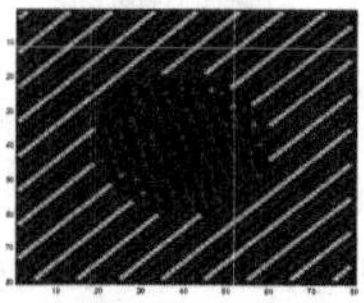

Fig. 3. The stopping function E of the first image calculated by using the intensity based definition $E(I) = \frac{1}{1+|\nabla I|^2}$

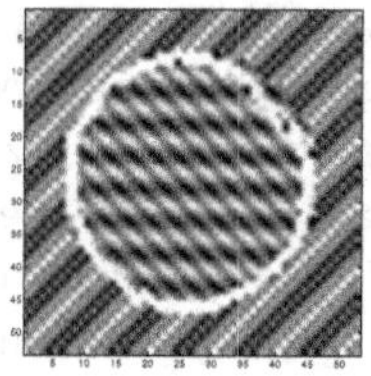

Fig. 4. The resultant boundary

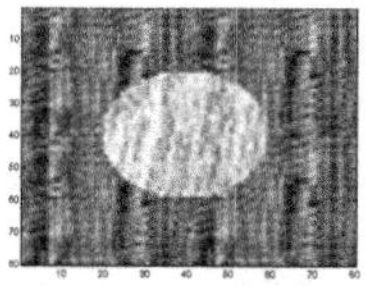

Fig. 5. An image comprised of two textures are taken from Brodatz album of textures [1]

We present the results of the 2D manifold approach for a synthetic image: the original image, the resulting stopping term E and the final boundary detected. We present some preliminary results for the 4D manifold approach with a Brodatz image: the resultant E is projected on the X-Y plane for each scale and orientation.

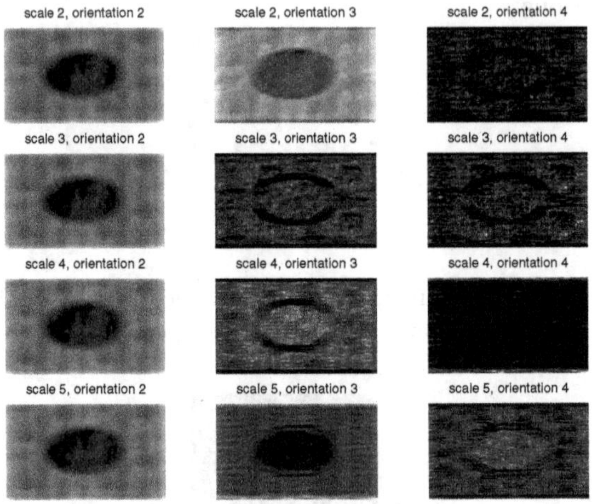

Fig. 6. The stopping function E for the Brodatz texture image, calculated by using the 4D manifold metric. For full size images see the web-page: **http://www-visl.technion.ac.il/gaborsnakes**

The first image (Fig. 1) is a synthesized texture composed of linear combination of spatial sinewave gratings of different frequencies and orientations. When the stopping term E is calculated using the $2D$ manifold metric, we obtain a clear picture of the texture gradients (i.e. where significant changes in texture occur) in the image (Fig. 2). So, our initial contour is drawn to the wanted boundary. As can be seen in figure (3), when E is calculated using intensity values only, $E(I) = \frac{1}{1+|\nabla I|^2}$, the texture gradients are not visible, and the resultant E will probably not attract the initial contour towards the boundary. Application of the geodesic snakes algorithm using the 2D manifold approach results in an accurate boundary, as can be seen in figure (4).

When we consider the entire Gabor spatial feature space, the stopping term E is a function of four variables x, y, θ, and σ. In more complex (texture-wise) images such as the Brodatz textures (Fig. 5), taken from [1], we may see the additional information that can be obtained. In figure (6) we present E as calculated for five scales and four orientations; however, only the components containing significant information is presented in the figure. We can see that information is preserved through scales. The E function contains more information when it is calculated by using the 4D manifold approach than the E function obtained by

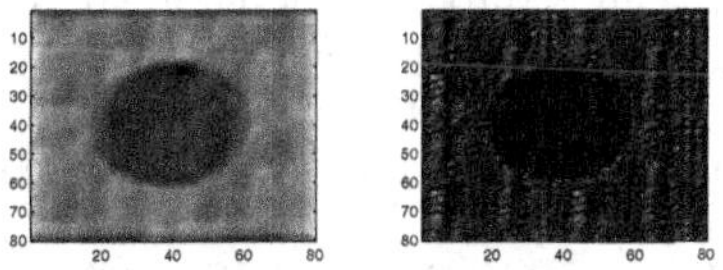

Fig. 7. The stopping function E used in the application of the proposed scheme to the Brodatz image using: (left) the 4D manifold approach incorporating specific scale and orientation (right) the intensity based definition $E(I) = \frac{1}{1+|\nabla I|^2}$

using the intensity based approach (Fig.7). In other words, we obtain a clear division of the image into two segments, which differ in their texture, and thereby get information about the relevant edges. As our main goal is to determine the boundaries in the image, we may deconvolve E for each scale and orientation with an appropriate gaussian function in order to obtain better spatial resolution.

The proposed texture segmentation scheme applies the geodesic active contours algorithm to the Gabor space of images, while the original geodesic snakes implements intensity gradients. The implementation of the feature space of images results in detection of texture gradients. We treat the Gabor transformed image as a 2D manifold embedded in a 6D space, or a 4D manifold embedded in a 6D space, and calculate the local metric on the manifold using the pull-back method. We then integrate the metric information to the geodesic snakes scheme. We have shown the feasibility of the proposed approach, and its advantages over the intensity geodesic snakes applied to multi-textured images. This is currently further extended by completing the application of geodesic snakes to a 4D manifold (x,y,θ,σ) embedded in a 6D space (R,J,x,y,θ,σ), and by the application of both schemes to medical images.

Acknowledgements

This research has been supported in part by the Ollendorff Minerva Center of the Technion, by the Fund for Promotion of Research at the Technion, and by the Consortium for Broadband Communication, administrated by the Chief Scientist of the Israeli Ministry of Industry and Commerce.

References

1. P. Brodatz, Textures: A photographic album for Artists and Designers, New York, NY, Dover, 1996.
2. V. Caselles and R. Kimmel and G. Sapiro, "Geodesic Active Contours", *International Journal of Conputer Vision*, 22(1), 1997, 61-97.
3. R. Conners and C. Harlow "A Theoretical Comparison of Texture Algorithms"*IEEE Transactions on PAMI*, 2, 1980, 204-222.

4. G. R. Cross and A. K. Jain, "Markov Random Field Texture Models" *IEEE Transactions on PAMI*, 5, 1983, 25-39.
5. J. G. Daugman, "Uncertainty relation for resolution in space, spatial frequency, and orientation optimized by two-dimensinal visual cortical filters", *J. Opt. Soc. Amer.* 2(7), 1985, 1160-1169.
6. D. Gabor "Theory of communication" *J. IEEE* , 93, 1946, 429-459.
7. S. Geman and D. Geman "Stochastic relaxation, Gibbs distribution and the Bayesian restoration of images", *IEEE Transactions on PAMI*, 6, 1984, 721-741.
8. B. Julesz "Texton Gradients: The Texton Theory Revisited", *Biol Cybern*, 54, (1986) 245-251.
9. M. Kaas, A. Witkin and D. Terzopoulos, "Snakes : Active Contour Models", *International Journal of Computer Vision*, 1, 1988, 321-331.
10. R. Kimmel, R. Malladi and N. Sochen, "Images as Embedding Maps and Minimal Surfaces: Movies, Color, Texture, and Volumetric Medical Images",*Proc. of IEEE CVPR'97*, (1997) 350-355.
11. R. Kimmel, N. Sochen and R. Malladi, "On the geometry of texture", Proceedings of the 4th International conference on Mathematical Methods for Curves and Surfaces, St. Malo, 1999.
12. T. S. Lee, "Image Representation using 2D Gabor-Wavelets", *IEEE Transactions on PAMI*, 18(10), 1996, 959-971.
13. T. S. Lee, D. Mumford and A. Yuille, "Texture segmentation by minimizing vector-valued energy functionals: the coupled-membrane model", *Lecture Notes in Computer Science, Computer Vision ECCV 92*, 588, Ed. G. Sandini, Springer-Verlag, 165-173.
14. L. M. Lorigo, O. Faugeras, W. E. L. Grimson, R. Keriven, R. Kikinis, "Segmentation of Bone in Clinical Knee MRI Using Texture-Based Geodesic Active Contours", *Medical Image Computing and Computer-Assisted Intervention*, 1998, Cambridge, MA, USA.
15. B. S. Manjunath and W. Y. Ma, "Texture features browsing and retrieval of image data", *IEEE Transactions on PAMI*, 18(8), 1996, 837-842.
16. S. Marcelja, "Mathematical description of the response of simple cortical cells", *J. Opt. Soc. Amer.*, 70, 1980, 1297-1300.
17. J. Morlet, G. Arens, E. Fourgeau and D. Giard, "Wave propagation and sampling theory - part 2: sampling theory and complex waves", *Geophysics*, 47(2), 1982, 222 - 236.
18. N. Paragios and R. Deriche, "Geodesic Active Regions for Supervised Texture Segmentation", *Proceedings of International Conference on Computer Vision*, 1999, 22-25.
19. M. Porat and Y. Y. Zeevi, "The generalized Gabor scheme of image representation in biological and machine vision", *IEEE Transactions on PAMI*, 10(4), 1988, 452-468.
20. M. Porat and Y. Y. Zeevi, "Localized texture processing in vision: Analysis and synthesis in the gaborian space", *IEEE Transactions on Biomedical Engineering*, 36(1), 1989, 115-129.
21. S. J. Osher and J. A. Sethian, "Fronts propagating with curvature dependent speed: Algorithms based on Hamilton-Jacobi formulations", *J of Computational Physics*, 79, 1988, 12-49.
22. N. Sochen, R. Kimmel and R. Malladi , "A general framework for low level vision", *IEEE Trans. on Image Processing*, 7, (1998) 310-318.
23. M. Zibulski and Y. Y. Zeevi, "Analysis of multiwindow Gabor-type schemes by frame methods", *Applied and Computational Harmonic Analysis*, 4, 1997, 188-221.

Color Image Enhancement by a Forward-and-Backward Adaptive Beltrami Flow

Nir A. Sochen[1,2], Guy Gilboa[1], and Yehoshua Y. Zeevi[1]

[1] Department of Electrical Engineering, Technion – Israel Institute of Technology
Technion City, Haifa 32000, Israel
gilboa@tx.technion.ac.il
zeevi@ee.technion.ac.il

[2] Department of Applied Mathematics, University of Tel-Aviv
Ramat-Aviv, Tel-Aviv 69978, Israel
sochen@math.tau.ac.il

Abstract. The Beltrami diffusion-type process, reformulated for the purpose of image processing, is generalized to an adaptive forward-and-backward process and applied in localized image features' enhancement and denoising. Images are considered as manifolds, embedded in higher dimensional feature-spaces that incorporate image attributes and features such as edges, color, texture, orientation and convexity. To control and stabilize the process, a nonlinear structure tensor is incorporated. The structure tensor is locally adjusted according to a gradient-type measure. Whereas for smooth areas it assumes positive values, and thus the diffusion is forward, for edges (large gradients) it becomes negative and the diffusion switches to a backward (inverse) process. The resultant combined forward-and-backward process accomplishes both local denoising and feature enhancement.

Keywords: scale-space, image enhancement, color processing, Beltrami flow, anisotropic diffusion, inverse diffusion.

1 Introduction

Image denoising, enhancement and sharpening are important operations in the general fields of image processing and computer vision. The success of many applications, such as robotics, medical imaging and quality control depends in many cases on the results of these operations. Since images cannot be described as stationary processes, it is useful to consider local adaptive filters. These filters are best described as solutions of partial differential equations (PDE).

The application of PDE's in image processing and analysis starts with the linear scale-space approach [,] which applies the heat equation by considering the noisy image as an initial condition. The associated filter is a Gaussian with a time varying scale. Perona and Malik [ ] in their seminal contribution, generalized the heat equation to a non-linear diffusion equation where the diffusion coefficient depends upon image features i.e. edges. This work paved the

G. Sommer and Y. Y. Zeevi (Eds.): AFPAC 2000, LNCS 1888, pp. 319– , 2000.

way for a variety of PDE based methods that were applied to various problems in low-level vision (see [ ] for an excellent introduction and overview).

The Beltrami framework was recently proposed by Sochen et al [ ] as a viewpoint that unifies many different algorithms and offer new possibilities of definitions and solutions of various tasks. Images and other vision objects of interest such as derivatives, orientations, texture, sequence of images, disparity in stereo vision, optical flow and more, are described as embedded manifolds. The embedded manifold is equipped with a Riemannian structure i.e. a metric. The metric encodes the geometry of the manifold. Non-linear operations on these objects are done according to the local geometry of the specific object of interest. The iterative process is understood as an evolution of the manifold. The evolution is a consequence of a non-linear PDE. No global (timewise) kernels can be associated with these non-linear PDE's. Short time kernels for these processes were derived recently in [ ].

We generalize the works of Perona and Malik [ ], Sochen et al [ ] and Weickert [ ] and show how one can design a structure tensor that controls the non-linear diffusion process starting from the induced metric that is given in the Beltrami framework. The proposed structure tensor is non-definite positive or negative and switches between them according to image features. This results in a forward-and-backward diffusion flow. Different regions of the image are forward or backwards diffused according to the local geometry within a neighborhood. The adaptive property of the process, that expresses itself in the local decision on the *direction* of the diffusion and on its *strength*, is the main novelty of this paper.

2 A Geometric Measure on Embedded Maps

2.1 Images as Riemannian Manifolds

According to the geometric approach to image representation, images are considered to be two-dimensional Riemannian surfaces embedded in higher dimensional spatial-feature Riemannian manifolds [, , , , , ,]. Let σ^μ, $\mu = 1, 2$, be the local coordinates on the image surface and let X^i, $i = 1, 2, \ldots, m$, be the coordinates of the embedding space than the embedding map is given by

$$(X^1(\sigma^1, \sigma^2), X^2(\sigma^1, \sigma^2), \ldots, X^m(\sigma^1, \sigma^2)). \tag{1}$$

Riemannian manifolds are manifolds endowed with a bi-linear positive-definite symmetric tensor which constitutes a *metric*. Denote by $(\Sigma, (g_{\mu\nu}))$ the image manifold and its metric and by $(M, (h_{ij}))$ the spatial-feature manifold and its corresponding metric. The induced metric can be calculated by $g_{\mu\nu} = h_{ij}\partial_\mu X^i \partial_\nu X^j$. The map $\mathbf{X} : \Sigma \to M$ has the following weight [ ]

$$E[X^i, g_{\mu\nu}, h_{ij}] = \int d^2\sigma \sqrt{g} g^{\mu\nu} (\partial_\mu X^i)(\partial_\nu X^j) h_{ij}(\mathbf{X}), \tag{2}$$

where the range of indices is $\mu, \nu = 1, 2$, and $i, j = 1, \ldots, m = \dim M$, and we use the Einstein summation convention: identical indices that appear one up and one down are summed over. We denote by g the determinant of $(g_{\mu\nu})$ and by $(g^{\mu\nu})$ the inverse of $(g_{\mu\nu})$. In the above expression $d^2\sigma\sqrt{g}$ is an area element of the image manifold. The rest, i.e. $g^{\mu\nu}(\partial_\mu X^i)(\partial_\nu X^j)h_{ij}(\mathbf{X})$, is a generalization of L_2. It is important to note that this expression (as well as the area element) does not depend on the choice of local coordinates.

The feature evolves in a geometric way via the gradient descent equations

$$X_t^i \equiv \frac{\partial X^i}{\partial t} = -\frac{1}{2\sqrt{g}} h^{il} \frac{\delta E}{\delta X^l}. \tag{3}$$

Note that we used our freedom to multiply the Euler-Lagrange equations by a strictly positive function and a positive definite matrix. This factor is the simplest one that does not change the minimization solution while giving a reparameterization invariant expression. This choice guarantees that the flow is geometric and does not depend on the parameterization.

Given that the embedding space is Euclidean, the variational derivative of E with respect to the coordinate functions is given by

$$-\frac{1}{2\sqrt{g}} h^{il} \frac{\delta E}{\delta X^l} = \Delta_g X^i = \frac{1}{\sqrt{g}} \partial_\mu(\sqrt{g} g^{\mu\nu} \partial_\nu X^i), \tag{4}$$

where the operator that is acting on X^i in the first term is the natural generalization of the Laplacian from flat surfaces to manifolds. In terms of the formalism implemented in our study, this is called *the second order differential parameter of Beltrami* [ ], or in short *Beltrami operator.*

2.2 The Metric as a Structure Tensor

There has been a few works using anisotropic diffusion processes. Cottet and Germain [ ] used a smoothed version of the image to direct the diffusion, while Weickert [,] smoothed also the structure tensor $\nabla I \nabla I^T$ and then manipulated its eigenvalues to steer the smoothing direction. Eliminating one eigenvalue from a structure tensor, first proposed as a color tensor in [ ], was used in [ ], in which the tensors are not necessarily positive definite. While in [,], the eigenvalues are manipulated to result in a positive definite tensor. See also [ ], where the diffusion is in the direction perpendicular to the maximal gradient of the three color channels (this direction is different than that of [ ]).

Let us first show that the diffusion directions can be deduced from the smoothed metric coefficients $g_{\mu\nu}$ and may thus be included within the Beltrami framework under the right choice of directional diffusion coefficients.

The induced metric $(g_{\mu\nu})$ is a symmetric uniformly positive definite matrix that captures the geometry of the image surface. Let λ_1 and λ_2 be the largest and the smallest eigenvalues of $(g_{\mu\nu})$, respectively. Since $(g_{\mu\nu})$ is a symmetric positive matrix its corresponding eigenvectors u_1 and u_2 can be chosen orthonormal. Let $U \equiv (u_1|u_2)$, and $\Lambda \equiv \begin{pmatrix} \lambda_1 & 0 \\ 0 & \lambda_2 \end{pmatrix}$, then we readily have the equality

$$(g_{\mu\nu}) = U \Lambda U^T. \tag{5}$$

Note also that

$$(g^{\mu\nu}) \equiv (g_{\mu\nu})^{-1} = U\Lambda^{-1}U^T = U \begin{pmatrix} 1/\lambda_1 & 0 \\ 0 & 1/\lambda_2 \end{pmatrix} U^T, \tag{6}$$

and that

$$g \equiv \det(g_{\mu\nu}) = \lambda_1 \lambda_2. \tag{7}$$

Our proposed enhancement procedure will control those eigenvalues adaptively so that only meaningful edges will be enhanced, where smooth areas will be denoised.

3 New Adaptive Structure Tensor

3.1 Changing the Eigenvalues

From the above derivation of the metric $g_{\mu\nu}$, it follows that the larger eigenvalue λ_1 corresponds to the eigenvector in the gradient direction (in the 3D Euclidean case: (I_x, I_y)). The smaller eigenvalue λ_2 corresponds to the eigenvector perpendicular to the gradient direction (in the 3D Euclidean case: $(-I_y, I_x)$). The eigenvectors are equal for both $g_{\mu\nu}$ and its inverse $g^{\mu\nu}$, whereas the eigenvalues have reciprocal values. We can use the eigenvalues as a means to control the Beltrami flow process. For convenience let us define $\lambda^1 \equiv \frac{1}{\lambda_1}$. As the first eigenvalue of $g^{\mu\nu}$ (that is λ^1) increases, so does the diffusion force in the gradient direction. Thus, by changing this eigenvalue we can reduce, eliminate or even reverse the diffusion process across the gradient.

What would be the best strategy to control the diffusion process via adjustment of the relevant parameters ? There are a few requirements that might be considered as guidelines :

- The enhancement should essentially be with relevance to the important features, while originally smooth segments should not be enhanced.
- The contradictory processes of enhancement and noise reduction by smoothing (filtering) should coexist.
- The process should be as stable as possible, though restoration and enhancement processes are inherently unstable.

Let us define $\hat{\lambda^1}(s)$ as a new adaptive eigenvalue to be put instead of the original λ^1. We propose that this new eigenvalue will be proportional to the combined gradient magnitude of the three channels (colors) $|\nabla I_\Sigma|$ (that is $\hat{\lambda^1} = \hat{\lambda^1}(|\nabla I_\Sigma|)$ in the following way:

$$\hat{\lambda^1}(s) = \begin{cases} 1 - (s/k_f)^n & , 0 \le s \le k_f \\ \alpha\left[((s-k_b)/w)^{2m} - 1\right] & , k_b - w \le s \le k_b + w \\ 0 & , \text{otherwise} \end{cases} \tag{8}$$

and its smoothed version:

$$\hat{\lambda^1}_\sigma(s) = \hat{\lambda^1}(s) \star G_\sigma(s) \tag{9}$$

where $|\nabla I_\Sigma| \equiv \left(\Sigma_i |\nabla I_i|^2\right)^{1/2}$, $\star$ denotes convolution, and $k_f < k_b - w$. We chose the exponent parameters n and m to be 4 and 1, respectively.

The new structure tensor has to be continuous and differentiable. In the discrete domain, ( ) could suffice (although it is only piecewise differentiable), whereas ( ) can fit the general continuous case. Other types of eigenvalue manipulation with similar nature may be considered.

The parameter k_f is essentially the limit of gradients to be smoothed out, whereas k_b and w define the range of the backward diffusion, and should assume values of gradients that we want to emphasize. In our formula the range is symmetric , and we restrain the width from overlapping the forward diffusion area. One way of choosing these parameters in the discrete case, is by calculating the mean absolute gradient (MAG).

The parameter α determines the ratio between the backward and forward diffusion . Under the condition of α that renders the backward diffusion process to become too dominant, the stabilizing forward process can no longer avoid oscillations. One can avoid the evolution of new singularities in smooth areas by bounding the maximum flux resulting from the backward diffusion to be smaller than the maximum affected by the forward one. Formally, we say:

$$\max_{s<k_f} \{s\lambda(s)\} > \max_{k_b-w<s<k_b+w} \{s\lambda(s)\} \tag{10}$$

In the case of our proposed eigenvalue, we get a simple formula for α, which just obeys this inequality by:

$$\alpha = k_f/2k_b \qquad \text{,for any } 0 < w < k_b - k_f \tag{11}$$

In practical applications, this bound can be doubled in value without experiencing major instabilities.

See [ ] for elaboration on the forward and backward diffusion for signal enhancement.

3.2 The Algorithm

The algorithm to implement the flow $\mathbf{I}_t = \Delta_{\hat{g}}\mathbf{I}$ for color image enhancement is as follows:

1. Compute the metric coefficients $g_{\mu\nu}$. For the N channel case (for color $N=3$) we have

$$g_{\mu\nu} = \delta_{\mu\nu} + \sum_{k=1}^{N} I^k_\mu I^k_\nu. \tag{12}$$

2. Diffuse the $g_{\mu\nu}$ coefficients by convolving with a Gaussian of variance ρ, thereby

$$\tilde{g}_{\mu\nu} = G_\rho * g_{\mu\nu}. \tag{13}$$

For $2D$ images $G_\rho = e^{-(x^2+y^2)/\rho^2}$.
3. Compute the inverse smoothed metric $\tilde{g}^{\mu\nu}$. Change the eigenvalues of the inverse metric λ^1, λ^2, $(\lambda^1 < \lambda^2)$, of $(\tilde{g}^{\mu\nu})$ so that $\lambda^1 = \hat{\lambda^1}(s)$ and $\lambda_2 = a$, $(a \geq 1)$. This yields a new inverse structure tensor $\hat{g}_{\mu\nu}$ that is given by:

$$(\hat{g}_{\mu\nu}) = \tilde{U} \begin{pmatrix} \hat{\lambda^1}(s) & 0 \\ 0 & a \end{pmatrix} \tilde{U}^T = \tilde{U}\hat{\Lambda}\tilde{U}^T. \tag{14}$$

4. Calculate the determinant of the new structure tensor. Note that $\hat{g}$ can now have negative values. In cases where the inverse eigenvalue $\hat{\lambda^1}$ is zero, the structure tensor determinant should assume a large value $M >> 1$.
$\hat{g} \equiv \det(\hat{g}_{\mu\nu}) = \hat{\lambda_1}\hat{\lambda_2} = \frac{1}{\hat{\lambda^1}\hat{\lambda^2}}$,

$$\hat{g} = \begin{cases} 1/a\hat{\lambda^1}(s) & , \hat{\lambda^1}(s) \neq 0 \\ M & , \text{otherwise} \end{cases} \tag{15}$$

5. Evolve the k-th channel via the Beltrami flow

$$I_t^k = \Delta_{\hat{g}} I^k \equiv \frac{1}{\sqrt{\hat{g}}} \partial_\mu \left(\sqrt{\hat{g}} \hat{g}^{\mu\nu} \partial_\nu I^k \right) \tag{16}$$

Remark: In this flow, we will not get imaginary values, though we have the term $\sqrt{\hat{g}}$ because in cases of negative $\hat{g}$ the constant imaginary term $i \equiv \sqrt{-1}$ will be canceled.

3.3 Variations to the Scheme

As the process involves inverse diffusion for enhancement - it is by definition not stable. To obtain a more stable process, which will denoise the image and preserve its edges, setting $\alpha = 0$ will remove the inverse diffusion part, and leave us with a coherent denoising scheme.

There are a few ways to increase regularity in this PDE-based approach. One can replace the proposed conductance coefficient Eq. ( ) by the smoothed one, Eq. ( ). As presented in the algorithm, convolving the metric with a smoothing kernel, before manipulating it, increases the stability of the process. It is possible also to smooth smaller scales in a noisy signal by preprocessing. As we enhance the signal afterwards, this smoothing process does not affect the end result that much and enables us to operate in an originally much noisier environment. Finally, operating in extremely noisy areas, when we know of the type of singularity, we can apply more pre-smoothing, and consider only the largest gradient within the backward diffusion range.

We can substitute the dependency of $\hat{\lambda^1}$ instead of on the gradient, on similar "edge detectors":

$$\hat{\lambda^1} = \hat{\lambda^1}(g) \tag{17}$$

or the smoothed version:

$$\hat{\lambda^1} = \hat{\lambda^1}(G_\rho \star g) \tag{18}$$

or on the original eigenvalue itself :

$$\hat{\lambda^1} = \hat{\lambda^1}(\lambda^1) \tag{19}$$

A local approach, that adjusts the parameters k_f, k_b, w to be of different values in different segments of the image, is currently investigated.

4 Results and Conclusion

The image feature enhancement procedure developed in the framework of geometry, incorporates a nonlinear adaptive structure tensor that controls the enhancement process along gradients. In other words, the structure tensor is locally adjusted according to a gradient-type measure. Whereas for smooth areas it assumes positive values, and thus the diffusion is forward, for edges it becomes negative and the diffusion switches to a backward (inverse) process. In this way we accomplish both of the conflicting tasks of local denoising and feature enhancement.

In Figure the left eye of the Mandrill image is shown, before and after the application of the adaptive Beltrami process. It depicts efficient denoising of the retina, with sharp edges somewhat enhanced. In Figure a blurred and noisy Tulip photo is processed, enhancing the center of the flower while denoising its background. In a detail enlargement of the same image (the flower's pattern in Fig.) one can see more clearly that the bright curly outline of the leaf is enhanced (brighter in its center), whereas smooth areas are denoised.
[For a closer look at the color images, please follow the web link: *http://www-visl.technion.ac.il/belt-fab*] .

Lastly, note that the general scheme can be easily degenerated into a coherent stable denoising scheme that preserves edges.

Acknowledgement

This research has been supported in part by the Ollendorf Minerva Center, by the Fund for the Promotion of Research at the Technion, and by the Consortium for Broadband Communication administered by the Chief Scientist of the Israeli Ministry of Industrial and Commerce.

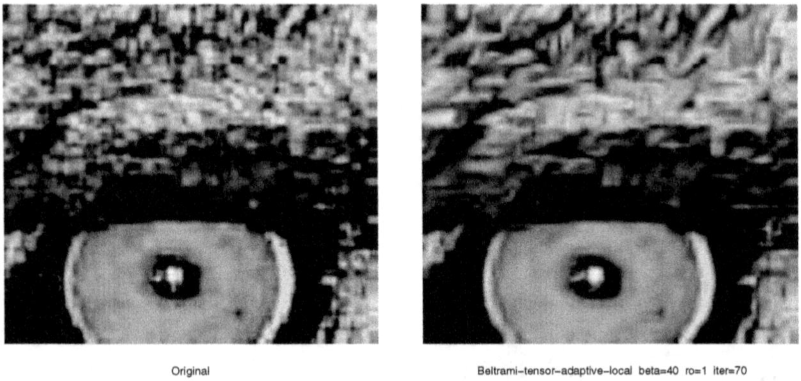

Fig. 1. Left - original eye image, right - enhanced and denoised eye image. 70 iterations, $[k_f, k_b, w] = [0.5, 4, 2] * MAG$, $\rho = 1$

Fig. 2. Left - original tulip, right - enhanced and denoised tulip. 40 iterations, $[k_f, k_b, w] = [0.7, 5, 3] * MAG$, $\rho = 1$

References

1. A. Chambolle, Partial Differential Equations and Image processing, Proceedings IEEE ICIP, 1 (1994) 16-20.
2. G. H. Cottet and L. Germain, Image processing through reaction combined with nonlinear diffusion, Math. Comp., 61 (1993) 659–673.
3. S. Di Zenzo, A note on the gradient of a multi image, Computer Vision, Graphics, and Image Processing, 33 (1986) 116–125.
4. G. Gilboa, N. Sochen, Y. Y. Zeevi, "Anisotropic selective inverse diffusion for signal enhancement in the presence of noise", to appear in IEEE ICASSP'2000, Istanbul, 2000.
5. R. Kimmel, N. Sochen and R. Malladi, "On the geometry of texture", Report, Berkeley Labs. UC, LBNL-39640, UC-405, November,1996.

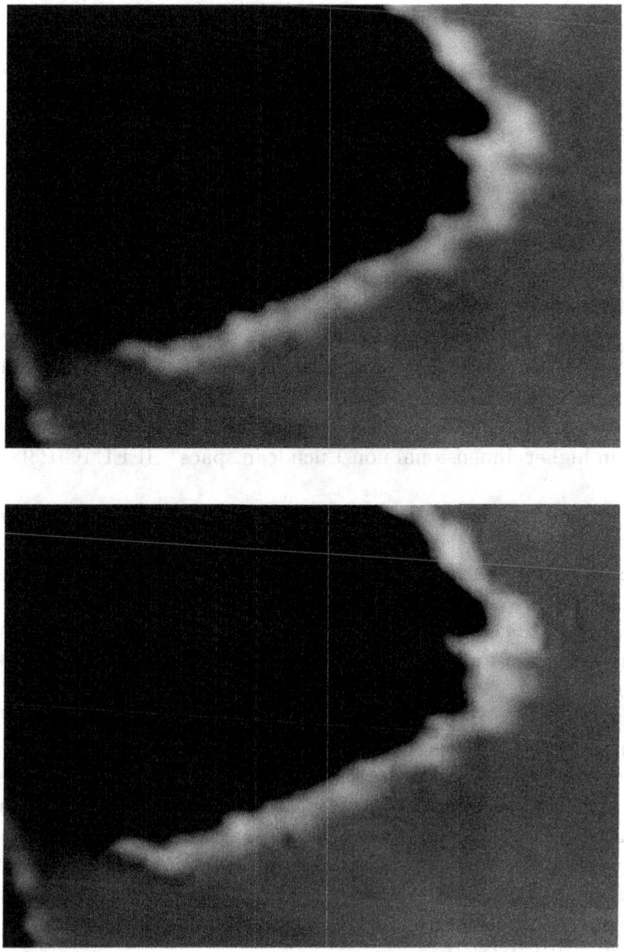

Fig. 3. Enlargement of Fig. – a pattern on the tulip: top – original, bottom – enhanced and denoised image. Note the white ridge at the center of the leaf's outline.

6. R. Kimmel, R. Malladi and N. Sochen, "Images as Embedding Maps and Minimal Surfaces: Movies, Color, Texture, and Volumetric Medical Images", *Proc. of IEEE CVPR'97*, (1997) 350-355.
7. R. Kimmel, N. Sochen and R. Malladi, "From High Energy Physics to Low Level Vision", *Lecture Notes In Computer Science:* 1252, First International Conference on Scale-Space Theory in Computer Vision, Springer-Verlag, 1997, 236-247.
8. J. J. Koenderink, "The structure of images", *Biol. Cybern.*, 50: 363-370, 1984.

9. E. Kreyszing, "Differential Geometry", Dover Publications, Inc., New York, 1991.

10. P. Perona and J. Malik, "Scale-space and edge detection using anisotropic diffusion", IEEE Trans. Pat. Anal. Machine Intel., vol. PAMI-12,no. 7, pp. 629-639, 1990.
11. A. M. Polyakov, "Quantum geometry of bosonic strings", *Physics Letters*, **103B** (1981) 207-210.
12. G. Sapiro and D. L. Ringach, Anisotropic Diffusion of multivalued images with applications to color filtering, IEEE Trans. Image Proc., 5 (1996) 1582-1586.
13. N. Sochen, R. Kimmel and R. Malladi, "A general framework for low level vision", *IEEE Trans. on Image Processing*, 7, (1998) 310-318.
14. N. Sochen and Y. Y. Zeevi, "Images as manifolds embedded in a spatial-feature non-Euclidean space", November 1998, EE-Technion report no. 1181.
15. N. Sochen, "Stochastic processes in vision I: From Langevin to Beltrami", EE-Technion technical report No. 245
16. N. Sochen and Y. Y. Zeevi, "Representation of colored images by manifolds embedded in higher dimensional non-Euclidean space", IEEE ICIP'98, Chicago, 1998.

17. B. M. ter Haar Romeny Ed., Geometry Driven Diffusion in Computer Vision, Kluwer Academic Publishers, 1994.
18. J. Weickert, "Coherence-enhancing diffusion of colour images", Image and Vision Comp., 17 (1999) 199-210.
19. J. Weickert, "Multiscale texture enhancement", Computer analysis of images and patterns; Lecture Notes in Computer Science, 970 (1995), pp. 230-237, Springer.

20. J. Weickert, "Anisotropic diffusion in image processing", Ph.D. Thesis, Kaiserslautern University, Germany, November 1995
21. J. Weickert, "Scale-space properties of nonlinear diffusion filtering with diffusion tensor", Report No. 110, Laboratory of Technomathematics, University of Kaiserslautern, 1994.
22. J. Weickert, Coherence-enhancing diffusion of colour images, Proc. VII National Symposium on Pattern Rec. and Image Analysis, Barcelona, 1 (1997) 239-244.
23. A. P. Witkin, "Scale space filtering", Proc. Int. Joint Conf. On Artificial Intelligence, pp. 1019-1023, 1983.

Point–Based Registration Assuming Affine Motion

Attila Tanács[1], Gábor Czédli[2], Kálmán Palágyi[1], and Attila Kuba[1]

[1] Department of Applied Informatics, University of Szeged,
H–6701 Szeged P.O.Box 652, Hungary
{tanacs,palagyi,kuba}@inf.u-szeged.hu
[2] Bolyai Institute, University of Szeged,
H–6720 Szeged, Aradi vértanúk tere 1, Hungary
czedli@math.u-szeged.hu

Abstract. Registration is a fundamental task in image processing. Its purpose is to find a geometrical transformation that relates the points of an image to their corresponding points of another image. The determination of the optimal transformation depends on the types of variations between the images. In this paper we propose a robust method based on two sets of points representing the images. One–to–one correspondence is assumed between these two sets. Our approach finds global affine transformation between the sets of points and can be used in any arbitrary dimension $k \geq 1$. A sufficient existence condition for a unique solution is given and proven. Our method can be used to solve various registration problems emerged in numerous fields, including medical image processing, remotely sensed data processing, and computer vision.

Keywords: registration problem; matching sets of points

1 Introduction

There is an increasing number of applications that require accurate aligning of one image with another taken from different viewpoints, by different imaging devices, or at different times. The geometrical transformation is to be found that maps a *floating image data set* in precise spatial correspondence with a *reference image data set*. This process of alignment is known as *registration*, although other words, such as *co–registration*, *matching*, and *fusion*, are also used. Examples of systems where image registration is a significant component include aligning images from different medical modalities for diagnosis, matching a target with a real–time image of a scene for target recognition, monitoring global land usage using satellite images, and matching stereo images to recover shape for autonomous navigation [,].

The registration technique for a given task depends on the knowledge about the characteristics of the type of variations. Registration methods can be viewed as different combinations of choices for the following four components [ ]:

G. Sommer and Y. Y. Zeevi (Eds.): AFPAC 2000, LNCS 1888, pp. 329– , 2000.

- *Search space* is determined by the type of transformation we have to consider, i.e., what is the class of transformations that is capable of aligning the images. Some widely used types are *rigid-body*, when translations and rotations are allowed only, *affine*, which maps parallel lines to parallel lines, and *nonlinear*, which can transform straight lines to curves.
- *Feature data set* describes what kind of image properties are used in matching.
- *Similarity measure* is a function of the transformation parameters which shows how well the floating and the reference image fit. The task of registration is to optimize this function.
- *Search strategy* determines what kind of optimization method to use.

Figure explains the major steps of a general registration process.

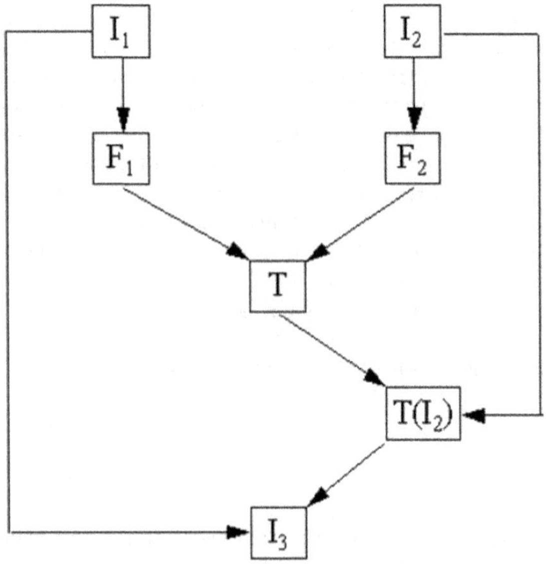

Fig. 1. Major steps of a general registration process. Feature data sets F_1 and F_2 are extracted from reference image I_1 and floating image I_2, respectively. Transformation T is calculated using F_1 and F_2. I_2 is aligned to I_1 by applying T. A brand new image I_3 can be calculated by fusing I_1 and $T(I_2)$

A general and robust solution for registration problems is selecting points as features. A general point–based method consists of three steps. First, the points are identified, then points in the floating image are corresponded with points in the reference image, finally a spatial mapping is determined. Point–based methods can be either *interactive* or *automatic*. Using an interactive point–based method, usually few pairs of points (4–20) are identified and corresponded by the user. Methods of this type are available for rigid–body [ ] and nonlinear [,] problems. Automatic determination of the features usually results huge amount of points. In this case finding correspondences can be rather difficult (e.g., the number of elements of the point sets is not necessarily the same) and require a special algorithm. Widely used methods are *head-hat* method [ ], *hierarchical Chamfer matching* [,], and *iterative closest point* [ ] method. These are used mainly for rigid–body problems, but extension to more general transformations is easy.

In this paper we propose an interactive point-based method.

2 Affine Method for Aligning Two Sets of Points

In this section we propose a robust method based on identified pairs of points, which assumes affine motion between the images. Let $k \geq 1$ denote the dimension of the images and let n be the number of pairs of points.

Our registration method is described by giving the following four components:

- *search space*
  Global transformation described by a $(k+1) \times (k+1)$ matrix $\mathcal{T}$ of the form

$$\mathcal{T} = \begin{pmatrix} t_{11} & t_{12} & \cdots & t_{1k} & t_{1,k+1} \\ t_{21} & t_{22} & \cdots & t_{2k} & t_{2,k+1} \\ \vdots & \vdots & \ddots & \vdots & \vdots \\ t_{k1} & t_{k2} & \cdots & t_{kk} & t_{k,k+1} \\ 0 & 0 & \cdots & 0 & 1 \end{pmatrix}$$

  is to be found. Given $\mathcal{T}$ and a point $x = (x_1, \ldots, x_k) \in \mathbb{R}^k$, the transformation sends x to $y = (y_1, \ldots, y_k) \in \mathbb{R}^k$ if and only if $(y_1, \ldots, y_k, 1)^T = \mathcal{T} \cdot (x_1, \ldots, x_k, 1)^T$ holds for the corresponding *homogeneous coordinates* [ ]. Notice that each affine transformation can be described this way (Fig.). This kind of transformation has $k \cdot (k+1)$ degrees of freedom according to the matrix elements to be determined.

- *feature data set*
  A set of n *reference points* $\{p_1, p_2, \ldots, p_n\}$, $p_i = (p_{i1}, \ldots, p_{ik}) \in \mathbb{R}^k$, and a set of n *floating points* $\{q_1, q_2, \ldots, q_n\}$, $q_i = (q_{i1}, \ldots, q_{ik}) \in \mathbb{R}^k$, are to be identified in the reference image and the floating image, respectively (Fig.). We assume that q_i is corresponded to p_i $(1 \leq i \leq n)$.

- *similarity measure*
  Suppose that we get point $\overline{q}_i = (\overline{q}_{i1}, \ldots, \overline{q}_{ik})$ when point q_i is transformed by matrix $\mathcal{T}$ $(1 \leq i \leq n)$:

$$\begin{pmatrix} \overline{q}_{i1} \\ \overline{q}_{i2} \\ \vdots \\ \overline{q}_{ik} \\ 1 \end{pmatrix} = \begin{pmatrix} t_{11} & t_{12} & \cdots & t_{1k} & t_{1,k+1} \\ t_{21} & t_{22} & \cdots & t_{2k} & t_{2,k+1} \\ \vdots & \vdots & \ddots & \vdots & \vdots \\ t_{k1} & t_{k2} & \cdots & t_{kk} & t_{k,k+1} \\ 0 & 0 & \cdots & 0 & 1 \end{pmatrix} \cdot \begin{pmatrix} q_{i1} \\ q_{i2} \\ \vdots \\ q_{ik} \\ 1 \end{pmatrix}.$$

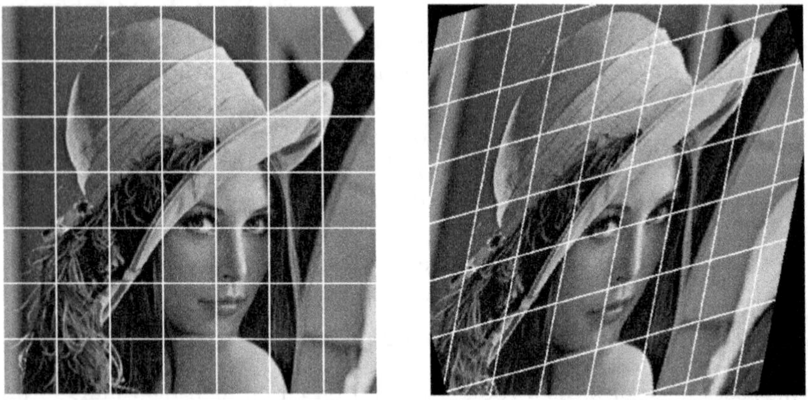

Fig. 2. Example of a 2D affine transformation: The original image (left) and the transformed one (right). Lines are mapped to lines, parallelism is preserved, but angles can be altered

Define the function $\mathcal{S}$ of $k \cdot (k+1)$ variables as follows:

$$\mathcal{S}(t_{11}, \ldots, t_{k,k+1}) = \sum_{i=1}^{n} \|\overline{q}_i - p_i\|^2 = \sum_{i=1}^{n}\sum_{j=1}^{k}(\overline{q}_{ij} - p_{ij})^2$$
$$= \sum_{i=1}^{n}\sum_{j=1}^{k}(t_{j1} \cdot q_{i1} + \ldots + t_{jk} \cdot q_{ik} + t_{j,k+1} - p_{ij})^2.$$

It can be regarded as the matching error.

– *search strategy*
The least square solution of matrix $\mathcal{T}$ is determined by minimizing function $\mathcal{S}$. Direct matching is applied. Function $\mathcal{S}$ may be minimal if all of the partial derivatives $\frac{\partial \mathcal{S}}{\partial t_{11}}, \ldots, \frac{\partial \mathcal{S}}{\partial t_{k,k+1}}$ are equal to zero. The required $k \cdot (k+1)$ equations:

$$\frac{\partial \mathcal{S}}{\partial t_{uv}} = 2 \cdot \sum_{i=1}^{n} q_{iv} \cdot (t_{u,k+1} - p_{iu} + \sum_{l=1}^{k} t_{ul} \cdot q_{il}) = 0$$
$$(1 \leq u, v \leq k),$$

$$\frac{\partial \mathcal{S}}{\partial t_{u,k+1}} = 2 \cdot \sum_{i=1}^{n}(t_{u,k+1} - p_{iu} + \sum_{l=1}^{k} t_{ul} \cdot q_{il}) = 0$$
$$(1 \leq u \leq k).$$

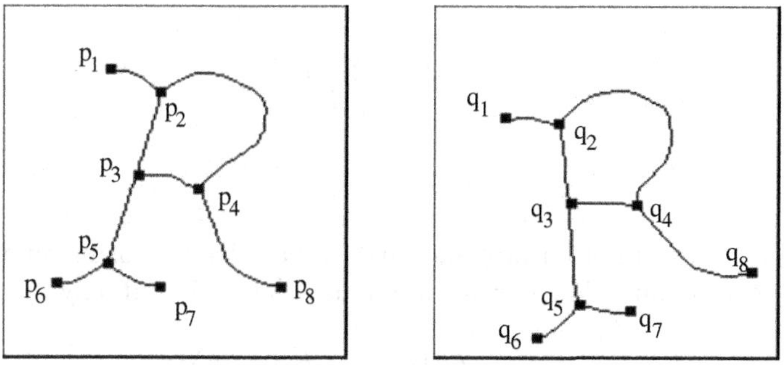

Fig. 3. Example of identified pairs of points in 2D. Eight pairs (p_i, q_i) of points are identified in the reference image (left) and in the floating image (right), respectively

We get the following system of linear equations:

$$\begin{pmatrix} \begin{matrix} a_{11} & \dots & a_{1k} & b_1 \\ \vdots & \ddots & \vdots & \vdots \\ a_{k1} & \dots & a_{kk} & b_k \\ b_1 & \dots & b_k & n \end{matrix} & & & \mathbf{0} \\ & \begin{matrix} a_{11} & \dots & a_{1k} & b_1 \\ \vdots & \ddots & \vdots & \vdots \\ a_{k1} & \dots & a_{kk} & b_k \\ b_1 & \dots & b_k & n \end{matrix} & & \\ & & \ddots & \\ \mathbf{0} & & & \begin{matrix} a_{11} & \dots & a_{1k} & b_1 \\ \vdots & \ddots & \vdots & \vdots \\ a_{k1} & \dots & a_{kk} & b_k \\ b_1 & \dots & b_k & n \end{matrix} \end{pmatrix} \cdot \begin{pmatrix} t_{11} \\ \vdots \\ t_{1k} \\ t_{1,k+1} \\ t_{21} \\ \vdots \\ t_{2k} \\ t_{2,k+1} \\ \vdots \\ t_{k1} \\ \vdots \\ t_{kk} \\ t_{k,k+1} \end{pmatrix} = \begin{pmatrix} c_{11} \\ \vdots \\ c_{1k} \\ d_1 \\ c_{21} \\ \vdots \\ c_{2k} \\ d_2 \\ \vdots \\ c_{k1} \\ \vdots \\ c_{kk} \\ d_k \end{pmatrix} ,$$

where

$$a_{uv} = a_{vu} = \sum_{i=1}^{n} q_{iu} \cdot q_{iv} ,$$

$$b_u = \sum_{i=1}^{n} q_{iu} ,$$

$$c_{uv} = \sum_{i=1}^{n} p_{iu} \cdot q_{iv} \ ,$$

$$d_u = \sum_{i=1}^{n} p_{iu}$$

$$(1 \leq u, v \leq k).$$

The above system of linear equations can be solved by using an appropriate numerical method. There exists a unique solution if and only if $det(M) \neq 0$, where

$$M = \begin{pmatrix} a_{11} & \dots & a_{1k} & b_1 \\ \vdots & \ddots & \vdots & \vdots \\ a_{k1} & \dots & a_{kk} & b_k \\ b_1 & \dots & b_k & n \end{pmatrix}.$$

3 Discussion

In this section we state and prove a sufficient existence condition for a unique solution.

By a hyperplane of the Euclidean space $\mathbb{R}^k$ we mean a subset of the form $\{a + x : x \in S\}$ where S is a $(k-1)$-dimensional linear subspace. Given some points $q_1, \dots, q_n$ in $\mathbb{R}^k$, we say that these points *span* $\mathbb{R}^k$ if no hyperplane of $\mathbb{R}^k$ contains them. If any $k+1$ points from $q_1, \dots, q_n$ span $\mathbb{R}^k$ then we say that $q_1, \dots, q_n$ are in general position.

Theorem. If $q_1, \dots, q_n$ span $\mathbb{R}^k$ then $\det(M) \neq 0$.

Proof. Suppose $\det(M) = 0$. Consider the vectors $v_j = (q_{1j}, q_{2j}, \dots, q_{nj})$ $(1 \leq j \leq k)$ in $\mathbb{R}^n$, and let $v_{k+1} = (1, 1, \dots, 1) \in \mathbb{R}^n$. With the notation $m = k+1$ observe that $M = \Big(\langle v_i, v_j \rangle\Big)_{m \times m}$ where $\langle \ , \ \rangle$ stands for the scalar multiplication. Since the columns of M are linearly dependent, we can fix a $(\beta_1, \dots, \beta_m) \in \mathbb{R}^m \setminus \{(0, \dots, 0)\}$ such that $\sum_{j=1}^{m} \beta_j \langle v_i, v_j \rangle = 0$ hold for $i = 1, \dots, m$. Then

$$0 = \sum_{i=1}^{m} \beta_i \cdot 0 = \sum_{i=1}^{m} \beta_i \sum_{j=1}^{m} \beta_j \langle v_i, v_j \rangle = \sum_{i=1}^{m} \beta_i \Big\langle v_i, \sum_{j=1}^{m} \beta_j v_j \Big\rangle =$$

$$\Big\langle \sum_{i=1}^{m} \beta_i v_i, \sum_{j=1}^{m} \beta_j v_j \Big\rangle = \Big\langle \sum_{i=1}^{m} \beta_i v_i, \sum_{i=1}^{m} \beta_i v_i \Big\rangle,$$

whence $\sum_{i=1}^{m} \beta_i v_i = 0$. Therefore all the q_j, $1 \leq j \leq n$, are solutions of the following (one element) system of linear equations:

$$\beta_1 x_1 + \dots + \beta_k x_k = -\beta_m. \tag{1}$$

Since the system has solutions and $(\beta_1, \ldots, \beta_m) \neq (0, \ldots, 0)$, there is an $i \in \{1, \ldots, k\}$ with $\beta_i \neq 0$. Hence the solutions of (1) form a hyperplane of $\mathbb{R}^k$. This hyperplane contains $q_1, \ldots, q_n$. Now it follows that if $q_1, \ldots, q_n$ span $\mathbb{R}^k$ then $\det(M) \neq 0$. Q.e.d.

4 Estimating Registration Error

Point–based registration might find imperfect matching due to the presence of error in localizing the points (note that points are often called *fiducials*). Maurer et al. [ ] proposed three types of measures of error:

- Fiducial localization error (FLE), which is the error in determining the positions of the fiducials.
- Fiducial registration error (FRE), which is the root mean square distance between corresponding points after registration. Note that point-based registration methods minimize this error measure.
- Target registration error (TRE), which is the distance between corresponding points representing ROIs (range–of–interest) after registration.

In real applications, only FRE is used, neither FLE nor TRE can be measured. Both FLE and TRE would require the knowledge of the exact spatial positions of the point pairs and in case we knew these, we would use these as pairs of points. But point selection is always prone to some error. The question is: if FRE is zero, does it really mean that the registration result is perfect? The answer is no, in a sense that the goal of registration is actually not the matching of the points, but the images in which the points are selected. Thus, using FRE as the measure of registration accuracy may be unreliable in some cases.

So, to what extent does the method tolerate the errors in selecting points, and how can we measure it? In real applications it can be estimated e.g., visually. In theory, we can make numerical simulations. In this case the exact spatial position of the points is well known, FLE can be modelled, and TRE can be calculated. In the last decade, investigations were focussed on TRE as a measure of theoretical accuracy of registration [,]. Note that each of these papers considers only rigid–body transformations.

There are two important results concerning registration errors [ ]:

- **Result 1.** For a fixed number of fiducials, TRE is proportional to FLE .
- **Result 2.** TRE is approximately proportional to $1/\sqrt{n}$ with n being the number of fiducials .

Fitzpatrick et al. [ ] gave an exact expression for approximating TRE assuming rigid–body transformations, thus proving both **Result 1** and **Result 2**.

In this paper we examine the dependence of TRE for our affine method via using numerical simulations.

4.1 Model for Numerical Simulations

Let $\mathcal{M} = \{(x, y, z) \mid x, y, z \in \mathbb{R}, 0 \leq x, y, z < 256\}$ be a cube–shaped region in the 3D Euclidean space. Let $P = \{p_1, p_2, \ldots, p_n\}$ be a set of n points used for modeling the fiducials identified in the reference image, where $p_i \in \mathcal{M}$ $(1 \leq i \leq n)$. A known affine transformation $\mathrm{T}_{\mathrm{known}}$ is chosen and the set $R = \{r_i \mid r_i = \mathrm{T}_{\mathrm{known}} \cdot p_i,\ i = 1, \ldots, n\}$ is calculated. Set R is corrupted by an n–dimensional noise vector $(\mu_1, \ldots, \mu_n)$ whose components are random variables having σ–Gaussian distribution. This is used for modeling the FLE. The set $Q = \{q_i \mid q_i = r_i + \mu_i,\ i = 1, \ldots, n\}$ is constructed, where pair (p_i, q_i) of points can be regarded as a pair of corresponding fiducials used for registration. It is assumed that the FLE is identically zero in the base image. The set $S = \{s_j \mid s_j \in \mathcal{M},\ j = 1, \ldots, m\}$ of m points is randomly selected to represent ROIs in the reference image. Note that the same $m = 20$ target points are used for our numerical simulations . Set S is also transformed to generate set of m points $U = \{u_j \mid u_j = \mathrm{T}_{\mathrm{known}} \cdot s_j,\ j = 1, \ldots, m\}$. The transformation $\mathrm{T}_{\mathrm{found}}$ is determined and it is applied to the set U to calculate the set of m points $V = \{v_j \mid v_j = \mathrm{T}_{\mathrm{found}} \cdot u_j,\ j = 1, \ldots, m\}$.

TRE is formulated as follows:

$$\sqrt{\frac{1}{m} \sum_{j=1}^{m} \|s_j - v_j\|^2}.$$

We repeated the iterations 10000 times.

4.2 Results

Figure shows that TRE is proportional to FLE, for a fixed number of fiducials. Therefore, **Result 1** holds for affine transformations, too.

Figure is to demonstrate how TRE depends on the number of fiducials, for a fixed FLE. Although **Result 2** does not hold, it can be seen that the TRE is inversely proportional to the number of fiducials.

5 Conclusions

In this paper we proposed a method capable of finding affine transformations based on selected pairs of points. We gave and proved a sufficient existence condition for a unique solution. We examined the theoretical registration accuracy of this method using numerical simulations.

In practice, we successfully use this method to register 3D MR brain studies, in which 12 anatomical landmarks were interactively identified.

Acknowledgement

This work was supported by OTKA T023804, OTKA T026243, OTKA T023186, and FKFP 0908/1997 grants.

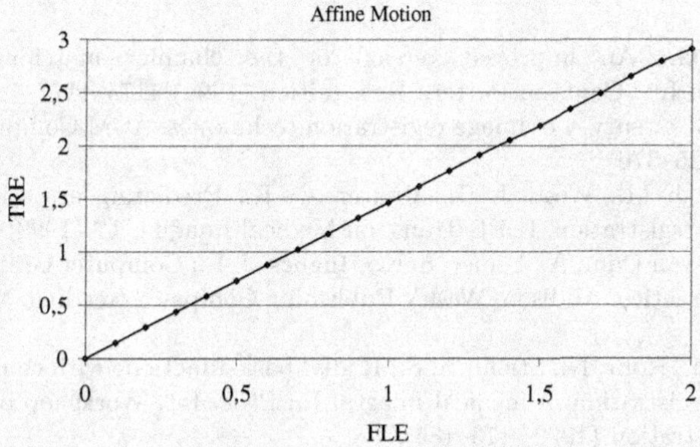

Fig. 4. TRE (for the 20 target points) as a function of FLE for 10 fiducials. It is confirmed that TRE is proportional to FLE

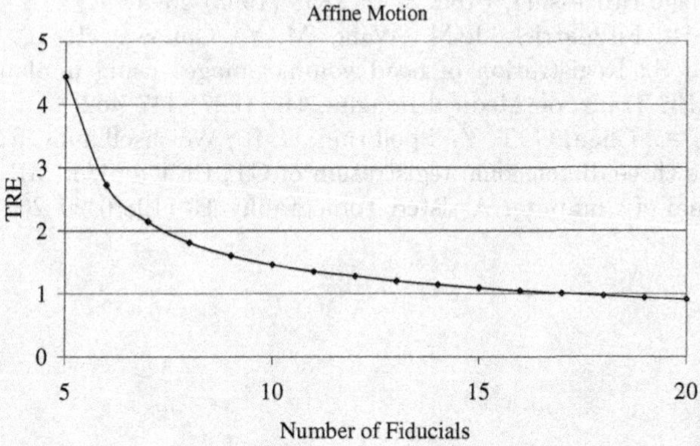

Fig. 5. TRE (for the 20 target points) as a function of the number of fiducials n, where $\sigma = 1$ Gaussian distribution is used for modelling FLE. TRE is inversely proportional to n

References

1. Arun, K. S., Huang, T. S., Blostein, S. D.: Least squares fitting of two 3-D point sets. IEEE Trans. Pattern Analysis and Machine Intelligence **9** (1987) 698–703
2. Barrow, H. G., Tenenbaum, J. M., Bolles, R. C., Wolf, H. C.: Parametric correspondence and chamfer matching: two new techniques for image matching. In: Proc. 5th Int. Joint Conf. Artificial Intelligence (1977) 659–663
3. Besl, P. J., McKay, N. D.: A method for registration of 3D shapes. IEEE Trans. Pattern Analysis and Machine Intelligence **14** (1992) 239–256

4. Bookstein, F. L.: Principal warps: thin–plate splines and the decomposition of deformations. IEEE Trans. Pattern Analysis and Machine Intelligence **11** (1989) 567–585
5. Borgefors, G.: An improved version of the chamfer matching algorithm. In: Proc. 7th Int. Conf. on Pattern Recognition (1994) 1175–1177
6. Brown, L. G.: A survey of image registration techniques. ACM Computing Surveys **24** (1992) 325–376
7. Fitzpatrick, J. M., West, J. B., Maurer, C. R.: Predicting error in rigid–body point-based registration. IEEE Trans. on Medical Imaging **17** (1998) 694–702
8. Foley, J. D., van Dam, A., Feiner, S. K., Hughes, J. F.: Computer Graphics — Principles and Practice. Addison–Wesley Publishing Company, Reading, Massachusetts (1991)
9. Fornefett, M., Rohr, K., Stiehl, H. S.: Radial basis functions with compact support for elastic registration of medical images. In: Proc. Int. Workshop on Biomedical Image Registration (1999) 173–185
10. Maintz, J. B. A., Viergever, M. A.: A survey of medical image registration. Medical Image Analysis **2** (1998) 1–36
11. Maurer, C. R., McCrory, J. J., Fitzpatrick, J. M.: Estimation of accuracy in localizing externally attached markers in multimodal volume head images. In: Medical Imaging: Image Processing, Proc. SPIE 1898 (1993) 43–54
12. Maurer, C. R., Fitzpatrick, J. M., Wang, M. Y., Galloway, R. L., Maciunas, R. J., Allen, G. S.: Registration of head volume images using implantable fiducial markers. IEEE Trans. on Medical Imaging **16** (1997) 447–462
13. Pelizzari, C. A., Chen, G. T. Y., Spelbring, D. R., Weichselbaum, R. R., Chen, C. T.: Accurate three dimensional registration of CT, PET and/or MR images of the brain. Journal of Computer Assisted Tomography **13** (1989) 20–26

Extended Kalman Filter Design for Motion Estimation by Point and Line Observations

Yiwen Zhang, Bodo Rosenhahn, Gerald Sommer

Institut für Informatik und Praktische Mathematik
Christian-Albrechts-Universität zu Kiel
Preußerstrasse 1-9, 24105 Kiel, Germany
yz,bro,gs@ks.informatik.uni-kiel.de

Abstract. The paper develops three extended Kalman filters (EKF) for 2D-3D pose estimation. The measurement models are based on three constraints which are constructed by geometric algebra. The dynamic measurements for these EKF are either points or lines. The real monocular vision experiments show that the results of EKFs perform more stable than that of LMS method.

1 Introduction and problem statement

The paper describes the design of EKFs which are used to estimate pose parameters of known objects in the framework of kinematics. Pose estimation in the framework of kinematics will be treated as nonlinear optimization with respect to geometric constraint equations expressing the relation between 2D image features and 3D model data.

The problem is described as follows. First, we make the following assumptions. The model of an object is given by points and lines in the 3D space. Further we extract line subspaces or points in an image of a calibrated camera and match them with the model of the object. The aim is to find the pose of the object from observations of points and lines in the images at different poses. Figure 1 shows the scenario with respect to observed line subspaces. The method of obtaining these is out of scope of this paper.

To be more detailed, in the scenario of figure 1 we describe the following situation: We assume 3D points $\{\boldsymbol{y}_i\}$ and lines $\{\boldsymbol{S}_i\}$, $i = 1, 2, ...,$ belonging to an object model. Further we extract points $\{\boldsymbol{b}_i\}$ and lines $\{\boldsymbol{l}_i\}$ in an image of a calibrated camera and match them with the model.

Three constraints can be depicted:

1. Point-line constraint: A transformed point, e.g. $\boldsymbol{x}_1$, of the model point $\boldsymbol{y}_1$ must lie on the projection ray $\boldsymbol{L}_{b_1}$, given by the optical center $\boldsymbol{c}$ and the corresponding image point b_1.
2. Point-plane constraint: A transformed point, e.g. $\boldsymbol{x}_1$, must lie on the projection plane $\boldsymbol{P}_{12}$, given by $\boldsymbol{c}$ and the corresponding image line $\boldsymbol{l}_1$.
3. Line-plane constraint: A transformed line, e.g. $\boldsymbol{L}_1$, of the model line $\boldsymbol{S}_1$ must lie on the projection plane $\boldsymbol{P}_{12}$, given by $\boldsymbol{c}$ and the the corresponding image line $\boldsymbol{l}_1$.

We want to estimate optimal motion parameters based on these three constraints which formally are written [1, 2] in motor algebra [3, 4, 5] as

G. Sommer and Y. Y. Zeevi (Eds.): AFPAC 2000, LNCS 1888, pp. 339-348, 2000.

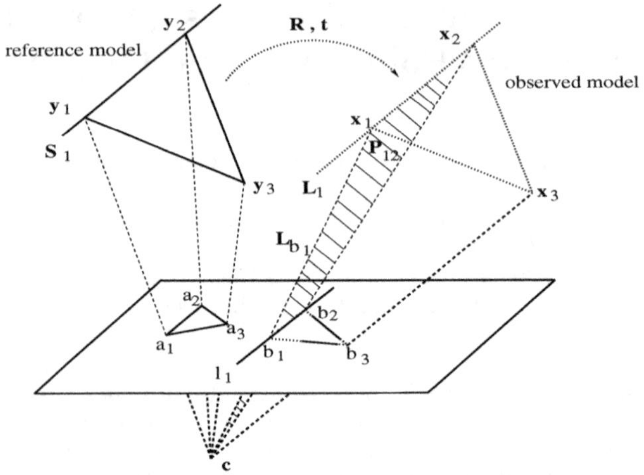

Fig. 1. The scenario. The solid lines at the left hand describe the assumptions: the camera model, the model of the object and the initially extracted lines on the image plane. The dashed lines at the right hand describe the actual pose of the model, which leads to the best fit of the object with the actual extracted lines.

Point-line constraint: $\boldsymbol{X_1 L}_{b_1} - \overline{\boldsymbol{L}_{b_1}\boldsymbol{X_1}} = \boldsymbol{I}(\boldsymbol{m_1} - \boldsymbol{n_1} \times \boldsymbol{x_1}) = \boldsymbol{0},$

Point-plane constraint: $\boldsymbol{P_{12}X_1} - \overline{\boldsymbol{X_1 P_{12}}} = \boldsymbol{I}(d_1 + \boldsymbol{p_1} \cdot \boldsymbol{x_1}) = \boldsymbol{0},$

Line-plane constraint: $\boldsymbol{L_1 P_{12}} + \boldsymbol{P_{12}}\overline{\boldsymbol{L}_1} = \boldsymbol{u_1} \cdot \boldsymbol{p_1} + \boldsymbol{I}(d_1 \boldsymbol{u_1} - \boldsymbol{p_1} \times \boldsymbol{v_1}) = \boldsymbol{0}.$

In above equations, we denote the point $\boldsymbol{X_1} = 1 + \boldsymbol{I x_1}$, the lines $\boldsymbol{L}_{b_1} = \boldsymbol{n_1} + \boldsymbol{I m_1}$ and $\boldsymbol{L_1} = \boldsymbol{u_1} + \boldsymbol{I v_1}$ and the plane $\boldsymbol{P_{12}} = \boldsymbol{p_1} + \boldsymbol{I} d_1$. More detailed derivation and interpretation of these constraints are described in [1, 2]. We use rotor algebra to describe points and their 3D kinematics and motor algebra to present lines and to model their kinematics. The reason we use rotor and motor algebra instead of matrix algebra is as follows. In EKF we define the state vector to be estimated as the parameter vector of rotation and translation. By rotor and motor algebra, there are 7 and 8 parameters, respectively. If we directly use matrix algebra, there will be up to 12 parameters (9 for rotation and 3 for translation). It is obviously that rotor or motor algebra will be more efficient. Moreover, using motor algebra we linearize the 3D Euclidean line motion model straightforwardly.

There are several approaches of optimal pose estimation based on least square methods [6]. Our preference is to use EKF for pose estimation because of their incremental, real-time potential and because of their robustness in case of noisy data. The robustness of the Kalman filter results from the fact that stronger modeling of the dynamic model is possible using additional priors compared to usual LMS estimators.

Because EKF means a general frame for handling nonlinear measurement models [7], the estimation of each considered constraint requires an individually designed EKF. The commonly known EKFs for pose estimation are related to 3D-3D point based measurements. The only EKF for line based measurements has been recently published by the authors [3]. But also that one has to estimate the motion of a line from 3D-3D measurements in motor algebra and not from 2D-3D measurements as in this paper. Zhang and Faugeras [8] used line segments in the frame of a point based standard EKF. Bar-Itzhack and Oshman [9] designed a quaternion EKF for point based rotation estimation.

The paper is organized as follows. After introduction and problem statement, in section two we will present three EKF approaches for motion estimation. In section three we compare the performance of different algorithms for constraint based pose estimation.

2 The Extended Kalman Filter for pose estimation

In this section we want to present the design of EKFs for estimating the pose based on three constraints. Because an EKF is defined in the frame of linear vector algebra, it will be necessary to map the estimation task from any chosen algebraic embedding to linear vector algebra (see e.g. [3]), at least so long no other solution exists. We present the design method in detail for constraint no.1 in subsection 2.1. The design results for constraints no.2 and 3 will be given in subsections 2.2 and 2.3, respectively.

2.1 EKF pose estimation based on point-line constraint

In case of point based measurements of the object at different poses, an algebraic embedding of the problem in the 4D linear space of the algebra of rotors $\mathcal{G}^+_{3,0,0}$, which is isomorphic to that one of quaternions $\mathbb{H}$, will be sufficient [4, 3]. Thus, rotation will be represented by a unit rotor $\boldsymbol{R}$ and translation will be a bivector $\boldsymbol{t}$. A point $\boldsymbol{y}_1$ transformed to $\boldsymbol{x}_1$ reads

$$\boldsymbol{x}_1 = \boldsymbol{R}\boldsymbol{y}_1\tilde{\boldsymbol{R}} + \boldsymbol{t}.$$

We denote the four components of the rotor as

$$\boldsymbol{R} = r_0 + r_1\sigma_2\sigma_3 + r_2\sigma_3\sigma_1 + r_3\sigma_1\sigma_2.$$

To convert a rotor $\boldsymbol{R}$ into a rotation matrix $\mathcal{R}$, simple conversion rules are at hand:

$$\mathcal{R} = \begin{pmatrix} r_0^2 + r_1^2 - r_2^2 - r_3^2 & 2(r_1r_2 + r_0r_3) & 2(r_1r_3 - r_0r_2) \\ 2(r_1r_2 - r_0r_3) & r_0^2 - r_1^2 + r_2^2 - r_3^2 & 2(r_2r_3 + r_0r_1) \\ 2(r_1r_3 + r_0r_2) & 2(r_2r_3 - r_0r_1) & r_0^2 - r_1^2 - r_2^2 + r_3^2 \end{pmatrix}.$$

In vector algebra, the above point transformation model can be described as

$$\mathbf{x_1} = \mathcal{R}\mathbf{y_1} + \mathbf{t}.$$

The projection ray $\mathbf{L}_{b_1}$ in the point-line equation is represented by Plücker coordinates $(\mathbf{n_1}, \mathbf{m_1})$, where $\mathbf{n_1}$ is its unit direction and $\mathbf{m_1}$ its moment. The point-line constraint equation in vector algebra of $\mathbb{R}^3$ reads

$$\mathbf{f_1} = \mathbf{m_1} - \mathbf{n_1} \times \mathbf{x_1} = \mathbf{m_1} - \mathbf{n_1} \times (\mathcal{R}\mathbf{y_1} + \mathbf{t}) = \mathbf{0}.$$

Let the state vector $\mathbf{s}$ for the EKF be a 7D vector, composed in terms of the rotor coefficients for rotation and translation,

$$\mathbf{s} = (\mathbf{R}^T, \mathbf{t}^T)^T = (r_0, r_1, r_2, r_3, t_1, t_2, t_3)^T.$$

The rotation coefficients must satisfy the unit condition

$$\mathbf{f_2} = \mathbf{R}^T\mathbf{R} - 1 = r_0^2 + r_1^2 + r_2^2 + r_3^2 - 1 = 0.$$

The noise free measurement vector $\mathbf{a_i}$ is given by the actual line parameters $\mathbf{n_i}$ and $\mathbf{m_i}$, and the actual 3D point measurements $\mathbf{y_i}$,

$$\mathbf{a_i} = (\mathbf{n_i}^T, \mathbf{m_i}^T, \mathbf{y_i}^T)^T = (n_{i1}, n_{i2}, n_{i3}, m_{i1}, m_{i2}, m_{i3}, y_{i1}, y_{i2}, y_{i3})^T.$$

For a sequence of measuremerts $\mathbf{a_i}$ and states $\mathbf{s_i}$, the constraint equations

$$\mathbf{f_i}(\mathbf{a_i}, \mathbf{s_i}) = \begin{pmatrix} \mathbf{f_{1i}} \\ \mathbf{f_{2i}} \end{pmatrix} = \begin{pmatrix} \mathbf{m_i} - \mathbf{n_i} \times (\mathcal{R}_\mathbf{i}\mathbf{y_i} + \mathbf{t_i}) \\ \mathbf{R_i}^T\mathbf{R_i} - 1 \end{pmatrix} = \mathbf{0}$$

relate measurements and states in a nonlinear manner. The system model in this static case should be

$$\mathbf{s_{i+1}} = \mathbf{s_i} + \boldsymbol{\zeta}_\mathbf{i},$$

where $\boldsymbol{\zeta}_\mathbf{i}$ is a v ector random sequence with knawn statistics,

$$E[\boldsymbol{\zeta}_\mathbf{i}] = 0,$$
$$E[\boldsymbol{\zeta}_\mathbf{i}^T \boldsymbol{\zeta}_\mathbf{k}] = \mathcal{Q}_\mathbf{i}\delta_\mathbf{ik},$$

where $\delta_\mathbf{ik}$ is the Kroneck er delta and the matrix $\mathcal{Q}_\mathbf{i}$ is assumed to be nonnegative definite.

We assume that the measurement system is disturbed by additive white noise, i.e., the real observed measurement $\mathbf{a}'_\mathbf{i}$ is expressed as

$$\mathbf{a}'_\mathbf{i} = \mathbf{a_i} + \boldsymbol{\eta}_\mathbf{i}.$$

The vector $\boldsymbol{\eta}_\mathbf{i}$ is an additive, random sequence with known statistics,

$$E[\boldsymbol{\eta}_\mathbf{i}] = 0,$$
$$E[\boldsymbol{\eta}_\mathbf{i}^T \boldsymbol{\eta}_\mathbf{k}] = \mathcal{W}_\mathbf{i}\delta_\mathbf{ik},$$

where the matrix $\mathcal{W}_\mathbf{i}$ is assumed to be nonnegative definite.

Since the observation equation is nonlinear (that means, the relationship betw een the measuremert $\mathbf{a}'_\mathbf{i}$ and state $\mathbf{s_i}$ is nonlinear), we expand $\mathbf{f_i}(\mathbf{a_i}, \mathbf{s_i})$ into a Taylor series about the $(\mathbf{a}'_\mathbf{i}, \hat{\mathbf{s}}_{/\mathbf{i-1}})$, where $\mathbf{a}'_\mathbf{i}$ is the real measurement and $\hat{\mathbf{s}}_{/\mathbf{i-1}}$ is the predicted state at situation i. By ignoring the second order terms, we get the linearized measurement equation

$$\mathbf{z_i} = \mathcal{H}_\mathbf{i}\mathbf{s_i} + \boldsymbol{\xi}_\mathbf{i},$$

where

$$\mathbf{z_i} = \mathbf{f_i}(\mathbf{a}'_\mathbf{i}, \hat{\mathbf{s}}_{/\mathbf{i-1}}) - \frac{\partial \mathbf{f_i}(\mathbf{a}'_\mathbf{i}, \hat{\mathbf{s}}_{/\mathbf{i-1}})}{\partial \mathbf{s_i}} \hat{\mathbf{s}}_{/\mathbf{i-1}}$$
$$= \begin{pmatrix} \mathbf{m}'_\mathbf{i} - \mathbf{n}'_\mathbf{i} \times (\hat{\mathcal{R}}_{\mathbf{i/i-1}}\mathbf{y}'_\mathbf{i} + \hat{\mathbf{t}}_{\mathbf{i/i-1}}) \\ \hat{\mathbf{R}}^T_{\mathbf{i/i-1}}\hat{\mathbf{R}}_{\mathbf{i/i-1}} - 1 \end{pmatrix} + \mathcal{H}_\mathbf{i}\hat{\mathbf{s}}_{/\mathbf{i-1}}.$$

The measurement matrix $\mathcal{H}_\mathbf{i}$ of the linearized measurement $\mathbf{z_i}$ reads

$$\mathcal{H}_\mathbf{i} = -\frac{\partial \mathbf{f_i}(\mathbf{a}'_\mathbf{i}, \hat{\mathbf{s}}_{/\mathbf{i-1}})}{\partial \mathbf{s_i}} = \begin{pmatrix} \mathcal{C}_{\mathbf{n}'_\mathbf{i}}\mathcal{D}_{\hat{\mathcal{R}}\mathbf{y}'} & \mathcal{C}_{\mathbf{n}'_\mathbf{i}} \\ \mathcal{D}_\mathbf{R} & 0_{1\times 3} \end{pmatrix},$$

where

$$\mathcal{D}_\mathbf{R} = \frac{\partial(\hat{\mathbf{R}}^T_{\mathbf{i/i-1}}\hat{\mathbf{R}}_{\mathbf{i/i-1}} - 1)}{\partial \mathbf{R_i}}$$
$$= \begin{pmatrix} -2\hat{r}_{(i/i-1)0} & -2\hat{r}_{(i/i-1)1} & -2\hat{r}_{(i/i-1)2} & -2\hat{r}_{(i/i-1)3} \end{pmatrix},$$

$$\mathcal{D}_{\hat{\mathcal{R}}\mathbf{y}'} = \frac{\partial(\hat{\mathcal{R}}_{i/i-1}\mathbf{y}_i')}{\partial \mathbf{R}_i} = \begin{pmatrix} d_1 & d_2 & d_3 & d_4 \\ d_4 & -d_3 & d_2 & -d_1 \\ -d_3 & -d_4 & d_1 & d_2 \end{pmatrix},$$

$$d_1 = 2(\hat{r}_{(i/i-1)0}y_{i1}' + \hat{r}_{(i/i-1)3}y_{i2}' - \hat{r}_{(i/i-1)2}y_{i3}'),$$
$$d_2 = 2(\hat{r}_{(i/i-1)1}y_{i1}' + \hat{r}_{(i/i-1)2}y_{i2}' + \hat{r}_{(i/i-1)3}y_{i3}'),$$
$$d_3 = 2(-\hat{r}_{(i/i-1)2}y_{i1}' + \hat{r}_{(i/i-1)1}y_{i2}' - \hat{r}_{(i/i-1)0}y_{i3}'),$$
$$d_4 = 2(-\hat{r}_{(i/i-1)3}y_{i1}' + \hat{r}_{(i/i-1)0}y_{i2}' + \hat{r}_{(i/i-1)1}y_{i3}').$$

The 3×3 matrix $\mathcal{C}_{\mathbf{n}_i'}$ is the sk ew-symmetric matrix of$\mathbf{n}_i'$. For any vector $\mathbf{y}$, we have$\mathcal{C}_{\mathbf{n}_i'}\mathbf{y} = \mathbf{n}_i' \times \mathbf{y}$ with

$$\mathcal{C}_{\mathbf{n}_i'} = \begin{pmatrix} 0 & -n_{i3}' & n_{i2}' \\ n_{i3}' & 0 & -n_{i1}' \\ -n_{i2}' & n_{i1}' & 0 \end{pmatrix}.$$

The measurement noise is given by

$$\boldsymbol{\xi}_i = -\frac{\partial \mathbf{f}_i(\mathbf{a}_i', \hat{\mathbf{s}}_{i/i-1})}{\partial \mathbf{a}_i}(\mathbf{a}_i - \mathbf{a}_i') = \frac{\partial \mathbf{f}_i(\mathbf{a}_i', \hat{\mathbf{s}}_{i/i-1})}{\partial \mathbf{a}_i}\boldsymbol{\eta}_i$$
$$= \begin{pmatrix} \mathcal{C}_{\hat{\mathbf{x}}_{i/i-1}} & \mathbf{I}_{3\times 3} & -\mathcal{C}_{\mathbf{n}_i'}\hat{\mathcal{R}}_{i/i-1} \\ \mathbf{0}_{1\times 3} & \mathbf{0}_{1\times 3} & \mathbf{0}_{1\times 3} \end{pmatrix}_{4\times 9} \boldsymbol{\eta}_i,$$

where $\mathbf{I}_{3\times 3}$ is a unit matrix and $\mathcal{C}_{\hat{\mathbf{x}}_{i/i-1}}$ is the skew-symmetric matrix of $\hat{\mathbf{x}}_{i/i-1}$ with

$$\hat{\mathbf{x}}_{i/i-1} = \hat{\mathcal{R}}_{i/i-1}\mathbf{y}_i' + \hat{\mathbf{t}}_{i/i-1}.$$

The expectation and the co variance of the new measurement noise $\boldsymbol{\xi}_i$ are easily derived from that of $\mathbf{a}_i'$ as

$$E[\boldsymbol{\xi}_i] = \mathbf{0},$$
$$E[\boldsymbol{\xi}_i^T\boldsymbol{\xi}_i] = \mathcal{V}_i = \left(\frac{\partial \mathbf{f}_i(\mathbf{a}_i', \hat{\mathbf{s}}_{i/i-1})}{\partial \mathbf{a}_i}\right)\mathcal{W}_i\left(\frac{\partial \mathbf{f}_i(\mathbf{a}_i', \hat{\mathbf{s}}_{i/i-1})}{\partial \mathbf{a}_i}\right)^T.$$

The EKF motion estimation algorithms based on point-plane and line-plane constraints can be derived in a similar way. We list the results below.

2.2 EKF pose estimation based on point-plane constraint

The projection plane $\mathbf{P}_{12}$ in the point-plane constraint equation is represented by $(d_1, \mathbf{p}_1)$, where d_1 is its Hesse distance and $\mathbf{p}_1$ its unit direction. The point-plane constraint equation in vector algebra of $\mathbb{R}^3$ reads

$$d_1 - \mathbf{p}_1{}^T(\mathcal{R}\mathbf{x}_1 + \mathbf{t}) = 0.$$

With the measurement vector $\mathbf{a}_i = (d_i, \mathbf{p}_i{}^T, \mathbf{y}_i{}^T)^T$ and the same state vector $\mathbf{s}$ as above, the measurement $\mathbf{z}_i$ of linearized measurement equation reads

$$\mathbf{z}_i = \begin{pmatrix} d_i' - \mathbf{p}_i'^T(\hat{\mathcal{R}}_{i/i-1}\mathbf{y}_i' + \hat{\mathbf{t}}_{i/i-1}) \\ \hat{\mathbf{R}}_{i/i-1}^T\hat{\mathbf{R}}_{i/i-1} - 1 \end{pmatrix} + \mathcal{H}_i\hat{\mathbf{s}}_{i/i-1}.$$

The measurement matrix $\mathcal{H}_i$ of the linearized measurement $\mathbf{z}_i$ now reads

$$\mathcal{H}_i = \begin{pmatrix} \mathbf{p_i}^T \mathcal{D}_{\hat{\mathcal{R}}\mathbf{y}'} & \mathbf{p_i}^T \\ \mathcal{D}_{\mathbf{R}} & \mathbf{0}_{1\times 3} \end{pmatrix}.$$

The measurement noise is given b y

$$\xi_i = \begin{pmatrix} 1 & -(\hat{\mathcal{R}}_{i/i-1}\mathbf{y}_i' + \hat{\mathbf{t}}_{i/i-1})^T & -(\mathbf{p}_i'^T \hat{\mathcal{R}}_{i/i-1}) \\ 0 & \mathbf{0}_{1\times 3} & \mathbf{0}_{1\times 3} \end{pmatrix}_{2\times 7} \eta_i.$$

2.3 EKF pose estimation based on line-plane constraint

Using the line-plane constraint, the reference model entity in $\mathcal{G}^+_{3,0,1}$ [3, 4] is the Plücker line $\boldsymbol{S_1} = \boldsymbol{n_1} + \boldsymbol{I}\boldsymbol{m_1}$. This line transformed by a motor $\boldsymbol{M} = \boldsymbol{R} + \boldsymbol{I}\boldsymbol{R}'$ reads

$$\boldsymbol{L_1} = \boldsymbol{M}\boldsymbol{S_1}\widetilde{\boldsymbol{M}} = \boldsymbol{R}\boldsymbol{n_1}\widetilde{\boldsymbol{R}} + \boldsymbol{I}(\boldsymbol{R}\boldsymbol{n_1}\widetilde{\boldsymbol{R}'} + \boldsymbol{R}'\boldsymbol{n_1}\widetilde{\boldsymbol{R}} + \boldsymbol{R}\boldsymbol{m_1}\widetilde{\boldsymbol{R}}) = \boldsymbol{u_1} + \boldsymbol{I}\boldsymbol{v_1}.$$

We denote the 8 components of the motor as

$$\begin{aligned} \boldsymbol{M} &= \boldsymbol{R} + \boldsymbol{I}\boldsymbol{R}' \\ &= r_0 + r_1\gamma_2\gamma_3 + r_2\gamma_3\gamma_1 + r_3\gamma_1\gamma_2 + \boldsymbol{I}(r_0' + r_1'\gamma_2\gamma_3 + r_2'\gamma_3\gamma_1 + r_3'\gamma_1\gamma_2). \end{aligned}$$

The line motion equation can be equivalently expressed by vector form,

$$\begin{aligned} \mathbf{u_1} &= \mathcal{R}\mathbf{n_1}, \\ \mathbf{v_1} &= \mathcal{A}\mathbf{n_1} + \mathcal{R}\mathbf{m_1}, \end{aligned}$$

with

$$\mathcal{A} = \begin{pmatrix} a_{11} & a_{12} & a_{13} \\ a_{21} & a_{22} & a_{23} \\ a_{31} & a_{32} & a_{33} \end{pmatrix},$$

$$\begin{aligned} a_{11} &= 2(r_0'r_0 + r_1'r_1 - r_2'r_2 - r_3'r_3), & a_{12} &= 2(r_3'r_0 + r_2'r_1 + r_1'r_2 + r_0'r_3), \\ a_{13} &= 2(-r_2'r_0 + r_3'r_1 - r_0'r_2 + r_1'r_3), & a_{21} &= 2(-r_3'r_0 + r_2'r_1 + r_1'r_2 - r_0'r_3), \\ a_{22} &= 2(r_0'r_0 - r_1'r_1 + r_2'r_2 - r_3'r_3), & a_{23} &= 2(r_1'r_0 + r_0'r_1 + r_3'r_2 + r_2'r_3), \\ a_{31} &= 2(r_2'r_0 + r_3'r_1 + r_0'r_2 + r_1'r_3), & a_{32} &= 2(-r_1'r_0 - r_0'r_1 + r_3'r_2 + r_2'r_3), \\ a_{33} &= 2(r_0'r_0 - r_1'r_1 - r_2'r_2 + r_3'r_3). \end{aligned}$$

The line-plane constraint equation in vector algebra of $\mathbb{R}^3$ reads

$$\begin{pmatrix} \mathbf{f_1} \\ \mathbf{f_2} \end{pmatrix} = \begin{pmatrix} \mathbf{p_1}^T\mathbf{u_1} \\ d_1\mathbf{u_1} + \mathbf{v_1} \times \mathbf{p_1} \end{pmatrix} = \begin{pmatrix} \mathbf{p_1}^T(\mathcal{R}\mathbf{n_1}) \\ d_1\mathcal{R}\mathbf{n_1} + (\mathcal{A}\mathbf{n_1} + \mathcal{R}\mathbf{m_1}) \times \mathbf{p_1} \end{pmatrix} = \mathbf{0}.$$

We use the 8 components of the motor as the state vector for the EKF,

$$\mathbf{s} = (r_0, r_1, r_2, r_3, r_0', r_1', r_2', r_3')^T$$

and these 8 components must satisfy both the unit and orthogonal conditions:

$$\begin{aligned} f_3 &= r_0^2 + r_1^2 + r_2^2 + r_3^2 - 1 = 0, \\ f_4 &= r_0r_0' + r_1r_1' + r_2r_2' + r_3r_3' = 0. \end{aligned}$$

The 10D noise free measurement vector $\mathbf{a_i}$ is given by the true plane parameters d_i and $\mathbf{p_i}$, and the true 6D line parameters $(\mathbf{n_i}, \mathbf{m_i})$,

$$\mathbf{a_i} = (d_i, \mathbf{p_i}^T, \mathbf{n_i}^T, \mathbf{m_i}^T)^T = (d_i, p_{i1}, p_{i2}, p_{i3}, n_{i1}, n_{i2}, n_{i3}, m_{i1}, m_{i2}, m_{i3})^T.$$

The new measurement in linearized equation reads

$$\mathbf{z_i} = \begin{pmatrix} \mathbf{p}_i'^T(\hat{\mathcal{R}}_{i/i-1}\mathbf{n}_i') \\ d_i'\hat{\mathcal{R}}_{i/i-1}\mathbf{n}_i' + (\hat{\mathcal{A}}_{i/i-1}\mathbf{n}_i' + \hat{\mathcal{R}}_{i/i-1}\mathbf{m}_i') \times \mathbf{p}_i' \\ \hat{\mathbf{R}}_{i/i-1}^T\hat{\mathbf{R}}_{i/i-1} - 1 \\ \hat{\mathbf{R}}_{i/i-1}^T\hat{\mathbf{R}}_{i/i-1}' \end{pmatrix} + \mathcal{H}_i\hat{\mathbf{s}}_{i/i-1}.$$

The measurement matrix $\mathcal{H}_i$ of the linearized measurement $\mathbf{z}_i$ reads

$$\mathcal{H}_i = \begin{pmatrix} -\mathbf{p}_i'^T \mathcal{D}_{\hat{\mathcal{R}}\mathbf{n}'} & \mathbf{0}_{1\times 4} \\ -d_i' \mathcal{D}_{\hat{\mathcal{R}}\mathbf{n}'} + \mathcal{C}_{\mathbf{p}_i'}(\mathcal{D}_{\hat{\mathcal{A}}\mathbf{n}'} + \mathcal{D}_{\hat{\mathcal{R}}\mathbf{m}'}) & \mathcal{C}_{\mathbf{p}_i'} \mathcal{D}_{\hat{\mathcal{R}}\mathbf{n}'} \\ \mathcal{D}_{\mathbf{R}} & \mathbf{0}_{1\times 4} \\ \mathcal{D}_{\mathbf{R}'} & \frac{1}{2}\mathcal{D}_{\mathbf{R}} \end{pmatrix},$$

where $\mathcal{D}_{\hat{\mathcal{R}}\mathbf{n}'} = \frac{\partial(\hat{\mathcal{R}}_{i/i-1}\mathbf{n}_i')}{\partial \mathbf{R}_i}$, $\mathcal{D}_{\hat{\mathcal{R}}\mathbf{m}'} = \frac{\partial(\hat{\mathcal{R}}_{i/i-1}\mathbf{m}_i')}{\partial \mathbf{R}_i}$, $\mathcal{D}_{\hat{\mathcal{A}}\mathbf{n}'} = \frac{\partial(\hat{\mathcal{A}}_{i/i-1}\mathbf{n}_i')}{\partial \mathbf{R}_i}$,
$\mathcal{D}_{\mathbf{R}'} = \frac{\partial(\hat{\mathbf{R}}_{i/i-1}^T \hat{\mathbf{R}}_{i/i-1}')}{\partial \mathbf{R}_i}$ and $\frac{1}{2}\mathcal{D}_{\mathbf{R}} = \frac{\partial(\hat{\mathbf{R}}_{i/i-1}^T \hat{\mathbf{R}}_{i/i-1}')}{\partial \mathbf{R}_i'}$

The 3×3 matrix $\mathcal{C}_{\mathbf{p}_i'}$ is the sk ew-symmetric matrix of$\mathbf{p}_i'$.

The measurement noise is given by

$$\xi_i = \begin{pmatrix} 0 & \mathbf{n}_i'^T \hat{\mathcal{R}}_{i/i-1} & \mathbf{p}_i'^T \hat{\mathcal{R}}_{i/i-1} & \mathbf{0}_{1\times 3} \\ \hat{\mathcal{R}}_{i/i-1}\mathbf{n}_i' & \mathcal{C}_{\hat{\mathbf{v}}_i} & d_i'\hat{\mathcal{R}}_{i/i-1} - \mathcal{C}_{\mathbf{p}_i'}\hat{\mathcal{A}}_{i/i-1} & -\mathcal{C}_{\mathbf{p}_i'}\hat{\mathcal{R}}_{i/i-1} \\ \mathbf{0}_{2\times 3} & \mathbf{0}_{2\times 3} & \mathbf{0}_{2\times 3} & \mathbf{0}_{2\times 3} \end{pmatrix} \eta_i$$

where $\mathcal{C}_{\hat{\mathbf{v}}_i}$ is sk ew-symmetric matrix of $\hat{\mathbf{v}}_i$, and $\hat{\mathbf{v}}_i$ is defined as

$$\hat{\mathbf{v}}_i = \hat{\mathcal{A}}_{i/i-1}\mathbf{n}_i' + \hat{\mathcal{R}}_{i/i-1}\mathbf{m}_i'.$$

Having linearized the measurement models, the EKF implementation is straightforward and standard. Further implementation details will not be repeated here [2, 7, 3, 8]. In next section, We will denote the EKF as **R**t EKF, if the state explicitly uses the rotor components of rotation **R** and of translation **t**, or MEKF, if motor components of motion **M** is used.

2.4 Some notes about the algorithms

Here we will give some specific notes on EKF algorithms.

The EKF algorithm requires an initial guess of motion not very far from the true one. One reasonable hypothesis is that the motion is "small". So w e can set the initial guess as "no motion": $\mathbf{s}_{1/0} = (1\ \ 0\ \ 0\ \ 0\ \ 0\ \ 0\ \ 0)^T$ and $\mathbf{s}_{1/0} = (1\ \ 0\ \ 0\ \ 0\ \ 0\ \ 0\ \ 0\ \ 0)^T$ for **R**tEKF and MEKF, respectively. In experiments, the estimate conv erges rapidly from the initial guess to near the true one within 4 or 5 runs, but for a qualified estimation, more than 15 runs are required.

In our experiments, we find, if the translation $||\mathbf{t}|| \gg 1$, the algorithms based on constraints no. 1 and no. 3 will frequently diverge. The reason is that these constraints contain cross product terms. Such situation can be analyzed by the equation of measurement noise $\boldsymbol{\xi}_i$. In constraint no. 1, suppose we set the origin of the coordinate system at somewhere on the reference model. If the estimated translator $||\hat{\mathbf{t}}_i|| \gg 1$, then, (usually) $||\hat{\mathbf{y}}_i|| \gg 1$. That will directly cause the components of the cov ariance matrix $\mathcal{V}_i$ to be far greater than that of the original co variance matrix $\mathcal{W}_i$. Such enlarged noise will easy make the EKF diverging. To solve this problem, we simply multiply the measurement function by a scalar as follows. At the beginning of the algorithm, we check the distance, $||\mathbf{m}_1||$, of the projection line $\mathbf{L}_{b_1}$. If $||\mathbf{m}_i|| > 1$, we can use a modified measurement equation

$$\mathbf{f}_i/||\mathbf{m}_i|| = \mathbf{m}_i/||\mathbf{m}_i|| - \mathbf{n}_i \times (\mathcal{R}_i\mathbf{y}_i/||\mathbf{m}_i|| + \mathbf{t}_i/||\mathbf{m}_i||) = 0.$$

At the end of the algorithm we multiply the estimated translation $\hat{\mathbf{t}}_{\mathbf{i}}^{*} = \hat{\mathbf{t}}_{\mathbf{i}}/||\mathbf{m}_{\mathbf{i}}||$ by $||\mathbf{m}_{\mathbf{i}}||$ to recover the true estimation $\hat{\mathbf{t}}_{\mathbf{i}}$. Similar analysis can be done for constraint no. 3. In case of constraint no. 3, we use the distance , d_i, of projection plane $\mathbf{P}_{\mathbf{ij}}$, that means, divide $\mathbf{f}_{\mathbf{2i}}$ by d_i. At the end of the algorithm, we multiply the estimated dual part $\hat{\mathbf{R}}_{\mathbf{i}}'^{*}$, $\hat{\mathbf{R}}_{\mathbf{i}}'^{*} = \hat{\mathbf{R}}_{\mathbf{i}}'/d_i$, by d_i to recover the true estimation $\hat{\mathbf{R}}_{\mathbf{i}}'$.

3 Experiments

In this section we present some experiments by real images. The aim of the experiments is to study the performance of the EKF algorithms for pose estimation based on geometric constraints. We expect that both the special constraint and the algorithmic approach of using it may influence the results. This behavior should be shown with respect to different qualities of data.

In our experimental scenario we took a B21 mobile robot equipped with a stereo camera head and positioned it two meters in front of a calibration cube. We focused one camera on the calibration cube and took an image. Then we moved the robot, focused the camera again on the cube and took another image. The edge size of the calibration cube is 46 cm and the image size is 384×288 pixel. Furthermore we defined on the calibration cube a 3D object model.

In these experiments we actually selected certain points by hand and from these the depicted lines are derived and, by knowing the camera calibration, the actual projection ray and projection plane parameters are computed.

The results of different algorithms for pose estimation are shown in table 1 In the second column of table 1 **Rt**EKF and **M**EKF denote the use of the EKF, MAT denotes matrix algebra, SVD denotes the singular value decomposition of a matrix to ensure a rotation matrix as a result. In the third column the used constraints, point-line (XL), point-plane (XP) and line-plane (LP) are indicated. The fourth column shows the results of the estimated rotation matrix $\mathcal{R}$ and the translation vector $\mathbf{t}$, respectively. The fifth column shows the error of the equation system. Since the error of the equation system describes the Hesse distance of the entities [1], the value of the error is an approximation of the squared average distance of the entities.

In a second experiment we compare the noise sensitivity of the various approaches for pose estimation. Matrix based estimations result in both higher errors and larger fluctuations in dependence of the noise level compared to EKF estimates. This is in agreement with the well known behavior of error propagation in case of matrix based rotation estimation. The EKF performs more stable. This is a consequence of the estimator themselves and of the fact that in our approach rotation is represented as rotors. The concatenation of rotors is more robust than that of rotation matrices.

4 Conclusions

In this paper we present three EKF algorithms for 2D-3D pose estimation. The aim of the paper is to design EKFs based on the geometric constraints. The model data are either points or lines. The observation frame is constituted by projection lines or projection planes. Any deviations from the constraint correspond the Hesse distance of the involved geometric entities. The representation

no.	$\mathcal{R}$ — t	Constraint	Experiment 1		Error
1	R EKF — R EKF	XL-XL	$\mathcal{R} = \begin{pmatrix} 0.986 & 0.099 & -0.137 \\ -0.127 & 0.969 & -0.214 \\ 0.111 & 0.228 & 0.967 \end{pmatrix}$	$\mathbf{t} = \begin{pmatrix} -35.66 \\ -203.09 \\ 156.25 \end{pmatrix}$	4.5
2	R EKF — R EKF	XP-XP	$\mathcal{R} = \begin{pmatrix} 0.986 & 0.115 & -0.118 \\ -0.141 & 0.958 & -0.247 \\ 0.085 & 0.260 & 0.962 \end{pmatrix}$	$\mathbf{t} = \begin{pmatrix} -37.01 \\ -198.55 \\ 154.86 \end{pmatrix}$	5.5
3	MEKF — MEKF	LP-LP	$\mathcal{R} = \begin{pmatrix} 0.985 & 0.106 & -0.134 \\ -0.133 & 0.968 & -0.213 \\ 0.108 & 0.228 & 0.968 \end{pmatrix}$	$\mathbf{t} = \begin{pmatrix} -53.30 \\ -214.37 \\ 138.53 \end{pmatrix}$	2.6
4	MEKF — MAT	LP-LP	$\mathcal{R} = \begin{pmatrix} 0.985 & 0.106 & -0.134 \\ -0.133 & 0.968 & -0.213 \\ 0.108 & 0.228 & 0.968 \end{pmatrix}$	$\mathbf{t} = \begin{pmatrix} -67.78 \\ -227.73 \\ 123.90 \end{pmatrix}$	2.7
5	SVD — MAT	LP-XP	$\mathcal{R} = \begin{pmatrix} 0.976 & 0.109 & -0.187 \\ -0.158 & 0.950 & -0.266 \\ 0.149 & 0.289 & 0.945 \end{pmatrix}$	$\mathbf{t} = \begin{pmatrix} -66.57 \\ -216.18 \\ 100.53 \end{pmatrix}$	7.1

Table 1. The results of experiment 1, depending on the used constraints and algorithms to evaluate their validity.

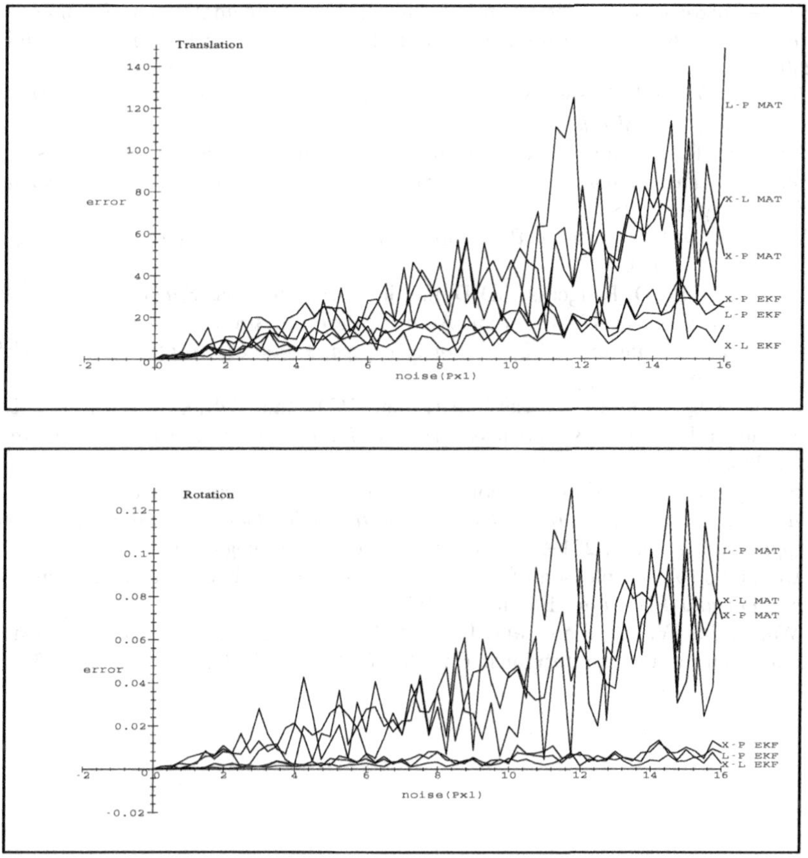

Fig. 2. Performance comparison with increasing noise. The EKFs perform with more accurate and more stable estimates than the matrix based method.

of both rotation and translation is assumed as rotors and motors in even geometric subalgebras of the Euclidean space. The experiments show advantages of that representation and of the EKF approach in comparison to normal matrix based LMS algorithms.

References

1. Rosenhahn B., Zhang Y. and Sommer G. Pose estimation in the language of Kinematics. *AFPA C*,(submitted), 2000.
2. Sommer G., Rosenhahn B. and Zhang Y. Pose Estimation using Geometric Constraints. *Techn. Rept. 2003, Institut für Informatik und Praktische Mathematik Christian-Albrechts-Universität zu Kiel*, 2000.
3. Zhang Y., Sommer G., and E. Bayro-Corrochano. The motor extended Kalman filter for dynamic rigid motion estimation from line observations. *In G. Sommer, editor, Geometric Computing with Clifford Algebra. Springer Verlag*, to be published, 2000.
4. Bayro-Corrochano E. The geometry and algebra of kinematics. *In Sommer G., editor, Geometric Computing with Clifford Algebra. Springer Verlag*, to be published, 2000.
5. Hestenes D. and G. Sobczyk. Clifford Algebra to Geometric Calculus. *D. Reidel Publ. Comp., Dordrecht*, 1984.
6. Carceroni R. L. and C. M. Brown. Numerical Methods for Model-Based Pose Recovery. *Techn. Rept. 659, Comp. Sci. Dept., The Univ. of Rochester, Rochester, N. Y.*, August 1998.
7. Andersson B. D. O. and J. B. Moore. Optimal Filtering. *Prentice Hall, Englewood Cliffs, N. J.*, 1979.
8. Zhang, Z. and O. Faugeras. 3D Dynamic Scene Analysis. *Springer Verlag*, 1992.
9. I. Y. Bar-Itzhack and Y. Oshman. Attitude determination from vector observations: quaternion estimation. *IEEE Trans. Aerospace and Electronic Systems*, AES-21: 128-136, 1985.
10. Phong T.Q., Horaud R., Yassine A. and P.D. Tao. Object pose from 2-d to 3-d point and line correspondences. *International Journal of Computer Vision*, 15: 225–243, 1995.
11. Sommer G. The global algebraic frame of the perception-action cycle. *In B. Jähne, H. Haussecker, and P.Geissler, editors, Handbook of Computer Vision and Applications*, Vol. 3: 221–264, Academic Press, San Diego, 1999.
12. Rooney J. A comparison of representations of general screw displacement. *Environment and Planning* , B5: 45-88, 1978.
13. Walker M. W., L. Shao, and R. A. Volz . Estimating 3-D location parameters using dual number quaternions. *CVGIP: Image Understanding*, 54: 358–367, 1991.

Author Index